Bacteria

IN BIOLOGY, BIOTECHNOLOGY AND MEDICINE

SIXTH EDITION

Bacteria
IN BIOLOGY, BIOTECHNOLOGY AND MEDICINE

SIXTH EDITION

Paul Singleton

John Wiley & Sons, Ltd

1st edition 1981 (reprinted 1985, 1987, 1989, 1991)   Japanese edition 1982
2nd edition 1992   French editions 1984, 1994, 1999
3rd edition 1995 (reprinted 1995, 1996)   German edition 1995
4th edition 1997 (reprinted 1997, 1998)   Polish edition 2000
5th edition 1999   Spanish edition 2004
6th edition 2004

Other Wiley Editorial Offices

John Wiley & Sons Inc., 111 River Street, Hoboken, NJ 07030, USA

Jossey-Bass, 989 Market Street, San Francisco, CA 94103–1741, USA

Wiley-VCH Verlag GmbH, Boschstr. 12, D-69469 Weinheim, Germany

John Wiley & Sons Australia Ltd, 33 Park Road, Milton, Queensland 4064, Australia

John Wiley & Sons (Asia) Pte Ltd, 2 Clementi Loop #02-01, Jin Xing Distripark, Singapore 129809

John Wiley & Sons Canada Ltd, 6045 Freemont Boulevard, Mississauga, Ontario L5R 4J3

Library of Congress Cataloging-in-Publication Data

Singleton, Paul.
    Bacteria in biology, biotechnology, and medicine / Paul Singleton. — 6th ed.
      p. ; cm.
    Includes bibliographical references and index.
    ISBN 0-470-09026-X (cloth : alk. paper) — ISBN 0-470-09027-8 (pbk. : alk. paper)
    1. Bacteria.   2. Medical bacteriology.   3. Biotechnology.   I. Title.
    [DNLM: 1. Bacteria.   2. Biotechnology.   QW 50 S617b 2004]
    QR74.8.S56   2004
    579.3 — dc22
    200414697

British Library Cataloguing in Publication Data

A catalogue record for this book is available from the British Library

ISBN 0 470 09026 X (cased)   0 470 09027 8 (paperback)

Typeset in 10.5/13pt Minion by Vision Typesetting Ltd, Manchester, UK
Printed and bound in Great Britain by Antony Rowe Ltd, Chippenham, Wiltshire

This book is printed on acid-free paper responsibly manufactured from sustainable forestry, in which at least two trees are planted for each one used for paper production.

Contents

Preface   ix

1   The bacteria: an introduction   1
1.1   What are bacteria?   1
1.2   Why study bacteria?   3
1.3   Classifying and naming bacteria   4

2   The bacterial cell   6
2.1   Shapes, sizes and arrangements of bacterial cells   6
2.2   The bacterial cell: a closer look   10
2.3   Trichomes and coenocytic bacteria   43

3   Growth and reproduction   44
3.1   Conditions for growth   44
3.2   Growth in a single cell   49
3.3   Growth in bacterial populations   51
3.4   Diauxic growth   68
3.5   Measuring growth   68

4   Differentiation   69
4.1   The life-cycle of *Caulobacter*   69
4.2   Swarming   72
4.3   Resting cells   73
4.4   Akinetes, heterocysts, hormogonia   77

5 Metabolism I: energy 81
5.1 Energy metabolism in chemotrophs 83
5.2 Energy metabolism in phototrophs 96
5.3 Other topics in energy metabolism 99
5.4 Transport systems 102

6 Metabolism II: carbon 117
6.1 Carbon assimilation in autotrophs 118
6.2 Carbon assimilation in heterotrophs 119
6.3 Synthesis, interconversion and polymerization of carbon compounds 122
6.4 Methylotrophy in bacteria 126

7 Molecular biology I: genes and gene expression 128
7.1 Chromosomes and plasmids 128
7.2 Nucleic acids: structure 132
7.3 DNA replication 138
7.4 DNA modification and restriction 145
7.5 RNA synthesis: transcription 147
7.6 Proteins: synthesis and other aspects 149
7.7 DNA monitoring and repair 160
7.8 Regulation of gene expression 163
7.9 RNA 188

8 Molecular biology II: changing the message 191
8.1 Mutation 191
8.2 Recombination 196
8.3 Transposition 199
8.4 Gene transfer 205
8.5 Genetic engineering/recombinant DNA technology 213

9 Bacteriophages 278
9.1 Virulent phages: the lytic cycle 280
9.2 Temperate phages: lysogeny 288
9.3 Androphages 291
9.4 Phage conversion 291
9.5 Transduction 292
9.6 How does phage DNA escape restriction in the host bacterium? 293

10 Bacteria in the living world 295

10.1 Microbial communities 295
10.2 Saprotrophs, predators, parasites, symbionts 300
10.3 Bacteria and the cycles of matter 302
10.4 Ice-nucleation bacteria 310
10.5 Bacteriology *in situ* – fact or fiction? 311
10.6 The greenhouse effect 312
10.7 Recombinant bacteria in the environment 314
10.8 Uncultivable/uncultured bacteria 314

11 Bacteria in medicine 316

11.1 Bacteria as pathogens 316
11.2 The routes of infection 317
11.3 Pathogenesis 330
11.4 The body's defences 340
11.5 The pathogen: virulence factors 358
11.6 Pathogen–host interactions: a new perspective 370
11.7 The transmission of disease 372
11.8 Laboratory detection and characterization of pathogens 373
11.9 Prevention and control of bacterial diseases 379
11.10 Notes on chemotherapy; phage therapy; biologicals 380
11.11 Some bacterial diseases 382

12 Applied bacteriology I: food 393

12.1 Bacteria in the food industry 393
12.2 Food preservation 395
12.3 Food poisoning and food hygiene 400

13 Applied bacteriology II: miscellaneous aspects 407

13.1 Feeding animals, protecting plants 407
13.2 Biomining (bioleaching) 409
13.3 Biological washing powders 411
13.4 Sewage treatment 411
13.5 Water supplies 415
13.6 Putting pathogens to work 422
13.7 Plastics from bacteria: 'Biopol' 422
13.8 Bioremediation 423
13.9 Biomimetic technology 424

14 **Some practical bacteriology   425**
    14.1   Safety in the laboratory   425
    14.2   Bacteriological media   426
    14.3   Aseptic technique   431
    14.4   The tools of the bacteriologist   432
    14.5   Methods of inoculation   435
    14.6   Preparing a pure culture from a mixture of organisms   436
    14.7   Anaerobic incubation   439
    14.8   Counting bacteria   441
    14.9   Staining   444
    14.10 Microscopy   446

15 **Man against bacteria   450**
    15.1   Sterilization   450
    15.2   Disinfection   454
    15.3   Antisepsis   457
    15.4   Antibiotics   457

16 **The identification and classification of bacteria   481**
    16.1   Identification   481
    16.2   Classification (taxonomy) of prokaryotes   500

**Appendix   Minidescriptions of some genera, families, orders and other
categories of bacteria   513**

Index   529

Preface

Studies on the structure and physiology of bacteria are finally dispelling the myth of 'the simple cell'; recent findings, reported in the 6th edition, include a cytoskeleton, mitotic activity, elaborate control systems and cross-talk with human cells – clearly, features of intricate, dynamic and interactive organisms whose complexity is only just beginning to be appreciated.

The text is extensively updated and many new references have been added. References are included in order to (i) indicate the source(s) of data and/or (ii) help readers to acquire further detail. (The references are enclosed by square brackets; abbreviated journal names are explained by the key inside the book's cover.)

Many of the recent advances have involved new or improved methods. The central role of methodology was recognized in earlier editions of this book by the inclusion of topics such as PCR, NASBA, cloning, mutagenesis, DNA sequencing, IVET and STM. In addition to these methods, this edition includes FRET, *in vitro* Tn5 transposition, Pyrosequencing™ and phage display.

I would like to acknowledge invaluable help from the Medical Library, University of Bristol.

Paul Singleton
Clannaborough Barton, Devon
March, 2004

1 The bacteria: an introduction

1.1 WHAT ARE BACTERIA?

Bacteria are minute organisms which occur almost everywhere. They sometimes reveal their presence – wounds 'go septic', milk 'sours', meat 'putrefies' – but usually we are unaware of them because their activities are less obvious and because they are so small. Indeed, the very existence of bacteria was unknown until the development of the microscope in the 17th century.

In most cases a bacterium is a single, autonomous cell. However, the bacterial cell has a *prokaryotic* organization and differs markedly from the *eukaryotic* cells of animals and plants; some of the differences are listed in Table 1.1 (Prokaryotic features referred to in the table are described in later chapters.) Eukaryotic organisms may have emerged about 2–3.5 billion years ago [Science (1996) *271* 470–477]; one hypothesis is that they arose from an energy-based relationship between two types of prokaryotic cell: one from the domain Bacteria and one from the domain Archaea (section 1.1.1) [the hydrogen hypothesis: Nature (1998) *392* 37–41; comment: Microbiology (2000) *146* 1019–1020]

Bacteria are included in the category 'microorganisms'; this contains several distinct types of organism – including algae, fungi, lichens, protozoa and viruses – as well as bacteria. Hence, while all bacteria are microorganisms, not all microorganisms are bacteria. The distinction between 'microbiology' (the study of microorganisms) and 'bacteriology' (the study of bacteria) should also be noted.

1.1.1 A note on the use of the term 'bacteria'

The term 'bacteria' has been widely used, in a general sense, to mean 'prokaryotic organisms' – and in this sense it included *all* prokaryotes. However, molecular studies have shown that the prokaryotes can be divided into two fundamentally different groups called *domains*: the domain Bacteria (note 'B', not 'b') and the domain Archaea [discussion: JB (1994) *176* 1–6]; hence, 'bacteria' is now used increasingly to refer

Table 1.1   Eukaryotic and prokaryotic cells: some major differences

Eukaryotic cells	Prokaryotic cells
The chromosomes are enclosed within a sac-like, double-layered nuclear membrane	There is no nuclear membrane: the chromosome(s) are in direct contact with the cytoplasm
Chromosome structure is complex; the DNA is usually associated with proteins called histones	Chromosome structure is relatively simple
The cytoplasmic membrane characteristically contains sterols	The cytoplasmic membrane usually lacks sterols (but sterols are present e.g. in *Mycoplasma*)
Cell division involves mitosis or meiosis	Meiosis does not occur. Mitotic activity has been reported (section 7.3.1)
The cell wall, when present, includes structural components such as cellulose or chitin, but never peptidoglycan	The cell wall, when present, usually contains peptidoglycan (a compound similar to peptidoglycan is found in some members of the Archaea) but never cellulose or chitin structural components
Mitochondria are generally present; chloroplasts occur in photosynthetic cells	Mitochondria and chloroplasts are never present
Each cell contains ribosomes of two types: a larger type in the cytoplasm and a smaller type in mitchondria and chloroplasts	Each cell contains ribosomes of only one size
Flagella, when present, have a complex structure	Flagella, when present, have a relatively simple structure

specifically to some or all members of the domain Bacteria. (Readers should be aware of the sense in which the term is used in any given context.)

In this book, 'bacteria' refers to members of the domain Bacteria; the book deals almost exclusively with the organisms in this group.

The domain Archaea (see Table 1.2) contains organisms that were previously classified as members of the kingdom Archaebacteria; this domain is apparently the smaller of the two – although we do not know how many archaeans (or indeed, bacteria) remain unidentified. Many species of the Archaea live in so-called 'extreme' environments characterized by high temperatures, high salinity etc. [for further information see e.g. *Extremophiles* (ISBN 0471 026182)]; however, archaeans are also found in relatively moderate environments such as soil [PNAS (1997) *94* 277–282] and marine habitats [Nature (2001) *409* 507–510] – while certain bacteria are known to occur in extreme environments [e.g. antarctic snow: AEM (2000) *66* 4514–4517].

Aspects of archaeal biology are mentioned in various parts of the book in order to give some idea of the way in which these organisms differ from the 'true bacteria'.

Table 1.2 The domains Bacteria and Archaea: some distinguishing features

Bacteria	Archaea
Distinct (bacterial) 16S rRNA	Distinct (archaeal) 16S rRNA
The larger domain (apparently)	The smaller domain (apparently)
Includes all the medically important prokaryotes	No species is known to be pathogenic
Relatively few species from extreme habitats	Many species live in extreme habitats
Some species are photosynthetic	No species is photosynthetic (but see section 5.2.2)
No species forms methane	All methane-formers are archaeans
Gene expression (transcription and translation) prokaryotic	Gene expression (in some archaeans, at least) closer to the eukaryotic pattern
Membrane lipids characteristically ester-linked	Membrane lipids characteristically ether-linked
Flagella grow at the tip	Flagella appear to assemble from the base; they are typically much thinner than bacterial flagella
Cell wall, when present, usually contains peptidoglycan	Cell wall, when present, may contain pseudomurein

1.2 WHY STUDY BACTERIA?

One important reason is the conquest of disease. Bacteria cause some major diseases as well as a number of minor ones; the prevention and control of these diseases depend largely on the efforts of medical, veterinary and agricultural bacteriologists. Pathogenic (i.e. disease-causing) bacteria and their activities are considered in Chapter 11.

Important though they are, the pathogenic bacteria are only a small proportion of the bacteria as a whole. Most bacteria do little or no harm, and many are positively useful to man. Some, for example, produce useful antibiotics, and some provide enzymes for biological washing powders. Some are used as microbial insecticides – protecting crops from certain insect pests – while others are employed in biomimetic technology (section 13.9) and biofuel cells. Bacteria are even used to leach out metals (including gold) from low-grade or refractory ores (section 13.2) and have been used to synthesize biodegradable plastics (section 13.7).

Perhaps surprisingly, bacteria contribute a lot to the food industry. We usually think of bacteria as a nuisance where food is concerned, causing spoilage and food poisoning, but certain types of bacteria are actually employed in the production of food. For example, in the manufacture of butter, cheese and yoghurt, bacteria are used to convert the lactose (in milk) to lactic acid; these bacteria also form compounds which give the products their characteristic flavours. Xanthan gum, a bacterial product, is widely used as a gelling agent and thickener in the food industry; gellan gum (another bacterial product) has similar uses (e.g. in jams and jellies). Vinegar is produced from alcohol (ethanol) by bacterial action. Bacteria are also involved in the manufacture of cocoa and coffee, and are constituents of *probiotics* (section 11.9).

Some of the activities described above can be understood by studying the chemical reactions (metabolism) of bacterial cells (Chapters 5 and 6). Chapters 12 and 13 look at the activities of some 'useful' bacteria.

Bacteria and their enzymes play central roles in biotechnology – with applications in medicine, industry and agriculture. Thus, for example, recombinant DNA technology has been used to produce agents such as streptokinase, a treatment for blood clots (thrombosis) (section 8.5.10).

Not least, bacteria have essential roles in the natural cycles of matter (Chapter 10) – on which, ultimately, all life depends. In the soil, bacteria affect fertility and structure – agricultural potential – so that a better understanding of bacterial activity will permit better management of land and crops; in the future this will be vital to the survival of our ever-expanding population.

From this brief summary it should be clear that the more we learn about bacteria the more effectively we can minimize their harmful potential and exploit their useful activities.

Some would exploit the *harmful* potential of bacteria in biological warfare (germ warfare), or terrorism. [Bioterrorism as a public health threat: EID (1998) *4* 488–494. Bioterrorism (special issue): EID (1999) *5* 491–565.]

1.3   CLASSIFYING AND NAMING BACTERIA

How is one type of bacterium distinguished from another, and how are bacteria classified? Bacteria may differ, for example, in their shape, size and structure, in their chemical activities, in the types of nutrients they need, in the form of energy they use, in the physical conditions under which they can grow, and in their reactions to certain dyes.

Features such as those listed above are widely used for classifying (and identifying) bacteria – such features being easily checked even in a modestly equipped laboratory. As in other areas of biology, bacteria are classified in a hierarchy of categories – e.g. families, genera, species; *species* which are sufficiently alike are placed in the same *genus*, and genera with a certain level of similarity are grouped into a *family*. A species may be subdivided into two or more *strains* – organisms which conform to the same species definition but which have minor differences. In general, members of (say) a bacterial family would have similar structure, would use the same form of energy, and would typically react in a similar way to certain dyes; the species in such a family may be grouped into genera on the basis of differences in chemical activities, nutrient requirements conditions for growth and (to some extent) shape and size.

Although useful for everyday purposes, the kind of classification described above does not necessarily indicate *evolutionary* relationships among the bacteria. In recent decades, classification (taxonomy) has been based increasingly on sequence data from bacterial *nucleic acids* (Chapter 7). Such classification, believed to reflect true evolutionary relationships, is considered in Chapter 16.

As in the case of animals and plants, each species of bacterium is given a name in the form of a Latin binomial. A binomial consists of (i) the name of the genus to which a given organism belongs, followed by (ii) the 'specific epithet' which acts as a label for one particular species; for example, *Escherichia coli* gives the name of the genus (*Escherichia*) and the specific epithet (*coli*). By convention, a Latin binomial is printed in italics, or is underlined once if handwritten; the name of a genus always begins with a capital letter, but a specific epithet always begins with a lower-case letter.

The name of a species may be abbreviated by abbreviating the name of the genus – e.g. *Escherichia coli* may be written '*E. coli*'; however, this should be done only when the full name of the genus has been mentioned earlier in the text so that the meaning of the abbreviation is clear.

The names of families and orders of bacteria are not printed in italics, but each has a capital initial letter. These names also have standardized endings, the name of a family always ending in '-aceae' (e.g. Enterobacteraceae) and the name of an order always ending in '-ales' (e.g. Actinomycetales).

The naming of bacteria is formally governed by various rules made by the International Committee on Systematics of Prokaryotes, formerly the International Committee on Systematic Bacteriology. The advantages of an internationally standardized system of naming are obvious, but the rules are not always adhered to in the literature – owing to a lack of awareness (of the rules, or of revised names) or to disagreement with published opinions.

2 The bacterial cell

2.1 SHAPES, SIZES AND ARRANGEMENTS OF BACTERIAL CELLS

2.1.1 Shape

Bacterial cells vary widely in shape, according to species. Rounded or 'spherical' cells – of any species – are called *cocci* (singular: *coccus*). Elongated, rod-shaped cells of any species are called *bacilli* (singular: *bacillus*), or simply *rods*. Cocci are not necessarily exactly spherical, and not all bacilli have exactly the same shape; for example, some cocci are more or less kidney-shaped, and some bacilli taper at each end (*fusiform* bacilli) or are curved (*vibrios*). Ovoid cells, intermediate in shape between cocci and bacilli, are called *coccobacilli* (singular: *coccobacillus*). There are also two types of spiral cell: those which are more or less rigid (*spirilla*, singular: *spirillum*; Plate 2.1, *bottom, left*), and those which are flexible (*spirochaetes*, singular: *spirochaete*; Plate 2.4). Then there are the so-called 'square bacteria' (flat, square bacteria) and 'box-like bacteria' (variously-shaped, angular bacteria). Finally, there are the *actinomycetes* – most species of which grow as fine, fungus-like threads called *hyphae* (singular: *hypha*; Fig. 2.1n); a group or mass of hyphae is called *mycelium*. Bacteria of various shapes are shown in Fig. 2.1 and in Plate 2.1.

As seen in the caption of Fig. 2.1, some of the names used to describe shapes of bacterial cells are also used as names for bacterial genera. Care should be taken, for example, not to confuse 'bacillus' with '*Bacillus*' (note that the latter has a capital 'B' and is printed in italics); some bacilli belong to the genus *Bacillus*, others do not!

Although the cells of a given species of bacterium are usually more or less uniform in shape, in some species the shape of the cell typically varies from one cell to another – sometimes quite markedly; this phenomenon is called *pleomorphism*.

L-form cells are irregularly-shaped or spherical cells which are produced spontaneously by some species of bacteria and which can be induced in other species e.g. by temperature shock and by other kinds of physicochemical stimulus; these cells were named after the Lister Institute of Preventive Medicine (London). [Induction of

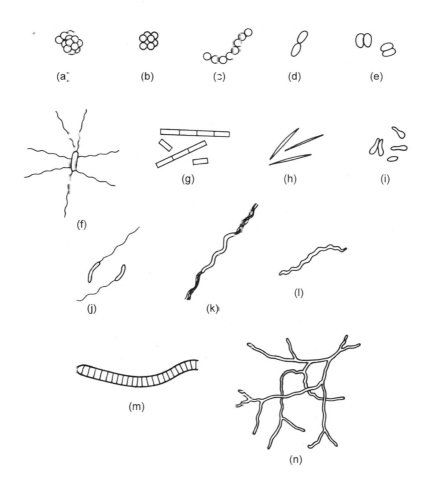

Figure 2.1 Shapes and arrangements of some bacteria with named examples (not drawn to scale). (a) Uniform spherical cells (cocci) in irregular clusters: *Staphylococcus aureus*. (b) Cocci, in regular packets of eight cells: *Sarcina ventriculi*. (c) Cocci in chains: *Streptococcus pyogenes*. (d) Slightly elongated cocci in pairs (diplococci): *Streptococcus pneumoniae*. (e) Pairs of cocci (diplococci) in which each cell is flattened or slightly concave on the side next to its neighbour: *Neisseria gonorrhoeae*. (f) Rod-shaped cell (bacillus): *Escherichia coli*. The lines arising from the bacillus represent fine, hair-like appendages called *flagella* which are described later in the chapter (g) Blunt-ended bacilli, singly and in chains: *Bacillus anthracis* (h) Bacilli with tapered ends (fusiform bacilli): *Fusobacterium nucleatum*. (i) Irregular-ly-shaped (pleomorphic) cells: *Corynebacterium diphtheriae*. (j) Curved bacilli (vibrios), each with one flagellum: *Vibrio cholerae*. (k) A rigid spiral cell (spirillum) with a tuft of flagella at each end: *Spirillum volutans*. (l) A flexible spiral cell (spirochaete): *Treponema pallidum*. (m) One end of a filament (*trichome*) of a cyanobacterium: *Oscillatoria limnetica*. Trichomes are discussed in section 2.3. (n) Thin, branched filaments (hyphae): *Streptomyces albus*.

L-form-like cells of *Bacillus subtilis* under microculture conditions: Microbiology (2003) *149* 2501–2511.]

Bacterial morphology in an evolutionary context. Studies on evolutionary aspects of bacterial morphology suggest that the coccus shape has arisen, independently, on various occasions (in different bacterial lineages) and that, when formed, it has usually persisted to the present day. By contrast, the spirochaetes and the (wall-less) mycoplasmas seem to have developed only once during the evolution of bacteria. Moreover, it seems probable that members of the domain Bacteria developed from a common rod-shaped ancestor because only rod-shaped organisms are found in the deepest lineages of the evolutionary tree; the coccus may be a degenerate form of the rod. It has been proposed that persistence in the types of morphology seen among today's bacteria reflects the biochemical and biophysical properties of the cell-wall polymer *peptidoglycan* (sections 2.2.9.1, 2.2.9.2; Fig. 2.7) as well as the genetic control involved in biosynthesis of peptidoglycan [Microbiology (1998) *144* 2803–2808].

Determinants of cell shape. Studies on the Gram-positive bacterium *Bacillus subtilis* have indicated that helical, actin-like filaments are contributory factors in determining the organism's rod shape. Two types of filament are involved. Filaments composed of the MreB protein influence the cell's width; these are short-pitch filaments that appear to encircle the cell, at the inner face of the cytoplasmic membrane, in the mid-cell region. Filaments composed of the Mbl protein are of longer pitch and they appear to run lengthwise in the cell; these filaments influence cell growth in the longitudinal axis.

Mutations in the *mreB* and *mbl* genes are generally associated with abnormal cell shape. Mutations in *mreB* are lethal in *B. subtilis* but not in *Escherichia coli*; those in *mbl* are apparently non-lethal.

MreB (like the ParM protein: section 7.3.1) exhibits some degree of relationship to the (eukaryotic) protein *actin*. Database searches have shown that homologues of the MreB and Mbl proteins occur in a wide range of bacteria (e.g. *Caulobacter, Haemophilus, Helicobacter, Pseudomonas, Treponema* and *Vibrio*) – and also in the archaean *Methanobacterium thermoautotrophicum*. Homologues were not found e.g. in various cocci or in species of *Corynebacterium, Neisseria, Mycobacterium* or *Mycoplasma*.

It was postulated that the role of the MreB/Mbl filaments is likely to involve guidance of the cell-wall-synthesizing machinery (rather than some form of a 'structural brace') [Cell (2001) *104* 913–922].

Later work on *B. subtilis* revealed that synthesis of peptidoglycan in the cylindrical part of the cell wall occurs in a helical pattern under the guidance of Mbl. Studies on rod-shaped bacteria which *lack* the MreB/Mbl proteins have indicated that these organisms can maintain their shape through an entirely different mechanism; for example, *Corynebacterium glutamicum* exhibits growth at both *poles* of the cell – so that the wall of a growing cell consists of a tube of inert wall material flanked on either side by a zone of active polar growth. [Control of cell morphogenesis in bacteria: Cell (2003) *113* 767–776.]

In most bacteria, the cell's shape depends on the presence of a cell wall (section 2.2.9) that contains the polymer peptidoglycan. Synthesis of peptidoglycan requires the activity of enzymes called penicillin-binding proteins (PBPs: section 2.2.8); a given species of bacterium typically contains a range of PBPs, each capable of specific role(s) in peptidoglycan synthesis. In rod-shaped bacteria, it appears that one set of PBPs are involved in wall synthesis (i.e. cell elongation: maintenance of rod shape) while another set deals with septation during cell division (section 3.2.1). Studies on *B. subtilis* have identified two PBPs – PBP2a (gene: *pbpA*) and PbpH (gene: *pbpH*, formerly *ykuA*) – whose function involves maintenance of the rod shape; each of these PBPs is redundant in the presence of the other [JB (2003) *185* 4717–4726].

Interestingly, if *E. coli*, which is normally rod-shaped, lacks a functional DacA protein (i.e. PBP5), the organism exhibits branching morphology [JB (2003) *185* 1147–1152].

2.1.2 Size

Bacterial cells are usually measured in *micrometres*, μm (formerly called microns, μ); 1 μm = 0.001 mm. Among the smallest bacteria are cells of *Chlamydia* (~0.2 μm); even smaller cells have been reported among the *nanobacteria* [PNAS (1998) *95* 8274–8279]. At the other end of the scale, some cells of *Spirochaeta* are about 250 μm in length, and the bacterium *Epulopiscium fishelsoni*, which inhabits the gut of the surgeon fish (*Acanthurus nigrofuscus*), is >600 μm [JB (1998) *180* 5601–5611]. These, however, are extreme cases; in most species the maximum dimension of a cell lies within the range 1–10 μm. Note that some small bacteria are of the same order of size as the limit of resolution of a good light microscope, which is about 0.2 μm.

2.1.3 Arrangements of bacterial cells

Under the microscope, bacteria of a given species may be seen as separate (individual) cells or as cells in characteristic groupings. According to species, cells may occur in pairs, in irregular clusters, in chains or filaments, in regular *packets* of four, eight or more cells, or in *palisade* form – a number of elongated cells side-by-side in a row with adjacent cells touching. The species *Pelodictyon clathratiforme* is unusual in that it forms three-dimensional networks of cells. In a number of species the cells form *trichomes* (section 2.3.1). Some arrangements of cells are shown in Fig. 2.1. These different arrangements of cells do not result from the aggregation of previously single cells; they occur because (i) cells of the different species divide (reproduce) in different ways, and (ii) two or more cells may remain attached after the process of cell division.

In nature, certain bacteria occur in stable, mixed-species groups of cells called *consortia* (singular: *consortium*) [see e.g. Microbial Ecology (1996) *31* 225–247]. Mixed bacterial–archaeal consortia have been reported in a terephthalate-degrading anaerobic sludge system [Microbiology (2001) *147* 373–382]. Consortia containing nitrogen-fixing species of *Clostridium* occur e.g. in wild rice plants [AEM (2004) *70* 3096–3102].

2.2   THE BACTERIAL CELL: A CLOSER LOOK

Are all bacteria basically similar in structure? No: cells of different species may differ greatly in both their fine structure (ultrastructure) and chemical composition; for this reason there is no 'typical' bacterium. Figure 2.2 shows a 'generalized' bacterium in a very diagrammatic way.

Figure 2.2 shows the cell's *chromosome* ('genetic blueprint') – a typically loop-like structure of DNA (Chapter 7) which is extensively folded to form a body called the *nucleoid*. Bathing the nucleoid is a complex fluid, the *cytoplasm*, which fills the interior of the cell. The cytoplasm contains *ribosomes*: minute bodies involved in the synthesis of proteins; sometimes there are also *storage granules* of reserve nutrients etc. The nucleoid, cytoplasm, ribosomes and storage granules all occur within the space bounded by a membranous sac, the *cytoplasmic membrane* (also called cell membrane or plasma membrane). The outermost layer in Fig. 2.2 is the tough (mechanically strong) *cell wall*. The cytoplasmic membrane and cell wall are referred to, jointly, as the *cell envelope*. Between the cytoplasmic membrane and the cell wall is the so-called *periplasmic region*. The *flagellum* is a thin, hair-like proteinaceous appendage (involved in cell motility – i.e. movement) which is attached by specialized structures to the cell envelope. Some cells have more than one flagellum, and some have many flagella; some cells have none. These and other features of the bacterial cell are considered in subsequent sections.

The bacterial cell is sometimes said to be 'simple' because it lacks the specialized structures (nucleus, mitochondrion etc.) characteristic of eukaryotic cells. At the *molecular* level, however, bacteria use subtle and sophisticated strategies that involve a

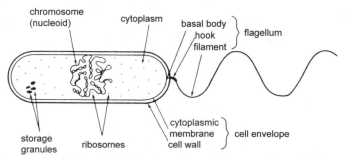

Figure 2.2   A generalized bacterium (diagrammatic). Note that some bacteria do not have all of the features shown in the diagram, and that some bacteria have structures not shown.

high degree of organization. For example, in some bacteria, groups of proteolytic enzymes undergo self-assembly to form hollow, cylindrical structures (*proteasomes*) whose active enzymatic sites occur on the inner surface; proteasomes of the 20S type resemble those in eukaryotes, and they probably degrade damaged or non-functional proteins. The breakdown of unwanted proteins must avoid uncontrolled intracellular proteolysis – hence the requirement for proteasomes with the *internal* active sites. Only unfolded proteins can enter the narrow opening of a proteasome; the docking, unfolding and translocation of proteins (in preparation for degradation) are probably ATP-dependent processes as proteasomes include ATPases. [Prokaryotic proteasomes: TIM (1999) 7 88–92.]

Another multicomponent complex, the *RNA degradosome*, has been demonstrated in *Escherichia coli*. A degradosome includes the enzyme ribonuclease E (RNase E) which degrades/processes RNA and is used e.g. to degrade *certain* types of mRNA [PNAS (2004) 101 2758–2763] and to form the 5S rRNA component of ribosomes (section 2.2.3). Degradosomes are associated with the cytoplasmic membrane via the N-terminal of RNase E. [Degradosomes in *E. coli*: PNAS (2001) 98 63–68.]

Bacterial sophistication is also exemplified by rudimentary 'cytoskeletal' structures such as the helical, proteinaceous filaments which, in some species, contribute to cell shape (section 2.1.1) – and by the FtsZ ring which appears, transiently, at the time of cell division (section 3.2.1 (septum formation) and Plate 3.1). Moreover, a 'mitosis-like' process appears to occur in the segregation of certain plasmids at cell division (section 7.3.1), and a similar process may occur in the segregation of chromosomes (section 3.2.1).

In general, the localization of proteins within a bacterial cell is now seen to be much more organized than was once supposed [Science (1997) 276 712–718]; for example, the activity of specific proteins at specific sites (and times) is crucial for various aspects of differentiation (e.g. section 4.1). We are finding that, in physiological processes, proteins are distributed (and become active) in ways that involve dynamic forms of regulation [Cell (2000) 100 89–98]. Moreover, the transcriptional and non-transcriptional regulatory systems in bacteria exhibit a fascinating degree of complexity [Science (2003) 301 1874–1877].

Although primarily single-celled organisms, some bacteria exhibit 'social' activity in that the behaviour of cells in *communities* can differ from that of individual, isolated cells (quorum sensing: section 10.1.2). Moreover, like higher organisms, some bacteria are influenced by circadian (diurnal) rhythms (section 7.8.12).

2.2.1 The nucleoid

In bacteria, the chromosome is typically a (closed) loop of *deoxyribonucleic acid* (DNA, described in Chapter 7); however, in some species (e.g. *Borrelia burgdorferi*) it is a linear molecule of DNA. In either case, the chromosome is extensively folded into a compact body, the *nucleoid*.

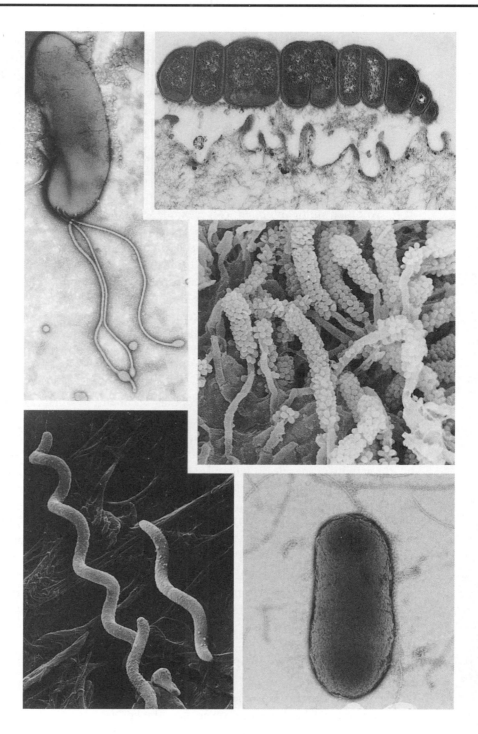

The DNA is associated with certain types of protein, some of which appear to be involved in folding the chromosome. In e.g. *Escherichia coli*, the nucleoid is associated with many copies of a small, basic, multi-functional protein – the HU protein – which binds strongly to angular/kinked double-stranded DNA. This protein is reported to have roles in DNA replication (Chapter 7), in some instances of transposition and site-specific recombination (Chapter 8), and in regulating gene expression [MM (1993) *7* 343–350; MM (2001) *39* 1069–1079]. Another small protein, the *H-NS protein*, appears to have a major role in condensing DNA to form the compact nucleoid [see e.g. NAR (2000) *28* 3504–3510].

In *Deinococcus radiodurans* the nucleoid has an unusual toroidal shape; this may contribute to the organism's extraordinary resistance to ionizing radiation [Science (2003) *299* 254–256].

Within the cell the nucleoid is associated with the cytoplasmic membrane. In *E. coli* the chromosome is about 1.3 mm long; the extent of folding is shown by the fact that all of this DNA fits into a cell of only a few micrometres in length – with room to spare!

A bacterial cell may contain more than one copy of the chromosome, according e.g. to conditions of growth; in *E. coli*, for example, cells growing rapidly have more chromosomes per cell than do those growing slowly (Chapter 3). However, in some

Plate 2.1   Some shapes, sizes and arrangements of bacteria. *Top left*. A typical view of *Helicobacter pylori* (×21000), a helical (twisted) bacillus from the human gastrointestinal tract. This cell has three flagella (Fig 2.2, section 2.2.14), each flagellum being covered by an extension of the cell's outer membrane (section 2.2.9.2); the bulbous structures at/near the ends of the flagella appear to be regions where the membrane has 'ballooned out' – due perhaps to the techniques used in electron microscopy. Each flagellum is about 3 μm long. *Top right*. A multicellular, filamentous bacterium, *Simonsiella*, which is present as part of the mouth microflora in about 25% of humans. This section, stained with ruthenium red, shows the filament adhering to the surface of a buccal epithelial cell. The filament is about 3.5 μm in length. *Centre*. In a 'filled' (i.e. repaired) tooth: microorganisms colonizing the small space between the filling material and the wall of the cavity. Each of the 'corn-cob' structures appears to be composed of a mass of small cocci (each less than 1 μm in diameter) clustered around the end of a hypha – perhaps indicating a symbiotic relationship between two types of organism. *Bottom left*. Cells of *Aquaspirillum peregrinum*, a motile, nitrogen-fixing bacterium found in various freshwater habitats. The cell on the right is about 7.3 μm in (axial) length, 0.6 μm in thickness. *Bottom right*. A single cell of *Escherichia coli*: a straight, round-ended bacillus about 2 μm in length.

The photograph of *Helicobacter pylori* is courtesy of Dr Alan Curry, PHLS, Withington Hospital, Manchester. *Simonsiella* is from *Letters in Applied Microbiology* (1988) *6* 125–128, courtesy of the author, Dr Caroline Pankhurst, King's College, London, with permission from Blackwell Scientific Publications. 'Corn-cob organisms' is courtesy of Prof. Dr Wolfgang Klimm, Medizinische Akademie 'Carl Gustav Carus', Dresden, Germany, reproduced from G Buchmann *et al. Microbial Ecology in Health and Disease* (1990) *3* 51–57, with permission from John Wiley & Sons Ltd. *Aquaspirillum peregrinum* is from *Journal of General Microbiology* (1986) *132* 877–881, by courtesy of Dr Hisanori Konishi, Yamaguchi University School of Medicine, Ube, Japan, and with permission of the Society for General Microbiology. *Escherichia coli* is courtesy of Dr Markus B. Dürrenberger, University of Zürich, Switzerland.

species (e.g. *Azotobacter vinelandii*, *Deinococcus radiodurans*) it appears that each cell normally contains several to many copies of the chromosome.

In *Vibrio cholerae* there are *two different types* of chromosome in each cell – chromosome 1 being the larger of the two [Nature (2000) *406* 477–483].

2.2.2   The cytoplasm

The cytoplasm is an aqueous (water-based) fluid containing ribosomes, nutrients, ions, enzymes, waste products and various molecules involved in synthesis, cell maintenance and energy metabolism; storage granules may be present under certain conditions. There is no equivalent of the endoplasmic reticulum found in eukaryotic cells.

Exceptionally (in a few cases), the cytoplasm may contain some unique and interesting items. In *Bacillus thuringiensis*, for example, it sometimes contains crystals that are poisonous for various insects; this species is used in agriculture for biological control (Chapter 13). Other bacteria, themselves parasites of protozoa, contain curious rolled-up ribbons of protein called *R bodies*. In some aquatic bacteria, tiny particles of magnetite (Fe_3O_4) called *magnetosomes* cause the cells to align in a magnetic field [ferric iron reductase from *Magnetospirillum*: JB (1999) *181* 2142–2147; magnetosome membrane from *Magnetospirillum*: AEM (2004) *70* 1040–1050].

2.2.3   Ribosomes

Ribosomes are minute, rounded bodies, each about 0.02 μm, made of RNA (a polymer similar to DNA – Chapter 7) and protein. They are sites where proteins are synthesized (Chapter 7), and the cytoplasm contains a large number of them (Plate 8.1, *top*).

Bacterial ribosomes are smaller than those in the cytoplasm of eukaryotic cells, but they are similar (in size) to those in the chloroplasts of plants and algae. Ribosomes are usually described not by their diameters but by their rate of sedimentation in an ultracentrifuge; measurement is made in Svedberg units (S) – the higher the value the more rapid is the rate of sedimentation. Each bacterial ribosome (70S) consists of one 50S subunit and one 30S subunit (yes, 50S + 30S = 70S!); the parts of a ribosome are held together probably by hydrogen bonding and by ionic and hydrophobic interactions – magnesium ions generally being important in maintaining the structure.

Crystallographic analysis of the 30S subunit of ribosomes from *Thermus thermophilus* has shown that, as predicted, most of the *interfacial* region of the subunit (in contact with the 50S subunit) consists of RNA [Nature (1999) *400* 833–840].

Most of a ribosome (about 70% by mass) is RNA (ribosomal RNA, or rRNA); a ribosome's function seems to depend primarily on its rRNA. Like the ribosomes themselves, rRNA molecules are also measured in Svedberg units (S). The 30S ribosomal subunit contains 16S rRNA, while the 50S subunit contains 5S and 23S rRNA.

rRNA is a polymer containing four different types of subunit (nucleotide), and in a given rRNA molecule the nucleotides are arranged in a definite and meaningful sequence. In rRNA molecules the *sequence* of nucleotides appears to remain relatively unchanged over evolutionary periods of time. Different sequences in related organisms may therefore indicate evolutionary divergence between such organisms; for example, in different species of the same genus, 16S rRNA typically differs by at least 1.5%. In recent years, the sequences of nucleotides in 16S rRNA have been used to classify prokaryotes into categories generally believed to reflect evolutionary relationships; for example, the difference in 16S rRNA was a major reason for distinguishing the domains Archaea and Bacteria.

Ribosomal proteins (r-proteins) account for ~30% by mass of a ribosome. In *E. coli*, the 30S subunit contains 21 different types of r-protein, while the 50S subunit has over 30 different types.

2.2.4 Storage granules

Under appropriate conditions, many bacteria produce polymers which are stored as granules in the cytoplasm; these compounds include poly-β-hydroxybutyrate and polyphosphate.

2.2.4.1 *Poly-β-hydroxybutyrate (PHB)*

PHB is a linear polymer of β-hydroxybutyrate (Fig. 2.3). In some species, granules of PHB accumulate when decreasing supplies of nutrients (other than carbon) restrict the cell's rate of growth; in these cells the PHB acts as a reservoir of carbon and energy (Chapters 5 and 6) – to be used when other nutrients become more plentiful. In the soil bacterium *Azotobacter beijerinckii*, PHB accumulates (up to 80% of the cell's mass) when oxygen is scarce; in this species PHB can replace oxygen as a source of oxidizing power.

Enzymes involved in the synthesis of PHB (and possibly also de-polymerizing enzymes) occur at the surface of the granules. Each mature granule is surrounded by a layer apparently <4 nm in thickness; this is believed to be a monolayer consisting mainly of *phasins* (molecules of amphiphilic protein) and phospholipids – all the molecules being orientated with their hydrophilic regions facing the cytoplasm [JBM (1997) *37* 45–52]. PHB granules can be stained *in situ* (i.e. in the cell) by dyes such as Nile blue A.

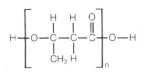

Figure 2.3 Poly-β-hydroxybutyrate: a common storage compound in bacteria.

PHB is the basis of a biodegradable plastic (Biopol: see section 13.7).

Other poly-β-hydroxyalkanoates (PHAs) occur in some bacteria; for example, granules of poly-β-hydroxyoctanoate form in *Pseudomonas oleovorans* when it is grown on *n*-octane. [Bacterial PHAs, including PHB and Biopol (multi-author symposium report): FEMSMR (1992) *103* 91–376.]

2.2.4.2 *Polyphosphate (polymetaphosphate; volutin)*

Polyphosphate (PO_3^{2-}-O-$[PO_3^-]_n$-PO_3^{2-}) granules occur in many types of bacteria. They are believed to act as reservoirs of phosphate and, in some cases, they appear to be involved in energy metabolism; polyphosphate may also serve to store or sequester cations.

When treated with certain dyes (e.g. polychrome methylene blue), polyphosphate granules develop a colour different to that of the dye used to stain them; this phenomenon is called *metachromacy*, and the granules are sometimes called *metachromatic granules*.

2.2.5 Gas vacuoles

Gas vacuoles (see Plate 2.2) occur only in certain (typically aquatic) prokaryotes; they are found e.g. in bloom-forming cyanobacteria (such as *Anabaena flos-aquae*) and in members of the Archaea (such as *Halobacterium* and *Methanosarcina*). Each vacuole consists of a cluster of tiny, elongated, hollow, gas-filled *vesicles*; each vesicle, which has a protein wall, is commonly about 60–250 nm in diameter. Until recently, only two types of protein had been detected in the vesicles of *Halobacterium*; five new proteins have now been identified [JB (2004) *186* 3182–3186].

In some cases gas vacuoles are formed constitutively, in others they are inducible.

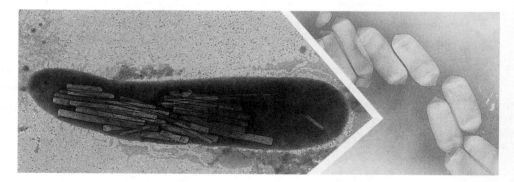

Plate 2.2   Gas vacuoles and vesicles (see section 2.2.5). *Left*. A single gas-vacuolated cell from the polar sea-ice. *Right*. Individual gas vesicles under higher magnification. Photographs courtesy of Dr John J. Gosink, University of Washington, Seattle, USA.

Gas vacuoles affect the buoyancy of free-floating cells; for cells in the aquatic environment, this will affect the intensity of received light – important in the ecology of photosynthetic organisms. Buoyancy can be regulated by at least two mechanisms. In one, increased intensity of light (at a location nearer the surface) stimulates photosynthesis and causes an accumulation of e.g. photosynthate, thus raising the cell's turgor pressure; in some species, the increased turgor pressure (above a certain level) causes gas vesicles to collapse, lowering the buoyancy. New gas vesicles form at a deeper level where the intensity of light is lower.

In an alternative mechanism gas vesicles do not collapse under high-intensity light but the formation of new vesicles is inhibited under such conditions; thus, existing vesicles are 'diluted out' during cell growth and division, leading to a loss of buoyancy. Formation of new vesicles is resumed at a lower depth where the intensity of light is lower. This occurs e.g. in the cyanobacterium *Oscillatoria agardhii*.

2.2.6 Carboxysomes

Carboxysomes are intracellular bodies, each about 100–500 nm in diameter, which are found in many *autotrophic* bacteria, i.e. bacteria which can use carbon dioxide for most or all of their carbon requirements (Chapter 6). Each carboxysome consists of a membranous sac or shell containing many copies of an enzyme (RuBisCO) involved in the 'fixation' of atmospheric carbon dioxide (Fig. 6.1).

2.2.7 Thylakoids

Thylakoids are flattened, intracellular membranous sacs which occur in most cyanobacteria; they usually occur close to, and parallel with, the cell envelope – but seem to be structurally distinct from the cytoplasmic membrane. Thylakoid membranes contain chlorophylls etc. and are the sites of photosynthesis (Chapter 5); in at least some cases they are also sites of respiratory activity (Chapter 5).

Structures similar to thylakoids (but called *chlorosomes* or *chlorobium vesicles*) are formed by bacteria of the suborder Chlorobiineae; they contain certain components of the photosynthetic apparatus.

2.2.8 Cytoplasmic membrane

The cytoplasmic membrane (CM) is a double layer of lipid molecules, about 7–8 nm thick, in which protein molecules are partly or wholly embedded – some proteins spanning the entire thickness of the membrane (Fig. 2.4); the arrangement of lipid molecules is such that the inner and outer faces of the membrane are hydrophilic ('water-loving') – stained darkly in Plate 2.3 (*top, left*) – while the interior is hydrophobic.

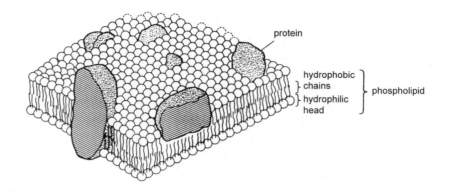

Figure 2.4 Cytoplasmic membrane (diagrammatic): protein molecules in a fluid bilayer of phospholipid molecules – the so-called 'fluid mosaic model'. Both surfaces of the membrane are hydrophilic; the interior is hydrophobic.

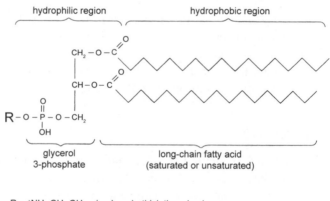

R = ⁺NH₃.CH₂.CH₂ – in phosphatidylethanolamine
R = CH₂OH.CHOH.CH₂ – in phosphatidylglycerol
R = ⁺N(CH₃)₃.CH₂.CH₂ – in phosphatidylcholine

Figure 2.5 Some of the diacyl glycerol phospholipids in bacterial cytoplasmic membranes. The bonds between glycerol 3-phosphate and the two long-chain fatty acid residues are ester bonds.

The CM is not a rigid structure: the lipid molecules are actually in a fluid state; it 'hangs together' as a result of inter-molecular forces.

The lipids are mainly phospholipids. One phospholipid, phosphatidylglycerol (Fig. 2.5), seems to occur in most bacteria; the presence of other types of lipid depends mainly on whether a given bacterium belongs to one or other of two broad categories: the Gram-positive (G +ve) and Gram-negative (G −ve) bacteria – see section 2.2.9. Phosphatidylethanolamine (Fig. 2.5) is generally more common and abundant in G −ve bacteria. Phosphatidylcholine (*lecithin*) (Fig. 2.5) occurs in some G −ve bacteria but not in G +ve ones. Small amounts of glycolipids are common in bacterial CMs; sphingolipids are rare, and sterols occur in bacteria of the family Mycoplasmataceae. In *E. coli* the main lipid is phosphatidylethanolamine; phosphatidylglycerol and diphosphatidylglycerol (DPG, *cardiolipin*) are minor components.

In some anaerobic bacteria the CM contains *plasmalogens*: glycerophopholipids in which the glycerol bears a 1-alkenyl ether group.

The lipids of the CM are often regarded simply as contributors to the selective permeability of the membrane – with proteins carrying out important functions such as energy conversion and sensing. However, CM phospholipids have essential roles as 'chaperones', i.e. molecules which help to ensure the correct *folding* of specific proteins (section 7.6); for example, in *E. coli*, phosphatidylethanolamine (Fig. 2.5) is required for the correct folding of the membrane protein LacY (product of the *lacY* gene: section 7.8.1.1) [EMBO Journal (1998) *17* 5255–5264].

The CM *proteins* include various enzymes, components of transport systems (see section 5.4), sensing systems (e.g. section 2.2.15.2) and energy-converting systems.

The enzymes generally inlcude *penicillin-binding proteins* (PBPs) which are involved in the synthesis of the cell envelope polymer *peptidoglycan* (sections 2.2.9.1, 3.2.1, 6.3.3.1; Fig. 2.7). The high-molecular-weight PBPs are transglycosylases (which synthesize glycan backbone chains) and/or transpeptidases (involved in synthesis of the peptide cross-links); the low-molecular-weight PBPs function e.g. as carboxypeptidases. (High-MW PBPs are generally more sensitive than low-MW PBPs to penicillins.) PBPs play a role in the determination of cell shape: see section 2.1.1.

There are also specialized enzymes which occur in only certain species. For example, in *Staphylococcus aureus* the enzyme *sortase* catalyses a reaction in which certain exported proteins are cleaved and covalently linked to peptidoglycan – such proteins thus being tethered to the cell envelope [sortase-catalysed anchoring of surface proteins to the cell wall of *S. aureus*: MM (2001) *40* 1049–1057]. [Sortase in *Listeria monocytogenes*: JB (2004) *186* 1972–1982.]

Components of energy-converting systems include ATPases and electron transport chains (Chapter 5).

Membrane permeability. The CM is not freely permeable to most molecules. Some small, uncharged or hydrophobic molecules – e.g. O_2, CO_2, NH_3 (but not NH_4^+) and water – can pass through more or less freely. Other molecules (including e.g. nutrients) and ions have to be transported across by special mechanisms, some of which require expenditure of energy by the cell; these mechanisms may allow the cell to accumulate a particular substance to a concentration far greater than that which occurs in the surrounding environment.

The permeability of the CM is important in *osmoregulation* (section 3.1.8). The CM contains so-called stretch-activated *mechanosensitive channels* which respond to increases in the cell's turgor pressure by increasing their pore size – thus facilitating efflux of water and particular solutes. An example of one such channel is the MscL channel in *E. coli*. [Mechanosensitive channels in *E. coli*: ARP (1997) *59* 633–657.] Such channels have been found in many types of bacteria, and they appear to have an important role in adaptation to osmotic stress; they open when the cell's turgor pressure is just below the level at which fatal disruption would occur [EMBO Journal (1999) *18* 1730–1737].

The MscL channel in *Mycobacterium tuberculosis* consists of a single protein (151 amino-acid residues) whose five identical transmembrane subunits are arranged symmetrically to form the 'wall' of the channel; the *E. coli* homologue (136 amino-acid residues) appears to be similar. At an appropriate level of tension in the membrane, the MscL structure (initially in the 'closed' state) appears to expand, in an iris-like fashion, and to become flattened; this has been calculated to form a non-selective pore of about 30 Å in diameter [Nature Structural Biology (2002) *9* 704–710].

In addition to mechanosensitive channels, the CM in some bacteria contains other molecular pathways which allow transmembrane passage of water and/or certain other uncharged molecules. This type of pathway, which consists of a single protein, was first described in red blood cells and designated aquaporin-1 (Aqp1). Subsequently it was shown that Aqp1 is related to (i) the major intrinsic protein (MIP) in mammalian lens fibre, (ii) tonoplast intrinsic proteins in plants, and (iii) the GlpF protein in *Escherichia coli*; proteins of this so-called MIP family (found in all classes of living organisms) share highly conserved regions.

The MIP family of proteins are of two main types: *aquaporins* and *glycerol facilitators*.

The transmembrane path formed by a protein of the MIP family is called a MIP channel. The MIP polypeptide is believed to pass through the CM a number of times in a series of loops; this structure apparently forms a continuous pathway for either (i) water (channel = *aquaporin*) or glycerol and/or other uncharged molecules (channel = *glycerol facilitator*). (In *Lactococcus lactis* the MIP channel transport water *and* glycerol.)

In *E. coli*, aquaporin AqpZ is encoded by gene *aqpZ*, and the glycerol facilitator GlpF is encoded by gene *glpF*. In *aqpZ* mutants growth is reported to be poor under low osmotic pressure, while in wild-type strains expression of *aqpZ* is stimulated under these conditions; this suggests that AqpZ has an important physiological role.

MIP channels have been found in some archaeans (e.g. *Archaeoglobus fulgidus*, *Methanobacterium thermoautotrophicum*) but are apparently absent in others (e.g. *Methanococcus jannaschii*); they also seem to be absent in certain bacteria (e.g. *Chlamydia trachomatis*, *Helicobacter pylori*, *Mycobacterium tuberculosis*, *Treponema pallidum*).

[The importance of aquaporin protein structures: EMBO Journal (2000) *19* 800–806. Microbial MIP channels: TIM (2000) *8* 33–38.]

Voltage-dependent (voltage-gated) ion channels occur in eukaryotic cell membranes but, until recently, have not been reported in prokaryotes; these channels open and close in response to changes in transmembrane voltage (see pmf in section

5.1.1.2). Recently, a voltage-dependent K^+ channel was described in an archaean (and a similar gene has been found in the bacterium *Pseudomonas aeruginosa*) [Nature (2003) *422* 180–185].

2.2.8.1   *Protoplasts*

If a cell loses its cell wall (Fig. 2.2) the resulting structure is called a *protoplast*. Although bounded only by the CM, a protoplast can survive (in the laboratory) and can carry out many of the processes of a normal living cell. However, if a protoplast be suspended in a medium more dilute than its cytoplasm, water will pass in through the CM (by osmosis) and the protoplast will swell and burst – an event known as *osmotic lysis*. In an intact bacterium, the delicate protoplast is usually saved from osmotic lysis by the mechanical strength of the cell wall (section 2.2.9).

2.2.8.2   *The cytoplasmic membrane in members of the Archaea*

In these organisms the CM contains lipids which do not occur in members of the Bacteria. Unlike the ester-linked glycerol–fatty acid bacterial lipids (Fig. 2.5), the archaeal lipids are characteristically ether-linked molecules which contain e.g. iso-prenoid or hydro-isoprenoid components. Some archaeal and bacterial CM lipids are structurally analogous – e.g. the di-ether and di-ester lipids, both types of molecule having a single polar end. However, some archaeal lipids (e.g. tetra-O-di(biphytanyl) diglycerol) contain two ether-linked glycerol residues – one at *each* end of the molecule; such molecules, having two polar ends, may span the width of the CM.

2.2.9   Cell wall

In most bacteria a tough outer layer – the cell wall (Fig. 2.2) – protects the delicate protoplast (section 2.2.8.1) from mechanical damage and osmotic lysis; it also deter-mines a cell's shape: an *isolated* protoplast is spherical, regardless of the shape of its original cell. Additionally, the cell wall acts as a 'molecular sieve' – a permeability barrier which excludes various molecules (including some antibiotics). However, the cell wall should not be thought of merely as an 'inert box' enclosing a living cell: it also plays an active role e.g. in regulating the transport of ions and molecules.

The cell walls of different species may differ greatly in thickness, structure and composition. However, there are only two major types of cell wall; whether a given cell has one or the other type of wall can generally be determined by the cell's reaction to certain dyes. Thus, when stained with crystal violet and iodine, cells with one type of wall retain the dye even when treated with solvents such as acetone or ethanol; cells with the other type of wall do not retain the dye (i.e. they become decolorized) under similar conditions. This important staining procedure was discovered empirically in the 1880s by the Danish scientist Christian Gram; the *Gram stain* is described in

Table 2.1 Some genera of Gram-negative and Gram-positive bacteria

Gram-negative	Gram-positive
Aeromonas	Bacillus
Bacteroides	Clostridium
Bordetella	Lactobacillus
Brucella	Listeria
Escherichia	Propionibacterium
Haemophilus	Staphylococcus
Pseudomonas	Streptococcus
Salmonella	Streptomyces
Thiobacillus	Streptoverticillium

section 14.9.1. Bacteria which retain the dye (and which have one type of wall) are called *Gram-positive* bacteria; those which can be decolorized (and which have the other type of wall) are called *Gram-negative* bacteria. Some named examples of Gram-negative and Gram-positive bacteria are listed in Table 2.1. The cell walls of Gram-positive and Gram-negative bacteria are described below. (In practice, some bacteria do not give a clear-cut result in the Gram stain: see section 14.9.1 for the meaning of *Gram-type-positive* and *Gram-type-negative*.)

2.2.9.1 *Gram-positive-type cell walls*

The Gram-positive-type wall is relatively thick (about 30–100 nm) and it generally has a simple, uniform appearance under the electron microscope. Some 40–80% of the wall is made of a tough, complex polymer, *peptidoglycan* (also called *murein*). Essentially, peptidoglycan consists of linear heteropolysaccharide chains that are cross-linked (by short peptides) to form a three-dimensional net-like structure (the *sacculus*) (Fig. 2.7) which envelops the protoplast.

In this type of wall the sacculus consists of *multilayered* peptidoglycan which, during growth, develops by the 'inside-to-outside' mechanism. In this process, new peptidoglycan is added to the inner (cytoplasmic) face of the wall; as growth continues, layers move outwards towards the cell surface, the oldest layers eventually being shed in fragments.

Covalently bound to peptidoglycan are compounds such as *teichoic acids*: typically, substituted polymers of glycerol phosphate or ribitol phosphate. [Teichoic acids in the actinomycete *Nocardiopsis*: FEMSMR (2001) *25* 269–283.] Some bacterial walls contains lipids; for example, *Mycobacterium* (and related bacteria) have *mycolic acids*: α-substituted, β-hydroxylated fatty acids (whose chain length, in *Mycobacterium*, is typically C_{50}–C_{90}). In some bacteria the wall contains carbohydrates (see e.g. *Streptococcus* in the Appendix).

The composition of the cell wall can vary with growth conditions; for example, in

Bacillus, the availability of phosphate affects the amount of teichoic acids associated with the wall.

Given the thickness of this type of wall – and the presence of hydrophobic components (e.g. mycolic acids) in some species – it may be thought that cells with Gram-positive-type walls would be ill-fitted to an aqueous environment (see section 3.1.3). It's therefore relevant that hydrophilic channels (also called 'porins') have been reported to run through the wall in *Streptomyces griseus* [MM (2001) *41* 665–673] and *Corynebacterium glutamicum* [JB (2003) *185* 4779–4786].

2.2.9.2 *Gram-negative-type cell walls*

The Gram-negative-type wall (20–30 nm thick) has a distinctly layered appearance under the electron microscope. The inner layer (Fig. 2.6) – nearest the cytoplasmic membrane – is widely believed to consist of a 'periplasmic gel', 15 nm thick, representing 1–10% of the dry weight of the cell; this view (based on e.g. electron microscopy) suggests that the peptidoglycan is multilayered. However, experiments on the turnover rate of peptidoglycan in growing cells of *Escherichia coli* suggest that the lateral cell wall may contain a *monolayer* of peptidoglycan [JB (1993) *175* 7–11].

The chemical composition of the peptidoglycan in *E. coli* is shown in Fig. 2.7, and its mode of synthesis is outlined in section 6.3.3.1. (Interestingly, although the cell wall of *Chlamydia trachomatis* lacks peptidoglycan, this organism was reported to contain the genes for peptidoglycan biosynthesis [but see AAC (1999) *43* 2339–2344].) Peptidoglycan is discussed further in sections 3.2.1, 5.4.8 and 5.4.9.

The outer layer of the wall, the so-called *outer membrane* (Fig. 2.6, Plate 2.3, *top*,

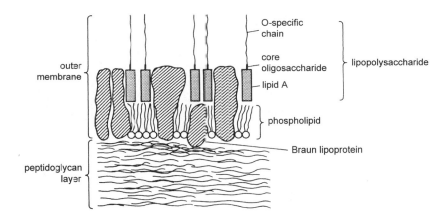

Figure 2.6 The Gram-negative-type cell wall (diagrammatic). O-specific chains form the outermost part of the wall. The outer membrane – an important permeability barrier that blocks e.g. certain antibiotics – is linked to peptidoglycan via the Braun lipoproteins. At the left-hand side of the outer membrane is shown a *porin* (section 2.2.9.2).

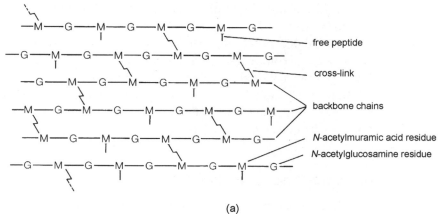

(a)

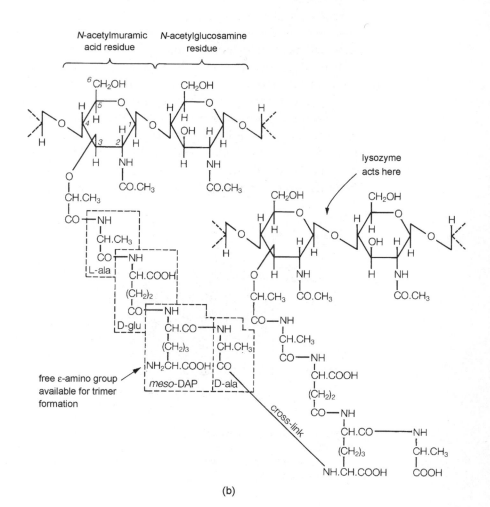

(b)

left), is essentially a protein-containing lipid bilayer – i.e. it resembles the cytoplasmic membrane. However, while the inward-facing lipids are phospholipids, the outward-facing lipids are macromolecules called *lipopolysaccharides* (LPS).

Lipid A (Fig. 2.6) is a glycolipid. In *E. coli* it consists of a phosphorylated disaccharide (two glucosamine residues) carrying six (lipophilic) acyl chains, four of the chains being on one of the glucosamine residues (to which is also linked the core oligosaccharide); all six acyl chains project into the lipophilic interior of the outer membrane.

Lipid A biosynthesis involves a precursor which is also used in the synthesis of peptidoglycan: UDP-*N*-acetylglucosamine (UDP-GlcNAc) (section 6.3.3.1). In lipid A biosynthesis, UDP-GlcNAc is initially mono-acylated by an enzyme encoded by gene *lpxA*. Then, de-acetylation (by a zinc-containing enzyme, de-acetylase, encoded by *lpxC*) is followed by a multi-step process resulting in the hexa-acylated disaccharide (lipid A). [A useful account of lipid A biosynthesis: TIM (1998) 6 154–159.]

Ideas for new candidate antibiotics have included compounds such as L-573,655 [Science (1996) *274* 980–982], directed against lipid A biosynthesis, and a synthetic

---

Figure 2.7 Peptidoglycan. The structures shown are typical of those in *Escherichia coli*; similar types of peptidoglycan are found in many other Gram-negative bacteria.

(a) The (three-dimensional) net-like peptidoglycan molecule: backbone chains of alternating residues of *N*-acetylglucosamine (G) and *N*-acetylmuramic acid (M) are held together by short peptides which link *N*-acetylmuramic acid residues. Each of the peptide bridges shown in the diagram links a given backbone chain to *one* other backbone chain. However, there are also peptide bridges which link together more than two backbone chains; the different types of peptide bridge in *E. coli* are described briefly in (b), below.

(b) Part of two adjacent backbone chains showing the mode of peptide linkage between them. (The numbers in italic are used to refer to particular carbon atoms within the molecule.) Each *N*-acetylmuramic acid residue bears a short *stem peptide* – in this example the tetrapeptide L-alanine–D-glutamic acid–*meso*-diaminopimelic acid(*meso*-DAP)–D-alanine (shown in dotted boxes); in this kind of peptide bridge there is a direct, covalent link between the D-alanine of one stem peptide and the ε-amino group of *meso*-DAP in the other stem peptide. As this particular type of peptide bridge consists of two stem peptides, each of four amino acid residues, it has been referred to as a *tetra-tetra dimer* (or a tetra-tetra). Another type of peptide bridge, the *tetra-tri dimer* (tetra-tri), is formed from one tetrapeptide and one tripeptide. A *trimer* peptide bridge is one which links together *three* of the glycan backbone chains; this occurs when a stem peptide on a *third* backbone chain forms a covalent linkage with the free ε-amino group (see diagram) of a dimer bridge. An appreciation of these different types of peptide bridge is essential for understanding a model for the replication of the peptidoglycan sacculus (section 3.2.1).

The enzyme *lysozyme* hydrolyses the *N*-acetylmuramyl-(1 → 4) linkages in the backbone chain (see diagram); such activity weakens the cell envelope.

In Gram-positive bacteria the peptidoglycan often differs from that shown above. For example, residues of *N*-glycollylmuramic acid occur in the backbone chains of *Mycobacterium*, and there are many differences in the types and positions of the cross-links. In the peptidoglycan of *Staphylococcus aureus* some of the muramic acid residues are not acetylated, and the stem peptides are linked by a penta- or hexa-glycine bridge; this bridge is susceptible to cleavage by the enzyme *lysostaphin* – an enzyme which can therefore lyse cells of this species.

Synthesis of peptidoglycan depends on the enzymic roles of the *penicillin-binding proteins* (PBPs) which occur in the cell envelope (sections 2.2.8 and 6.3.3.1); PBPs can be inactivated by penicillin and by related *β*-lactam antibiotics (section 15.4.1).

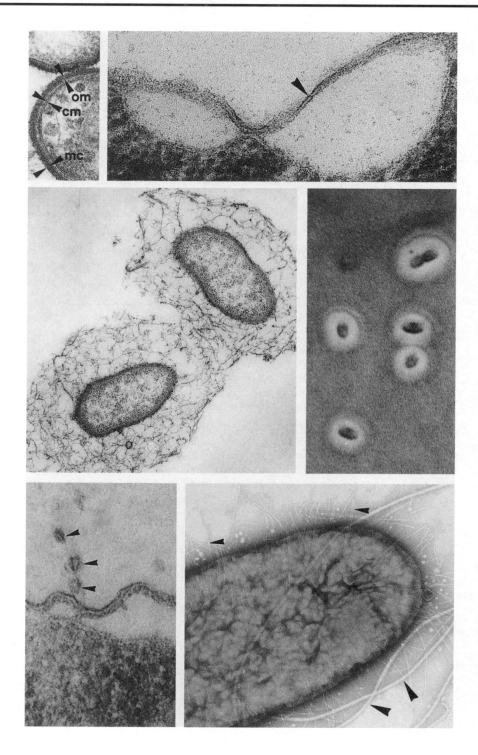

analogue of lipid A, E5531, which has useful activity against endotoxic shock (section 11.3.4) in humans [TIM (1998) 6 154–159].

The core oligosaccharide (in e.g. *E. coli* and related bacteria) contains glucose and galactose residues and substituted residues of other sugars, including heptose phosphate.

The O-specific chains, which form the outermost part of the cell wall, consist of linear or branched chains of oligosaccharide subunits; the chemical composition of the O-specific chain can vary from strain to strain, and this is exploited in the identification of particular strains by serology (section 16.1.5.1). (*Note.* Some bacteria (e.g. *Neisseria meningitidis*) lack the *E. coli*-type O-specific chain; in these bacteria the structure analogous to LPS is called *lipo-oligosaccharide* (LOS).) [LOS: CRM (1998) 24 281–334.] Variability of LOS chemotypes in *Neisseria gonorrhoeae* appears to involve changes in transcription within the *IgtABCDE* gene cluster [JB (2004) 186 1038–1049].

About half the mass of the outer membrane consists of proteins; in e.g. *E. coli* and related bacteria these include thousands of molecules of the *Braun lipoprotein* (Fig. 2.6) each of which is linked to the underlying peptidoglycan. There are also enzymes, proteins involved in specific uptake ('transport') mechanisms, and *porins*. A porin is a protein molecule which forms part, or all, of a water-filled channel ('pore') that spans the thickness of the outer membrane and allows the passage of certain ions and molecules. Porins are commonly trimeric (i.e. three porin molecules form the channel – as in *E. coli*) but sometimes they are dimeric [as in *Paracoccus denitrificans*: JB (1985) 162 430–433] or monomeric [as in cyanobacteria: JB (2000) 182 1191–1199]. (*Note.*

Plate 2.3 The cell envelope and surface structures in some Gram-negative bacteria. *Top left.* The outer membrane (om) and cytoplasmic membrane (cm), and a microcapsule (mc), in a thin section of *Bacteroides fragilis* (×52000). The microcapsule seen in this electronmicrograph would not be seen by light microscopy. *Top right.* The cell envelope in a thin section of a plasmolysed cell of *Escherichia coli* (×100 000). (Plasmolysis, in which water is withdrawn from the cell, causes the cytoplasmic membrane to shrink away from the outer membrane.) Here, gaps are seen between the outer membrane (arrowhead) and the cytoplasmic membrane below it. Near the centre of the photograph, the outer membrane is joined to the cytoplasmic membrane; this 'fused' region is called an adhesion site. *Centre left.* Two capsulated cells of *B. fragilis* as seen by electron microscopy. The macrocapsules have been stained with ruthenium red. *Centre right.* Light microscopy of cells of *B. fragilis* from the same culture as those seen to the left (approx. ×1000). Here, the background has been stained darkly with eosin–carbolfuchsin ('negative staining') so that each capsule can be seen as a bright 'halo' surrounding its cell. The cell at the top is dividing. *Bottom left.* An F pilus (a specialized, hair-like appendage) arising from the cell envelope in *E. coli* (×150000). The F pilus itself (less than 10 nm in diameter) is barely visible; however, in order to make it detectable, it has been 'labelled' with a particular bacteriophage (MS2) which binds specifically to the sides of these pili. Here, three MS2 bacteriophages (each about 25 nm in diameter) have bound, close together, along part of the length of the F pilus; each bacteriophage is indicated by an arrowhead. *Bottom right.* Part of an *E. coli* cell specially stained to show the large number of fimbriae (indicated by small arrowheads) (×54000). The large arrowheads point to fragments of flagella.

Photographs of *B. fragilis* courtesy of Dr Sheila Patrick, Queen's University of Belfast. Adhesion site and F pilus in *E. coli* courtesy of Dr Manfred E. Bayer, Fox Chase Cancer Center, Philadelphia. *E. coli* fimbriae courtesy of Dr Anne Field, Public Health Laboratory Service, London.

The term 'porin' is also used to refer to the *complete* pore structure, whether it be composed of one, two or three porin proteins.)

In *E. coli* and related bacteria, the outer membrane contains several different types of porin which are designated e.g. OmpC, OmpF etc. (Omp = outer membrane protein); the relative proportions of these porins can vary according to the cell's environment (see e.g. section 7.8.6). The number of *types* of porin in a cell can also vary; for example, in *E. coli* the PhoE porin (concerned with the uptake of phosphate) is synthesized when levels of phosphate are low.

Variation in a cell's porins may be associated with an altered level of suceptibility to specific antibiotics [see e.g. Microbiology (1998) *144* 3003–3009].

As well as their role as channel components, porins may act e.g. as receptor sites for phages (Chapter 9); for example, in *E. coli* the OmpC porin is a receptor for phages T4 adn TuIb.

During infection by Gram-negative bacteria, porins may induce the release of certain *cytokines* (Chapter 11) from monocytes and lymphocytes.

Components of the outer membrane are held together by ionic and other interactions. Adjacent core oligosaccharides appear to be linked by divalent cations, particularly Mg^{2+} and Ca^{2+}; experimental removal of these cations (by ion chelators) can disrupt the outer membrane. Lipid A is hydrophobically bound to the fatty acid residues of phospholipids. Some proteins appear to be linked to the core oligosaccharides.

The outer membrane is generally permeable to small ions and to small hydrophilic molecules – but much less permeable (or impermeable) to hydrophobic or amphipathic molecules. Permeability is increased (with adverse effects) by certain antibiotics, e.g. polymyxins (section 15.4.5), and is likely to be increased by those agents which inhibit lipid A biosynthesis (section 15.4.11).

Cyanobacteria are generally categorized as Gram-type-negative bacteria. However, in many cases the cell wall in these organisms exhibits features of both Gram-negative and Gram-positive species [JB (2000) *182* 1191–1199].

2.2.9.3 *Cell walls in members of the Archaea*

In some of these organisms (e.g. species of *Halobacterium*, *Methanococcus* and *Thermoproteus*) the wall consists mainly or solely of a so-called 'S layer' (section 2.2.12) closely associated with the cytoplasmic membrane. In e.g. *Methanobacter* the wall contains *pseudomurein*: a peptidoglycan-like polymer in which the backbone chains contain *N*-acetyl-D-glucosamine (and/or *N*-acetyl-D-galactosamine, depending on species) and *N*-acetyl-L-talosaminuronic acid. Unlike peptidoglycan, pseudomurein is not cleaved by the enzyme *lysozyme*.

2.2.9.4   Layers external to the cell wall

In some bacteria there are one or more layers external to the wall; these layers include e.g. capsules, S layers and M proteins (see later).

2.2.9.5   Wall-less bacteria

Prokaryotic cells which lack cell walls are found among members of the Bacteria (e.g. *Mycoplasma*) and the Archaea (e.g. *Thermoplasma*).

2.2.10   Adhesion sites (Bayer's patches) in Gram-negative bacteria

Adhesion sites are localized 'fusions' of the outer membrane and cytoplasmic membrane (Plate 2.3: *top*, *right*). They may be seen under conditions of plasmolysis and may give important clues about the cell's physiology [MM (1994) *14* 597–607]. Suggested functions of these sites include the export of components of the outer membrane, and secretion of proteins. In some cases, these sites may form during infection of a bacterial cell by certain types of bacteriophage (Chapter 9), the fusion being created by proteins ejected from the phage [see e.g. MM (2001) *40* 1–8].

2.2.11   Capsules and slime layers

In some bacteria the outer surface of the cell wall is covered with a layer of material called a *capsule*. A *macrocapsule*, or 'true' capsule, is thick enough to be seen (in suitable preparations) under the ordinary light microscope (i.e. thicker than about 0.2 μm), while a *microcapsule* can be detected only by electron microscopy or e.g. serological techniques (Plate 2.3: *centre*, and *top left*). A *slime layer* is a watery secretion which adheres loosely to the cell wall; it commonly diffuses into the medium when a cell is growing in a liquid environment.

Capsules are composed mainly of water; the organic part is usually a homopolysaccharide (e.g. cellulose, dextran) or a heteropolysaccharide (e.g. alginate, colanic acid [genes for colanic acid in *E. coli*: JB (1996) *178* 4885–4893], hyaluronic acid), but in some strains of e.g. *Bacillus anthracis* the capsule is a homopolymer of D-glutamic acid. Species of *Xanthobacter* can form an α-polyglutamine capsule together with copious polysaccharide slime. Capsule-to-wall binding may be ionic and/or covalent.

Capsules have various functions. For example, they may (i) help to prevent desiccation; (ii) act as a permeability barrier to toxic metal ions; (iii) prevent infection by bacteriophages (Chapter 9); (iv) act as a nutrient reserve; (v) promote adhesion – important e.g. in those bacteria which form dental plaque; and (vi) help the cell to avoid phagocytosis. In pathogenic bacteria, capsule formation often correlates with pathogenicity (i.e the ability to cause disease): in a given, normally capsulated pathogen, capsule-less strains are typically non-pathogenic. (See also sections 11.3.2.1 and 11.5.1.)

Capsules can mask *adhesins* (section 11.5.6) [JB (2004) *186* 1249–1257].

Some secreted polysaccharides are used industrially. For example, the hetero-polysaccharide *xanthan gum* (produced by strains of *Xanthomonas campestris*) is used in the food industry as a gelling agent, gel stabilizer, thickener, and inhibitor of crystallization.

2.2.12   S layers

Some cells have a so-called S layer – usually as the outermost layer of the cell; an S layer consists of a repeating pattern of protein or glycoprotein subunits arranged e.g. in squares or hexagons. In those bacteria which have an S layer, the S layer usually overlays the cell wall – e.g. in Gram-negative bacteria it covers the outer membrane. S layers occur e.g. in strains of *Aeromonas*, *Campylobacter*, *Clostridium*, *Pseudomonas* and *Treponema*. Double S layers, containing the same or different subunits, occur e.g. in strains of *Aquaspirillum*, *Bacillus* and *Corynebacterium*. [Secretion of S layers in Gram-negative bacteria: FEMSMR (2000) *24* 21–44 (35–38).]

In a strain of *Bacillus stearothermophilus*, variation in the S layer has been found to involve a chromosome–plasmid re-arrangement of DNA [JB (2001) *183* 1672–1679].

S layers also occur in archaeans (e.g. strains of *Desulfurococcus*, *Halobacterium*, *Methanococcus* and *Sulfolobus*). In some archaeans (e.g. *Methanococcoides*) the S layer *is* the cell wall, i.e. it overlays the cytoplasmic membrane.

2.2.13   M proteins

Molecules of M protein form a thin layer on the cell wall of the pathogen *Streptococcus pyogenes* – a Lancefield group A streptococcus (see *Streptococcus* in the Appendix) that causes e.g. 'sore throat', scarlet fever and necrotizing fasciitis; M protein makes the bacterium less susceptible to phagocytosis, thus promoting virulence. M protein can also act as an *adhesin* (section 11.5.6).

The composition of M protein differs slightly in different strains of *S. pyogenes*. This enables strains to be classified by serological tests (section 16.1.5.1) into about 60 groups (*Griffith's serogroups*) – all the strains in a given group having a similar type of M protein; this is one example of *typing* (section 16.1.5) that is useful in epidemiology.

2.2.14   Flagella, fimbriae and pili

In many bacteria there are fine, hair-like proteinaceous filaments extending from the cell surface; these filaments can be divided into three main types: flagella, fimbriae and pili.

2.2.14.1 *Flagella*

Flagella (singular: *flagellum*) enable a cell to swim through a liquid medium, i.e. they are involved in cell motility (section 2.2.15). Depending on species, a cell may have a single flagellum (*monotrichous* arrangement); one flagellum at each end (*amphitrichous* arrangement); a tuft of flagella at one or both ends (monopolar or bipolar *lophotrichous* arrangement); or flagella which arise at various points on the cell surface (*peritrichous* arrangement) (see Fig. 2.1)

Each flagellum consists of a *filament, a hook* and a *basal body* (Fig. 2.2). The filament is helical, $5-20 \times 0.02$ μm, and is composed of eleven protein fibrils arranged like the strands of a rope; a fine channel (about 70 Å in diameter) runs through the axis of the filament. (In some species of e.g. *Pseudomonas* and *Vibrio*, and in *Helicobacter pylori*, the filament is sheathed, i.e. covered by an extension of the outer membrane – see e.g. Plate 2.1 *top, left.*) The hook is a hollow, flexible, proteinaceous structure (Fig. 2.8). Flagellar *filaments* can be seen by light microscopy, but only after special staining. [Simple flagellar stain: JCM (1989) *27* 2612–2615.]

Each flagellum originates at a *basal body* (= *flagellar motor*): a complex energy-converting mechanism which spans the thickness of the cell envelope. Essentially, the motor consists of a number of ring-shaped structures mounted coaxially on a hollow, rod-shaped shaft, together with associated components in the cytoplasmic membrane (Fig. 2.8); precise details (e.g. number of rings) differ according to species.

The flagellum *rotates on its axis*. Rotation has been demonstrated by anchoring the free end of the flagellum to a glass surface and observing rotation of the cell body. The current view is that (in e.g. *E. coli* and other enteric bacteria), the MS ring, together with the C ring, rotates relative to surrounding stationary components, and that this causes rotation of the shaft and, hence, filament; the L and P rings are believed to be anchored in the cell envelope and to form a 'bush' for the rotating shaft (Fig. 2.8).

Flagellar rotation needs energy in the form of an ion gradient across the cytoplasmic membrane (see Chapter 5) and commonly involves a flow of protons through the flagellar motor. Sodium motive force (section 5.3.6) drives the polar flagellum in *Vibrio cholerae* [JB (1999) *181* 1927–1930]. [Models for the flagellar motor (theoretical considerations): PTRSLB (2000) *355* 491–501.]

Assembly of the flagellum. Flagellar components are assembled in strict sequence. The following refers to assembly in *E. coli* and related bacteria. First, the MS ring is inserted into the cytoplasmic membrane; the C ring is then added. For further assembly, many of the components pass *through* the MS ring in the order in which they are incorporated into the flagellum: first, subunits for the rod (shaft), then those of the hook, and finally subunits of the filament. Thus, during the final stage of assembly, the protein (*flagellin*) subunits for the (elongating) filament pass through the filament's axial channel and are added sequentially to the growing tip.

Clearly, an open channel from the cytoplasm (where components are synthesized)

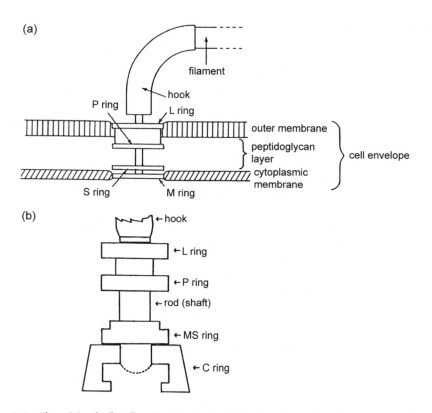

Figure 2.8 The origin of a flagellum (section 2.2.14.1) in the cell envelope of a Gram-negative enteric bacterium (e.g. *E. coli*) (diagrammatic, not drawn to scale).

(a) An earlier model.
(b) A more recent view [based on e.g. Kihara *et al.* (1996) JB *178* 4582–4589; Katayama *et al.* (1996) JMB *255* 458–475]. The L and P rings are associated with the outer membrane and peptidoglycan layer, respectively. S and M rings are now seen as a single structure, the MS ring, in the cytoplasmic membrane; it comprises multiple copies of the FliF protein (Table 7.3).

The C ring ('switch'), bound to the MS ring, consists of proteins FliG, FliM and FliN (Table 7.3). FliG adjoins the MS ring; FliM and FliN are distal components.

Encircling the MS ring in the cytoplasmic membrane are the MotA/MotB proteins; these are stationary, MotB being linked to peptidoglycan. The L, P, MS and C rings, the shaft, and the MotA/MotB proteins comprise the *flagellar motor* (=*basal body*). Torque generation (flagellar rotation) appears to result from interaction between FliG and MotA, with energy provided by an inward flow of protons through the motor. The C ring is also involved in chemotaxis (Fig. 7.13).

Katayama *et al.* (cited above) mention a putative 'export apparatus' within the C ring (dotted line).

to the cell's exterior (where the filament is assembled) could not be tolerated because it would compromise the structural integrity of the cell. Hence, a so-called *export apparatus* (Fig. 2.8) must perform a 'gatekeeper' function by permitting only appropriate proteins to pass through the MS ring. The export apparatus and the C ring may be assembled with, or soon after, the MS ring. Moreover, the distal (growing) end of the filament has a *cap* (composed of FliD protein) which remains in place while flagellin subunits are being added to the growing filament; incorporation of the flagellin subunits may be facilitated by *rotation* of the flagellar cap [Science (2000) *290* 2148–2152].

Unlike other flagellar proteins, those which form the L and P rings have signal sequences (section 7.6) and are apparently transported across the cytoplasmic membrane via the general secretory pathway (section 5.4.3) rather than through the MS ring.

Interestingly some proteins of the basal body exhibit homology with proteins of type III secretory systems (section 5.4.4).

Genetic control of flagellar assembly is outlined in section 7.8.2.7.

[Genetics and assembly of flagella in Gram-negative bacteria: FEMSMR (2000) *24* 21–44 (22–27).]

Flagella in spirochaetes. Spirochaetes are Gram-negative bacteria which have a unique structure. One (or usually many) flagella originate from *each* end of the cell and extend – between the peptidoglycan layer and outer membrane – towards the centre of the cell (where, in some species, flagella from both ends overlap). The number of these so-called *periplasmic flagella*, per cell, varies with species. For example, in *Leptonema illini* a single flagellum arises at each end but there is no overlap at the centre of the cell; in *Borrelia burgdorferi* the seven or so flagella from each end do overlap. The periplasmic flagella appear to *rotate* in a way similar to that of other flagella [JB (1992) JB *174* 832–840]. The periplasmic flagella of *Treponema pallidum* (the causal agent of syphilis) are shown in Plate 2.4. [Structure/motility of spirochaetes: JB (1996) *178* 6539–6545.]

The flagellum in members of the Archaea. The archaeal flagellum differs markedly from that of bacteria in composition, structure and apparent mode of assembly. For example, the archaeal filament subunit protein, flagellin, is typically glycosylated, and the filament itself is typically much thinner than the bacterial filament. Moreover, archaeal flagellin contains a signal sequence (section 7.6), suggesting passage into the membrane, unlike bacterial flagellin (which passes through the hollow structures of the developing flagellum); this feature, and certain similarities between archaeal flagellins and type 4 fimbriae have suggested that the archaeal flagellum is actually assembled from the *base* (in contrast to the 'tip growth' in bacterial flagella) [JB (1996) *178* 5057–5064].

Genes homologous to those which encode bacterial flagellar structures have not been detected in archaeans, even in those whose complete genome sequence is known. This has suggested that the archaeal flagellum is a unique form of organelle [FEMSMR (2001) *25* 147–174].

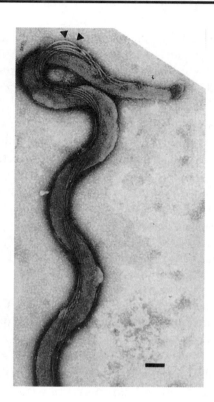

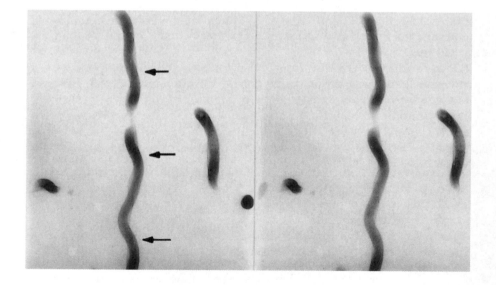

2.2.14.2 *Fimbriae*

Fimbriae (singular: *fimbria*) are quite common on Gram-negative bacteria (including cyanobacteria), less common on Gram-positive species. This account refers specifically to the fimbriae of Gram-negative bacteria.

Fimbriae are functionally distinct from flagella and pili – but are often (confusingly) called pili (see section 2.2.14.3). Fimbriae may occur all over the cell (as in Plate 2.3, *bottom right*) or they may be localized to particular parts of the cell surface. Different types of fimbria may occur on the same cell.

Each fimbria is essentially a rod-like structure of uniform width, typically 2–8 nm in diameter and from 0.1 μm to several micrometres in length. While a fimbria may contain more than one type of component, the main structure is composed of a single type of protein subunit (called either *fimbrillin* or *pilin*). Each fimbria is formed by a linear sequence of identical protein subunits which, in some cases, are associated with a small number of different (often specialized) subunits; in at least some fimbriae (e.g. the 'common' type I fimbriae of *E. coli*) the linear sequence of subunits appears as a tightly coiled helix under the electron microscope. Fimbrial subunits are often rich in non-polar amino acids; consequently, fimbriate cells are commonly more hydrophobic than afimbriate cells (hydrophobicity being indicated e.g. by the tendency of cells to congregate at an air–water or oil–water interface).

Some type of fimbria promote cell–cell adhesion or adhesion between the bacterium and substratum. The adhesive fimbriae of some pathogenic bacteria are important factors in the process of infection (sections 11.2.1 and 11.5.6); for example, in *Neisseria* spp the fimbrial tips have specific protein *adhesins* which promote binding between these pathogens and the host cells.

In the laboratory, bacteria which have certain types of adhesive fimbria can be

---

Plate 2.4 Spirochaetes. *Top*. Part of a cell of *Treponema pallidum* showing the periplasmic flagella (arrowheads). The scale bar is 0.1 μm.

Bottom. Stereoelectronmicrographs of part of a cell of *Borrelia burgdorferi* taken by high-voltage (1 MV) electron microscopy with a tilt angle of ~10° (scale: 13 mm = 1 μm). Arrows indicate regions where the flagellar bundle is on the 'observer-side' of the cell. The lower part of the cell is seen 'flat-on', i.e. a view which shows the *flattened* waveform maximally. An axial twist is located near the centre of the photograph; part of the cell which curved outwards (towards the observer) has been sliced off during preparation of the section – showing that the flagellar bundle runs underneath.

Photograph of *T. pallidum* reprinted from *Medical Microbiology* 14th edn, Greenwood, D. *et al.* (eds), 1992, p. 419, by courtesy of Dr Charles W. Penn, University of Birmingham, UK, and with permission from the copyright owner, Elsevier.

Photograph of *B. burgdorferi* reprinted from *Journal of Bacteriology* (1996) *178* 6539–6545 with permission from the American Society for Microbiology, and by courtesy of the authors: Dr S. Goldstein (University of Minnesota), Dr Karolyn Buttle (New York State Department of Health) and Professor N. W. Charon (West Virginia University).

detected by their ability to cause *haemagglutination*, i.e. the binding together of erythrocytes (red blood cells) into visible clumps; if this effect is inhibited by the presence of D-mannose it is called mannose-sensitive haemagglutination (MSHA). Other types of fimbria may cause mannose-resistant haemagglutination (MRHA) – which is not affected by the presence of D-mannose.

The type, number and composition of fimbriae on a cell depend on various factors. Clearly, for a given type of fimbria to be present its protein subunits must be encoded by the cell's chromosome(s) or plasmid(s) (Chapter 7). Even then, however, factors such as temperature may be important; for example, K88 and K99 fimbriae are not formed if the temperature is below ~18°C. (See also section 7.8.11.) Moreover, in some cases, a genetic switching mechanism can determine either the presence/absence or composition of a given type of fimbria. For example, the common type I fimbriae of *E. coli* can be switched 'on' (i.e. present) or 'off' (i.e. absent) by inversion of a particular region of DNA (Chapter 7) in a relevant gene; in principle, this switching mechanism resembles *phase variation* in the flagella of *Salmonella* – see Fig. 8.3(c). Again, in the pathogen *Neisseria gonorrhoeae*, the major subunit of the fimbriae is encoded by a gene which can undergo repeated *recombination* (section 8.2) with another gene in the same chromosome – giving rise to a very large number of different forms of the subunit (so that fimbriae can vary in composition in different cells of *N. gonorrhoeae*) [MM (1996) *21* 433–440].

Interestingly, there is evidence of 'cross-talk' between the genetic systems of two types of fimbria: common type I fimbriae and the so-called P fimbriae. Thus, a protein involved in the regulation of P fimbriae has been reported to inhibit the expression of common type I fimbriae [EMBO Journal (2000) *19* 1450–1457].

The fimbriae of Gram-negative bacteria can be classified according to criteria such as (i) the homology of amino acid sequences in the major subunit proteins, (ii) physical, antigenic and adhesive properties, and (iii) the mechanism of secretion and assembly. Examples of different types of fimbria are given in Table 2.2.

Secretion and assembly of fimbriae. Unlike flagella, most or all fimbriae appear to be assembled from the *base*. The common type I fimbriae of *E. coli* are assembled as follows. Subunits are synthesized in the cytoplasm, each with a signal seqeuence (section 7.6), and are transported across the cytoplasmic membrane in a *sec*-dependent manner (section 5.4.3). In the periplasmic region (section 2.2) each subunit is bound by a *chaperone* (the FimC protein) which folds the subunit protein and apparently directs it to the outer membrane.

In the outer membrane, a number of molecules of the *usher protein* (FimD) form a ring-shaped structure through which the oligomerized (linked) subunits of the nascent fimbria are translocated as a linear (not helical) filament. The first subunits to be externalized are those forming the distal end of the fimbria (a specialized region

Table 2.2 Fimbriae of Gram-negative bacteria: some examples

Fimbria	Species/strain	Comments
Type I (= type 1) (a) ('common pili' *or* 'common type I fimbriae' *or* F1 fimbriae)	*Escherichia coli* and other enterobacteria	Chromosomally encoded (by the *fim* operon); subject to on/off switching (see text). Each fimbria is ~7 nm in diameter. MSHA (see text) with e.g. guinea-pig RBCs. The fimbriae can bind to *uroplakins* (proteins on the luminal surface of mammalian bladder) – a factor relevant to colonization of the bladder by UPEC (uropathogenic *E. coli*) [PNAS (2000) *97* 8829–8835]. The fimbriae (specifically, FimH: see text) can also promote *invasion* of human urinary tract epithelium [EMBO Journal (2000) *19* 2803–2812]. [FimH-mediated invasion in urinary tract infections: PNAS (2000) *97* 8829–8835.]
(b) P fimbriae	Strains of *E. coli* associated with cystitis and pyelonephritis	Also called 'Pap pili'. Chromosomally encoded by the *pap* operon (similar to the *fim* operon); expression is temperature-regulated (see section 7.8.11). P fimbriae mediate MRHA (see text). Receptors for P fimbriae occur on uroepithelium (see section 11.2.1). Reported to be a highly potent elicitor of the *cytokine* IL–6 (interleukin–6) which can stimulate synthesis of immunoglobulins and promote the release of acute-phase proteins. PapB, a regulatory protein of the *pap* operon, is reported to inhibit expression of *fim*-encoded fimbriae [EMBO Journal (2000) *19* 1450–1457].
(c) K88 (F4 fimbriae)	*E. coli* strains pathogenic for piglets	Plasmid-encoded. The major subunit is an adhesin. The (short) fimbriae seen under the electron microscope may be broken fimbriae [FEMSMR (1996) *19* 25–52 (35)].
(d) K99 (F5 fimbriae)	*E. coli* strains pathogenic for calves and lambs	Plasmid-encoded. The major subunit has adhesive properties.
Type II (= type 2)	Enterobacteria including strains of *Salmonella*	Similar to type I fimbriae but lack the adhesive and haemagglutinating properties.
Type III (= type 3)	Enterobacteria	Mannose-insensitive adhesion. Ox RBCs are agglutinated only when previously treated with tannic acid ('tanned ox' haemagglutination). [Type III fimbriae in *Klebsiella* spp: JMM (1985) *20* 203–214.]
Type IV (= type 4)	Various Gram-negative pathogens, including *Neisseria gonorrhoeae, N. meningitidis, Vibrio cholerae* (TCP: 'toxin co-regulated pili'), *Pseudomonas aeruginosa* and some strains of *E. coli*	Chromosomally encoded in most cases; the 'bundle-forming pili' of EPEC (see Table 11.2) are plasmid-encoded. All type IV fimbriae have a similar mode of secretion and assembly. Each fimbria is typically 5–6 nm in diameter, 1–2 μm in length; a plasmid-encoded type IV fimbria known as 'longus' (formed by strains of ETEC) can reach 20 μm in length [MM (1994) *12* 71–82]. Retractable (at approx. 1 μm/sec) [Nature (2000) *407* 98–102] and involved in twitching motility (see section 2.2.15.1). [Force of retraction > 100 pN: PNAS (2002) *99* 16012–16017.] Type IV fimbriae are important virulence factors in certain pathogens. In e.g. *Neisseria gonorrhoeae* the type IV fimbriae are subject to antigenic variation. [Assembly of type IV fimbriae: FEMSMR (2000) *24* 21–44 (31–35). Role of PilC in retraction in pathogenic *Neisseria*: EMBO Journal (2004) *23* 2009–2017.]

known as the *fibrillum*): the protiens FimF, FimG and FimH. FimH is a mannose-binding *adhesin* which apparently forms the most distal part of the fimbria; FimG may have an important role in 'nucleating' the fibrillum [PNAS (2000) *97* 9240–9245]. Subsequently, FimA subunits, which form the major part of the fimbria, are translocated through the usher pore and added sequentially to the base of the growing fimbria; these subunits form a helical structure *after* they have passed through the usher pore.

Genes encoding the structure and assembly of fimbriae occur in *operons* (section 7.8.1); for example, the operon for common type I fimbriae includes genes *fimA* (major subunit), *fimC* (chaperone) and *fimD* (usher).

[Fimbriae in *E. coli*; FEMSMR (1996) *19* 25–52. Secretion and assembly of type I fimbriae: FEMSMR (2000) *24* 21–44 (27–31) and PNAS (2000) *97* 9240–9245.]

Unlike type I fimbriae, type IV fimbriae are assembled from a complex in the cytoplasmic membrane, i.e. the base of the fimbria is linked to the cytoplasmic membrane. [Model for the assembly of type IV fimbriae in *Neisseriea gonorrhoeae*: FEMSMR (2000) *24* 21–44 (31–35). Relationship between type II protein secretion (section 5.4.3) and the formation of type IV fimbriae: EMBO Journal (2000) *19* 2221–2228.]

2.2.14.3 *Pili*

Pili (singular: *pilus*) are elongated or hair-like proteinaceous structures which project from the cell surface; they are found specifically on those Gram-negative cells which have the ability to transfer DNA to other cells by *conjugation* (Chapter 8) – a process in which (in at least some cases) the pili themselves play an essential role. The genes encoding pili occur in genetic elements called *plasmids* (Chapter 7). Commonly, only one or a few pili occur on a given cell.

Note. The appendages described in this section are consistently referred to (by all authors) as 'pili' – but the appendages described in section 2.2.14.2 are sometimes referred to as 'fimbriae' and sometimes as 'pili'. (Some authors use the terms interchangeably and randomly, as though they were synonyms.) However, a logical use of language would require that different entities be referred to by different terms; this book follows the usage in *Dictionary of Microbiology and Molecular Biology* [ISBN 0-471-49064-4], i.e. the terms 'pili' and 'fimbriae' are used with distinct and mutually exclusive meanings.

The various types of pili differ e.g. in size and shape. For example, some are long, thin and flexible, while others are short, rigid and nail-like; the type of pilus correlates with the physical conditions under which conjugation can take place (Chapter 8).

The best-studied flexible pilus, the *F pilus*, is one to several micrometres long, 8–9 nm in diameter; a 'labelled' F pilus is seen in Plate 2.3 (*bottom, left*). The F pilus, a tubular structure with an axial canal about 2.5 nm in diameter, is made (mainly or solely) from a single type of subunit called *pilin*, a 70-amino-acid molecule which contains one residue of D-glucose and two phosphate residues. The way in which these subunits are assembled to form the pilus is not known. [Structure of F-pilin and a model for the organization of pilin subunits in the F pilus: MM (1997) *23* 423–429.]

2.2.15 Motility and chemotaxis

Many bacteria are *motile*, i.e. they can actively move about in liquid media. Not only can they move, they can also move towards better sources of nutrients and away from harmful substances; such a directional response is called *chemotaxis*.

2.2.15.1 *Motility*

In most cases motility is due to the possession of one or more flagella (section 2.2.14.1). Flagellar motility involves energy-requiring *rotation* of the flagellum from the basal body.

In peritrichously flagellate bacteria (e.g. *Salmonella*, *E. coli*), the flagella rotate independently of one another [Cell (1983) *32* 109–117]. Each flagellum rotates counterclockwise (CCW) for most of the time (about 95%) and clockwise (CW) for the rest of the time; the timing of the switch from CCW to CW rotation depends on the given flagellum. When most of the flagella are rotating CCW they bunch together at one end of the cell so that the cell moves forward with the opposite end leading. Under uniform conditions such 'smooth swimming' is interrupted about once per second by *tumbling*: a brief random movement of the cell (lasting about 0.1 second) caused by a switch from CCW to CW rotation in some of the bunched flagella; when this switch occurs, the bundle of flagella is disrupted, causing the random movement. Smooth swimming is resumed when most of the flagella are again rotating CCW.

Alternate swimming and tumbling results in a three-dimensional *random walk*: the cell moves in a series of straight lines in randomly determined directions.

In general, monotrichously flagellated cells (e.g. those of *Pseudomonas aeruginosa*) can reach speeds of about 70 μm/s, compared with about 30 μm/s for peritrichously flagellated bacteria (though the peritrichously flagellated cells of *Thiovulum majus* may reach speeds >600 μm/s [Microbiology (1994) *140* 3109–3116]). In monotrichously flagellated cells the flagellum rotates clockwise and counterclockwise for roughly equal periods of time; during change-over, randomization of direction (similar to that caused by tumbling in peritrichous cells) may be due to Brownian motion (see section 16.1.1.3).

Motility in spirochaetes is associated with rotation of their periplasmic flagella. In *Leptonema illini*, counterclockwise flagellar rotation causes the anterior (forward-pointing) end of the cell to assume a helical shape and the cell to rotate clockwise about its axis; the cell thus appears to move forward with a screw-like action. *Borrelia burgdorferi* swims as a flattened waveform, rather than as a helical form, although 'axial twists' along the length of the cell mean that different sections of the cell are in different planes; as in *L. illini*, the cell body appears to rotate clockwise about its axis. [Structure/motility of spirochaetes: JB (1996) *178* 6539–6545.]

Motility without flagella. Several types of flagellum-less motility can be distinguished. These are outlined below.

Gliding motility occurs in certain bacteria which lack flagella – e.g. species *Beggiatoa*, *Myxococcus* and *Oscillatoria*. It is a smooth form of motion which appears to occur only when cells are in contact with a solid surface, the gliding cell leaving a 'slime trail' as it moves.

Gliding may occur when nutrient levels are low, being suppressed by high levels. It often appears to require the presence of calcium ions and is inhibited by chelators of Ca^{2+} (e.g. EGTA). In bacteria of the *Cytophaga–Flexibacter* group gliding appears to require the presence of sulphonolipids in the cell envelope [Nature (1986) *324* 367–369].

Non-motile forms occur in the life cycle of some gliding bacteria (e.g. the myxospores of myxobacteria). In *Flexibacter elegans* (cells ~1–50 μm in length) gliding occurs only in cells longer than ~5 μm. In *Myxococcus xanthus* two types of motility have been distinguished: the so-called 'adventurous' motility of individual cells and the 'social' gliding motility of groups of cells; the rate of social motility has been found to decrease with increase in length of filaments [PNAS (1999) *96* 15178–15183].

In some cyanobacteria gliding has been associated with a so-called *junctional pore complex* (JPC); this is a channel in the cell envelope through which a carbohydrate mucilage is secreted. Each organism has a number of JPCs, and it has been suggested that secretion via JPCs may provide the thrust required for locomotion [JB (2000) *182* 1191–1199].

The motility apparatus of *Myxoccocus xanthus* has been investigated by shock-freezing and freeze-drying. This has shown a structure resembling a helical 'continuum' or band (of about 170–380 nm in width) apparently wrapped around the protoplast (i.e. within the periplasmic space) forming a closed loop; 'nodes', associated with the continuum, may travel along trichomes (section 2.3.1) in a wave-like manner (see Plate 2.5), each node possibly corresponding to a region of close contact between the trichome and substratum [Microbiology (2001) *147* 939–947].

Twitching motility is a jerky form of movement which is seen when certain polarly fimbriate bacteria are in contact with a solid surface. It occurs in various species – e.g. *Acinetobacter calcoaceticus*, *Moraxella* spp, *Neisseria* spp and certain species of *Pseudomonas*, including *P. aeruginosa*. Once regarded as a passive form of motion, twitching motility has been reported to involve retraction of type IV fimbriae, at approx. 1 μm/second, and to require protein synthesis [Nature (2000) *407* 98–102].

Vibrio cholerae exhibits a flagellum-independent *spreading* in which surface tension may be involved [JB (2001) *183* 3784–3790].

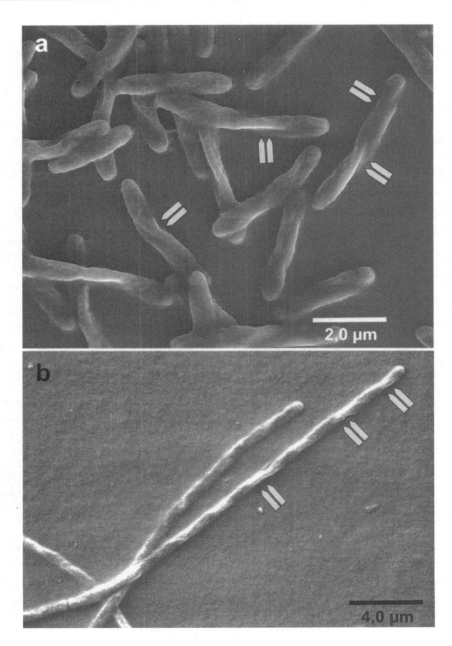

Plate 2.5 Shock-frozen cells of gliding bacteria (see 'gliding motility' in section 2.2.15.1). (a) Cells of *Myxococcus xanthus*, some of which show evidence of a helical 'continuum' along part or all of the cell's length. (b) Gliding filaments of *Flexibacter filiformis*. Double arrows indicate locations of nodes.

[For further information see: Microbiology (2001) *147* 939–947.]

Photographs provided by courtesy of Dr H. Lürsdorf, Gesellschaft für Biotechnologische Forschung mbH, Bereich Mikrobiologie, Braunschweig, Germany.

Actin-based motility is exhibited by certain pathogenic bacteria *within* (eukaryotic) host cells (see e.g. section 11.2.2.1).

Flagellum-less but *calcium*-dependent swimming (mechanism unknown) occurs e.g. in strains of the cyanobacterium *Synechococcus* [JB (1997) *179* 2524–2528].

2.2.15.2 *Chemotaxis*

Chemotaxis means directional movement in response to a chemical concentration gradient. In a chemically uniform environment, flagellated cells typically adopt a 'random walk' (section 2.2.15.1). Suppose, however, that the concentration of nutrients increases in a certain direction; can a cell – which changes direction *randomly* – travel towards the higher concentration of nutrients? It can, simply by tumbling less frequently when moving towards and more frequently when moving away from the higher concentration. In this way, more time is spent going in the 'right' direction – so that the cell's overall (net) movement is towards the higher concentration of nutrients.

Substances which attract a cell are called *chemoattractants*, while those which cause a cell to move away are called *chemorepellents*; the general term *chemoeffector* is used to refer to both of these categories.

How does the cell 'sense' different concentrations, and how does it control its rate of tumbling? If a cell swims towards an increasing or decreasing concentration of chemoeffector it can detect the *change* in concentration e.g. by means of receptors in the cell envelope; depending on the direction of the gradient, a receptor may (in a given time) be either more or less likely to bind a molecule of the chemoeffector.

When a chemoattractant binds to, or leaves, a receptor, the effect is to inhibit or promote certain intracellular signals; these signals, which originate at the receptor complex in the cell envelope, are aimed at the flagellar motor(s) via a so-called *signal transduction pathway* (section 7.8.6; Fig. 7.13). For example, the release of a chemoattractant molecule by an MCP receptor enhances the signal for CW rotation (favouring tumbling); that is, tumbling occurs more often when the cell is moving into a decreasing concentration of the chemoattractant. On the other hand, movement towards a higher concentration of chemoattractant will increase the proportion of time spent in smooth swimming. Thus, concentration gradients in the environment produce a *biased random walk* by appropriately modifying the rate of tumbling when the cell is moving in a given direction. On moving into a uniform (high or low) concentration of chemoeffector the 'routine' rate of tumbling is resumed.

Aspects of chemosensing, including those in non-enteric bacteria, are discussed in a minireview [JB (1998) *180* 1009–1022].

[Signalling components in bacterial locomotion and sensory reception: JB (2000) *182* 1459–1471. Signalling in bacterial chemotaxis (review): JB (2000) *182* 6865–6873.]

Aerotaxis is a particular type of chemotaxis in which (motile) cells respond to a concentration gradient of dissolved oxygen; cells may move to a higher or lower concentration of oxygen according to that which is optimal for their growth (section 3.1.6). What is the mechanism of aerotaxis? In at least some cases, the signal for aerotactic movement may involve changes in energy status (section 5.3.5).

[Sensing the environment by ways other than transmembrane signalling, including monitoring oxygen levels via energy status: JB (2001) *183* 4681–4686.]

2.3   TRICHOMES AND COENOCYTIC BACTERIA

Earlier it was mentioned that most bacteria can live as single, autonomous cells. Thus, for example, each cell in a chain of streptococci leads an essentially independent life – except, of course, in that each cell shares a micro-environment with its neighbour(s). Some bacteria, however, normally exist in *trichomes* or as *coenocytic* organisms.

2.3.1   Trichomes

A trichome is a row of cells which have remained attached to one another following successive cell divisions; the cells are separated by *septa* (cross-walls, singular: *septum*), but, in at least some trichomes, adjacent cells communicate with one another via small pores (*microplasmodesmata*). (In a simple 'chain' of cells – as formed e.g. by some streptococci – such pores are not formed.) The positions of the septa may or may not be obvious (as constrictions) from the outside of the trichome. The cells of a trichome may or may not be covered by a common sheath. Trichomes are formed by many cyanobacteria and e.g. by species of *Beggiatoa*.

2.3.2   Coenocytic bacteria

The filamentous actinomycete *Streptomyces*, and some other bacteria, form tube-like hyphae which lack septa, the cytoplasm being continuous from one nucleoid to the next. Such a multinucleate organism is called a *coenocyte*.

3 Growth and reproduction

In a bacterial cell growth involves a *co-ordinated* increase in the mass of its constituent parts. Growth is not simply an increase in total mass: that could be due, for example, to the accumulation of a storage compound within the cell.

Usually, growth leads to the division of a cell into two similar or identical cells. Thus, growth and reproduction are closely linked in bacteria, and the term 'growth' is generally used to cover both processes.

3.1 CONDITIONS FOR GROWTH

Bacteria grow only if their environment is suitable; if it's not optimal, growth may occur at a lower rate or not at all – or the bacteria may die, depending on species and conditions.

Essential requirements for growth include (i) a supply of suitable nutrients; (ii) a source of energy; (iii) water; (iv) an appropriate temperature; (v) an appropriate pH; (vi) appropriate levels (or the absence) of oxygen. Of course, none of these factors operates in isolation: an alteration in one may enhance or reduce the effects of another; for example, the highest temperature at which a bacterium can grow may well be lowered if the pH of the environment is made non-optimal.

3.1.1 Nutrients

Cells need nutrients as raw materials for growth, maintenance and division. As a group, the bacteria use an enormous range of compounds as nutrients; these include various sugars and other carbohydrates, amino acids, sterols, alcohols, hydrocarbons, inorganic salts and carbon dioxide. However, no individual bacterium can use all of these compounds: it hasn't the range of enzymes to deal with them all, and (in any case) its cell envelope does not have uptake ('transport') systems for all of them. A given type of bacterium typically uses only a relatively limited range of compounds.

Whatever the organism, cells need sources of carbon, nitrogen, phosphorus, sulphur and other materials from which living matter is made. Some bacteria obtain all nutritional requirements from simple inorganic salts and substances such as carbon dioxide and ammonia; others need – to varying extents – more or less complex organic compounds derived from other organisms. Some important aspects of nutrition are discussed in Chapters 6 and 10.

3.1.2 Energy

Energy is needed for most of the essential chemical reactions which go on in a living cell; it is also needed e.g. for flagellar motility and for the uptake of various nutrients. All of this energy is obtained from sources in the environment. In phototrophic species energy is derived mainly or solely from light, while chemotrophic species obtain energy by processing chemicals taken from the environment. Some species can use both methods. Energy is discussed in Chapter 5.

3.1.3 Water

A high proportion of the mass of a bacterium is water, and, during growth, nutrients and waste products enter and leave the cell, respectively, *in solution*; hence, bacteria can grow only in or on materials which have adequate free (available) water. (Not all the water in a given material is necessarily available for bacterial growth; some, for example, may be bound by hydrophilic gels or by ions in solution.)

An extreme lack of water (desiccation) is tolerated to different degrees by different species of bacteria, though many species do not survive for long in the air-dried state. [Desiccation tolerance of prokaryotes: MR (1994) *58* 755–805.]

3.1.4 Temperature

Generally, for a given type of bacterium, growth proceeds most rapidly at a particular temperature: the *optimum growth temperature*; the rate of growth tails off as temperatures increase or decrease from the optimum. For any given bacterium there are maximum and minimum temperatures beyond which growth will not occur.

Thermophilic bacteria are those whose optimum growth temperature is >45°C. These *thermophiles* occur e.g. in composts and hot springs and near hydrothermal vents on the ocean floor; they include species of *Thermobacteroides* (opt. 55–70°C) and *Thermomicrobium* (opt. 70–75°C).

Among the Archaea, *Pyrodictium* has an optimum growth temperature of 105°C. Even more remarkable, an archaean referred to as 'strain 121' – isolated from water near an ocean-floor hydrothermal vent (temperature: 300°C) – has been reported to grow between 85 and 121°C [Science (2003) *301* 934]. This archaean was not killed by autoclaving (see sections 15.1 and 15.1.1.3); in fact, at 121°C it grew with a generation time (doubling time) of 24 hours.

Thermophiles may exhibit special features associated with adaptation to growth at high temperatures; such adaptation may be seen in the composition of their cell components (e.g. enzymes [TIM (1998) *6* 307–314] or cytoplasmic membranes) and, in some cases, even in their mode of energy metabolism (see section 5.3.6).

Thermoduric bacteria can survive – though not necessarily grow – at temperatures which would normally kill most other vegetative (i.e. growing) bacteria. In dairy bacteriology, 'thermoduric' bacteria are those which survive pasteurization (section 12.2.1.1).

Mesophilic bacteria grow optimally at temperatures between 15 and 45°C. The *mesophiles* live in a wide range of habitats, and they include all those bacteria which cause disease in man and other animals.

Psychrophilic bacteria grow optimally at or below 15°C, do not grow above about 20°C, and have a lower limit for growth of 0°C or below. The *psychrophiles* occur e.g. in polar seas. At −2°C to −20°C, association with particles or surfaces is important for bacterial activity [AEM (2004) *70* 550–557].

Psychrotrophic bacteria can grow at low temperatures (e.g. 0–5°C), but they grow optimally above 15°C, with an upper limit for growth >20°C.

3.1.5 pH

Most bacteria grow best at or near pH 7 (neutral), and the majority cannot grow under strongly acidic or strongly alkaline conditions. However, some (found e.g. in mine drainage and in certain hot springs) not only tolerate but actually 'prefer' acidic or highly acidic conditions; these *acidophiles* include *Thiobacillus thiooxidans*, a species whose optimal pH is 2–4. Among the Archaea, species of *Sulfolobus* grow at pH 1–5, while *Thermoplasma acidophilum* has an optimum pH of 0.8–3.

Alkalophiles grow optimally under alkaline conditions – typically above pH 8. *Thermomicrobium roseum* (opt. pH 8.2–8.5) occurs e.g. in hot springs, and *Exiguobacterium aurantiacum* (opt. pH 8.5 and 9.5) has been found in potato-processing effluents; other alkalophiles occur in natural alkaline lakes.

Acidophiles and alkalophiles may grow slowly – or not at all – at pH 7.

3.1.6 Oxygen

Some bacteria need oxygen for growth. Others need the *absence* of oxygen for growth. Yet others can grow regardless of the presence or absence of oxygen.

Bacteria which *must* have oxygen for growth are called 'strict' or 'obligate' *aerobes* to emphasize their absolute need for oxygen.

Strict or obligate *anaerobes* grow only when oxygen is absent; these organisms occur e.g. in river mud and in the rumen.

Bacteria which normally grow in the presence of oxygen but which can still grow under anaerobic conditions (i.e. in the absence of oxygen) are called *facultative anaerobes*. Similarly, those which normally grow anaerobically but which can grow in the presence of oxygen are called *facultative aerobes*.

Microaerophilic bacteria generally grow best when the concentration of oxygen is (usually much) lower than it is in air.

Some bacteria exhibit *aerotaxis* (section 2.2.15.2).

The definition of anaerobe is likely to need revision given reports that the 'strict anaerobe' *Desulfovibrio gigas* (a sulphate-reducing bacterium) has an oxygen-reducing respiratory chain [FEBS Letters (2001) *496* 40–43], and that the anaerobe *Bacteroides fragilis* encodes a cytochrome *bd* oxidase and can benefit from nanomolar concentrations of oxygen [Nature (2004) *427* 441–444].

3.1.7 Inorganic ions

All bacteria need certain inorganic ions in low concentrations, higher concentrations usually inhibiting growth. The ions have various functions – e.g. magnesium in the outer membrane (section 2.2.9.2), iron in cytochromes (section 5.1.1.2) and in a range of enzymes, and manganese and nickel in enzymes or enzyme systems. (See also section 11.4.1.3.)

Some bacteria (the *halophiles*) – e.g. the actinomycete *Actinopolyspora* – grow only in the presence of a high concentration of electrolyte (usually NaCl). Certain members of the Archaea (e.g. *Halobacterium salinarium*), which occur in salt lakes and salted fish etc., are examples of extreme halophiles: they need at least 1.5 M NaCl for growth, 3–4 M NaCl for good growth. The electrolyte serves to maintain the structure of e.g. ribosomes and the cell envelope; in dilute solutions the cells of some species break open due to weakening of the cell envelope.

Halotolerant bacteria are non-halophiles which can grow in electrolyte up to about 2.5 M; they include many strains of *Staphylococcus*.

Inorganic ions can also play a *regulatory* role in various aspects of bacterial physiology. For example, Ca^{2+} ions have known or suspected roles in e.g. transformational competence (section 8.4.1; [JB (1995) *177* 486–490]), chemotaxis (section 2.2.15.2) and sporulation in *Bacillus subtilis* (section 7.8.6.1) [calcium signalling in bacteria: JB (1996) *178* 3677–3682]. For an L-form (section 2.1.1), calcium was a specific requirement for growth [JB (2000) *182* 1419–1422].

3.1.8 Response to changed conditions: adaptation

Bacteria which live in highly specialized environments – e.g. at extreme temperatures, or within the cells of other organisms – usually cannot tolerate other types of environment. By contrast, many bacteria continue to grow (or, at least, survive) even when major changes occur in their environment; such continued growth (or survival) means that these organisms are either unaffected by the particular change(s) or are able to *adapt* (i.e. change themselves) so that they can then either tolerate or take advantage of the new conditions.

Adaptation often requires the synthesis of new types of protein (for specific functions) and suppression of other types of protein; this involves changes in the expression of various *genes* (Chapter 7). One example is the adaptation of *Salmonella typhimurium* to strongly acidic conditions (section 7.8.2.6).

Bacteria respond in various ways to low pH (acidity). The acid tolerance response has been studied in various enterobacteria (section 7.8.2.6). Gram-positive bacteria use mechanisms such as proton pumps, production of alkali, changes in the cytoplasmic membrane, and strategies for protecting/repairing particular macromolecules [MMBR (2003) *67* 429–453].

Another example is adaptation to changes in osmotic pressure in the environment or medium; such adaptation is called *osmoregulation*. If, for example, the osmotic pressure increases it tends to withdraw water from a cell – decreasing the cell's turgor pressure and increasing the concentrations of intracellular solutes to levels that may inhibit growth.

Osmoregulation may involve the activity of specific structures in the cell envelope – such as mechanosensitive channels and aquaporins (see section 2.2.8). In *E. coli*, a significant up-shift in environmental osmotic pressure causes a change in the relative proportions of certain pore-forming proteins in the outer membrane, thereby reducing the average pore size (section 7.8.6).

E. coli reacts immediately to an increase in the environmental osmotic pressure by taking up potassium ions (K^+) and synthesizing glutamate (as counter ion); this serves to counteract a high external osmotic pressure by increasing the cell's internal osmotic pressure. In *E. coli* uptake of K^+ can occur via several distinct types of *transport system* (section 5.4): e.g. the Trk system (which is *constitutive*, i.e. always present), and the Kdp system (which is *induced* by high external osmotic pressure: see section 7.8.6). The Kdp system appears to be triggered by decreased turgor pressure (or by a related parameter).

Osmotic up-shift also promotes uptake (or *de novo* synthesis) of so-called *compatible solutes*. These are certain small molecules that can be accumulated intracellularly to high concentrations without adversely affecting cell viability. The accumulation of a compatible solute increases the cell's osmotic pressure, the solute thus being able to replace at least some of the accumulated potassium ions (which are physiologically less friendly at high concentrations).

Compatible solutes include *glycine betaine* $(CH_3)_3.N^+.CH_2COO^-$. This molecule, which is common in nature, is taken up efficiently by *E. coli* via a binding-protein-dependent transport system (section 5.4.1.1) that is designated ProU [FEMSMR (1994) *14* 3–20]. Uptake of glycine betaine is common in both Gram-negative and Gram-positive bacteria (*Staphylococcus aureus* has two transport systems for it).

Interestingly, glycine betaine (and related compounds) also give protection against heat stress in *Bacillus subtilis* [JB (2004) *186* 1683–1693].

The bacterium *Sinorhizobium meliloti* metabolizes (i.e. does not accumulate) most of the known osmoprotectants, including glycine betaine. In this organism, osmotic up-shift triggers the accumulation of *endogenous* solutes (e.g. glutamate and *N*-acetylglutaminylglutamine amide), accumulation being stimulated by e.g. exogenous sucrose [JB (1998) *180* 5044–5051].

[Regulation of compatible solute accumulation in bacteria: MM (1998) *29* 397–407.]

Osmotic *down-shift* in the environment triggers efflux systems – e.g. K$^+$ and glutamate are rapidly released by *E. coli*, and specific compatible solutes are released by other organisms. Efflux is apparently not dependent on metabolic energy; this contrasts e.g. with uptake of glycine betaine via the ProU transport system (which is ATP-dependent). Much of the K$^+$ released by *E. coli* during osmotic down-shift passes through mechanosensitive channels (section 2.2.8) – e.g. the MscL channel; this was deduced from studies on cells with mutations in the *mscL* gene [JBC (1997) *272* 32150–32157].

Many facultatively anaerobic bacteria (section 3.1.6) can adapt to anaerobiosis (or to a return to aerobiosis) by switching to an alternative form of energy metabolism (Chapter 5), this requiring e.g. the synthesis of new (protein) enzymes.

Genetic/regulatory aspects of adaptation are discussed in sections 7.8.2.3–7.8.2.6 and 7.8.6.

3.2   GROWTH IN A SINGLE CELL

In a growing cell there is a co-ordinated increase in the mass of the component parts. 'Co-ordinated' does not mean that all the parts are made simultaneously: some are synthesized more or less continually, but others are made in a definite sequence during certain fixed periods. The cycle of events in which a cell grows, and divides into two daughter cells, is called the *cell cycle*.

3.2.1   The cell cycle

Most of the studies discussed below have been carried out on *Escherichia coli* (a Gram-negative bacillus); some were carried out on *Bacillus subtilis* (a Gram-positive bacillus). Much of the recent research on control of the cell cycle has focused on the Gram-negative bacterium *Caulobacter crescentus*.

The cell's dimensions during growth. Rapidly growing cells are larger (in mass and volume) than those growing slowly. When the doubling time (section 3.2.3) is less than 1 hour, the *C* and *D* periods of the cell cycle (Fig. 3.2, *lower part*) are practically constant, and, at the start of each *C* period, the cell's mass (per chromosome origin) is also practically constant *for a given rate of growth*; hence, under these conditions, the faster a cell grows (i.e. the shorter the I period) the greater will be the increase in cell mass and volume by the end of the (constant) $C + D$ interval of time before cell division.

When growing at a *constant* rate, the cell's increase in mass involves an increase in length only, the diameter remaining unchanged. If the growth rate increases the cells become larger: the cell's diameter increases, slowly, but its length increases more rapidly; in fact, the length initially 'overshoots' (i.e. increases more than is appropriate for the new rate of growth). Subsequently, the diameter increases to its new value, and the length decreases to its final value (which is still higher than the original value). A fall in growth rate causes an initial 'undershoot' in cell length, the cell's diameter, again, adjusting more slowly than its length to the new growth rate [JGM (1993) *139* 2711–2714].

Synthesis of the cell envelope. One early study [MR (1991) *55* 649–674] concluded that peptidoglycan is incorporated diffusely in the lateral wall of the cell, while a study on peptidoglycan turnover [JB (1993) *175* 7–11] suggested that the lateral wall contains a monolayer of peptidoglycan. In *E. coli*, synthesis of peptidoglycan in the lateral wall does not continue at a uniform rate throughout the cell cycle; as discussed later, some time prior to cell division, synthesis is switched from the lateral wall to a narrow region at mid-cell in preparation for septation [JB (2003) *185* 1125–1127].

Höltje [Microbiology (1996) *142* 1911–1918] described a '3-for-1' model for the biosynthesis of the peptidoglycan sacculus. This model is consistent with the observed features of biosynthesis and accounts for the different types of peptide bridge that are now known to occur in the sacculus (see Fig. 2.7, legend). The model is outlined in Fig. 3.1. To co-ordinate the various synthetic and lytic processes involved, Höltje postulated the existence of two types of holoenzyme: one active at the cell's wall, the other active during formation of the septum; both types of holoenzyme would involve penicillin-binding proteins (PBPs, section 2.2.8) and 'lytic transglycosylases', i.e. enzymes which can degrade a glycan strand processively (that is, moving and cutting).

Studies on cell shape determination in *Bacillus subtilis* (section 2.1.1) have shown that peptidoglycan in the cylindrical region of the cell wall is synthesized in a helical pattern under the guidance of the Mbl protein [Cell (2003) *113* 767–776].

During growth, the turnover of peptidoglycan involves the re-cycling of peptides (section 5.4.8).

It has been assumed that the cytoplasmic membrane develops passively, i.e. simply 'keeping in step' with the expanding peptidoglycan sacculus. However, peptidoglycan synthesis seems to be dependent on the synthesis of membrane phospholipids – suggesting a novel form of regulation for the growth of the cell envelope [Microbiology (1996) *142* 2871–2877]. In this context it is interesting that the cytoplasmic membrane phospholipid phosphatidylethanolamine (Fig. 2.5) is needed for the correct folding

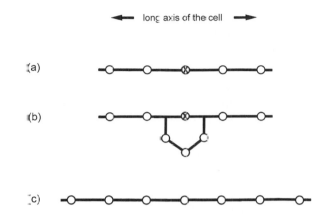

Figure 3.1 Outline of the 3-for-1 model describing the mode of growth of the peptidoglycan sacculus in *E. coli* [Microbiology (1996) *142* 1911–1918]. In peptidoglycan, the glycan backbone chains (Fig. 2.7a) are *perpendicular* to the long axis of the cell; the diagram shows the *end-on* view of backbone chains (circles) joined by peptide bridges (straight lines).

(a) A peptidoglycan monolayer; the monolayer is a stress-bearing structure and is under considerable tension. The chain in the centre (⊗) – called the 'docking' chain – is to be replaced by three chains (hence 3-for-1).

(b) A triplet of three linked chains is located below (i.e. on the cytoplasmic side of) the peptidoglycan monolayer, with the flanking chains of the triplet covalently bound either side of the docking chain; for this to happen, a stem peptide (Fig. 2.7b) from each flanking chain has bound to the ε-amino group of a dimeric peptide bridge to form a *trimeric* bridge (Fig. 2.7b).

(c) The docking chain has been removed by enzymic action, and – because of the tension in the monolayer – the incoming triplet of chains has taken up its position in the sacculus; note that this contributes to cell *elongation*. The model assumes that the central chain in the triplet will, when incorporated in the sacculus, function as a new docking chain; for this reason, the dimeric bridge to each flanking chain is assumed to have a free ε-amino group located in the stem peptide of the flanking chain.

For cell length to *double* during the cell cycle, the model assumes that (i) docking and non-docking chains alternate in the sacculus, and (ii) only those docking chains which are present at the *start* of the new cell cycle will be replaced by an incoming triplet; that is, 'new' docking chains inserted *during* the current cell cycle will not be replaced until the next cell cycle. For this purpose, the growth mechanism must be able to distinguish 'old' from 'new' docking chains. This is possible because newly synthesized peptidoglycan is linked to the sacculus solely by tetra-tetra peptide bridges (Fig. 2.7b); it is therefore postulated that *new* chains (identified by their tetra-tetra links) are not recognized as docking chains, but that at the start of each new cell cycle, all tetra-tetra links will have been changed enzymically (by an LD-carboxypeptidase) to tetra-tri links (Fig. 2.7b) and that this allows recognition of the docking chains.

The successive addition of triplets at the mid-point of the cell – perhaps in association with the FtsZ ring (section 3.2.1, septum formation) – might account for the ingrowth of peptidoglycan during the development of the septum.

(and function) of the LacY protein (section 7.8.1.1, Fig. 7.11) [EMBO Journal (1998) *17* 5255–5264]; it is therefore conceivable that a similar function may be carried out by membrane lipids for protein(s) involved in the synthesis of peptidoglycan.

Chromosome (DNA) replication. Two different mechanisms have been proposed for chromosomal replication in *Escherichia coli* and *Bacillus subtilis*. In one model, the multi-component DNA-replicating machinery (the 'replication factory' or *replisome*) is localized at the mid-cell position; in this model the chromosomal DNA would move through a stationary replisome – replicated DNA being extruded from the complex in both directions [Science (1998) *282* 1516–1519; Biochimie (1999) *81* 803–810; GD (2001) *15* 2031–2041]. The other model proposes that *two* replication complexes migrate, in opposite directions, from their initial central position in the cell; the proponents of this model have reported corresponding movement of the β-subunit of DNA polymerase III [JB (2002) *184* 867–870].

Interestingly, in the first model of chromosome replication (above), the two copies of the *origin* (section 7.3) move rapidly apart to one-quarter and three-quarter positions in the cell, while in the second model of replication, the (paired) *replication complexes* move rapidly apart to one-quarter and three-quarter positions in the cell.

The necessary co-ordination between chromosomal replication (DNA replication: section 7.3) and cell division is described in section 3.2.1.1 and Fig. 3.2.

What controls the *initiation* of a round of DNA replication in the cell cycle? The answer is not known, but initiation – though precisely timed – can occur at different times after the start of the cycle; moreover, when there are multiple chromosomes in a given cell, initiation occurs at all chromosomal origins at more or less the same time [MM (1993) *10* 457–463]. That the timing of initiation is related to growth rate can be seen in Fig. 3.2: compare the top and lower sets of $I + C + D$ sequences, and note the timing of initiation (start of the C period) at the different growth rates. (The reader may find it useful to draw another set of $I + C + D$ sequences with I half the length of C, noting the timing of initiation.)

Certain factors seem to be important in the initiation of replication. One factor is the intracellular concentration of DnaA protein (or, perhaps, of an activated form of it); thus, before replication can begin, the strands of (double-stranded) DNA must separate at the chromosome's *origin* (section 7.3) – this requiring the prior binding of many molecules of DnaA in this region of the chromosome. Insufficient DnaA may therefore delay initiation. Studies on replication systems have identified mutant DnaA proteins that are defective in their ability to open the DNA duplex at *oriC* [MM (2000) *35* 454–462].

Another factor is the degree of methylation (sections 7.4 and 7.8.8) at the origin, *oriC*; the timing of initiation may involve delayed or gradual methylation at *oriC* – which could occur e.g. if the relevant sites in the DNA were not freely accessible to the methylating enzymes.

A further factor apparently involved in initiation is the Fis protein (product of the *fis* gene). Like DnaA, Fis has binding sites in the *oriC* region – and it also appears to be involved in regulating genes that encode (i) DnaA and (ii) a subunit of DNA polymerase III (the main DNA polymerizing enzyme in *E. coli*) [JB (1996) *178* 6006–6012].

Finally, initiation is linked to growth rate. One suggestion for a mechanism involves the small molecule guanosine 5'-diphosphate 3'-diphosphate ('guanosine tetraphosphate'; ppGpp) whose intracellular concentration is inversely proportional to growth rate; at low rates of growth, the high level of ppGpp can e.g. inhibit synthesis of the DnaA protein – which, in turn, may delay initiation (see above). (See also section 7.8.2.5.) Interestingly, ppGpp has also been linked to the expression of the FtsZ protein involved in septation (see below).

If DNA replication is blocked then progress through the cell cycle is affected (see section 3.2.1.3).

Chromosome decatenation and partition. During cell division, daughter cells each receive the same number of chromosomes. When synthesized, chromosomes are initially *catenated* (interlocked, like the links of a chain) and they have to be separated; such *decatenation* involves enzymes, called *topoisomerases*, which can break DNA strand(s), transiently, and pass DNA strand(s) through the gap. If an essential topoisomerase is not functional in the cell (e.g. through mutation), the process of decatenation is prevented; for example, certain mutations in the *E. coli parC* gene (which encodes one subunit of topoisomerase IV) prevent decatenation and give rise to filamentous cells (containing interlinked chromosomes) and anucleate cells (containing no chromosome). In fact, topoisomerase IV (whose subunits are encoded by the *parC* and *parE* genes in *E. coli*) is reported to be the sole enzyme responsible for decatenation in this organism [GD (2001) *15* 748–761].

As well as decatenation, the chromsomes have to be separated (= segregated, partitioned) and folded so that each subsequently forms a compact *nucleoid* (section 2.2.1) within a daughter cell.

The mechanism by which chromosomes are partitioned to daughter cells is still under investigation. An earlier view was that partition may involve e.g. (i) attachment of nucleoids to the cytoplasmic membrane, the nucleoids being drawn apart as the cell elongates; (ii) mutual repulsion between nucleoids; and/or (iii) the action of a putative force-generating protein (product of the *mukB* gene) [JB (1992) *174* 7883–7889]. A passive mechanism, as in (i), is no longer given credence. Moreover, the MukB protein (a homologue of the Smc protein in *Bacillus subtilis* – see below) is now thought to contribute to the formation of a compact chromosome by constraining supercoils.

In *Bacillus subtilis*, normal partitioning of chromosomes is dependent on the *spo0J* gene product. Following initiation of replication, Spo0J–origin complexes are formed and subsequently move apart within the cell; such movement is not driven by cell elongation and is presumed to be involved in the movement of daughter chromosomes to opposite poles of the cell. Thus, partitioning appears to be a dynamic process [GD (1997) *11* 1160–1168].

Cohesion within SpoOJ–origin complexes apparently requires the Soj protein, which seems to be an ATPase that moves from one origin to another during replication. In *B. subtilis* the Smc (= SMC) protein (encoded by gene *smc*) plays an important role in nucleoid structure and (hence) is needed for normal partitioning of the nucleoids. A mutation in *smc* may cause abnormal (often less compact) nucleoids which may be imperfectly folded. [Bacterial chromosome segregation: Microbiology (2001) *147* 519–526.]

Homologues of Smc occur e.g. in *E. coli* (the MukB protein) and in the archaeans *Methanococcus jannaschii* and *Archaeoglobus fulgidus*.

One current model for partitioning in bacteria is the *extrusion–capture* model. In this scheme, which is predicated on the presence of a central, stationary replisome, bi-directional extrusion of DNA from the replisome provides the energy for partition [GD (2001) *15* 2031–2041].

The report of mitosis-like partitioning of the R1 plasmid in *E. coli* [EMBO Journal (2002) *21* 3119–3127] (section 7.3.1) raised the possibility that actin-like filaments of MreB protein may be involved in the partitioning of bacterial chromosomes. It was subsequently reported that dysfunctional MreB protein gives rise to aberrant partition of chromosomes in *E. coli*, and this has suggested that filaments of the MreB protein are involved in directional movement and segregation of chromosomes [EMBO Journal (2003) *22* 5283–5292].

Septum formation. The growing cell eventually divides into two cells by the development of a cross-wall *(septum)*. An early indication of septation is the appearance of a ring-shaped aggregate of FtsZ molecules (encoded by the *ftsZ* gene) circumferentially on the inner surface of the cytoplasmic membrane, mid-way along the cell. (A fluorescent image of the FtsZ ring (= 'Z ring') is shown in Plate 3.1.) The FtsZ ring determines the plane of the forthcoming septum.

Binding between FtsZ and a cytoplasmic membrane protein (ZipA) appears to be necessary for eventual cell division [Cell (1997) *88* 175–185]. [Interaction between ZipA and a C-terminal fragment of FtsZ (X-ray crystallography): EMBO Journal (2000) *19* 3179–3191.]

The coupling between the assembly of the FtsZ ring and the cell cycle is not understood. Studies on *ftsZ* mutants of *E. coli* have suggested a role for ppGpp in the control of FtsZ expression during cell division [JB (1998) *180* 1053–1062].

The mid-cell positioning of the FtsZ ring depends on the products of the *min* genes (see section 3.2.1.2).

Interestingly, anucleate cells can also develop an FtsZ ring, indicating that the early division apparatus can form without a chromosome [MM (1998) *29* 491–503].

The FtsZ ring is also found in members of the Archaea, suggesting that the cell

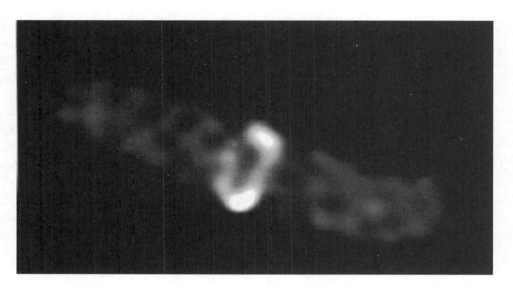

Plate 3.1 The FtsZ ring of *Escherichia coli* (section 3.2.1). Molecules of the FtsZ protein form a transient circumferential ring, at the mid-cell position, prior to septation. This image was obtained by fusing the gene for FtsZ with the gene for *green fluorescent protein* (GFP) (see section 8.5.2); the resulting hybrid protein, FtsZ–GFP, has the fluorescence characteristics of GFP, so that detection of GFP (by fluorescence microscopy) reveals the intracellular location of the FtsZ protein. Photograph reproduced, with permission, from *Proceedings of the National Academy of Sciences of the USA* (1996) **93** 12998–13003 (Copyright 1996 National Academy of Sciences, USA) by courtesy of the authors: Dr Xiaolan Ma, Dr William Margolin (University of Texas Medical School) and Dr David Ehrhardt (Howard Hughes Medical Institute).

division apparatus was similar in a common prokaryotic ancestor [MM (1996) *21* 313–319].

Assembly of the FtsZ ring appears to start from a so-called *nucleation site* and proceeds, bi-directionally, around the circumference of the cell. Formation of the nucleation sites themselves may involve the Era protein (encoded by the *era* gene) – a GTPase of the Ras superfamily; Era may be an important factor in the control of cell division [MM (1998) *29* 19–26].

Subsequently, in the plane of the FtsZ ring, peptidoglycan grows inwards to form the septum.

At a very early stage of septation in *E coli* there is a sharp fall in the addition of peptidoglycan to the lateral cell wall – synthesis of new peptidoglycan being largely confined to a narrow mid-cell region at the site of the future septum. Interestingly, synthesis of this 'pre-septal' peptidoglycan occurs even in the absence of a (functional) penicillin-binding protein 3 (PBP3) – which normally mediates ingrowth of the septum; under these conditions, *E. coli* forms aseptate filaments, although new peptidoglycan continues to be synthesized, at regularly spaced sites, in narrow zones that correspond to aborted division sites.

Synthesis of pre-septal peptidoglycan seems to require the FtsZ protein, suggesting a role in the switch from cell elongation to pre-septal synthesis. Other proteins (e.g. ZipA) may be needed for the switch and/or for later synthesis. FtsEX proteins, homologous to ABC transporters (section 5.4.1), are also reported to be needed for division [JB (2004) *186* 785–793].

[The bacterial cell division site (new insights): JB (2003) *185* 1125–1127.]

Studies on cells which lack a functional DacA protein (PBP5) have found that branching arises at, or close to, the pre-septal sites [JB (2003) *185* 1147–1152].

The fully formed septum is split (by an enzyme, EnvA) into two layers, each layer forming a new pole (end) of one of the daughter cells. Inward growth of the outer membrane completes the cell envelope of each daughter cell.

On cell division in *E. coli*, daughter cells usually separate immediately. In other species, separation may not occur immediately – leading to one or other of the groupings mentioned in section 2.1.3.

3.2.1.1   *The Helmstetter–Cooper model*

During the cell cycle it's clearly essential that chromosome replication and cell division be properly co-ordinated. Chromosome replication within the cell cycle is described by the Helmstetter–Cooper model (Fig. 3.2). During slow growth, one complete duplication of the chromosome occurs for each division of the cell – so that each new daughter cell receives only one chromosome; during faster growth, a new round of chromosome replication starts before the previous round has been completed, so that each daughter cell receives more than one chromosome – about $1\frac{1}{2}$ chromosomes in Fig. 3.2. This helps to explain why it is that the rate of growth can affect the number of chromosomes in a cell (section 2.2.1).

3.2.1.2   *Genes involved in the cell cycle: the morphogenes*

Genetic control of the cell cycle involves apppropriate functioning of the so-called *morphogenes* and their products. These genes have to ensure not only that appropriate products are formed at the right time but also that particular gene products are targeted to the correct positions within the cell. For example, following DNA replication and chromosome segregation, the septum must develop mid-way along the length of the cell – rather than at alternative *polar* sites in the cell. This means that assembly of the FtsZ ring (section 3.2.1) must somehow be guided to the correct mid-cell division site; such guidance involves the products of the *min* genes: *minC* (encoding MinC, which inhibits assembly of the FtsZ ring), *minD* (encoding MinD, an ATPase) and *minE* (encoding MinE, which regulates the intracellular position of the MinCD complex). One model for this guidance system in *E. coli* [Science (2002) *298* 1942–1946] is described below.

Initially, molecules of the MinCD complex form a lining on the inner surface of the cytoplasmic membrane in one half of the cell; development of this 'polar cap' of MinCD molecules requires binding of ATP by MinD – MinD–ATP molecules polymerizing on the membrane. MinE molecules form a circumferential, annular band at mid-cell. The MinE ring then moves into that half of the cell containing the MinCD polar cap; in so doing, MinE promotes hydrolysis of MinD-associated ATP, and this is accompanied by expulsion of the MinCD proteins – which move to the opposite pole of the cell and again form a polar cap. MinE proteins again form an annular band at mid-cell; this time the band moves into the other half of the cell (i.e. that containing the new polar cap) and, again, ejects MinCD proteins. Thus, the MinE ring oscillates between the cell's poles, forcing the MinCD proteins to occupy mainly *polar* locations; consequently, the influence of MinC is minimal at the mid-cell location, permitting assembly of the FtsZ ring. At the alternative (polar) sites of septation, polymerization of FtsZ is prevented by MinC, i.e. the FtsZ ring does not form at these sites [PNAS (1999) *96* 14819–14824].

The product of morphogene *envA* is required to split the septum into the two new poles of the daughter cells.

Other morphogenes include those encoding the penicillin-binding proteins.

3.2.1.3   *Control of the cell cycle*

In the cell cycle, the occurrence of any event could be directly dependent on the occurrence of the previous event. Alternatively, events could be controlled independently, and co-ordinated. Early arguments for independent control [e.g. MM (1991) *5* 769–774] included the following. (i) Some cell-cycle events can be inhibited without inhibiting others. For example, cell division *can* occur without prior replication of the chromosome, or without proper partition of the nucleoids; in such cases there would be no initiating signal from the (normally) preceding event. (ii) There is no evidence for *direct* coupling between chromosome replication and cell division, but indirect co-ordination clearly does occur. For example, damage to DNA (which may inhibit chromosome replication) triggers the SOS system (section 7.8.2.2) – a *general* regulatory mechanism which, among other effects, inhibits cell division; such inhibition operates via the SOS gene *sulA* whose product inhibits the activity of FtsZ, a protein needed for the development of the septum (section 3.2.1).

More recent studies on bacteria have shown that growth, differentiation and the cell cycle include so-called *checkpoints* at which, normally, a given event is completed before other(s) can begin. [Checkpoints in prokaryotes: JB (2003) *185* 1128–1146 (1135–1136).] (Under abnormal conditions the sequence of events in the cell cycle can be upset; for example, mutation (Chapter 8) can lead to the formation of *minicells*: small, cell-like bodies, lacking a chromosome, which are formed when the septum develops in a polar location rather than at the centre of a dividing cell.)

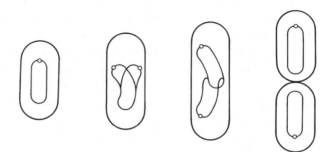

Slow growth

Rapid growth

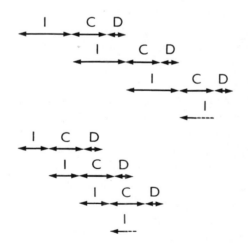

In at least some bacteria (perhaps all) certain events in the cell cycle are controlled by *two-component regulatory systems* (section 7.8.6). This type of system is involved in the regulation of a wide range of activities, including e.g. environmental sensing, but *essential* two-component regulatory systems (e.g. those involved in cell-cycle control) are necessary for cell viability.

Control of the cell cycle in the Gram-negative bacterium *Caulobacter crescentus* is being actively researched. In this organism (section 4.1, Fig. 4.1) the cell cycle is divided into three phases:

- G1 phase   This includes all of the motile (swarm cell) stage and ends with the transition stage in which the flagellum has been lost and biogenesis of the stalk has begun.
- S phase   This follows the G1 phase and extends to the pre-division cell. In this phase the cell grows, the chromosome is replicated, and components of the flagellum are synthesized.
- G2 phase   This phase – the pre-division cell – is characterized by chromosome segregation, septum formation, and an external flagellum; it ends with cell division. (G2 phase is shown in Fig. 4.1 at stage 3.)

Figure 3.2   Chromosome replication during growth at different rates. Cells and chromosomes are represented diagrammatically. Replication begins at a specific location in the chromosome – the *origin* (section 7.3) – which is shown as a small circle. During slow growth (*top*) each new daughter cell has exactly one chromosome because, following duplication of the chromosome, a new round of chromosomal replication does not begin until after completion of cell division. During faster growth (*below*) a new round of replication has begun before the previous round has been completed – i.e. well before cell division; consequently, each daughter cell in the diagram has about $1\frac{1}{2}$ chromosomes.

The cell division cycle may be represented as a linear sequence of three periods: I, C and D. C is the period during which chromosome replication occurs D is the period in which the septum forms, cell division occurring at the end of the D period. I is the period between each successive *initiation* of chromosomal (i.e. DNA) replication; it is also equal to the doubling time (section 3.2.3). When one I period ends the next I period begins.

In each set of $I+C+D$ sequences in the diagram, the relationship between a given $I+C+D$ sequence and the one immediately above it is that of a daughter cell to its parent cell. The upper set of $I+C+D$ sequences shows successive cell cycles during slow steady growth; notice that when $I=C+D$, a new round of chromosome replication does not start until after the D period, i.e. until after completion of cell division. The lower set of $I+C+D$ shows successive cell cycles during faster steady growth; here, with I shorter than C, chromosome replication is initiated before the previous round has been completed. Hence, at the end of D the new round of replication is already well advanced. Thus, in faster-growing cells, the stage which chromosome replication has reached in a new daughter cell is determined by the timing of initiation – i.e. start of the C period – in the parent cell.

In rapidly growing cells of *Escherichia coli* (generation times <60 minutes) the C and D periods are more or less constant for a given strain (inter-strain variation occurs in the length of these periods, particularly in the D period); the C and D periods are reported to increase linearly as the generation time increases above 60–70 minutes [Microbiology (2003) *149* 1001–1010].

In *C. crescentus*, CtrA is a response regulator (section 7.8.6) involved in the direct or indirect control of about one-quarter of the cell-cycle-regulated genes [Science (2000) *290* 2144–2148]; it also has a major role in regulating the initiation of DNA replication. During the cell cycle the CtrA protein undergoes phosphorylation (to form CtrA~P) and proteolysis, at specific times and locations, in order to carry out its functions. The timing of activation of CtrA is controlled in a way which is not understood; it is believed to involve a membrane-associated histidine kinase, CckA, and probably a tyrosine kinase, DivL, and the response regulator DivK [Science (2003) *301* 1874–1877].

In the *swarm* cell (Fig. 4.1), CtrA/CtrA~P binds to the chromosome at the origin of replication (Chapter 7), blocking the *initiation* of DNA replication in that cell until it eventually becomes a stalked (mother) cell [PNAS (1998) *95* 120–125]. The protein also blocks transcription of the 'early' cell division gene *ftsZ* (section 3.2.1) in swarm cells.

Before the end of G2 phase the *stalked* (mother) cell has lost its CtrA/CtrA~P: the protein is dephosphorylated/proteolysed in *that* compartment of the pre-division cell just before cell separation. (Proteolysis of CtrA involves e.g. the DivK protein, another response regulator.) The absence of CtrA/CtrA~P in the stalked cell at this time allows immediate initiation of DNA replication and permits transcription of the *ftsZ* gene (*ftsZ* is repressed by CtrA).

Once DNA replication has started in the stalked cell there is no need to exclude CtrA. In fact, at some time after the initiation of replication CtrA is required for various functions in the growing cell. What decides when the synthesis of CtrA can begin again? During DNA replication (section 7.3), as the *replication fork* (Fig. 7.8) passes through the P_1 promoter of the *ctrA* gene the DNA of the promoter becomes transiently hemi-methylated (see sections 7.4 and 7.8.8); because only the hemi-methylated form of this promoter can be activated, the hemi-methylation of P_1 *per se* acts as a signal for the renewed synthesis of CtrA [MM (2003) *47* 1279–1288].

Near the end of DNA replication CtrA accumulates in the stalked cell, transcription of *ctrA* being upregulated (at the P_2 promoter) by CtrA~P in a positive feedback loop. CtrA is needed at this stage e.g. for activation of 'late' genes *ftsQ* and *ftsA* from promoter P_{QA}; expression of these genes is required for cell division in *C. crescentus*. Experimental inhibition of DNA replication (with hydroxyurea) inhibits the (normal) synthesis of CtrA; this, in turn, prevents transcription of *ftsQA* from P_{QA}, blocking cell division. It was therefore suggested that CtrA mediates a cell-cycle checkpoint at which cell division is blocked in the absence of DNA replication [EMBO Journal (2000) *19* 4503–4512].

An *in vivo* genomic analysis found at least 95 genes whose regulatory sites are bound directly by CtrA [PNAS (2002) *99* 4632–4637].

A recent review on bacterial cell-cycle control pointed out that, in addition to transcriptional regulatory circuits, important roles are played by non-transcriptional pathways and by the temporal and spatial organization of specific regulatory proteins [Science (2003) *301* 1874–1877].

Essential two-component regulatory systems have also been reported in e.g. *Bacillus subtilis* [Microbiology (2000) *146* 1573–1583] and *Mycobacterium tuberculosis* [JB (2000) *182* 3832–3838].

3.2.1.4 *The cell cycle in members of the Archaea*

[The cell cycle in archaeans: MM (1998) *29* 955–961.]

3.2.2 Modes of cell division

The division of one cell into two (typically similar or identical) cells by the formation of a septum (as in *E. coli*) is called *binary fission*; it is the commonest form of cell division in bacteria. 'Asymmetrical' binary fission, in which daughter cells are not similar to the parent cell, occurs e.g. in *Caulobacter* (Chapter 4).

Multiple fission involves repeated binary fission of cells within a common bag-like structure; it occurs e.g. in certain cyanobacteria.

In *ternary fission*, three cells are formed from one; it occurs e.g. in *Pelodictyon* (which forms three-dimensional networks of cells).

Budding is a form of cell division in which a daughter cell develops from the mother (parent) cell as a localized outgrowth (*bud*); it occurs e.g. in species of *Blastobacter*, *Hyphomicrobium* and *Nitrobacter*.

3.2.3 Doubling time

The time taken for one complete cell cycle, the *doubling time*, varies with species and with growth conditions. For a minimum doubling time, optimum growth conditions are necessary. In *E. coli* the minimum doubling time is about 20 minutes, and in some species of *Mycobacterium* (for example) it is many hours; it has been estimated that *Mycobacterium leprae* (the causal agent of leprosy) has a doubling time of about 2 weeks in infected tissues.

3.3 GROWTH IN BACTERIAL POPULATIONS

Following cell division, each daughter cell can itself grow and divide – so that one cell can quickly give rise to a large population of cells if conditions are favourable. Given suitable conditions, such populations may develop either on solid surfaces or within the body of a liquid. In bacteriology, any solid or liquid specially prepared for bacterial growth is called a *medium* (section 14.2)

3.3.1   Growth on a solid medium

One common type of 'solid' medium, widely used in bacteriological laboratories, is a jelly-like substance (an *agar* gel) containing nutrients and other ingredients. Suppose that a single bacterial cell is placed on the surface of such a medium and given everything necessary for growth and division. The cell grows, divides into two cells, and each daughter cell does the same. If growth and division continue, the progeny of the original cell eventually reach such immense numbers that they form a compact heap of cells that is usually visible to the naked eye; this mass of cells is called a *colony*.

Typically, under given conditions, each species forms colonies of characteristic size, shape (Fig. 3.3), colour and consistency; different types of colony may be formed when growth occurs on different media or when other factors differ. The size of a colony may be limited e.g. by local exhaustion of nutrients (due to the colony's own growth); for this reason, crowded colonies are generally smaller than well-spaced ones. The rate at which a colony increases in size depends on temperature and other factors. Bacteria which produce pigments generally form brightly coloured colonies (e.g. red, yellow, violet) while colonies of non-pigmented bacteria usually look grey, whitish or cream-coloured. In consistency, a colony may be mucoid (viscous, mucus-like), butyrous (butter-like), friable (crumbly) etc. and its surface may be smooth or rough, glossy or dull etc.

Instead of starting with a single cell, suppose that a very large number of cells is spread over the surface of the medium. In this case there may not be enough space for individual colonies to develop; accordingly, the progeny of all these cells will form a continuous layer of bacteria which covers the entire surface of the medium. Such continuous growth is called *confluent growth*. Confluent growth may also result when one or a few cells of a *motile* bacterium are deposited on the medium; following

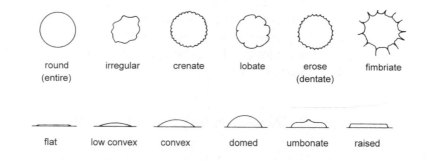

| round | irregular | crenate | lobate | erose | fimbriate |
| (entire) | | | | (dentate) | |

| flat | low convex | convex | domed | umbonate | raised |

Figure 3.3   Shapes of bacterial colonies. The upper row shows the outline or *edge* of some colonies as seen from above. The lower row shows the *elevation* of some colonies as seen from one side. A colony with, say, a round (entire) outline may have any one of the elevations, depending e.g. on the species of bacterium.

growth, the numerous progeny of these cells may swim through the surface film of moisture and eventually cover the whole surface of the medium.

3.3.2 Growth in a liquid medium

Bacteria can move freely through a liquid medium either by diffusion or, in motile species, actively by locomotion; thus, as cells grow and divide, the progeny are commonly dispersed throughout the medium. Usually, as the concentration of cells increases, the medium becomes increasingly turbid (cloudy). Certain bacteria are exceptional in that they tend to form a layer (a *pellicle*) at the surface of the medium; below the pellicle the medium may be almost free of cells. Some pellicles include bacterial products as well as the bacteria themselves; for example, a tough cellulose-containing pellicle is formed by strains of *Acetobacter xylinum*.

3.3.2.1 Batch culture

Suppose that a few bacterial cells are introduced into a suitable *liquid* medium which is then held at the optimum growth temperature for that species. At regular intervals a small volume of the medium can be withdrawn and a count made of the cells it contains (counting methods: section 14.8). In this way we can follow the development of a population, i.e. the increase in cell numbers with time. By plotting the number of cells against time we obtain a *growth curve* which, for a given species growing under given conditions, has a characteristic shape.

By growing bacteria in or on a medium we produce a *culture*; thus, a culture is a liquid or solid medium containing bacteria which have grown (or are still growing) in or on that medium. The process of maintaining a particular temperature (and/or other desired conditions) for bacterial growth is called *incubation*. The initial process of adding the cells to the medium is called *inoculation*.

When bacteria are introduced into a fresh liquid medium, cell division may not begin immediately: there may be an initial *lag phase* in which little or no division occurs. During lag phase the cells are adapting to their new environment – for example, by making enzymes to utilize the newly available nutrients. The length of the lag phase will depend largely on the conditions under which the cells existed *before* they were introduced into the medium. A long lag phase will often occur if the cells had previously existed under harsh conditions, or had been growing with different nutrients or at a different temperature; the lag phase will be short (or even absent) if the cells had been growing in a similar or identical medium at the same temperature.

During the (adaptive) lag phase, molecules are being synthesized, but the increase in total mass of the cell population is not matched by an increase in cell numbers; the cells are said to be undergoing *unbalanced growth*.

Once adapted to the new medium, the cells begin to grow and divide at a rate which is maximum for the species under the existing conditions; this is the *logarithmic* (= *log*) *phase* or *exponential phase* of growth. In this phase, cell numbers double at a constant rate (Table 3.1) – as does the mass of the population; this indicates *balanced growth*.

In the log phase of growth, a plot of cell numbers versus time gives a sharply rising curve on a simple arithmetical scale (Fig. 3.4a); clearly, such a scale would not be adequate for large numbers of cells. Is there a better way of plotting growth in the log phase? Table 3.1 (bottom row) shows that cell numbers can be expressed as powers of 2; for example, the 8 cells at 60 minutes can be written as 2^3 cells (in which 3 is the *index*). Each of the indices in Table 3.1 is, of course, the *logarithm* (to base 2, i.e. $\log_2$) of the corresponding number of cells. Now, if – instead of plotting cell numbers directly – we plot the $\log_2$ of each number, the result is a straight-line graph (Fig. 3.4b). In such a graph each unit on the $\log_2$ scale represents a doubling in cell numbers; the *doubling time* (here, the time, in minutes, needed for a doubling in cell numbers) can therefore be read off directly from the time-scale of the graph. The doubling time is also called the *generation time*.

Usually, it's more convenient to use $\log_{10}$ rather than $\log_2$ when constructing a

Table 3.1 Increase in cell numbers with time for *Escherichia coli* growing under optimal conditions in the logarithmic phase

Time (minutes)	0	20	40	60	80	100	200	300
Number of generations (i.e., rounds of cell division)	0	1	2	3	4	5	10	15
Number of cells	1	2	4	8	16	32	1024	32768
Number of cells as a power of 2	2^0	2^1	2^2	2^3	2^4	2^5....	2^{10}....	2^{15}

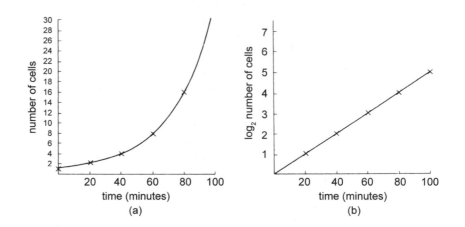

Figure 3.4 The logarithmic (exponential) phase of growth in *Escherichia coli*. Cell numbers are plotted on (a) a simple arithmetical scale, and (b) a logarithmic ($\log_2$) scale. (Compare with Table 3.1.)

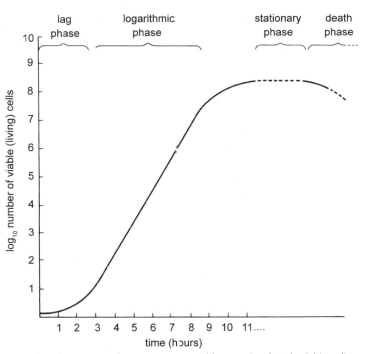

Figure 3.5 Batch culture. A growth curve constructed for a strain of *Escherichia coli* growing in nutrient broth at 37°C.

growth curve. The $\log_{10}$ and $\log_2$ of any number can be interconverted by using the formula: $\log_{10}N = 0.301 \log_2N$; the log phase of the graph will still be a straight line – only the *slope* of the graph will change.

As they grow, the cells use nutrients and they also produce waste products which accumulate in the medium. Eventually, therefore, growth slows down and stops due either to a lack of nutrients or to the accumulation of waste products (or both); the phase in which there is no overall increase in the number of living cells is called the *stationary phase*. (A minireview entitled 'Life after log' is worth reading [JB (1992) *174* 345–348].) (See also Rsd protein in section 7.8.9.) The stationary phase leads eventually to the *death phase* in which the number of living cells in the population progressively decreases.

Figure 3.5 shows the phases of growth during batch culture. 'Batch culture' is so-called because growth from lag phase to death phase occurs in the same batch of medium.

3.3.2.2 Cells in different phases of growth

Cells can undergo marked changes in their general biology and metabolism during the different phases of growth. For example cells which – when growing – are normally killed by penicillin (section 15.4.1) are resistant to this antibiotic when growth ceases.

During periods of non-growth or growth-limitation, by-products of primary (growth-directed) metabolism may be used by the cell to synthesize so-called 'secondary metabolites' (*idiolites*) which are not used for growth. In some species, idiolites include important antibiotics or toxins. [Function of secondary metabolites: ARM (1990) *44* 395–427.] Leptomycin B, an agent which blocks the cell cycle in eukaryotic cells, is a secondary metabolite of *Streptomyces*. [Detecting secondary metabolites in microbial extracts: AEM (2001) *67* 371–376.]

The production of substances during a particular phase of growth can be optimized by adding nutrient(s) at appropriate times; such *fed batch culture* is used in some industrial processes.

3.3.2.3   *Continuous culture*

When bacteria are grown in a fixed volume of liquid medium (as in batch culture) the composition of the medium continually changes as nutrients are used up and waste products accumulate. Batch culture is suitable for many types of study, but sometimes it is preferable that cells be grown under constant, controlled, defined conditions. This is achieved by using *continuous culture* (*continuous-flow culture, open culture*). In this process, bacteria are grown in a liquid medium within an apparatus called a *chemostat*; during growth, there is a continual inflow of fresh, sterile medium, and a simultaneous outflow – at the same rate – of culture (i.e. medium + cells). Constant and thorough agitation of the medium is necessary to ensure rapid mixing of the inflowing fresh medium with culture in the chemostat. Under these conditions, cells can exhibit continual logarithmic growth, i.e. balanced growth, for an extended period of time. Because growth is occurring under constant and defined conditions, this form of culture is useful e.g. for studies on bacterial metabolism.

Under steady-state conditions in a chemostat, the increase in cell numbers through growth is exactly balanced by the loss of cells from the chemostat; the mass of cells (the *biomass*) in the chemostat therefore remains constant. To achieve a steady state, the dilution rate is made equal, numerically, to the specific growth rate. The *dilution rate* (*D*) is given by F/V, where F is the rate at which the medium enters the chemostat (in litres per hour) and V is the volume of culture. The *specific growth rate* (μ) is the number of grams of biomass formed, per gram of biomass, per hour. In the chemostat, growth is normally kept at a *sub*maximal rate because instability tends to occur in the system at or near the maximum growth rate; the growth rate is controlled by controlling the concentration of an essential nutrient in the inflowing medium. The specific growth rate and the concentration of the growth-limiting nutrient often have the relationship predicted by the Monod equation:

$$\mu = \mu_{max} \frac{s}{k_s + s}$$

in which μ_{max} is the maximum growth rate, s is the concentration of the growth-limiting nutrient, and k_s is the concentration of the growth-limiting nutrient when $\mu = 0.5\,\mu_{max}$.

Clearly, D should not exceed the *critical dilution rate*, D_c, which corresponds to μ_{max}. If it does, the culture becomes progressively diluted to extinction and is said to have undergone 'wash-out'.

In practice, ideal (predictable) operation of a chemostat is not always achieved – due e.g. to less than perfect (instantaneous) mixing, or to factors such as 'wall growth' (organisms growing on the walls of the culture vessel and forming a separate, static (unmixed) population of cells). Another type of problem involves the emergence of *mutants* (Chapter 8) during the long periods of growth; mutant cells may e.g. grow more rapidly (than cells of the original population) so that the steady state will not be maintained.

3.3.2.4 Synchronous growth

In a population of growing bacteria, all the cells do not divide at the same instant. However, in the laboratory, we can obtain a population in which all the cells divide at approximately the same time; in such *synchronous growth* the logarithmic portion of the growth curve (Fig. 3.5) appears as a series of steps, each step representing an abrupt doubling of cell numbers.

One method for obtaining a synchronous culture – *selection synchrony* – depends on the fact that the mass:volume ratio (i.e. density) of a bacterium varies during the cell cycle, being highest in (i) cells about to divide, and (ii) newly formed daughter cells. In a given non-synchronous population, cells having the *lowest* density are likely to be at a similar stage of the cell cycle; this subpopulation of (least-dense) cells can be separated from the rest of the population by using the principle of isopycnic centrifugation (section 8.5.1.4). This purely physical method has the advantage that it avoids disturbance to the cells' metabolism.

3.3.2.5 Growth rate versus temperature – Arrhenius plots and Ratkowsky plots

Changes in temperature markedly affect growth rate because they affect the cell's growth-directed chemical reactions (metabolism – Chapters 5 and 6).

Over a limited range of temperatures, the relationship between growth rate and temperature is similar to that between chemical reaction rates and temperature; this has been expressed in Arrhenius plots – in which the logarithm of growth rate is plotted against the reciprocal of absolute temperature ($1/K$). However, over an extended temperature range such plots are non-linear. Ratkowsky and colleagues [JB (1983) *154* 1222–1226] showed that, for various species of bacteria, a linear relationship could be demonstrated over the full biokinetic temperature range by plotting the square root of growth rate against absolute temperature.

3.4 DIAUXIC GROWTH

If a bacterium is given a mixture of two different nutrients, it may use one in preference to the other – utilization of the second beginning only after the first has been exhausted. For example, given a mixture of glucose and lactose, *Escherichia coli* will use the glucose first, starting on the lactose only when all the glucose has been used. During the transition from one nutrient to the other, growth may slow down or even stop; this pattern of growth is called *diauxie* (or diauxy). The mechanism of diauxie is discussed in section 7.8.2.1.

3.5 MEASURING GROWTH

Growth (change in cell numbers, or biomass, with time) can be measured e.g. by (i) counting cells (section 14.8); (ii) determining the increase in dry weight of biomass formed in a given time interval; (iii) monitoring the uptake (or release) of a particular substance; (iv) measuring the amount of a radioactive substance incorporated in biomass in a given time, and (v) *nephelometry*: measuring the increase in scattered light from a beam passing through a liquid culture (light scattering increases as cell numbers increase).

Flow cytometry is a rapid and convenient method for counting bacteria in suspension (e.g. in a liquid medium). In this method, cells are counted automatically as they pass through a small hole or tube in the instrument (flow cytometer). One commercial product (Bacteria Counting Kit; Molecular Probes Inc., Eugene, Oregon, USA) contains a fluorescent dye (SYTO BC) which has a high affinity for nucleic acid and which penetrates and stains both Gram-positive and Gram-negative cells; excitation (within the instrument) produces a strong green fluorescent signal which is detected by the counting device. Flow cytometers should be checked and calibrated regularly; this can be easily done by using known numbers of synthetic (polystyrene) fluorescent microspheres of uniform size and density.

4 Differentiation

In most species of bacteria, major changes in form or function do not occur: progeny cells are more or less identical, in both appearance and behaviour, to their parental cells. In some bacteria, however, one type of cell can give rise to a markedly different type of cell; the timing of such *differentiation* is often related to conditions in the cell's environment. In the next few pages we look at some diverse examples of bacterial differentiation.

4.1   THE LIFE-CYCLE OF *Caulobacter*

Caulobacter is a Gram-negative, strictly aerobic bacterium found in soil and water. It forms two distinctly different types of cell, and the change from one type to the other is an essential part of the life-cycle (Fig. 4.1). Having a swarm cell in the life-cycle is advantageous because it allows the organism to spread to, and colonize, different locations. The motile daughter cell must lose its flagellum and develop a stalk (called a *prostheca)* before it can divide; it may thus be regarded as immature. The stalked (mature) mother cell can produce swarm cells but cannot itself become a swarm cell.

Clearly, for *Caulobacter* to grow normally, the flagellum and prostheca must develop at the correct times, and in the correct places, to allow *asymmetric* division into stalked and swarm cells. How is this done? Answers to some of the questions on development and differentiation – in *Caulobacter* and other bacteria – have been obtained through studies on the intracellular *localization* of particular proteins using techniques such as GFP fusion (section 8.5.2) and FRET (section 8.5.16); in some cases the presence or activation of a specific protein at a particular location within the cell has been linked to a particular stage or function in cell division or differentiation.

Control of the cell cycle in *C. crescentus* is discussed in section 3.2.1.3.

The development of the flagellum in *C. crescentus* has been extensively studied, and aspects of this process are outlined, below, to exemplify some of the features of differentiation in this organism.

Although the flagellum occurs on the swarm cell, its production begins in the

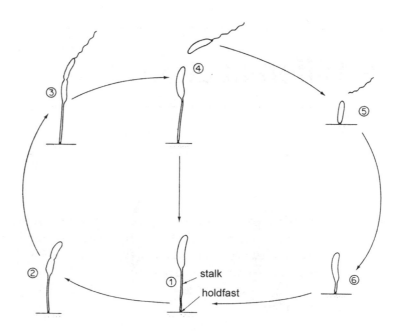

Figure 4.1 The life-cycle of *Caulobacter*. **1**. A mature, stalked cell attached to a surface by an adhesive 'holdfast'. **2**. The stalked cell begins to divide. **3**. A flagellum develops at the free end of the daughter cell. **4**. Asymmetric binary fission is complete, and the motile daughter cell (*swarm cell*) swims away. The stalked cell can continue to grow and produce more swarm cells. **5**. The swarm cell loses its flagellum and becomes attached by its holdfast. **6**. A stalk develops, and the daughter cell matures into a new stalked mother cell.

Stage 3 shows the *pre-division cell* (see section 3.2.1.3). The holdfast and stalk develop at the pole which loses the flagellum.

stalked cell during the S phase of the cell cycle (section 3.2.1.3); by the time the cells separate flagellar rotation has already begun.

As in other bacteria, flagellar assembly in *C. crescentus* involves a number of genes (~40) encoding structural and regulatory functions. Moreover, as in e.g. *Escherichia coli* (section 7.8.2.7), flagellar genes are expressed in a hierarchical or cascade fashion that reflects the order in which their products or functions are required during flagellar assembly. (In some cases the designation of a gene that encodes a particular flagellar component is the same for both *E. coli and C. crescentus*; for example, in both species the MS ring (Fig. 2.8) is encoded by *fliF*, the switch/C ring by *fliG, fliM* and *fliN*, the L ring by *flgH* and the P ring by *flgI*. By contrast, the filament subunit (flagellin) in *E. coli* is encoded by *fliC* but in *C. crescentus* the genes *fljM, fljN* and *fljO* encode flagellins.)

As in *E. coli* (Table 7.3), the flagellar genes of *C. crescentus* are categorized into classes; however, there are three classes (I–III) in *E. coli* and four (I–IV) in *C. crescentus* – in the latter organism *ctrA* (see section 3.2.1.3) is regarded as a class I flagellar gene.

In *E. coli*, *fliF*, *flgH* and *flgI* are all class II genes, while in *C. crescentus fliF* is class II but *flgH* and *flgI* are class III genes; this reflects a difference in the number of *checkpoints* (see section 3.2.1.3) during flagellar assembly in the two species. Thus, in *E. coli* there is a single checkpoint: the anti-sigma factor FlgM (Table 7.3) blocks expression of class III genes until completion of the hook region of the flagellum (section 7.8.2.7). In *C. crescentus* there are two checkpoints: (i) assembly of the MS ring and C ring/switch is required before further (class III) components can be added, and (ii) assembly of the L and P rings and the hook region is required before the filament (encoded by class IV genes) can be added.

As indicated earlier (section 3.2.1.3) checkpoints can be bypassed as a result of mutation. In *C. crescentus*, so-called *bfa* (bypass of flagellar assembly) mutant class IV (filament) genes are transcribed but not translated [JB (1995) *177* 3176–3184].

Recent work has shown that transcription of some flagellar genes occurs earlier than expected, e.g. *fliE* (thought to be a class III gene according to order of assembly) is transcribed with class II genes [JB (2001) *183* 725–735]; this, and the unexpected pattern of binding of the gene regulator CtrA, has suggested that regulation of the flagellar genes in *C. crescentus* may be more complex than once thought [PNAS (2002) *99* 4632–4637].

Temporal/spatial regulation in the assembly of the flagellum. The early (class II) flagellar genes encode those components of the basal body (e.g. MS ring: Fig. 2.8) that are associated with the cytoplasmic membrane; expression of these genes in the stalked cell (during the S phase) initiates flagellar assembly. Class II genes are activated by the CtrA protein; synthesis of CtrA starts in the developing stalked cell during DNA replication when the replication fork (Fig. 7.8) forms a transiently hemi-methylated (activated) *ctrA* gene [MM (2003) *47* 1279–1288].

FliF protein is targeted (by an unknown mechanism) to a specific polar site in the developing swarm cell, i.e. the MS ring develops prior to septum formation.

The operon (section 7.8.1) containing the class II gene *fliF* also contains the regulatory genes *flbD* (encoding a non-active gene-regulator protein) and *flbE* (encoding a kinase that is able to phosphorylate FlbD). *After* septum formation in the pre-division cell, FlbD is phosphorylated (activated) by the kinase FlbE *in the swarm cell compartment*. FlbD~P then promotes transcription of the late flagellar genes in that compartment, permitting development of the swarm cell's flagellum. Notice the requirement for activation of FlbD at a specific stage and localization to a particular compartment of the pre-division cell. (Analogously, during sporulation in *Bacillus subtilis*, localization of the phosphatase SpoIIE to a specific side of the asymmetric septum is an important factor in development of the endospore – see legend of Fig. 7.14.)

Many of the mechanisms that regulate development and differentiation in *C. crescentus* – including flagellar synthesis and assembly – have been at least partially characterized, but the mechanism that guides polar development of the flagellum is still unknown [Science (2003) *301* 1874–1877].

In the swarm cell, CtrA/CtrA~P binds to the chromosome's *origin* (section 7.3) and represses the initiation of DNA replication until the cell eventually loses its flagellum and becomes a stalked cell. The pole which loses the flagellum develops the stalk.

[Signal transduction mechanisms in *Caulobacter crescentus* development and control: FEMSMR (2000) *24* 177–191.]

Differentiation from non-motile to motile cells, and vice versa, also occurs in species of *Hyphomicrobium* and *Rhodomicrobium*.

4.2 SWARMING

Proteus is a Gram-negative bacillus found e.g. in the intestines of man and other animals. If cells of *P. mirabilis* (or of another species, *P. vulgaris*) are incubated on a suitable solid medium, the first progeny cells are short, sparsely flagellated bacilli about 2–4 μm in length; these cells form a colony in the usual way. However, after several hours of growth, some of the cells around the edge of the colony grow to lengths of 20–80 μm and develop numerous additional flagella; these cells are called *swarm cells*. The swarm cells swim out to positions a few millimetres from the colony's edge, and, there, each divides into several short bacilli – similar to those in the original colony; these cells grow and divide normally for a number of generations, forming a ring of heavy growth which surrounds (and is concentric with) the original colony. Later, another generation of swarm cells is formed at the outer edge of the ring and the cycle is repeated. In this way the entire surface of the medium becomes covered by concentric rings of growth (Plate 4.1). This phenomenon is called *swarming*.

Swarming is not essential for *Proteus*, and it doesn't happen on all types of medium. It occurs typically on rather moist surfaces; under such conditions, swarm cells help the organism to spread to new sources of nutrients.

Swarming also occurs in various other bacteria, both Gram-negative (including e.g. *Serratia marcescens* (Plate 4.2) and *E. coli*) and Gram-positive; it is often demonstrated by using an *agar* medium (section 14.2) containing a concentration of agar lower than that in normal growth media. Unlike *P. mirabilis*, which forms concentric rings of growth (due to discrete periods of swarming), some bacteria swarm continuously, forming a thin layer of growth, while in others swarming gives rise to individual microcolonies. [Swarming (review): MM (1994) *13* 389–394.] Interestingly, work on *E. coli* indicates that, although swarming involves components of the chemotaxis system (section 2.2.15.2), the substances/signals that provoke swarming differ from the known chemoeffectors in chemotaxis [PNAS (1998) *95* 2568–2573].

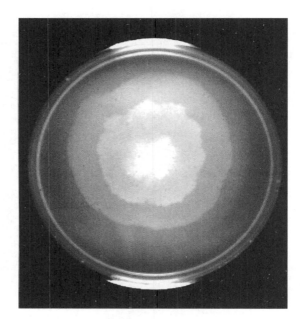

Plate 4.1   Swarming: concentric rings of growth on a suitable agar medium (section 4.2). Photograph courtesy of Dr Rasika M. Harshey, University of Texas at Austin, Texas, USA.

4.3   RESTING CELLS

In some bacteria differentiation can result in the formation of a resting cell – either a *spore* or a *cyst*. Resting cells may function as disseminative units (helping to spread the organism) and/or as dormant cells which are capable of surviving in a hostile environment. Under suitable conditions a spore or cyst *germinates* to form a new vegetative cell.

4.3.1   Endospores

Endospores have been studied more thoroughly than any other type of bacterial spore; they are formed by species of *Bacillus*, *Clostridium*, *Coxiella*, *Desulfotomaculum*, *Thermoactinomyces* and a few other genera. An endospore is formed *within* a cell as a response to starvation – specifically, a shortage of carbon, nitrogen and/or phosphorus. It exists in a state of dormancy: few, if any, of the chemical reactions in a vegetative (growing) cell take place in the mature endospore. Dormancy can persist for long periods: the age of some specimens may be measurable on a geological scale [Microbiology (1994) *140* 2513–2529]. The endospore is highly resistant to many hostile factors: extremes of temperature and pH, desiccation, radiation, various chemical agents and physical damage; in fact, irreversible inactivation of endospores

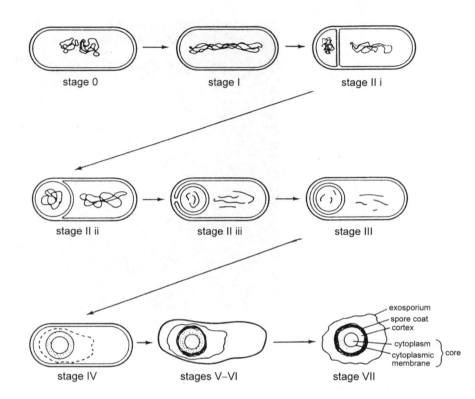

Figure 4.2 Endospore formation (diagrammatic). **Stage 0**. A vegetative cell containing two chromosomes is about to sporulate. **Stage I**. An *axial filament*, composed of the two chromosomes, is formed. **Stage II**. The development of an asymmetrical septum divides the protoplast into two unequal parts (Plate 4.3). The septum is considerably thinner than that formed during normal cell division because it contains much less peptidoglycan; subsequently the peptidoglycan is hydrolysed (removed). The smaller of the two protoplasts will become the endospore, and is called the *forespore* or *prespore*. The cytoplasmic membrane of the larger protoplast invaginates to engulf the forespore. Stage II is subdivided into parts i, ii and iii, as shown. When completely engulfed, the forespore is bounded by two membranes (**stage III**). **Stage IV**. Modified peptidoglycan is laid down between the two membranes of the forespore to form a rigid layer called the *cortex*. A loose protein envelope called the *exosporium* may begin to develop at about this time. **Stages V–VI**. A multilayered protein *spore coat* is deposited outside the outermost membrane (stage V), and the spore matures (stage VI) to develop its characteristic resistance to heat and its bright, refractile appearance under the light microscope; during this time, calcium dipicolinate accumulates in the 'core' – i.e. the spore's protoplast (bounded by the inner membrane). **Stage VII**. The completed spore is released by disintegration of the mother cell. (Note that an exosporium is not present on all endospores.)

How is the asymmetric (polar) septum formed? In *vegetative* (growing) cells of the endospore-forming species *Bacillus subtilis*, septation during cell division occurs at the mid-cell position; as in *E. coli* this is determined by the position of the FtsZ ring (section 3.2.1) which, in turn, is guided to this site by the *min* gene products (section 3.2.1.2). In a *sporulating* cell, an FtsZ ring initially forms at the mid-cell site; however, the ring later develops into a helical structure that extends into both poles of the can be guaranteed only by the harsh treatment of a *sterilization* process (section 15.1).

The formation of an endospore is described in Fig. 4.2, and its structure is shown in Plate 4.2. The heat-resistance of the endospore is believed to be due to its low content of water. Calcium dipicolinate (in the core) may act as a secondary stabilizer; diplicolinic acid (pyridine-2,6-dicarboxylic acid) is synthesized by a branch of the diaminopimelic acid pathway. [Mutant spores of *Bacillus subitlis* which lack dipicolinic acid: JB (2000) *182* 5505–5512.]

Under suitable conditions, an endospore *germinates*, i.e. it becomes metabolically active. The endospores of some species need to be 'activated' before they can germinate; activation may consist e.g. of sublethal heating or exposure to certain chemicals. Germination is promoted by chemicals called *germinants*; according to species, germinants may include e.g. L-alanine, some purine nucleosides, various ions or certain sugars. Germination may be initiated by the binding of a germinant to an inner membrane receptor. The transition from a germinated endospore to a vegetative cell is called *outgrowth*.

Genetic control of the initiation of endospore formation is described in section 7.8.6.1.

Note. 'Endospore' is often abbreviated to 'spore'. However, the endospore should not be confused with other types of bacterial spore (see section 4.3.2).

4.3.2   Other bacterial spores

In many of the hypha-forming actinomycetes, *exospores* are produced by septation and fragmentation of the hyphae (Fig. 4.3). These spores lack specialized structures (such as cortex and spore coat) but they do show some resistance e.g. to dry heat, desiccation and certain chemicals. The spores of *Streptomyces* are metabolically less active than the vegetative hyphae, though they are not completely dormant.

---

cell – finally giving rise to an FtsZ ring at *both* polar sites in the cell. (Progress from a single, mid-cell FtsZ ring to two polar rings involves upregulation of levels of FtsZ and the regulatory influence of the SpoIIE protein.) Septation occurs at only one of the two polar sites: septation at one inhibits septation at the other.

Interestingly, even when the septum is nearly complete the forespore still contains only a portion of its chromosome; the rest of the chromosome (in the mother cell) is fed into the forespore by septum-associated SpoIIIE (a DNA translocase).

Activation of transcription factor σ^F (which regulates gene expression in the forespore) is determined partly by the SpoIIE protein, which localizes *on the forespore side* of the asymmetric septum. Activation of σ^F is also promoted by the (initial) *absence* of an anti-sigma factor, SpoIIAB, whose gene occurs in that part of the chromosome initially excluded from the forespore (see above). These are two examples of the way in which an aspect of differentiation (here, the activation of σ^F) can be regulated by the spatial and temporal distribution of specific proteins and genes (SpoIIE and *spoIIAB*, respectively).

[Generating and exploiting polarity in bacteria: Science (2002) *298* 1942–1946 (1944–1945).]

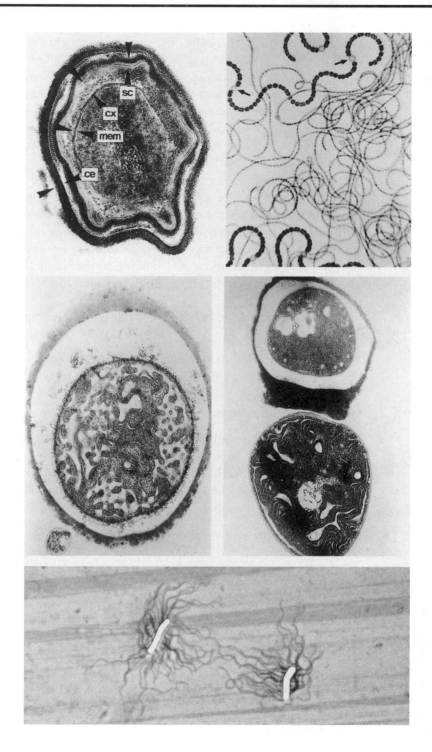

In the actinomycetes *Actinoplanes* and *Pilimelia*, motile (flagellated) *zoospores* are formed inside a closed sac called a *sporangium*; sporangia develop from the vegetative hyphae.

4.3.3   Bacterial cysts

Cysts are formed e.g. by the soil bacterium *Azotobacter vinelandii*. The desiccation-resistant cysts of this organism are dormant, and they can survive in dry soil for years. Encystment may be triggered by changes in the levels of carbon and nitrogen in the environment; it involves loss of flagella and the development of a complex cyst wall containing the polysaccharide *alginate* [alginate biosynthesis (review): Microbiology (1998) *144* 1133–1143], protein and lipid. Typically, PHB (section 2.2.4.1) accumulates in the cyst.

[Alginate formation in *A. vinelandii* in the stationary phase: Microbiology (2001) *147* 483–490.]

4.4   AKINETES, HETEROCYSTS, HORMOGONIA

These structures are formed by various filamentous (trichome-forming) cyanobacteria – photosynthetic organisms which occur e.g. in soil, in natural waters and in symbiotic associations with certain eukaryotes.

Plate 4.2   Differentiation.
Top left. Transmission electronmicrograph of an endospore of *Bacillus subtilis* within the mother cell (approx. 1 μm across). The 'core' (protoplast) of the endospore is bounded by the membrane (mem). Between the membrane and the multilayered spore coat (sc) is the cortex (cx). Surrounding the endospore is the cell envelope (ce) of the mother cell.

Top right. Light-micrograph of *Anabaena spiroides* (large filaments) and *Anabaena circinalis* (fine filaments) from Hebgen Lake, Montana, USA. Each arrow points to a heterocyst; these heterocysts have developed at intercalary positions among the vegetative cells of the filament.

Centre, left. Transmission electronmicrograph of a heterocyst of *Anabaena oscillarioides*. Note the thick envelope. The very fine black dots (seen most clearly with a hand lens) indicate locations of the enzyme *nitrogenase* (section 10.3.2.1); each dot is a particle of colloidal gold which is attached, as an electron-opaque label, to a molecule which binds specifically to nitrogenase.

Centre, right. Transmission electronmicrograph of an *Anabaena oscillarioides* heterocyst (above) and vegetative cell (below).

Bottom. Swarm cells of *Serratia marcescens* (see section 4.2). Note the large number of flagella on each cell.

Endospore courtesy of Dr John Coote, University of Glasgow, Scotland. Heterocysts courtesy of Associate Professor John C. Priscu, Montana State University, Bozeman, Montana, USA. Swarm cells courtesy of Dr Rasika M. Harshey, University of Texas at Austin, Texas, USA.

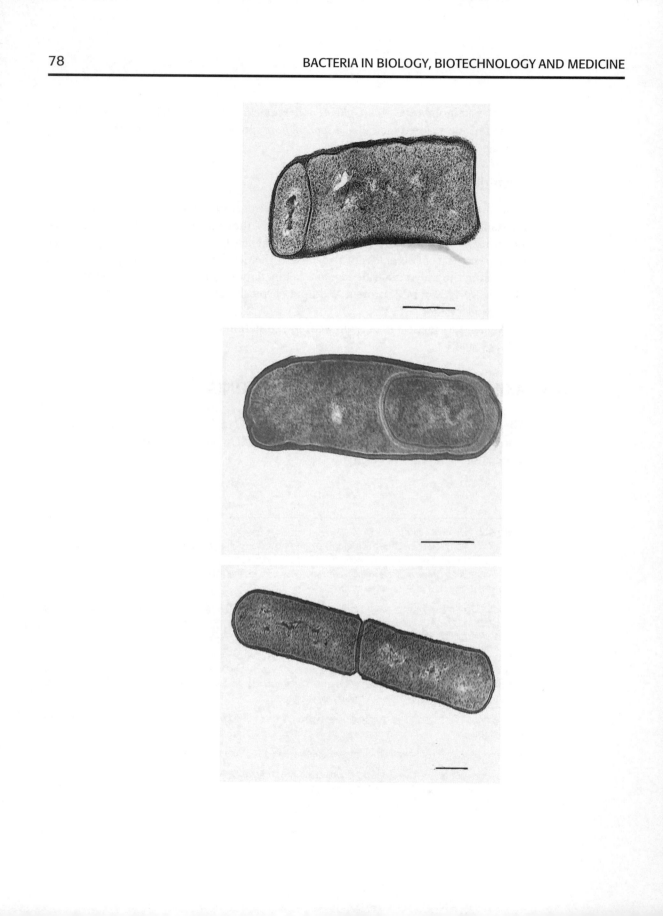

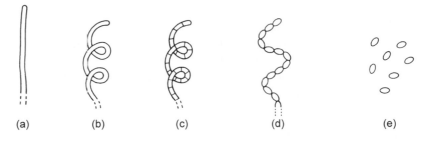

Figure 4.3 Exospore formation in a species *Streptomyces* (an actinomycete). (a) The tip of a vegetative aerial hypha. (b) The tip of the hypha becomes coiled. (c) Septa develop along the length of the coiled hypha. (d) The walls of the developing spores thicken, and each spore becomes rounded. (e) The spores are released.

4.4.1 Akinetes

Akinetes are differentiated cells produced by some species e.g. under starvation conditions; each has a thickened cell wall and a cytoplasm rich in storage compounds (such as glycogen). Akinetes are usually larger than vegetative cells, and they have a lowered rate of metabolism; they show some resistance to desiccation and cold, and may function as overwintering and/or disseminative units.

4.4.2 Heterocysts

Heterocysts are formed by some species when there is a shortage of usable nitrogen compounds. Under such conditions, some of the cells in a trichome undergo differentiation, each forming a *heterocyst*: a specialized compartment within which atmospheric (gaseous) nitrogen can be 'fixed' – i.e. converted to a usable nitrogen compound (section 10.3.2.1). The process of differentiation includes e.g. development of a thick envelope, re-arrangement of the thylakoids, cessation of (photosynthetic) oxygen evolution, and synthesis of *nitrogenase* (an enzyme used in nitrogen fixation). The thick envelope (see Plate 4.2) seems to protect (oxygen-sensitive) nitrogenase against

Plate 4.3 Sporulation versus cell division in *Bacillus subtilis*.
Top. A cell at an early stage of sporulation (Fig. 4.2, stage IIi) showing the asymmetrical septum. Bar=0.3 μm.
Centre. The developing endospore (*forespore*) at a later stage. Bar=0.3 μm.
Below. A cell undergoing vegetative cell division. The septum is located mid-cell and is much thicker than the asymmetric septum. Bar=0.3 μm.

Photographs courtesy of Dr Imrich Barák, Institute of Molecular Biology, Slovak Academy of Sciences, Bratislava, Slovak Republic.

atmospheric oxygen; *Anabaena flos-aquae* forms a thicker envelope when grown under higher partial pressures of oxygen [JGM (1992) *138* 2673–2678].

Communication between a heterocyst and an adjacent vegetative cell occurs via fine pores (*microplasmodesmata*) in their contiguous cytoplasmic membranes; during nitrogen fixation, fixed nitrogen is transferred to the vegetative cell which, in turn, transfers carbon and other materials to the heterocyst.

4.4.3 Hormogonia

A hormogonium is a short trichome, lacking both akinetes and heterocysts, formed e.g. from a vegetative trichome; the cells of a hormogonium may be smaller than those of the parent trichome. Typically, hormogonia exhibit gliding motility. In some species (e.g. *Nostoc muscorum*) only the hormogonia contain gas vascuoles – reinforcing the idea that these short trichomes have a primarily disseminative role.

5 Metabolism I: energy

'Metabolism' refers, collectively, to all the chemical reactions that occur in a living cell or organism; as a result of these reactions, molecules are built up (*anabolism*), broken down (*catabolism*) or changed from one type to another, atoms and ions are oxidized or reduced, and various entities are transported from one site to another.

All living cells require energy, in one form or another, and a major function of metabolism is to provide the cell with a usable form of energy. However, although this chapter focuses on energy, it must be remembered that, in a living cell, the various aspects of metabolism are integrated (see e.g. section 6.2).

A metabolic reaction usually needs a specific protein catalyst (*enzyme*). A given type of enzyme will function only within a limited range of physicochemical conditions – outside this range enzymic activity may be temporarily or permanently lost. The enzymes of extremophiles (section 1.1.1) work under extraordinary conditions of e.g. temperature or pH. These enzymes appear to have unique structures and/or may be stabilized by factors such as thermoprotectants; they are useful for studying enzyme stability and have potential applications in biotechnology [TIM (1998) *6* 307–314].

As well as enzymes, other types of molecule (both protein and non-protein) carry out specialized roles in particular types of energy metabolism. These molecules (discussed in the following pages) include the iron–protein *cytochromes* (involved in oxidation–reduction reactions), magnesium-containing *chlorophylls* (in photosynthetic species), and various energy-carrying nucleotides.

Within the cell, a sequence of metabolic reactions, in which one substance is converted to another (or others), is called a *metabolic pathway*; in such a pathway the *substrate* (e.g. a nutrient) is converted, often via one or more *intermediates*, to so-called *end-product(s)*. Many of the metabolic pathways in bacteria do not occur in eukaryotic cells.

Metabolic reactions are commonly *endergonic*, i.e. they require energy. Energy is also needed e.g. for the uptake ('transport') of nutrients, for some aspects of osmo-regulation (section 3.1.8), and (in motile species) for locomotion.

What sources of energy are used by bacteria? The main sources of energy are: (i)

(a)

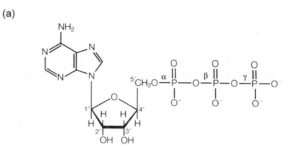

(b)

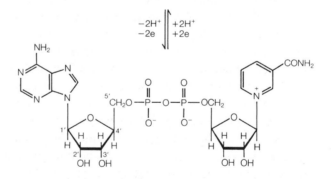

Figure 5.1 Some energy currency molecules. (a) Adenosine 5'-triphosphate (ATP). When donating energy, the γ-bond is broken and the terminal phosphate group is lost; the resulting molecule, adenosine 5'-diphosphate (ADP), must be phosphorylated to regenerate ATP.

(b) Nicotinamide adenine dinucleotide (NAD) showing the reduced (upper) and oxidized (lower) forms; e = electron. Strictly, the oxidized form should be written NAD$^+$ but is often written NAD for convenience; similarly, the reduced form should be written NADH + H$^+$ but is often written NADH. NAD phosphate (NADP) is NAD 2'-phosphate – the phosphate group being at the 2'-position of the (left-hand) sugar (D-ribose) molecule.

chemicals which yield energy when metabolized, and (ii) light (radiant energy). On this basis, two broad categories of bacteria can be distinguished:

- *chemotrophs* obtain energy by metabolizing chemicals from the environment
- *phototrophs* use radiant energy

These categories are not mutually exclusive: some bacteria can use either source of energy – use of a given source depending on conditions.

Neither of the two primary sources of energy can be used *directly* by a cell to fuel its energy requirements. Consequently, every bacterial cell has the ability to convert a primary source of energy into form(s) of energy usable by the cell. What *is* a usable form of energy? Given certain substances, or light, bacteria make various high-energy compounds that can be used, directly, to satisfy their requirements. These compounds include adenosine 5'-triphosphate (ATP), phosphoenolpyruvate (PEP), acetyl phosphate and acetyl-CoA. Such compounds have been called *energy currency molecules* because the cell can use them (rather than the primary form of energy) just as we spend coins and banknotes rather than gold bars. Some of these currency molecules are shown in Fig. 5.1.

In a cell, ATP (Fig. 5.1a) yields usable energy when its terminal phosphate bond is broken; accordingly, as molecules of ATP are used up (supplying the cell's energy needs) molecules of adenosine 5'-diphosphate (ADP) are formed. Hence, the primary (environmental) source of energy is used in such a way that ATP is re-synthesized by phosphorylation of ADP.

Another energy currency molecule, nicotinamide adenine dinucleotide (NAD) (Fig. 5.1b), carries energy in the form of 'reducing power': it accepts and yields energy by being (respectively) reduced and oxidized. Other carriers of reducing power include NAD phosphate (NADP) and flavin adenine dinucleotide (FAD).

In many types of cell, a primary source of energy can also be converted to an electrochemical form; this consists of a gradient of ions (usually protons: H^+) between the two surfaces of the cytoplasmic membrane. The energy of the ion gradient can be used e.g. for transport (see later), for driving flagellar rotation (section 2.2.14.1) – and for the synthesis of high-energy compounds!

How *does* a bacterium form high-energy compounds, or an ion gradient, from a primary source of energy? Different strategies are used by chemotrophs and phototrophs, as discussed below.

5.1 ENERGY METABOLISM IN CHEMOTROPHS

Chemotrophs metabolize chemicals for energy. Those chemotrophs which use organic chemicals are called *chemoorganotrophs*; those which use inorganic chemicals are called *chemolithotrophs*. (The *uptake* of chemicals by chemotrophs is discussed in section 5.4.)

5.1.1   Energy metabolism in chemoorganotrophs

A given chemoorganotroph can exhibit one or both of two main types of metabolism:

- *fermentation* (metabolism of a substrate *without* an external oxidizing agent)
- *respiration* (metabolism of a substrate *with* an external oxidizing agent)

Fermentation and respiration are both forms of *energy-converting metabolism*, i.e. metabolism in which one form of energy is converted to another; both can be used to convert a primary source of energy (i.e. a given chemical substrate) to a form of energy which the cell can use.

Some chemoorganotrophs can carry out only one of these forms of metabolism; others can carry out either, depending on conditions.

5.1.1.1   Fermentation

Fermentation is a type of energy-converting metabolism in which the substrate is metabolized *without the involvement of an exogenous* (i.e. *external*) *oxidizing agent*. (*Note.* Fermentation typically – but not necessarily – occurs anaerobically, i.e. in the absence of oxygen, but this is *not* the distinguishing feature of fermentation: as we shall see later, respiration also can occur anaerobically.) Because no external oxidizing agent is used, the products of fermentation – collectively – are neither more nor less oxidized than the substrate; that is, the oxidation of any intermediate in a fermentation pathway is balanced by the reduction of other intermediate(s) in that pathway. This is illustrated diagrammatically in Fig. 5.2.

These ideas can be understood more easily by looking at some real metabolic pathways. In many bacteria the fermentation of glucose begins with a pathway known as *glycolysis* or as the *Embden–Meyerhof–Parnas pathway* (EMP pathway) (Fig. 5.3). In this pathway, 1 molecule of glucose yields (via a number of intermediates) 2 molecules of the end-product, pyruvic acid. At two places in this pathway (Fig. 5.3) energy from *exergonic* (i.e. energy-yielding) reactions is used to phosphorylate ADP – that is, to synthesize the energy currency molecule ATP from ADP. When energy from a

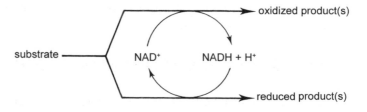

Figure 5.2 A fermentation pathway (diagrammatic). Here, the (organic) substrate gives rise to intermediates which undergo *mutual* oxidation and reduction; taken together, the products have the same oxidation state as the original substrate. (Compare with Fig. 5.4.)

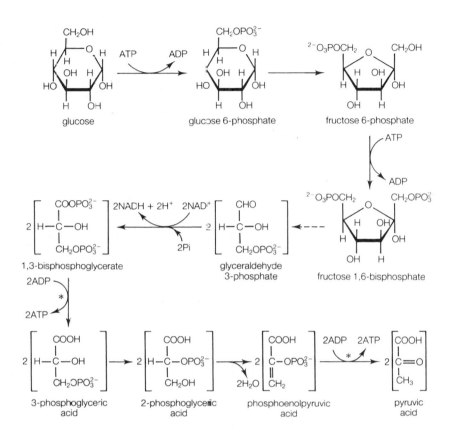

Figure 5.3 The Embden–Meyerhof–Parnas pathway. The broken arrow from fructose 1,6-bisphosphate indicates a simplified section of the pathway. Each of the two substrate-level phosphorylations is marked with an asterisk; in each case, phosphate is transferred to ADP from an energy-rich organic phosphate. Although two reactions in the pathway actually require ATP, there is nevertheless a *net* gain of 2ATP for each molecule of glucose metabolized (*Note*. Pyruvic acid and phosphoenolpyruvic acid are often referred to as 'pyruvate' and 'phosphoenolpyruvate', respectively.) Most of the reactions in this pathway are reversible.

chemical reaction is used, directly, for the synthesis of ATP from ADP the process is known as *substrate-level phosphorylation*.

In terms of energy currency molecules, the EMP pathway can be summarized as follows:

$$2ATP + 4ADP + 2NAD \rightarrow 2ADP + 4ATP + 2NADH$$

Thus, each molecule of glucose metabolized gives 2NADH and a *net* yield of 2ATP.

For the metabolism of further molecules of glucose there is clearly a need for fresh supplies of both ADP and NAD. ADP is regenerated from ATP when the latter is used to supply energy. However, as no external oxidizing agent is used in fermentation, how is NAD regenerated from NADH? Earlier it was said that the fermentation of glucose may *begin* with the EMP pathway; in fact, the EMP pathway is only the 'front end' of a number of different pathways: what happens to the NADH (and the pyruvic acid) depends on subsequent reactions. In the simplest case, NADH donates its reducing power to the pyruvic acid, i.e. the oxidation of NADH to NAD is coupled with the reduction of pyruvic acid to lactic acid:

$$\text{pyruvic acid} + \text{NADH} \rightarrow \text{lactic acid} + \text{NAD}$$

This reaction completes one possible fermentation pathway: a so-called *lactic acid fermentation*; a pathway in which lactic acid is the only (or predominant) product is called a *homolactic fermentation*, while a *heterolactic fermentation* yields a mixture of lactic acid and other products. Lactic acid – a waste product – is released by the cells.

The so-called 'lactic acid bacteria' include e.g. species of *Lactobacillus* and *Leuconostoc* (see Appendix); they are widely used in the manufacture of fermented foods (e.g. some cheeses and other dairy products, salami, sauerkraut and sourdough bread). [Genetics, metabolism and applications of lactic acid bacteria (symposium): FEMSMR (1993) *12* 1–272.]

Notice that lactic acid ($C_3H_6O_3$) has the same oxidation state as glucose ($C_6H_{12}O_6$) – i.e. no net oxidation or reduction has occurred; although glyceraldehyde 3-phosphate undergoes oxidation, this is balanced by the reduction of pyruvic acid to lactic acid (Fig. 5.4).

Other fermentation pathways which have the EMP pathway as their 'front end' include the mixed acid fermentation (Fig. 5.5) and the butanediol fermentation (Fig. 5.6). In these pathways the pyruvic acid is metabolized to several end-products, the relative proportions of which can vary according to conditions of growth.

The *mixed acid fermentation* (Fig. 5.5) occurs e.g. in *Escherichia coli* and in species of *Proteus* and *Salmonella*. Under acidic conditions, *E. coli*, and other species which have the formate hydrogen lyase enzyme system, can split the formic acid into carbon dioxide and hydrogen; thus, these organisms carry out an *aerogenic* (gas-producing) fermentation. Organisms which lack the enzyme system carry out the fermentation *anaerogenically* (i.e. without forming gas); they include species of *Shigella*. (*Shigella* is a reminder that gas is not *necessarily* produced during fermentation.)

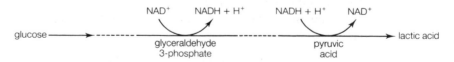

Figure 5.4 Homolactic fermentation: a diagrammatic view of the fermentation of glucose to lactic acid via the Embden–Meyerhof–Parnas pathway. Note that the oxidation is balanced by an equivalent reduction. The homolactic fermentation is carried out by some of the so-called 'lactic acid bacteria' which are used in the food industry.

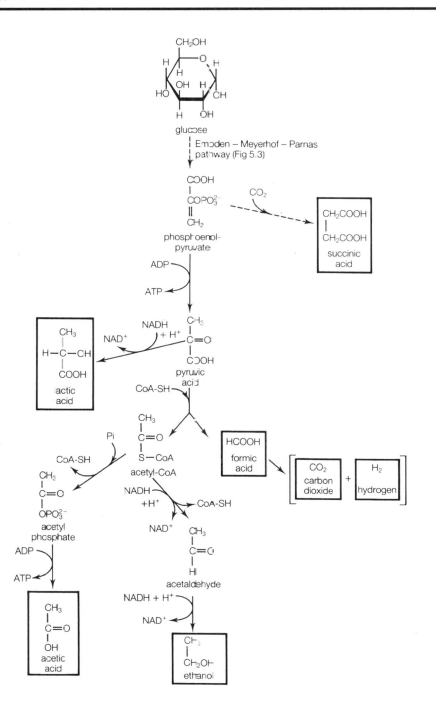

Figure 5.5 The mixed acid fermentation. CoA = coenzyme A; Pi = inorganic phosphate.

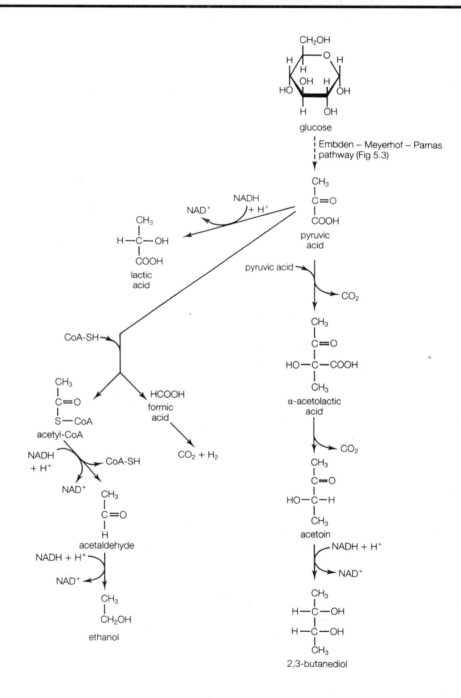

Figure 5.6 The butanediol fermentation. (See also section 16.1.2.5 – the Voges–Proskauer test.)

The *butanediol fermentation* (Fig. 5.6) occurs e.g. in species of *Enterobacter*, *Erwinia*, *Klebsiella* and *Serratia*. The amount of acid formed in this fermentation is generally much less than that formed in the mixed acid fermentation. Some strains form small amounts of diacetyl ($CH_3.CO.CO.CH_3$) from the acetolactic acid under certain conditions.

Although the mixed acid and butanediol fermentations are more complicated than the homolactic fermentation, they nevertheless conform to the same basic principles; while the relative proportions of end-products may vary, the formation of products more oxidized than glucose (e.g. formic acid) is always balanced by the formation of products more reduced than glucose (e.g. ethanol).

NADH and other compounds formed during fermentation can be used in various ways. Some of the NADH will be oxidized in biosynthetic reactions (rather than in the formation of waste products such as lactic acid); this also allows compounds such as pyruvate to be used as precursor molecules for biosynthesis (section 6.3.1). In both respiration (section 5.1.1.2) and fermentation, there is (in chemoorganotrophs) a close link between energy metabolism and carbon metabolism (section 5.3.1).

As well as high-energy compounds, such as ATP and NADH, bacteria growing fermentatively may use energy in the form of an ion gradient, mentioned earlier. In these bacteria, this type of energy is usually obtained by converting some of the ATP at specific membrane-associated enzyme systems called *ATPases* (see section 5.1.1.2 for details); some fermentative bacteria can also generate ion-gradient energy from end-product efflux (section 5.3.3).

Note. Before leaving the topic of fermentation it is worth mentioning that, in industry, the term 'fermentation' is commonly used for *any* chemical process mediated by microorganisms – even for those processes in which fermentation is not involved.

5.1.1.2   *Respiration*

Respiration is a type of energy-converting metabolism in which the substrate is metabolized *with the involvement of an exogenous (i.e. external) oxidizing agent* (compare fermentation: section 5.1.1.1). Respiration can occur in the presence of oxygen (oxygen itself serving as external oxidizing agent) – but it can also occur anaerobically (when other inorganic or organic oxidizing agents are used in place of oxygen). As we are still talking about chemoorganotrophs, the *substrate* is always an organic compound, even though the oxidizing agent may be inorganic or organic.

Because an external oxidizing agent is used, the substrate undergoes a *net* oxidation (Fig. 5.7); glucose, for example, can be oxidized to carbon dioxide and water. The oxidation of a substrate provides more energy than that obtainable – from the same substrate – by fermentation.

How is the substrate oxidized, and how is usable energy obtained? The typical mode

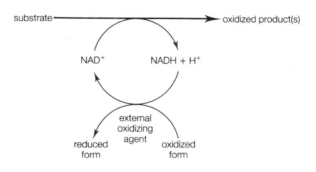

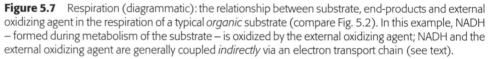

Figure 5.7   Respiration (diagrammatic): the relationship between substrate, end-products and external oxidizing agent in the respiration of a typical *organic* substrate (compare Fig. 5.2). In this example, NADH – formed during metabolism of the substrate – is oxidized by the external oxidizing agent; NADH and the external oxidizing agent are generally coupled *indirectly* via an electron transport chain (see text).

of oxidation of an *organic* substrate is shown in Fig. 5.7: oxidation is coupled with the reduction of NAD, the resulting NADH being oxidized by an external oxidizing agent. NADH and the external oxidizing agent usually interact indirectly via an *electron transport chain* (ETC) located in the cytoplasmic membrane. An ETC is a chain of specialized molecules (redox agents) which form a conducting path for electrons; the sequence of the (different) redox agents is such that electrons can flow down a redox gradient (towards the more positive end) in a series of oxidation/reduction reactions. Electrons from NADH flow down the gradient to an external oxidizing agent; when the latter is oxygen, the situation can be summarized as in Fig. 5.8. The final recipient of the electrons (in this case oxygen) is called the *terminal electron acceptor*.

Electron flow of this kind necessarily yields energy because the electrons are moving from high-energy to lower-energy locations. Typically, this liberated energy is used by the cell for pumping protons (hydrogen ions: H^+) across the cytoplasmic membrane – from the inner to the outer surface; this creates an imbalance of electrical charge (and pH) between the two surfaces of the membrane, and the tendency of protons to move back across the membrane (and thus abolish the imbalance) constitutes a form of energy known as *proton motive force*. Proton motive force (pmf) is one of the cell's most important and versatile forms of energy. It can be used – directly – to satisfy several types of energy demand. Thus, it can drive flagellar rotation (section 2.2.14.1). It can provide energy for the transport (uptake) of various ions; ion uptake is an energy-requiring process because the cytoplasmic membrane is ordinarily impermeable to ions (section 2.2.8). It can provide energy for the transport of certain substrates across the cytoplasmic membrane – e.g. lactose uptake in *E. coli*. Pmf can also provide energy for the phosphorylation of ADP to ATP at enzyme complexes (*ATPases*) located in the cytoplasmic membrane; thus, protons can pass *through* the ATPase (from the outer to the inner surface of the membrane) and, in doing so, provide energy necessary for the release of ATP from catalytic sites on the cytoplasmic side of the

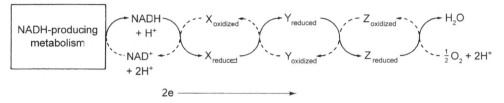

Figure 5.8 Respiration (diagrammatic): an electron transport chain (ETC) consisting of three components (X, Y and Z), with oxygen as the terminal electron acceptor, in the respiration of a typical *organic* substrate. The cell's ongoing metabolism produces NADH, from which NAD must be regenerated by oxidation; some NADH is oxidized to NAD in biosynthetic reactions, and some is oxidized via the ETC. In the ETC, NADH is oxidized by transferring electrons to (and thus reducing) the oxidized form of X. The reduced form of X is oxidized by transferring electrons to the oxidized form of Y – and so on. The final step is the reduction of oxygen to water. The solid curved lines indicate the path of electron flow. Oxidation of NADH (enzyme: NADH dehydrogenase) and reduction of oxygen (at a cytochrome oxidase) both appear to occur at the *inner* (i.e. cytoplasmic) face of the cytoplasmic membrane.

ATPase. In respiration, when pmf is used as a source of energy for the synthesis of ATP from ADP the process is called *oxidative phosphorylation* (compare 'substrate-level phosphorylation' in section 5.1.1.1). Interestingly, membrane ATPases can also catalyse the *hydrolysis* of ATP to ADP, the liberated energy being used to pump protons (outwards) across the membrane – i.e. to augment pmf; thus, the energy in ATP and pmf is interconvertible! These ideas are summarized in Fig. 5.9.

The ATPase itself consists of two main parts. The F_0 domain is a proton channel that spans the width of the membrane; the F_1 domain interacts with the cytoplasmic side of F_0 and contains the catalytic sites at which ATP is synthesized. (This type of ATPase is called an (F_0F_1)-type ATPase.) The F_0 domain appears to *rotate*, relative to the F_1 domain, during translocation of protons. [Energy transduction in ATP synthase: Nature (1998) *391* 510–513.]

In a bacterium, the electron transport chain involved in respiration (the *respiratory chain*) occurs in the cytoplasmic membrane (and, in some cases, in the thylakoids); its components vary from species to species, and variations may occur even in a given species growing under different conditions. Components found in respiratory chains include: (i) *cytochromes*: iron-containing proteins which receive and transfer electrons by the alternate reduction and oxidation of the iron atom; (ii) *iron–sulphur proteins* such as the *ferredoxins*; and (iii) *quinones*: aromatic compounds which can undergo reversible reduction. In the respiration of a typical organic substrate (Fig. 5.8), the oxidation of NADH and the reduction of the terminal electron acceptor (oxygen in Fig. 5.8) both appear to occur at the *inner* (i.e. cytoplasmic) face of the cytoplasmic membrane.

In Fig. 5.8 the source of NADH is given simply as 'NADH-producing metabolism'. We can now look at some actual pathways used in bacterial respiration. In many

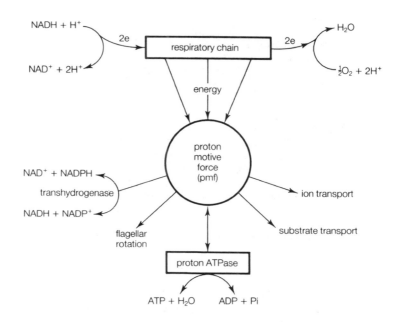

Figure 5.9 Some of the roles of proton motive force. 'Respiratory chain' is the term used for an electron transport chain involved in respiratory metabolism (i.e. respiration). 'Proton ATPase' is an enzyme system (in the cytoplasmic membrane) which catalyses the pmf-dependent phosphorylation of ADP to ATP as well as the hydrolysis of ATP to ADP and inorganic phosphate (Pi); pmf is *used* for the synthesis of ATP but is *augmented* by the (energy-yielding) hydrolysis of ATP. Pmf also controls the (reversible) reduction of NADP by NADH; NADPH is used e.g. for some of the cell's biosynthetic reactions (e.g. ammonia assimilation – section 10.3.2).

bacteria, the respiratory metabolism of glucose starts with the Embden–Meyerhof–Parnas pathway (EMP pathway, Fig. 5.3). Beyond pyruvic acid, however, the pathways of fermentation and respiration are completely different. In respiration, pyruvic acid is often converted to acetyl-CoA and fed into a cyclical pathway known as the *tricarboxylic acid cycle* (TCA cycle, Fig. 5.10) – also known as the *Krebs cycle* or the *citric acid cycle*.

Figure 5.10 shows that, in the TCA cycle, acetyl-CoA and oxaloacetic acid (OAA) combine to form citric acid; in the subsequent reactions, the original molecule of pyruvic acid is, in effect, oxidized to carbon dioxide. For each molecule of pyruvic acid oxidized, 4 molecules of NAD(P) and 1 of FAD are reduced, 1 molecule of ATP is synthesized, and 1 molecule of OAA is regenerated. NADH and $FADH_2$ can be oxidized via a respiratory chain, the resulting pmf being used e.g. for the synthesis of ATP at a membrane ATPase. In terms of energy yield, it should now be clear that respiration is much more efficient than fermentation: in respiration, the oxidation of NADH can lead, via pmf, to the synthesis of ATP, whereas in fermentation (where there is no

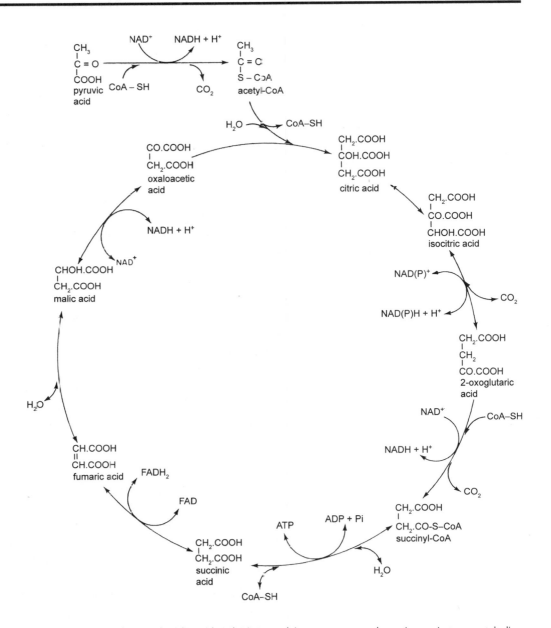

Figure 5.10 The tricarboxylic acid cycle (TCA cycle): a common pathway in respiratory metabolism (section 5.1.1.2). The reaction isocitric acid→2-oxoglutaric acid is catalysed by the enzyme isocitrate dehydrogenase which, in most bacteria, is specific for NADP rather than NAD; NADPH is used e.g. in biosynthetic reactions. FAD (flavin adenine dinucleotide) – like NAD – carries energy in the form of 'reducing power'; FADH$_2$ can be oxidized via a respiratory chain, oxidation yielding pmf. In many bacteria the step from succinyl-CoA to succinic acid involves the synthesis of guanosine 5'-triphosphate (GTP) rather than ATP; GTP is used as an energy currency molecule e.g. in the binding of the aminoacyl-tRNA to the 'A' site in protein synthesis (Fig. 7.9).

external oxidizing agent) the cell has to get rid of NADH by synthesizing waste products such as lactic acid.

Anaerobic respiration. In principle, anaerobic respiration is the same as aerobic respiration: both use an external oxidizing agent. Under anaerobic conditions, however, oxidizing agents such as nitrate, sulphate and fumarate are used in place of oxygen; pmf can be generated by an anaerobic respiratory chain. The substrate (electron *donor*) used by chemoorganotrophs in anaerobic respiration may be any of various organic compounds, depending on species and conditions.

In *nitrate respiration* nitrate is used as the terminal electron acceptor, nitrate being reduced to nitrite, nitrous oxide, nitrogen or ammonia – depending on species. When the nitrate is reduced mainly to nitrogen and/or nitrous oxide (i.e. gases) the process is called *denitrification* because it results in a loss of nitrogen to the atmosphere. This process can be a problem in agriculture because it can lower soil fertility (section 10.3.2.2). However, denitrification can be useful for eliminating the nitrogenous content of sewage in waste-water treatment plants (section 13.4).

Bacteria capable of denitrification include strains of *Alcaligenes faecalis*, *Bacillus licheniformis*, *Paracoccus denitrificans* and *Pseudomonas stutzeri*. It was once thought that denitrification occurs only under anaerobic or microaerobic conditions. We now know that oxygen affects different denitrifying bacteria in different ways. For some of these bacteria anaerobiosis *is* necessary for denitrification. Others can denitrify in the presence of oxygen – in some cases consuming both oxygen and nitrate simultaneously – though the rate of denitrification typically decreases with increasing concentrations of oxygen.

Dissimilatory reduction of nitrate to ammonia (DRNA) is carried out by certain species under appropriate conditions (section 10.3.2).

The *sulphate-* and *sulphur-reducing bacteria* use sulphate and sulphur, respectively, as terminal electron acceptors; during anaerobic respiration, sulphate or sulphur is reduced to sulphide (Fig. 10.3). This type of anaerobic respiration is carried out e.g. by species of *Desulfovibrio* (sulphate reducers) and *Desulfuromonas* (sulphur reducer); these organisms typically occur in anaerobic mud and soil, and they are responsible for much of the hydrogen sulphide found in organically polluted waters.

In *fumarate respiration* the terminal electron acceptor is exogenous fumarate – which is reduced to succinate. Fumarate respiration is carried out by a range of bacteria, including *E. coli*, under appropriate conditions.

Selenate is the terminal electron acceptor in a form of anaerobic respiration that occurs in certain (Gram-negative and Gram-positive) bacteria – e.g. *Thauera selenatis* [IJSB (1993) *43* 135–142]. [Arsenic and selenium respiration: FEMSMR (1999) *23* 615–627. Arsenate as terminal electron acceptor: AEM (2004) *70* 2741–2747.]

Other terminal electron acceptors reported to be used in bacterial respiration include manganese (IV) and uranium (VI).

5.1.2   Energy metabolism in chemolithotrophs

Chemolithotrophs use *inorganic* substrates for energy metabolism – e.g. sulphide, elemental sulphur, ammonia, hydrogen, ferrous ions etc. Metabolism commonly involves aerobic or anaerobic respiration: electrons from the substrate are transferred to the external oxidizing agent, pmf being generated (see also section 5.3.4). Owing to the nature of the energy substrate, this type of metabolism *usually* does not involve NAD (compare with chemo*organo*trophs: Figs 5.2 and 5.7); chemolithotrophic reduction of NAD is carried out e.g. by those hydrogen-oxidizing bacteria which have a 'hydrogen dehydrogenase' (NAD-reducing hydrogenase).

Species of *Thiobacillus* occur e.g. in soil and marine mud. Typically they oxidize substrates such as sulphide and sulphur aerobically, although *T. denitrificans* can metabolize these substrates anaerobically using nitrate as terminal electron acceptor (nitrate respiration). *T. denitrificans* can use Fe (II) (ferrous ions) as electron donor (i.e. substrate) in anaerobic nitrate-dependent growth (the nitrate being reduced primarily to gaseous nitrogen) [AEM (1996) *62* 1458–1460]. *T. ferrooxidans*, a strict aerobe, can oxidize ferrous ions as well as sulphur substrates. Thiobacilli have roles in the sulphur cycle (section 10.3.3).

Nitrifying bacteria are obligately aerobic respiratory organisms which live in soil and aquatic environments. They obtain energy by *nitrification*: the oxidation of ammonia to nitrite (e.g. by species of *Nitrosococcus* and *Nitrosomonas*), and the oxidation of nitrite to nitrate (e.g. by species of *Nitrobacter* and *Nitrococcus*). *Anaerobically*, some bacteria oxidize ammonium to nitrogen in the *anammox* reaction [FEMSMR (1999) *22* 421–437] in which nitrite is terminal electron acceptor; this occurs in wastewater treatment plants and anoxic marine habitats [Nature (2003) *422* 608–611]. (See also section 10.3.2.)

Some bacteria oxidize arsenite (AsIII) to arsenate [JB (2004) *186* 1614–1619]. [The microbial arsenic cycle: FEMSME (2004) *48* 15–27.]

5.1.2.1   *Inorganic fermentation*

Until 1987 it was thought that no *inorganic* substrate could be fermented. Then, Bak and Cypionka [Nature (1987) *326* 891–892] described a type of energy metabolism in which the substrate (sulphite or thiosulphate) underwent 'disproportionation' to yield sulphide and sulphate. This 'chemolithotrophic fermentation' occurs e.g. in *Desulfovibrio sulfodismutans*.

5.1.2.2   *Methane production*

Methane (CH_4) is a by-product of an energy-yielding process in certain obligately anaerobic members of the Archaea – the *methanogens*; apparently, no member of the

Bacteria produces methane. Methanogens live e.g. in river mud and in the rumen of cows and other ruminants (environments characterized by an E_h below about -330 mV).

Some methanogens (e.g. *Methanobacterium*, *Methanobrevibacter*, *Methanococcus*, *Methanothermus*) produce methane by a complex series of reactions in which hydrogen is involved in the reduction of CO_2. Essentially, the carbon atom of CO_2 is bound, sequentially, to each of a series of C_1 carrier molecules and undergoes progressive reduction via $-CH_2-$ and CH_3 to methane (CH_4); the carrier molecules include methanofuran, tetrahydromethanopterin and several coenzymes. The final step of the reaction is thermodynamically favourable ($\Delta G^{0'} = -112.5$ kJ) and seems to be coupled to the generation of pmf.

Some methanogens (e.g. *Methanococcoides*, *Methanolobus*) can form methane from e.g. acetate or methanol. Growth on acetate is slower than that on carbon dioxide and hydrogen (the $\Delta G^{0'}$ of the reaction $CH_3COOH \rightarrow CH_4 + CO_2$ is about -36 kJ). Essentially, cleavage of acetate yields CO and CH_3, the latter being reduced to methane while CO is oxidized by water (enzyme: CO dehydrogenase) to CO_2 and 2H. In marshy soils, acetate may be the preferred substrate at lower temperatures [FEMSME (1997) *22* 145–153], and studies on the degradation of rice straw in anoxic paddy soil indicate that a high proportion of the methane is synthesized from acetate under those conditions [FEMSME (2000) *31* 153–161]. Nevertheless, despite the overall picture, the contribution of hydrogen to methanogenesis, locally, may be 70–100% [FEMSME (1999) *28* 193–202].

Methane production in anoxic soils may be suppressed e.g. by nitrate – this apparently being due to the effects of intermediates of the denitrification process [FEMSME (1999) *28* 49–61].

[Enzymes in methanogenesis: FEMSMR (1999) *23* 13–38.]

5.2 ENERGY METABOLISM IN PHOTOTROPHS

Phototrophs obtain energy from sunlight – in most cases by *photosynthesis*.

5.2.1 Photosynthesis

In photosynthesis, the energy in light is absorbed by specialized pigments and is used to form energy currency molecules and/or pmf; in all cases, photosynthesis occurs in membranes containing *chlorophylls*, accessory pigments and electron transport chain(s). Chlorophylls are green, magnesium-containing pigments. Cyanobacteria contain chlorophyll *a* (which also occurs in algae and higher plants); the other photosynthetic bacteria contain one or more *bacteriochlorophylls* – pigments which are similar to the chlorophylls. The so-called *light reaction* of photosynthesis refers to all the (photochemical) events involved in the conversion of light energy to pmf or chemical energy; the *dark reaction* (= light-independent reaction) refers to the cell's use of its photosynthetically-derived energy for the synthesis of carbon compounds.

Photosynthetic bacteria can be divided into two main categories: (i) those which carry out photosynthesis aerobically (and which produce oxygen as a by-product), and (ii) those which carry out photosynthesis anaerobically (and which do not produce oxygen).

5.2.1.1 *Oxygenic (oxygen-producing) photosynthesis in bacteria*

This process – which closely resembles photosynthesis in green plants and algae – occurs in the cyanobacteria. Because the cyanobacteria carry out eukaryotic-type photosynthesis they were, for many years, regarded not as bacteria but as 'blue-green algae'; today, the bacterial nature of these organisms is not in doubt.

In almost all cyanobacteria photosynthesis occurs in the thylakoid membranes (section 2.2.7). (In strains of *Gloeobacter*, which lack thylakoids, the photosynthetic components appear to occur in the cytoplasmic membrane.)

Chlorophylls occur in so-called *reaction centres* to which light is channelled by specialized protein–pigment *light-harvesting complexes* (LHCs); the composition of the LHCs can be influenced by environmental factors such as light quality and the availability of nitrogen and sulphur [JB (1993) *175* 575–582]. When the chlorophyll receives light energy it ejects highly energized electrons; these electrons can flow down an electron transport chain and provide energy for (i) pmf generation and/or (ii) the direct reduction of NADP. (The cell uses NADPH e.g. for various biosynthetic reactions.)

Oxygenic photosynthesis is generally represented by the *Z scheme* (Fig. 5.11). Energized electrons ejected from photosystem II (PSII) flow down an electron transport chain to photosystem I (PSI), and this electron flow generates pmf across the thylakoid membrane; when such photosynthetically-derived pmf is used for the synthesis of ATP (at a membrane-bound ATPase) the process is called *photophosphorylation*. Electrons ejected from PSI have sufficient energy to reduce NADP to NADPH. The flow of electrons (from left to right in Fig. 5.11) requires an input of electrons to PSII. This is achieved by the oxidation of water, oxygen being formed as a by-product. A revised eukaryotic Z scheme includes the reduction of NADP by PSII [TIBS (1996) *21* 121–122].

Some cyanobacteria (e.g. *Oscillatoria limnetica*) can – as an alternative – carry out photosynthesis anaerobically, using sulphide instead of water as an electron donor; the sulphide is oxidized to elemental sulphur. Some cyanobacteria can even grow as chemoorganotrophs.

A small group of prokaryotic organisms (Prochlorophyta; prochlorophytes) – related to the cyanobacteria – contain both of the eukaryotic-type chlorophylls: *a* and

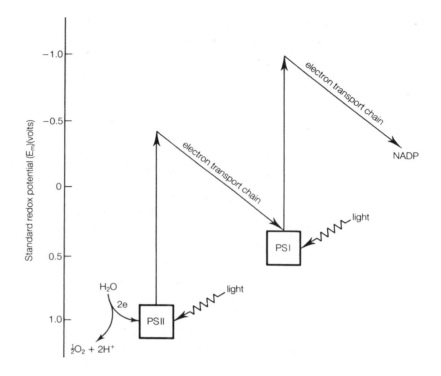

Figure 5.11 Oxygenic photosynthesis: simplified Z scheme (scale approximate). Light energy causes the reaction centres (PSI, PSII) to eject energized electrons whose energy levels and destinations are shown by the arrows; electron flow from PSII to PSI generates pmf. Electrons ejected from PSII are replaced by the oxidation of water, oxygen being liberated.

b. Only three species are currently known: *Prochloron didemni*, *Prochlorothrix hollandica* and *Prochlorococcus marinus*. [Photosynthetic machinery in prochlorophytes: FEMSMR (1994) *13* 393–414.]

5.2.1.2 *Anoxygenic photosynthesis*

Anoxygenic photosynthesis (in which oxygen is not produced) is carried out anaerobically by bacteria of the order Rhodospirillales. In the so-called 'purple' photosynthetic bacteria (suborder Rhodospirillineae), all the photosynthetic components occur in intracellular membranes which are continuous with the cytoplasmic membrane. In the 'green' photosynthetic bacteria (suborder Chlorobiineae), components of the light-harvesting complexes occur in *chlorosomes* (section 2.2.7) while the reaction centres occur in the cytoplasmic membrane.

In the purple bacteria, electrons ejected from a reaction centre follow a *cyclic* path –

via an electron transport chain – back to the reaction centre; the pmf which is generated can be used e.g. for the synthesis of ATP (i.e. photophosphorylation).

In green bacteria the electron flow generates pmf (for e.g. ATP synthesis) and it can also be used for the direct reduction of NAD to NADH. Such *non-cyclic* electron flow needs a supply of electrons from an exogenous donor; electron donors used by the green photosynthetic bacteria include e.g. sulphide and thiosulphate – but never water, because in these organisms photosynthesis is always anoxygenic.

Some bacteria of the Rhodospirillales (including many purple bacteria) can live as chemoorganotrophs under aerobic or microaerobic conditions.

5.2.1.3 *Electron donors in photosynthesis*

Inorganic electron donors in photosynthesis include e.g. water (used only by cyanobacteria), sulphide, sulphur and hydrogen; organic donors include e.g. formic acid and methanol. Phototrophs which use inorganic electron donors are called *photolithotrophs*; those which use organic donors are called *photoorganotrophs*.

5.2.2 The purple membrane

In the Archaea, certain strains of *Halobacterium salinarium* (an extreme halophile – section 3.1.7) can use the energy in sunlight even though they cannot carry out photosynthesis. In these strains, a differentiated, purple-pigmented region of the cytoplasmic membrane – the *purple membrane* – develops under anaerobic or microaerobic conditions in the light; the main purple pigment is called *bacteriorhodopsin*. When a molecule of bacteriorhodopsin absorbs light energy it passes rapidly through a cycle of states (photointermediates), all within a few thousandths of a second; during this time protons are pumped outwards across the membrane, i.e. the 'photocycling' in bacteriorhodopsin generates pmf. Unlike photosynthesis, pmf generation in the purple membrane appears to involve the *direct* pumping of protons across the membrane, and chlorophyll is not used. [Bacteriorhodopsin (overview and various aspects): Nature (2000) *406* 569–570, 645–657.]

5.3 OTHER TOPICS IN ENERGY METABOLISM

5.3.1 The yield of ATP

In a living cell, no pathway or process occurs in isolation but rather forms part of a complex whole. Because of this, the maximum theoretical yield of ATP from the respiration of (say) one molecule of glucose may not be achieved owing to interactions between the respiratory pathway and other pathways in the cell. For example, not all the energy liberated (as pmf) by electron transport may be used for oxidative phosphorylation: some may be needed e.g. for ion transport or flagellar motility. Again, some of the NADH generated during respiration may be used for biosynthesis; *this* NADH is clearly not available for oxidation (via the respiratory chain) so that it cannot contribute to pmf and, hence, ATP synthesis. Finally, intermediates in energy metabolism may be drawn off to supply the cell with 'building blocks' for biosynthesis; thus, e.g. pyruvic acid is used in the synthesis of the amino acids alanine, valine and leucine. The withdrawal of intermediates necessarily sacrifices the energy which would otherwise have been obtained by their metabolism.

5.3.2 Reverse electron transport

Bacteria need reducing power – e.g. NADH and/or NADPH – for biosynthesis. For fermentative bacteria this is never a problem, while the cyanobacteria and 'green' photosynthetic bacteria obtain these reduced molecules by direct reduction using non-cyclic electron flow. However, the 'purple' photosynthetic bacteria cannot do this: electrons ejected from their reaction centres do not have enough energy to reduce NAD. Instead, these organisms use *reverse electron transport*; in this process, pmf is used to drive electrons 'uphill' to a membrane-bound enzyme, NAD dehydrogenase, where NAD is reduced to NADH. This requires an input of electrons from an external electron donor; the purple bacteria typically use organic electron donors, but some can use inorganic donors such as sulphide. Reverse electron transport is also used by the nitrifying bacteria and other chemolithotrophs; note that chemolithotrophs (section 5.1.2) typically do not form NADH during metabolism of the energy substrate.

In *Thiobacillus ferrooxidans* reduction of pyridine nucleotides is reported to involve uphill electron flow through an oxidoreductase complex working in reverse [JB (2000) *182* 3602–3606].

5.3.3 End-product efflux

Fermentative bacteria make a great deal of NADH; in fact, they have to get rid of some of it by synthesizing waste products such as lactic acid (section 5.1.1.1). Making a virtue of necessity, some bacteria (e.g. *Lactococcus cremoris*) gain energy by linking protons to their waste lactate; when lactic acid passes outwards across the cytoplasmic membrane its 'passenger' protons automatically augment pmf!

5.3.4 Extracytoplasmic oxidation

Complex energy substrates are generally transported across the cytoplasmic membrane before being metabolized. However, some simple energy substrates (such as H_2 and Fe^{2+}) appear to be oxidized at the *outer* face of the cytoplasmic membrane or in the periplasmic region; such *extracytoplasmic oxidation* of a substrate is characterized by:

(i) release of protons from the substrate and/or water extracytoplasmically, (ii) a transmembrane flow of electrons (from the substrate) to the cytoplasmic (inner) face of the membrane, and (iii) interaction of these electrons with protons and a terminal electron acceptor (e.g. oxygen). The net result is pmf generation. For example, the strict aerobe *Thiobacillus ferrooxidans* gains energy from the oxidation of Fe^{2+} (to Fe^{3+}) at low pH; according to one scheme, Fe^{2+} (produced extracytoplasmically) reacts with water to yield ferric hydroxide and protons, while electrons (from Fe^{2+}) reduce a terminal electron acceptor (oxygen?) at the cytoplasmic face of the membrane.

Some bacteria (e.g. *Pseudomonas aeruginosa*) can generate pmf by the extracytoplasmic oxidation of glucose to gluconate; in these bacteria the enzyme glucose dehydrogenase – with its cofactor *pyrroloquinoline quinone* (PQQ) – occurs bound to the outer surface of the cytoplasmic membrane. The gluconate can be transported across the membrane and phosphorylated to 6-phosphogluconate, which can enter the Entner–Doudoroff pathway (Fig. 6.2).

E. coli (and many related bacteria) normally form a membrane-bound glucose dehydrogenase which lacks the necessary PQQ; such organisms can carry out extracytoplasmic oxidation only if they are provided with PQQ [IJSB (1989) *39* 61–67]. However, a mutant strain of *E. coli* with a non-functional PTS system (section 5.4.2) can synthesize PQQ and carry out extracytoplasmic oxidation – suggesting that PQQ-encoding genes may be present but not expressed under normal conditions [JGM (1991) *137* 1775–1782].

5.3.5   Pmf and aerotaxis

Surprisingly, the respiratory-type organism *Azospirillum brasilense* – which uses oxygen as terminal electron acceptor – is microaerophilic (section 3.1.6); to reach optimal conditions, the organism exhibits aerotaxis (section 2.2.15.2), moving towards regions where the oxygen concentration is only 3–5 μM. In *A. brasilense* pmf increases when the cell swims towards the preferred concentration of oxygen, and decreases when the cell swims away from it; it has been suggested that this change in the level of pmf acts as the signal which regulates aerotactic movements [JB (1996) *178* 5199–5204].

The ability of *A. brasilense* to carry out aerobic respiratory metabolism under such low levels of oxygen is believed to be due to a highly efficient *oxidase* (oxygen-reducing enzyme). Interestingly, for *A. brasilense* (a nitrogen-fixing organism), these oxygen-poor conditions are useful for carrying out nitrogen fixation (section 10.3.2.1).

In *E. coli*, a sensor protein, Aer, mediates aerotaxis by responding to redox changes in the electron transport chain. The Tsr protein (an MCP: Fig. 7.13) was thought to be another, independent sensor for aerotaxis that responds to changes in pmf [PNAS (1997) *94* 10541–10546].

The type of *energy taxis* referrred to above may be common in the microbial world. [Ecological role of energy taxis in microorganisms: FEMSMR (2004) *28* 113–126.]

5.3.6   Sodium motive force (smf)

Smf is the energy associated with an electrochemical gradient of sodium ions between the inner and outer surfaces of the cytoplasmic membrane; it is analogous to pmf.

In some cases a cell can generate smf by using pmf: the entry of protons into the cytoplasm is linked to the exit of sodium ions across the cytoplasmic membrane ($= Na^+/H^+$ *antiport*). In *E. coli*, such antiport is used to provide a source of energy (smf) suitable e.g. for the uptake ($=$ *transport*, section 5.4) of the sugar melibiose; thus, melibiose and sodium ions are jointly transported into the cytoplasm ($= Na^+/melibiose$ *symport*), i.e. melibiose uptake consumes smf.

In some marine bacteria (e.g. *Vibrio alginolyticus*) energy from respiration can be used directly for the outward pumping of Na^+ via a *sodium pump*; this organism can use smf e.g. for the uptake of various amino acids. Smf is also used to energize rotation of the polar flagellum in *Vibrio cholerae* [JB (1999) *181* 1927–1930].

Interestingly, external and internal concentrations of sodium ions appear to influence the expression of virulence factors in *V. cholerae* [PNAS (1999) *96* 3183–3187].

At the higher range of growth temperatures, the inherent permeability of the cytoplasmic membrane for both protons and sodium ions tends to increase; however, the increase in permeability is greater for protons than it is for sodium ions. A pmf-based organism will need to use extra energy for generating a given level of pmf simply in order to compensate for the higher inward diffusion of protons at these higher temperatures. In some organisms this problem is minimized because the (special) composition of their cytoplasmic membrane is such that raised temperatures have less effect on proton permeability. However, some thermophiles (section 3.1.4) have abandoned the use of protons in their energy metabolism; for example, *Clostridium fervidus* uses Na^+ as the sole energy-coupling ion (i.e. pmf is not involved in this species). [Ion permeability at high temperatures: FEMSMR (1996) *18* 139–148.]

Enzymes that translocate sodium ions: BBA (1997) *1318* 11–51].

5.4   TRANSPORT SYSTEMS

The cytoplasmic membrane (section 2.2.8) is an efficient barrier which stops most molecules (and all ions) from passing freely into and out of the cytoplasm. (Of course, this is essential: the cell must be able to control its own internal environment.) In addition to the cytoplasmic membrane, Gram-negative bacteria also have the extra barrier of the outer membrane (section 2.2.9.2). However, during ongoing metabolism the cell must be able to take up various substrates (for energy and other purposes) and get rid of waste products; these functions are carried out by various *transport systems.* Transport systems are also used e.g. for recycling certain cell components (e.g. section 5.4.8) and for combatting antibiotics (e.g. the TET protein, section 15.4.11).

A given transport system may be specific for a particular substrate, or for a number of related substrates. A given susbstrate may have more than one type of transport system, even in the same cell.

Transport typically requires energy. The energy is usually supplied in the form of pmf and/or a high-energy phosphate (such as ATP); in some cases the energy is supplied as sodium motive force (section 5.3.6). Pmf can be used, directly, as energy for the transport of both charged and uncharged solutes. In *E. coli*, for example, *proton/lactose symport* refers to the uptake of lactose at the expense of pmf, lactose and protons being transported *jointly* through the membrane. Also in *E. coli*, sodium motive force provides the energy in Na⁺ melibiose symport – in which the sugar melibiose is transported through the membrane at the expense of smf. The ABC transporters (section 5.4.1) provide examples of transport energized by ATP.

Energy for the transport of *ions* may be provided by hydrolysis of ATP at a membrane-bound ATPase. For example, in *E. coli*, potassium ions (K⁺) can be taken up by the so-called Kdp system (section 7.8.6) which involves a specialized K⁺-ATPase (*potassium pump*); hydrolysis of ATP at the pump allows uptake of K⁺ against a concentration gradient.

PTS (section 5.4.2) is used for importing certain substrates, particularly sugars and related compounds. As well as a transport mechanism, PTS also has a role in chemotaxis (sections 2.2.15.2, 7.8.6; Fig. 7.13) and is associated with catabolite repression (section 7.8.2.1).

The transport of *proteins* across the bacterial cell envelope has been well studied because pathogenic bacteria often cause disease by secreting protein toxins. In fact, Gram-negative bacteria have at least six distinct pathways for the export/secretion of proteins. (*Export* means transmembrane translocation of a protein toward the cell's exterior without subsequent release of the protein; in *secretion*, the translocated protein is released to the external environment.) The following pathways for export/secretion of proteins occur in Gram-negative bacteria: type I (section 5.4.1.2), type II (section 5.4.3), type III (section 5.4.4), type IV (section 5.4.5), type V (section 5.4.6), and the Tat export system (section 5.4.7).

5.4.1 ABC transporters

An ABC transporter is a transport system, found in both Gram-positive and Gram-negative bacteria, which consists of a multi-protein complex in the cell envelope – two of the proteins having a specific ATP-binding site called the *ATP-binding cassette* (ABC) on their inner (cytoplasmic) surface; a protein which has an ABC site is referred to here as an 'ABC protein'. Hydrolysis of ATP at an ABC site provides the energy for transport.

There are many different ABC transporters. Each one imports or exports/secretes one or more types of molecule or ion; collectively, they transport a wide range of substrates, from inorganic ions to sugars and proteins.

Interestingly, as well as transporting specific solute(s), an ABC transporter may regulate channels through which *other* solutes are translocated. For example, in *Salmonella typhimurium*, the transporter encoded by genes *sapABCDF* imports certain toxic peptides (apparently for detoxification) but it *also* helps to regulate a channel for K^+ ions – ions which are necessary for bacterial resistance to these peptides [see e.g. Cell (1995) *82* 693–696].

[The ATP-hydrolysing region of ABC transporters: FEMSMR (1998) *22* 1–20.]

5.4.1.1 ABC importers (binding-protein-dependent transporters; periplasmic permeases)

Importers mediate the uptake of e.g. certain amino acids, sugars and ions. Typically, an importer consists of (i) two different proteins (in the cytoplasmic membrane), (ii) an associated *dimer* of ABC proteins, and (iii) a substrate-specific *periplasmic binding protein* which binds substrate and transfers it to the membrane complex; all importers include the binding protein.

One example of an ABC importer is the ProU system for uptake of the osmoregulatory molecule glycine betaine (section 3.1.8). Another example is the histidine uptake system in *Salmonella typhimurium*; in this system the periplasmic protein binds histidine and passes it to the membrane complex which, in turn, transfers it to the cytoplasm [see e.g. JBC (1997) *272* 859–866]. [Crystal structure of HisP, the ABC protein of the *Salmonella typhimurium* histidine permease system: Nature (1998) *396* 703–707.]

In *Escherichia coli*, vitamin B_{12} is taken up in a two-stage transport process: (i) pmf-dependent translocation across the outer membrane, and (ii) binding-protein-dependent transport across the cytoplasmic membrane in which the binding protein BtuF [PNAS (2002) *99* 16642–16647] transfers the vitamin to the two ABC proteins (BtuCD); negatively charged glutamate residues on BtuF may interact with positively charged arginine residues on the two ABC proteins.

5.4.1.2 ABC exporters

Collectively, ABC exporters mediate the export/secretion of a range of proteins – including enzymes, haemolysins (section 16.1.4.1) and peptide antibiotics – and, in some bacteria, the polysaccharide components of the capsule. It seems that, typically, a given exporter can transport various related or similar molecules. [ABC transporters (bacterial exporters): MR (1993) *57* 995–1017.]

In *Lactococcus lactis* the LmrA transporter mediates an efflux system that extrudes amphiphilic compounds; it appears to be functionally identical to the mammalian P-glycoprotein (= MDR (multiple drug resistance) protein) which mediates multi-drug resistance, including resistance to anti-cancer drugs [Nature (1998) *391* 291–295].

Proteins secreted by pathogenic bacteria include some important virulence factors (section 11.5). In *E coli*, an ABC exporter mediates one-step transport of the α-haemolysin from the cytoplasm to the cell's exterior. Other proteins secreted via ABC exporters include the cyclolysin of *Bordetella pertussis* (causal agent of whooping cough); the cyclolysin, a ~180 kDa protein with e.g. pore-forming and adenylate cyclase activity, may inhibit phagocytes (e.g. macrophages) by raising their intracellular levels of cyclic AMP (cAMP). In *Streptomyces antibioticus* an ABC exporter secretes the antibiotic *oleandomycin* (related to erythromycin).

ABC exporters are categorized as *type I secretory systems* (Fig. 5.13).

Typically, proteins exported by ABC exporters lack an N-terminal signal sequence (compare section 5.4.3) but have a C-terminal *secretion sequence* that may interact directly with the ABC protein.

In Gram-negative bacteria, an exporter may include e.g. (i) ABC proteins, (ii) a so-called *membrane fusion protein* (MFP) in the cytoplasmic membrane, but partly in the periplasm, and (iii) a component in the outer membrane; the MFP may link the cytoplasmic and outer membranes, facilitating transport. Exporters that transport molecules into the periplasm or outer membrane as the *final* target site may have fewer protein components than those exporters which *secrete* proteins.

It appears that, in at least some cases, components of an exporter assemble in a definite sequence to form the complete transport system, and that assembly is promoted/initiated by the binding of substrate (substrate = molecule to be transported) to the ABC protein; thus, on binding, the ABC protein interacts with MFP – the latter then binding to the outer membrane component [EMBO Journal (1996) *15* 5804–5811].

5.4.2 PTS: the phosphoenolpyruvate-dependent phosphotransferase system

This transport system is used by various Gram-negative and Gram-positive bacteria for the uptake of certain sugars (e.g. glucose, fructose, lactose), sugar alcohols (e.g. mannitol) and substituted sugars (e.g. N-acetylglucosamine).

The source of energy for the PTS is phosphoenolpyruvate (PEP; for the formula see Fig. 5.3).

In the simplest case, PEP supplies phosphate and energy for the sequential phosphorylation of two soluble (cytoplasmic) energy-coupling proteins (designated I and

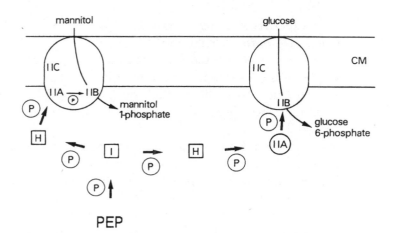

Figure 5.12 Transport by the phosphoenolpyruvate-dependent phosphotransferase system (PTS) (simplified, diagrammatic). PTS (section 5.4.2) is shown here transporting mannitol and glucose across the cytoplasmic membrane (CM) from the periplasm (above) into the cytoplasm (below) in *Escherichia coli*. Note that glucose and mannitol are taken up via different membrane-associated systems ('permeases'), as described below.

The source of energy for transport is phosphoenolpyruvate (PEP). Energy is initially fed into the system by the sequential transfer of a phosphate group ℗ from the 'high-energy phosphate' PEP to enzyme I (designated I) and then to the HPr protein (designated H). Phosphate is then transferred to a *permease* (=enzyme II complex, designated II). (Note that the H and I molecules are not specific to any one permease, i.e. they can phosphorylate any of the permeases.) The permease for mannitol (left) is entirely membrane bound, and consists of three domains: IIA, IIB and IIC. IIC is a hydrophobic, transmembrane domain which binds the substrate and appears to contain, or to contribute to, a transmembrane channel; the IIA and IIB domains each contain a phosphorylation site, but phosphorylation of the substrate seems directly to involve only the IIB domain.

In the glucose permease the IIA domain is a separate protein, but the phosphorylation sequence is from IIA to IIB, and phosphorylation of the substrate involves IIB.

In all cases, phosphorylation of the substrate molecule is an integral part of its transfer into the cytoplasm.

In some PTS permeases the components are grouped in other ways. For example, the fructose permease of enteric bacteria consists of a membrane-bound protein, containing IIC and IIB, and a separate protein which includes both IIA and H functions. In the fructose permease of *Rhodobacter capsulatus*, a single, separate protein includes the functions of IIA, H and I.

According to a proposed nomenclature for PTS permeases [JB(1992) *174* 1433–1438], the mannitol permease of *E. coli* is designated IICBA^{Mtl,Eco}, and the glucose permease of *E. coli* is designated IICB^{Glc,Eco}+IIA^{Glc,Eco}.

In addition to their roles in *transport*, the H, IIA and IIB components of the PTS system are involved in the regulation of certain catabolic operons (section 7.8.2.1).

H); H transfers phosphate and energy to a sugar-specific permease associated with the cytoplasmic membrane (the so-called II complex) which binds the substrate and then phosphorylates and *concomitantly* transports it into the cytoplasm (Fig. 5.12: uptake of mannitol). Note that phosphorylation of the substrate is an integral part of the transport process.

As sugars are often metabolized via their phosphate derivatives (e.g. glucose 6-phosphate in Figs 5.3 and 6.2), phosphorylation during uptake is a positive aspect of PTS transport.

A recently identified PTS allows *Escherichia coli* to grow on *N*-acetylmuramic acid as the sole source of carbon and energy [JB (2004) *186* 2385–2392].

The PTS is involved not only in transport: the cell can also respond *chemotactically* to sugars that are taken up by the PTS. In one possible mechanism, particular components of PTS interact, via unknown intermediate(s), with the CheY protein of the MCP system (see Fig. 7.13), thus influencing flagellar rotation. PTS-mediated chemotaxis differs from the MCP system in several ways [JCB (1993) *51* 1–6]: (i) the substrate must be *transported*, not simply bound; (ii) adaptation does not involve methylation; and (iii) the PTS does not respond to repellents.

As well as *transporting* substrates, the PTS is involved in some aspects of carbon *utilization*. Thus, PTS phosphate carriers (Fig. 5.12) regulate the activity of certain operons (section 7.8.1) involved in the uptake/catabolism of particular sources of carbon; in some cases the PTS phosphate carriers control operons by phosphorylating transcription regulator proteins at a site designated 'PTS-regulator domain' (PRD) [MM (1998) *28* 865–874].

5.4.3 The general secretory pathway for proteins (*sec*-dependent pathway; type II secretory system) in Gram-negative bacteria

Many proteins which are *secreted* – e.g. toxins, certain enzymes – are transported from the cytoplasm to the cell's exterior via the two-stage *general secretory pathway* (GSP); such transport requires energy: pmf as well as ATP. The initial stage(s) of the GSP are also used by some proteins whose *final* destination is the cytoplasmic membrane, the periplasm or the outer membrane.

Among the various pathways for protein secretion, the GSP is defined by certain genes (analogous to the *sec* genes in *E. coli*) whose products form the apparatus for transporting proteins into and through the cytoplasmic membrane (CM).

Proteins secreted via the GSP are synthesized with a special N-terminal *signal sequence* (section 7.6) which facilitates their passage into/through the CM. The signal sequence is enzymically cleaved during passage through the membrane.

First stage. Either during translation (section 7.6) or post-translationally, the given protein (to be transported) is carried to the CM by a so-called *chaperone* protein, SecB, which seems to prevent folding of the nascent protein (folding would inhibit trans-membrane translocation via this pathway). SecB is targeted to an ATPase, SecA, at the CM. SecA associates with a membrane-embedded protein complex (SecYEG) [JB (1997) *179* 5699–5704], termed the *translocon*, which forms a secretory channel through the CM. (SecYEG is represented in the Archaea, and in eukaryotes, as well as in bacteria.) [Three-dimensional structure of the *E. coli* SecYEG complex at an in-plane resolution of 8 Å: Nature (2002) *418* 662–665. Interactions between SecA and SecYEG: EMBO Journal (2003) *22* 4375–4384.] (In *Methanococcus jannaschii* the heterotrimeric protein complex SecY is reported to form a protein-conducting channel [Nature (2004) *427* 36–44].) SecB Is released, and energy from hydrolysis of ATP at SecA is believed to drive the protein through the translocon; it is not known how the energy from ATP hydrolysis is coupled to protein translocation.

For some proteins, targeting from the cytoplasm to the CM occurs, during translation (co-translationally), via a different pathway. One scheme for this pathway in *E. coli* is as follows. The nascent polypeptide (with its ribosome) first binds to a so-called *signal recognition particle* (SRP) that consists of (i) an RNA molecule and (ii) a 48 kDa multifunctional protein (designated P48 or Ffh) which has GTPase activity. (It is not known whether polypeptide–SRP binding results in inhibition of further translation on the ribosome (as occurs in eukaryotes); in prokaryotes, such arrest may not be necessary – the SRP pathway may simply facilitate the association of nascent polypeptide chains with the translocon.) The complex then binds a free, cytosolic protein, FtsY. The FtsY–SRP–ribosome–polypeptide entity binds to the CM, transiently, via FtsY; hydrolysis of GTP energizes dissociation of the complex, and the polypeptide (with its ribosome) binds at a translocon. [Model for the SRP pathway in *E. coli*: EMBO Journal (1998) *17* 2504–2512.] (Interestingly, in *E. coli*, reduced levels of SRP lead to induction of the heat-shock response (section 7.8.2.3); this may help to maintain the cell's viability by increasing its capacity to degrade mis-localized proteins [JB (2001) *183* 2187–2197].)

What decides whether a protein is transported via the SecB–SecA pathway or the SRP pathway? It appears that the choice of pathway reflects the *hydrophobicity* of the protein's N-terminal sequence. Thus, the SRP pathway may be used mainly for targeting proteins of the CM whose N-terminal sequences are characteristically highly hydrophobic; for these proteins, the SRP pathway may be advantageous in that completing the translation process at a membrane-associated site could avoid the risk of aggregation of the hydrophobic domains of a protein in the (hydrophilic) cytoplasm. Studies in which the signal peptide sequence was replaced, or modified, indicate that the targeting pathway of *E. coli* pre-secretory and integral CM proteins is specified by the hydrophobicity of the targeting signal [PNAS (2001) *98* 3471 –3476].

The export of certain proteins via the SRP pathway is sensitive to inhibition of SecA, suggesting that SecA may be required for both the SecB and SRP pathways [JB (2003) *185* 5706–5713].

Second stage. Transport from the periplasm to the cell's exterior (via the outer membrane) involves a number of specific proteins. This stage of the GSP apparently does not occur in *E. coli*, i.e. *E. coli* appears not to *secrete* proteins via the GSP.

However, this stage of the pathway is used by other enterobacteria. For example, *Klebsiella oxytoca* secretes a lipoprotein enzyme, *pullulanase*, whose translocation across the outer membrane requires the products of 14 *pul* genes. (If all the *pul* genes from *K. oxytoca*, including that encoding pullulanase, are expressed in *E. coli* then this organism can secrete pullulanase.) Again, the plant-pathogenic species *Erwinia chrysanthemi* secretes the enzymes pectinase and cellulase via the GSP.

The complex of proteins which, together, permit secretion of proteins across the outer membrane has been termed the *secreton*. Some of the secreton proteins from different organisms are similar or identical, and it has also been found that certain of these proteins are closely related to those involved in the secretion of subunits of certain fimbriae; thus, in at least some cases, the secreton may be a protein complex similar to the apparatus involved in the formation of certain fimbriae [EMBO Journal (2000) *19* 2221–2228].

The GSP is a major route for protein secretion in many types of Gram-negative bacteria.

5.4.4 Type III protein secretory systems in Gram-negative bacteria

In type III systems, proteins are secreted via a pore or channel that crosses the cell envelope from cytoplasm to cell surface; unlike the pores in ABC transporters (section 5.4.1), these systems may involve 20 or more different proteins. The components of type III systems are similar (homologous) in various species; moreover, some type III proteins are similar to the components of other protein-secreting systems. (*Note.* Type III *secretion* is distinct from type II transport, but the latter is apparently used for exporting certain outer membrane components of the type III transport system.)

Type III systems are characteristically found in pathogenic bacteria – e.g. EPEC (section 11.2.1), *Salmonella*, *Shigella*, *Yersinia*; genes for a type III system have been reported in *Chlamydia psittaci* [MM (1997) *25* 351–359]. The proteins secreted by these systems are typically active in some way against (eukaryotic) target cells. Type III systems occur in plant-pathogenic bacteria [JB (1997) *179* 5655-5662] as well as in human and animal pathogens.

Some type III systems are encoded by genes on pathogenicity islands (section 11.5.7) [JMM (2001) *50* 116–126].

In some type III systems, protein secretion may be activated by contact with a eukaryotic cell; this appears to occur e.g. in EPEC [MM (1998) *28* 143–155] and *Yersinia*. In such cases, proteins may be secreted *directly* into the eukaryotic cell from the bacterium; in support of this notion, type III-secreted proteins typically have little or no effect on eukaryotic host cells, *in vitro*, in the absence of the secreting bacterium, and at least some proteins secreted by type III systems have been demonstrated inside eukaryotic host cells. In the *Yersinia* system, secreted proteins (Yops) have been linked to specific effects within the eukaryotic target cells (section 11.5.1); in *Yersinia* spp, a type III system and a number of associated virulence proteins (Yops) constitute a so-called *virulon* [see e.g. PNAS (2000) *97* 8778–8783]. (Analogous systems also occur in certain other bacteria, including *Salmonella* and *Shigella*.)

Expression of type III systems in certain pathogenic strains of *E. coli* (EHEC, EPEC) may be regulated by quorum sensing (section 10.1.2) [PNAS (1999) *96* 15196–15201].

Proteins secreted by type III systems have an N-terminal sequence which may be targeted to the pore or channel but which (unlike the siganl sequence in type II systems) is not cleaved.

In at least some cases the type III secretory structure includes a multiprotein *needle complex* which consists of a base, spanning the cell envelope, and a needle-like extension from the surface of the bacterium; this complex resembles a straight version of the basal body–hook portion of a flagellum (Fig. 2.8). In *Salmonella typhimurium* the main component of the needle-like part of the type III system is the product of the *prgI* gene; the length of the needle is reported to be regulated by the *invJ* gene product [PNAS (2000) *97* 10225–10230].

In *Yersinia enterocolitica*, isolated needles are reported to be ~60–68 nm in length and 6–7 nm in width, with an axial channel of ~2 nm diameter; polymerization of a particular 6 kDa protein may (i) result in assembly of the needle, and (ii) provide the force necessary for the needle to puncture the cytoplasmic membrane of a eukaryotic cell [PNAS (2001) *98* 4669–4674].

[Structure and compositon of the needle complex in *Shigella flexneri*: MM (2001) *39* 652–663.]

Proteins secreted by type III systems include e.g. the IpaB protein of *Shigella*, the EspB and EspE proteins by EPEC, the various Yops of *Yersinia*, and the *harpins* of plant-pathogenic bacteria.

The energy requirements for type III secretion may include ATP hydrolysis; in a type III system of *Yersinia* a pore-associated protein (YscN) may act as an ATPase.

5.4.5   Type IV (autotransporter) protein secretion in Gram-negative bacteria

A protein secreted by this pathway is synthesized as part of a *larger protein* consisting of: (i) an N-terminal leader peptide, (ii) an α-domain, and (iii) a C-terminal β-domain. The mechanism of secretion is not fully understood, but one model is as follows. The leader peptide acts as a *signal sequence* (section 7.6) – so that the protein first crosses the cytoplasmic membrane via a *sec*-dependent (type II) mechanism (section 5.4.3), with cleavage of the signal sequence. The β-domain then forms a 'β-barrel' pore in the outer membrane – through which the α-domain (of the same protein) passes to the cell's surface. The protein may then e.g. (i) remain intact in the outer membrane; (ii) undergo *autocatalytic* cleavage, the α-domain being secreted; (iii) undergo cleavage by *another* outer membrane protein (such as the OmpT protease in *E. coli*), the α-domain being secreted; (iv) undergo cleavage, the α-domain remaining attached non-covalently to the β-domain. The α-domain can thus act as a 'passenger' which may be secreted via the β-domain *of the same protein.*

Another name for the α-domain is *autotransporter* but, confusingly, this name is also used to refer to the complete (three-part) precursor protein described above – and also to the β-domain.

The energy requirement for type IV secretion is not clear; some authors have assumed that secretion across the outer membrane is an energy-independent process.

The first known autotransporter was the IgA1 protease of *Neisseria gonorrhoeae*: a (secreted) enzyme which cleaves a subset of IgA antiboides (Fig. 11.2).

Another example is the (cell-surface-retained) IcsA protein which is responsible for the invasiveness of *Shigella* in dysentery (section 11.3.3.2).

In *Neisseria meningitidis*, fragments of an autotransported cell-surface protein (AspA) are shed into the supernatant [INFIM (2002) *70* 4447–4461]. Strains of *E. coli* that cause acute pyelonephritis may use an autotransporter for the secretion of serine proteases [INFIM (2004) *72* 593–597].

The *E. coli* AIDA-I autotransporter system (AIDA = adhesin involved in diffuse adhesion) is an adhesin which appears to occur in multiple copies on the cell surface. The AIDA-I system has been used for the *display* of heterologous proteins (e.g. the cholera toxin B subunit) on the surface of OmpT-deficient (mutant) strains of *E. coli*; in these studies, the gene for the heterologous protein was fused (*in vitro*) with DNA encoding the AIDA-I β-domain so that, when expressed in *E. coli*, the heterologous (passenger) protein was exported to (and exposed at) the cell surface (section 8.5.13).

Note. The designation 'type IV' secretion system has been widely used for the system described above [e.g. MMBR (1997) *61* 136–169; TIM (1998) *6* 370–378; *Cellular Microbiology* (1999, ISBN 0471–98681-X) page 42; *Bacteria* (1999, ISBN 0471–98880–4) page 94; *Dictionary of Microbiology and Molecular Biology* (2001, ISBN 0471–49064–4) page 618]. However, some authors use the designation 'type V' secretion system for autotransporters [see e.g. INFIM (2001) *69* 1231–1243]; these authors use the designation 'type IV' secretion system for the mode of secretion described in the next section (5.4.6).

5.4.6   Type V protein secretion in Gram-negative bacteria

The type V system involves a number of proteins that form (i) a channel across the cell envelope, and (ii) a tubular extension of this channel that projects from the surface of the bacterial cell; the latter feature thus resembles the 'needle' of the type III system

Protein export/secretion in Gram-negative bacteria

TYPE I SYSTEMS (section 5.4.1.2)

• Energized by ATP hydrolysis at a specific ABC site
• One-step translocation via a channel between the cytoplasm and the cell's surface – no periplasmic intermediate
• Secreted proteins characteristically lack an N-terminal signal peptide but have a C-terminal secretion sequence
• Analogous transport systems found in Gram-positive bacteria and in eukaryotic cells; these systems include the mammalian P-glycoprotein (a molecular pump which mediates resistance to anti-cancer drugs)

TYPE II SYSTEMS (section 5.4.3)

• Energized by ATP hydrolysis and pmf (GTP involved in some cases)
• Secreted proteins have an N-terminal signal sequence which is cleaved on passage through the cytoplasmic membrane
• Secretion involves two-step translocation, the first step being dependent on genes analogous to the *sec* genes of *Escherichia coli*. The first step involves a protein complex in the cytoplasmic membrane (the *translocon*) and the second step involves a protein complex in the outer membrane (the *secreton*)

TYPE III SYSTEMS (section 5.4.4)

• Energy requirements may include ATP hydrolysis
• Secreted proteins have an N-terminal sequence which is not cleaved
• Secretion involves one-step translocation through a multi-protein complex in the cell envelope; in at least some cases the transmembrane channel is extended by a tubular *needle complex* which extends from the cell surface
• Found in pathogenic bacteria, including EPEC (section 11.2.1) and species of *Salmonella*, *Shigella* and *Yersinia*
• In some pathogenic bacteria, type III systems may be regulated by quorum sensing (section 10.1.2)
• In some type III systems protein secretion may be activated by contact with a target cell

TYPE IV SYSTEMS * (section 5.4.5)

• Energy requirements unknown; energy-independent secretion across the outer membrane has been suggested
• Secretion involves two-step translocation, the secreted protein being translated as part of a larger protein which embodies the secretion mechanism; *sec*-dependent transport across the cytoplasmic membrane is involved
• Secreted proteins include some virulence factors (e.g. the IcsA protein of *Shigella*)
• The secreted protein may be cleaved from the parent protein autocatalytically or by a separate protease in the outer membrane
• The cleaved protein may be secreted or may remain bound, non-covalently, at the cell surface

TYPE V SYSTEMS * (section 5.4.6)

- ATP hydrolysis may be a source of energy
- Morphological features in common with type III systems
- Found in pathogenic bacteria – e.g. *Agrobacterium tumefaciens*, *Bordetella pertussis*, *Helicobacter pylori*, *Rickettsia prowazekii*
- In *A. tumefaciens* protein secretion is controlled by a two-component regulatory system
- In some cases type V secretion may depend on initial contact with a target cell

THE Tat SYSTEM (section 5.4.7)

- Energized by pmf
- Exports *folded* proteins across the cytoplasmic membrane
- Most (not all) proteins exported with a bound cofactor
- Export pathway for e.g. energy-associated proteins and outer-membrane components
- Signal peptide characteristically contains two consecutive arginine residues
- Found in Gram-positive as well as Gram-negative bacteria

* See note on nomenclature in section 5.4.5.

Figure 5.13 Protein export/secretion in Gram-negative bacteria: some characteristics of six distinct pathways. See text for further details.

(section 5.4.4). (See section 5.4.5 for a note on the naming of 'type IV' and 'type V' systems.)

Certain proteins at the base of the transmembrane channel (near the cytoplasmic membrane) are reported to have ATPase activity, suggesting that ATP hydrolysis may be a source of energy for type V transport.

Type V transport systems occur in a range of Gram-negative bacteria, including species pathogenic for plants and animals – for example, *Agrobacterium tumefaciens, Bordetella pertussis, Brucella suis, Helicobacter pylori* and *Rickettsia prowazekii.*

A. tumefaciens causes tumours (*crown gall*) in a variety of plants. Using type V secretion, *A. tumefaciens* injects single-stranded DNA from a Ti (tumour-inducing) plasmid directly into a plant cell; a protein, covalently attached to the DNA, appears to carry so-called nuclear localization signals targeted to the eukaryotic nucleus. Transfer of DNA from the bacterium is controlled by a two-component regulatory system (section 7.8.6) [FEMSMR (1998) *21* 291–319]; the transferred DNA (referred to as T-DNA) inserts into the plant's nuclear DNA, and plant hormones – encoded by the T-DNA – are apparently responsible for tumorigenesis.

In *Bordetella pertussis* (the causal agent of whooping cough) type V transport is used to secrete pertussis toxin into the extracellular environment.

In *Helicobacter pylori* a type V system is encoded by a so-called *pathogenicity island* (section 11.5.7) designated *cag* [see e.g. Science (1999) *284* 1328–1333].

In some cases (e.g. *H. pylori*), secretion of proteins by the type V system may depend on initial contact with a target cell (a further point of similarity with type III systems).

5.4.7   Exporting folded proteins: the Tat system

Unlike the *sec* (type II) transport system (section 5.4.3), the Tat system exports *folded* proteins across the cytoplasmic membrane (i.e. proteins which have been folded in the cytoplasm prior to transport). The typical protein which uses this pathway has a role in energy metabolism (e.g. respiration) or is a component of the outer membrane; however, proteins with other functions (including virulence factors) are also transported by the Tat system. Most (not all) proteins which use this pathway bind a cofactor *prior to* export across the CM; for example, proteins involved in energy metabolism may bind an iron–sulphur or other redox-associated complex.

The Tat system consists of some three or four proteins encoded by the *tatABCD* and *tatE* genes. TatA may form a transmembrane channel; the other gene products may have roles in recognizing the signal peptide on a protein targeted for export via the Tat pathway.

The (N-terminal) Tat signal peptide resembles the *sec* peptide but it differs e.g. in being longer and in containing a conserved motif that characteristically includes two consecutive arginine residues (hence Tat: *twin arginine translocation*).

Export via the Tat pathway is dependent on pmf.

The Tat pathway is found in Gram-positive bacteria (e.g. *Bacillus subtilis*) as well as in Gram-negative bacteria (e.g. *Escherichia coli, Pseudomonas aeruginosa*).

[Tat transport (review): Microbiology (2003) *149* 547–556.]

5.4.8 Salvage transport

In some cases a cell can re-cycle part(s) of macromolecules – rather than lose them to the environment. This makes sense not only in terms of conserving (possibly) scarce materials but also in saving the energy that would otherwise be needed for synthesis. For example, in *E. coli* there is an ongoing turnover of peptidoglycan (Figs 2.7 and 3.1), which is partly re-cycled. In a current model, peptidoglycan is degraded (by periplasmic enzymes) to disaccharide–tripeptide subunits. The subunits complex with a periplasmic binding protein (designated MppA) and are taken across the cytoplasmic membrane by a specific transport system (encoded by genes *oppBCDF*); within the cytoplasm, the tripeptide (L-alanyl-γ-D-glutamyl-*meso*-diaminopimelate – Fig. 2.7) is cleaved from the subunit by an enzyme (amidase) and is joined to UDP-MurNac (section 6.3.3.1) by another enzyme (a *ligase* – product of gene *mpl* [JB (1996) *178* 5347–5352]) – thus re-entering the pathway of cell wall biosynthesis [JB (1998) *180* 1215–1223].

5.4.9 Transport through the peptidoglycan barrier

An intact cell envelope is needed to prevent osmotic lysis. However, the peptidoglycan layer cannot *completely* enclose the protoplast because various large molecules (e.g. the *E. coli* α-haemolysin – section 5.4.1.2) are transported through the envelope, while the assembly of fimbriae (and other appendages) requires the outward transport of protein components. In certain types of conjugation (section 8.4.2) a large DNA–protein complex is transported from donor to recipient cell via the envelopes of both cells. Moreover, the peptidoglycan layer must be breached during the assembly of the flagellar basal body, and adhesion sites (section 2.2.10) are likely to involve the linkage of membranes at regions of discontinuity in the peptidoglycan.

The assembly of transport-specific or other structures which cross the cell envelope is likely to involve enzyme(s) which locally modify peptidoglycan while preserving structural integrity (cf. Fig. 3.1). The assembly of some of these structures has indeed been linked with peptidoglycan metabolism [review: JB (1996) *178* 5555–5562].

5.4.10 Transport of proteins across archaeal cytoplasmic membranes

Two pathways of protein secretion, resembling those in bacteria, have been found in members of the Archaea. One pathway resembles the *sec*-dependent pathway in *Escherichia coli* (section 5.4.3); however, archaeal systems are reported to lack a component with significant homology to the ATPases associated with the translocon. The other system resembles the Tat system (section 5.4.7) for exporting folded proteins.

[Protein transport across archaeal cytoplasmic membranes: FEMSMR (2004) *28* 3–24.]

6 Metabolism II: carbon

Why carbon? Simply because virtually all the compounds which comprise – and which are formed by – living organisms are carbon compounds. In this chapter we consider (i) the carbon requirements of bacteria, (ii) the ways in which different carbon compounds are assimilated, and (iii) the synthesis, interconversion and polymerization of carbon compounds. The metabolism of nitrogen and sulphur is considered in Chapter 10.

What sort of carbon compounds do bacteria need for growth? Some bacteria are able to use carbon dioxide for most or all of their carbon requirements; such bacteria are called *autotrophs*, and the use of carbon dioxide as the sole (or main) source of carbon is called *autotrophy*. In some autotrophic bacteria autotrophy is optional; in others it is obligatory: *only* carbon dioxide can be used as a source of carbon – even when glucose or other substrates are freely available. The *obligate* autotrophs use simple energy substrates (Chapter 5) as well as a simple carbon source: they are either chemolithotrophs (i.e. *chemolithoautotrophs*) or photolithotrophs (i.e. *photolitho-autotrophs*).

Chemolithoautotrophic metabolism occurs in relatively few bacteria (and it also occurs in some members of the Archaea): this type of metabolism is unique in the living world, being found only in these specialized prokaryotes and in no other type of organism. The chemolithoautotrophs have important roles e.g. in the nitrogen cycle (Chapter 10).

Photolithoautotrophic metabolism is found e.g. in the cyanobacteria – and, of course, in green plants and algae.

Most bacteria are *not* autotrophic: they cannot use carbon dioxide as a major source of carbon, and their growth depends on a supply of complex carbon compounds derived from other organisms; bacteria which need complex carbon compounds are called *heterotrophs*. Chemoorganotrophic heterotrophs are *chemoorganoheterotrophs*. Collectively, the heterotrophs can use a vast range of carbon sources – including sugars, fatty acids, alcohols and various other organic substances. Heterotrophic bacteria are widespread in nature, and they include (for example) all those species which cause disease in man, other animals and plants.

6.1   CARBON ASSIMILATION IN AUTOTROPHS

In autotrophs, carbon dioxide from the environment is used to form complex organic compounds; when carbon dioxide is incorporated into such compounds it is said to have been 'fixed'. Different autotrophs have different pathways for carbon dioxide fixation, but there are two very common pathways: the Calvin cycle and the reductive TCA cycle.

6.1.1   The Calvin cycle

This pathway (also called the *reductive pentose phosphate cycle*) is used by a wide range of autotrophs, including some anoxygenic photosynthetic bacteria and most or all cyanobacteria. Part of the Calvin cycle is shown in Fig. 6.1. Each turn of the Calvin cycle requires a considerable input of both ATP and reducing power, i.e. carbon dioxide fixation needs a lot of energy. The key enzymes in this pathway are ribulose 1,5-bisphosphate carboxylase–oxygenase (RuBisCO – see also section 2.2.6) and phosphoribulokinase; these enzymes are found only in cells which fix carbon dioxide via the Calvin cycle.

6.1.2   The reductive TCA cycle

This pathway is used for carbon dioxide fixation e.g. by phototrophic bacteria of the family Chlorobiaceae, and by the chemotrophic archaean *Sulfolobus*. Essentially, the pathway resembles the TCA cycle (Fig. 5.10) operating in reverse, with its one-way reactions (such as oxaloacetic acid→citric acid) being modified by different enzymes/reaction sequences. As in the Calvin cycle, carbon dioxide fixation requires a great deal of energy.

6.1.3   Carboxydobacteria

Carboxydobacteria can use carbon *monoxide* as the sole source of carbon and energy, i.e. they do not conform to the strict definition of 'autotroph'. However, they oxidize carbon monoxide to carbon dioxide, aerobically, and they assimilate carbon dioxide via the Calvin cycle. These organisms, which occur in soil, polluted waters and sewage, include e.g. *Bacillus schlegelii* and *Pseudomonas carboxydovorans*. [Growth of mycobacteria on carbon monoxide: JB (2003) *185* 142–147.]

Carbon monoxide is oxidized to carbon dioxide by an enzyme which is sometimes referred to as 'CO dehydrogenase'; this enzyme operates in a respiratory chain in which the cytochrome oxidase is insensitive to inhibition by carbon monoxide. In at least some cases it appears that the enzyme is functional only when attached to a

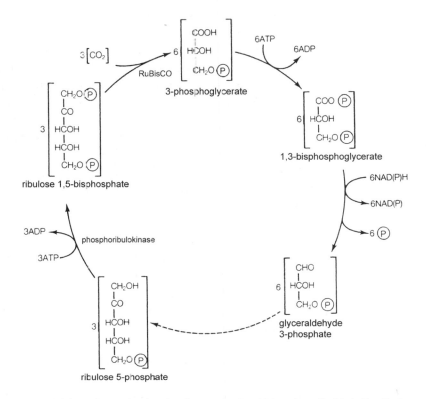

Figure 6.1 Part of the Calvin cycle, showing the reaction in which carbon dioxide is 'fixed'. The purpose of the pathway is to enable carbon dioxide to be used for the synthesis of the complex carbon compounds which make up the cell itself. Essentially, carbon enters the cycle by the fixation of carbon dioxide, and carbon leaves the cycle when intermediates are withdrawn for use in biosynthesis. For example, 3-phosphoglycerate is used for the synthesis of certain amino acids (such as glycine and serine) and also for the formation of pyruvate – itself a precursor of alanine, leucine and other amino acids; note that obligate autotrophs must synthesize all the amino acids necessary for protein synthesis. Continued operation of the Calvin cycle demands a continual supply of ribulose 1,5-bisphosphate; hence, the amount of carbon removed from the cycle must not exceed that put into the cycle. (The encircled 'P' represents phosphate.) RuBisCO is the enzyme ribulose 1,5-bisphosphate carboxylase–oxygenase. The many interconnecting pathways in the Calvin cycle have been omitted for clarity.

membrane-bound electron acceptor. Two distinct groups of these organisms have been identified [AEM (2003) *69* 7257–7265].

6.2   CARBON ASSIMILATION IN HETEROTROPHS

Collectively, the heterotrophs can assimilate a vast range of carbon sources; only a few examples are given below.

Many heterotrophs can use glucose, and its mode of assimilation will depend on the pathways and enzyme systems in any given organism. For example, many bacteria (including *E. coli*) can assimilate glucose via 'energy' pathways such as the EMP pathway (Fig. 5.3) and – subsequently – the TCA cycle (Fig. 5.10). In *Pseudomonas aeruginosa* glucose can be assimilated via extracytoplasmic oxidation (section 5.3.4). These pathways provide both energy and the compounds used as starting points in biosynthesis.

Many bacteria (and some members of the Archaea) can assimilate glucose, and e.g. gluconates, via the *Entner–Doudoroff* (ED) *pathway* (Fig. 6.2); this resembles the EMP pathway (Fig. 5.3) in that both pathways start with the phosphorylation of a hexose and both can yield e.g. pyruvic acid. The ED pathway provides NAD(PH) for bio-synthesis as well as a number of useful precursor molecules; moreover, in organisms with both EMP and ED pathways, glyceraldehyde 3-phosphate can feed into the EMP pathway and yield energy via substrate-level phosphorylation. In *E. coli* the ED pathway may be important in carbon assimilation, and substrates such as gluconates (found e.g. in the mammalian intestine) can be metabolized even when glucose is present. [ED pathway in *E. coli*: JB (1998) *180* 3495–3502.]

The *hexose monophosphate pathway* (HMP pathway) (= *pentose phosphate pathway*) provides a source of ribulose 5-phosphate (Fig. 6.2) – an important precursor of nucleotides (Chapter 7) and of the amino acid histidine. It also provides the NADPH needed as a source of reducing power for various biosynthetic reactions – such as the synthesis of fatty acids. Moreover, the HMP pathway is a source of erythrose 4-phosphate, a precursor molecule required in the synthesis of aromatic amino acids such as L-tryptophan and L-phenylalanine. The HMP pathway is widely distributed in nature, occurring also in eukaryotic microorganisms and in animals and plants.

In *E. coli*, the sugar lactose is split into glucose and galactose. The glucose is metabolized by the EMP pathway (Fig. 5.3), and the galactose is metabolized by the Leloir pathway. In the *Leloir pathway*, galactose is 1-phosphorylated (by ATP), and the galactose 1-phosphate is converted, enzymatically, via glucose 1-phosphate to glucose 6-phosphate – which enters the EMP pathway.

Cellulose, a polymer of glucose, is used as a carbon source by a number of actinomycetes (bacteria found in soil and compost heaps etc.) and e.g. by certain bacteria in the *rumen* (the grass-digesting part of the alimentary canal in cows and other ruminants). These bacteria produce enzymes which can degrade certain types of cellulose – outside the cell – into products which include e.g. the disaccharide *cellobiose*. Cellulolytic (i.e. cellulose-degrading) bacteria include species of *Cellulomonas* and e.g. *Clostridium thermocellum* and some strains of *Pseudomonas* and *Ruminococcus*.

It should be noted that the *availability* of a usable substrate does not necessarily lead to its uptake and utilization: see catabolite repression (section 7.8.2.1).

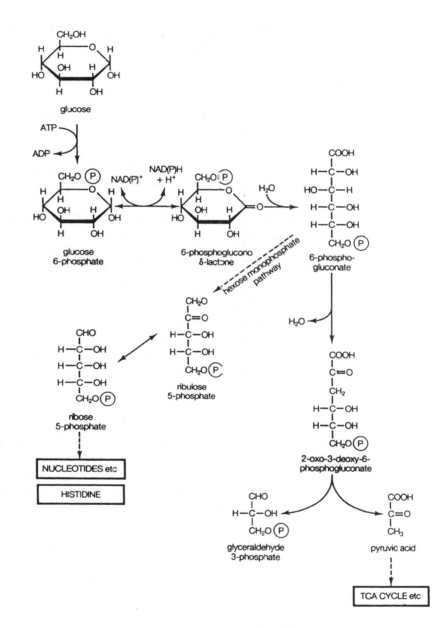

Figure 6.2 The Entner–Doudoroff pathway for assimilation of carbon in various bacteria (and some species of Archaea). The *hexose monophosphate pathway* (=*pentose phosphate pathway*) is similar as far as 6-phosphogluconate. Glyceraldehyde 3-phosphate may be converted to pyruvic acid (as in the EMP pathway: Fig. 5.3) or (e.g. in pseudomonads) re-cycled – re-entering the Entner–Doudoroff pathway via gluconeogenesis (section 6.3.2).

6.3 SYNTHESIS, INTERCONVERSION AND POLYMERIZATION OF CARBON COMPOUNDS

6.3.1 Synthesis of carbon compounds

Synthesis of the vast range of molecules which form the structure of a living cell clearly demands an enormously complex (and rigorously controlled) network of chemical reactions. Moreover, as well as 'structural' molecules the cell needs energy with which to carry out the various anabolic reactions; carbon and energy metabolism are usually closely inter-linked (section 5.3.1). Here, we have space to look only briefly at some of the generalities of biosynthesis.

Many compounds in the initial assimilative pathways can be used more or less immediately as starting points for biosynthesis; from the cell's point of view this makes good sense: if lengthy metabolism were needed to make the carbon available, any early breakdown in the carbon pathway could be a serious problem for biosynthesis. Thus, for example, in the Calvin cycle (Fig. 6.1) the very first product of carbon dioxide fixation, 3-phosphoglycerate, can be used to synthesize the amino acids cysteine, glycine and serine (themselves components of proteins – section 7.6) as well as other compounds. Pyruvic acid (from the EMP and Entner–Doudoroff pathways) is a precursor of e.g. various amino acids, and ribose 5-phosphate (from the HMP pathway) is used e.g. to synthesize nucleotides (components of DNA and RNA – Chapter 7).

Intermediates in the TCA cycle (Fig. 5.10) can also be used for biosynthesis. For example, oxaloacetic and 2-oxoglutaric acids are precursors of a range of amino acids; acetyl-CoA can be used e.g. for fatty acid synthesis; and succinyl-CoA is used for the synthesis of porphyrins – components of chlorophylls and cytochromes. Because intermediates withdrawn from the TCA cycle cannot be used to regenerate oxaloacetic acid (OAA), this compound must be generated in some other way if the cycle is to continue. There are various reactions for achieving this – for example, under certain conditions, OAA may be formed directly by the carboxylation of pyruvic acid or of phosphoenolpyruvate; such 'replenishing' reactions are called *anaplerotic sequences*.

In general, the assimilative pathways in both autotrophs and heterotrophs yield a range of 3-, 4- and 5-carbon molecules which are useful precursors in biosynthesis.

In some cases, *de novo* biosynthesis of cell components can be avoided by re-cycling materials through salvage pathways. One example is the re-use of tri-peptide units from peptidoglycan (section 5.4.8).

Each step in a biosynthetic reaction is normally enzyme-mediated, very few reactions occurring spontaneously. As well as enzymes, there are commonly other requirements. Reductive steps typically involve NAD(P)H or $FADH_2$, oxidative steps NAD or FAD, and phosphorylation usually involves ATP, GTP or phosphoenolpyruvate. Coenzymes have important roles in bacterial metabolism; they include e.g.

coenzyme A (a carrier of acetyl and other acyl groups) and thiamine pyrophosphate (TPP) (involved e.g. in various decarboxylation reactions). The way in which a diverse range of metabolic activities can depend on a given coenzyme is well illustrated by the roles of tetrahydrofolate (THF), a complex compound derived from a pteridine derivative, p-aminobenzoic acid and glutamic acid. Some of the roles of THF are shown in Table 6.1; notice that these THF-dependent functions involve essential aspects of cell growth: the synthesis of dTMP and purines (and, hence, the synthesis of DNA and RNA – Chapter 7) and the synthesis of proteins. THF itself is synthesized by the reduction of its precursor, dihydrofolic acid (DHF); in some strains of bacteria the synthesis of DHF (and hence THF) can be inhibited by sulphonamide antibiotics (section 15.4.9) – which thus inhibit the THF-dependent functions.

Some examples of simple biosynthetic pathways are shown in Fig. 6.3.

6.3.2 Interconversion of carbon compounds

In many pathways, individual reactions – or even sequences of reactions – can go in either direction, depending on conditions. Moreover, intermediates (as well as end-products) can pass from one pathway to another, so that the flow of carbon into various products can be regulated according to the cell's requirements. This versatility is indicated by the fact that, in many cases, the great diversity of structural molecules can be synthesized from a single carbon source. Only a few brief examples of interconvertibility can be mentioned here.

The interconvertibility of carbon compounds is well illustrated by *gluconeogenesis*: the synthesis of glucose 6-phosphate from non-carbohydrate substrates such as acetate, glycerol or pyruvate. Essentially, this is achieved by conversion of the substrate (where necessary) to an intermediate in the EMP pathway (Fig. 5.3) – followed by reversal of that pathway; non-reversible reactions in the EMP pathway are by-passed

Table 6.1 The coenzyme tetrahydrofolate (THF): examples of roles in bacterial metabolism

THF derivative	Function
N^{10}-formyl-THF	Initiation of protein synthesis (formation of N-formylmethionine – section 7.6)
N^5-methyl-THF	Biosynthesis of L-methionine
N^5, N^{10}-methylene-THF	Glycine ↔ serine interconversion
N^5, N^{10}-methylene-THF	Biosynthesis of deoxythymidine monophosphate (dTMP) (Fig. 6.3)
N^5, N^{10}-methenyl-THF	Biosynthesis of purine bases (found in DNA and RNA)

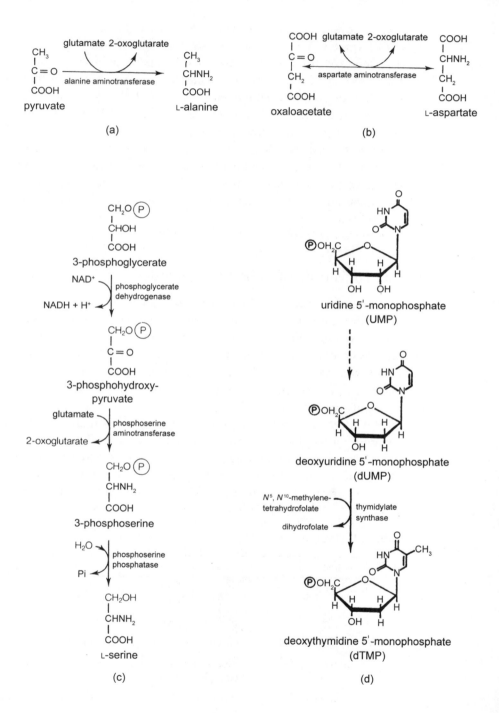

(a)

(b)

(c)

(d)

by other enzymes. Bacteria capable of gluconeogenesis (e.g. *E. coli*) can, if necessary, use intermediates (as well as the end-product) in the reversed EMP pathway.

Where the hexose monophosphate and Entner–Doudoroff pathways occur in the same cell, the common intermediate 6-phosphogluconate (Fig. 6.2) may be metabolized by either pathway. Increased metabolism via the HMP pathway may be needed e.g. for chromosome replication – DNA synthesis requiring increased production of ribose 5-phosphate, a component of both purine and pyrimidine nucleotides (section 7.2).

6.3.3   Polymerization of carbon compounds

Polymerization involves the chemical linkage of small molecules to form a large, often chain-like molecule called a *polymer*; a *homopolymer* is formed when all the small molecules are similar, and a *heteropolymer* results when the small molecules are not all alike. In bacteria, polymerization is involved in the synthesis of certain storage compounds and in the formation of various cell wall and capsular structures. Two (hetero)polymers of major importance – proteins and nucleic acids – are considered in the next chapter.

6.3.3.1   Peptidoglycan synthesis

In *E. coli*, *N*-acetylglucosamine and *N*-acetylmuramic acid (Fig. 2.7) are synthesized in the cytoplasm as their UDP (uridine 5'-diphosphate) derivatives – here abbreviated to UDP-GlcNAc and UDP-MurNAc, respectively. A chain of five amino acids is then added to UDP-MurNAc to form the so-called 'Park nucleotide'. The Park nucleotide is transferred (with release of UMP) to a long-chain lipophilic molecule (a *bactoprenol*) in the cytoplasmic membrane; subsequently, UDP-GlcNAc (with release of UDP) is added to form a bactoprenol–disaccharide–pentapeptide subunit. Subunits are then transferred (with release of bactoprenol) from the membrane to the periplasm; here, they are presumably joined together by *transglycosylation* reactions to form glycan chains (nascent strands of peptidoglycan). Incorporation of this new peptidoglycan

Figure 6.3   Some simple biosynthetic pathways: synthesis of the amino acids L-alanine (a), L-aspartate (b) and L-serine (c). (*Note*. Biochemists often give the *formula* of an organic acid in the un-ionized form but *name* the compound as though it were the salt – e.g $CH_3.CO.COOH$ = pyruvate.) In these particular examples, the reaction with glutamate introduces nitrogen into the (non-nitrogenous) precursor molecule; glutamine carries out this function in the synthesis of L-tryptophan. Note that each reaction is catalysed by a specific enzyme. An encircled 'P' represents a phosphate group; 'Pᵢ' is inorganic phosphate. (d) Uridine 5'-monophosphate (UMP) (top) has been synthesized via a series of reactions (not shown) from the simple initial precursor molecules aspartate and carbamoyl phosphate. Here, UMP first undergoes a reaction at the 2' position of the ribose (see Fig. 7.2), forming 2'-deoxyribose. In the next reaction, a coenzyme, N^5, N^{10}-methylene-THF (see section 6.3.1 and Table 6.1), donates a methyl group (-CH_3) to the 5-position of uracil (Fig. 7.3) – thus converting uracil to thymine.

into the existing sacculus may proceed according to the 3-for-1 mechanism (section 3.2.1, Fig. 3.1), peptide cross-linking involving *transpeptidation* reactions. Transglycosylation and transpeptidation are mediated by the enzymic action of penicillin-binding proteins (section 2.2.8).

In *E. coli* a significant proportion of peptidoglycan is broken down in each generation; however, through an efficient re-cycling system, constituent tripeptides are salvaged and used in the synthesis of new peptidoglycan (section 5.4.8).

Certain antibiotics inhibit peptidoglycan synthesis. Cycloserine blocks synthesis of the Park nucleotide. *β*-Lactams (see section 15.4.1) inhibit the penicillin-binding proteins. *Vancomycin* blocks the transfer of the subunit from the membrane to the periplasm by binding to the pentapeptide part of the bactoprenol–disaccharide–pentapeptide. Vancomycin, a complex glycopeptide, is bactericidal for a range of Gram-positive bacteria but is excluded from most Gram-negative cells; it is clinically useful e.g. against MRSA (section 15.4.11) and is also used for the treatment of pseudomembranous colitis associated with *Clostridium difficile*. *Teicoplanin* is structurally and functionally similar to vancomycin but is more active against enterococci.

6.3.3.2   *Synthesis of poly-β-hydroxybutyrate*

When non-carbon nutrients become scarce, many bacteria form intracellular granules of poly-*β*-hydroxybutyrate (PHB) – degrading these granules when conditions return to normal; PHB thus acts as a reserve of carbon and/or energy. Typically, PHB (section 2.2.4.1; Fig. 2.3) is synthesized from acetyl-CoA via acetoacetyl-CoA and *β*-hydroxybutyryl-CoA; the polymerizing enzyme, PHB synthetase, occurs at the surface of the granules.

The synthesis of PHB is described in a review on the production of polyhydroxy-alkanoates [Biotechnology (1995) *13* 142–150].

6.4   METHYLOTROPHY IN BACTERIA

Methylotrophy is a type of aerobic metabolism characterized by (i) the obligate or facultative use of certain 'C$_1$ compounds' for carbon and energy, and (ii) the assimilation of formaldehyde – a product of methylotrophic metabolism – as a major source of carbon. The C$_1$ compounds used as substrates in methylotrophy include methanol, methylamine and methane; some methylotrophic bacteria can use methyl groups cleaved from certain larger molecules. The *methylotrophs* (organisms which are capable of methylotrophy) include species of *Hyphomicrobium*, *Methylococcus*, *Methylomonas* and *Methylophilus*. Those bacteria which use carbon monoxide as the sole source of carbon and energy (the carboxydobacteria – section 6.1.3) were once classified as methylotrophs, but are now excluded because they do not produce/assimilate formaldehyde.

The obligate methylotroph *Methylophilus methylotrophus* was once grown commercially (on methanol) for use as an animal feed; thousands of tons of the product ('Pruteen') were made each year – until the price of protein fell and the process was discontinued.

In methylotrophs, the formaldehyde produced metabolically is assimilated by the ribulose monophosphate pathway (RuMP pathway) or the serine pathway. Both are cyclic pathways. In the RuMP pathway, formaldehyde condenses with ribulose monophosphate to form hexulose-6-phosphate; subsequent reactions regenerate the acceptor molecule (RuMP) and allow pyruvate (for example) to be withdrawn for biosynthesis. In the serine pathway, formaldehyde reacts with glycine, forming serine; subsequent reactions regenerate glycine, 2-phosphoglycerate being withdrawn for biosynthesis.

Methanotrophic bacteria [review: MR (1996) *60* 439–471] are those methylotrophs which are able to use methane as sole source of carbon and energy; most are *obligately* methanotrophic. The methanotrophic bacteria, which include species of *Methylococcus* and *Methylomonas*, occur e.g. in soil, water and sediments; they use oxygen (O_2) and the enzyme *methane monooxygenase* (MMO) to oxidize methane to methanol and water. Further oxidation leads to the production of formaldehyde; some formaldehyde is assimilated by the RuMP or serine pathway, while some is oxidized to carbon dioxide to yield energy – common electron acceptors being NAD and NADP.

The oxidation of methane by methanotrophs significantly affects the global methane budget in that much less methane reaches the atmosphere; hence, methanotrophs help to reduce the contribution of methane to global warming.

The ecology of methanotrophic species has been studied by molecular methods [FEMSME (1998) *27* 103–114].

[Detection and classification of atmospheric methane-oxidizing bacteria in soil: Nature (2000) *405* 175–178.]

7 Molecular biology I: genes and gene expression

7.1 CHROMOSOMES AND PLASMIDS

7.1.1 Chromosomes

The chromosome (section 2.2.1) consists mainly of a polymer called *deoxyribonucleic acid* (DNA). This polymer is built up of subunits called *nucleotides*. The *number* of nucleotides in the polymer varies in different chromosomes – e.g. the chromosome of *Buchnera* (one of the smallest bacterial chromosomes) contains $\sim 640 \times 10^3$ nucleotides, while that of *Mycobacterium tuberculosis* contains $\sim 4.4 \times 10^6$. (The smallest prokaryotic genome reported so far is that of the archaean *Nanoarchaeum equitans* (490885 nucleotides) [PNAS (2003) *100* 12984–12988].) The nucleotides in DNA are of four different types (see section 7.2); the *sequence* in which these nucleotides occur in chromosomal DNA encodes all the information needed to specify the structure and behaviour of a given bacterium. (DNA and the related polymer, *ribonucleic acid* (RNA: section 7.9), are both classified as *nucleic acids*; both types of molecule carry information in the sequence of their nucleotides. Within the bacterium, RNA is needed to convert the information in DNA into normal physiological processes – see later.)

Chromosomal DNA dictates the life of a cell e.g. by encoding all the enzymes (thus controlling structure and metabolism) and by encoding various RNA molecules (involved in protein synthesis and certain control functions). Information carried by DNA regulates and co-ordinates growth and differentiation. Moreover, DNA is also a self-monitoring system: there are various mechanisms for detecting and repairing damaged or altered DNA.

All of the information carried by DNA is passed to daughter cells when the parent cell divides (section 3.2.1). During replication, DNA is normally copied very accurately so that the characteristics of the species remain stable from one generation to the next; the tendency of daughter cells to inherit parental characteristics – heredity – is the main focus of *genetics*.

Clearly, nucleic acids determine both the 'routine' life of a cell and the process of heredity; both of these roles are studied in *molecular biology*.

In most bacteria the long DNA molecule of the chromosome forms a closed *loop*; in others (e.g. *Borrelia burgdorferi, Streptomyces* spp) the DNA molecule is linear. The reason for this is not known. Either way, the chromosome forms a compact body called the *nucleoid* (section 2.2.1).

Each bacterial cell may contain one or more copies of the chromosome – depending e.g. on growth conditions (section 3.2.1.1; Fig. 3.2) and on species. Some species normally have multiple copies; for example, *Deinococcus radiodurans* has a minimum of about four, while *Desulfovibrio gigas* may have more than ten copies.

A bacterium typically contains only one *type* of chromosome. Interestingly, each cell of *Vibrio cholerae* contains two *dissimilar* circular chromosomes which, together, form the organism's genetic complement [Nature (2000) *406* 477–483].

As mentioned above, the information in DNA is encoded in the *sequence* in which the four different types of subunit (four types of nucleotide) are joined together. In a given species of bacterium, the DNA contains a distinct and highly characteristic sequence of nucleotides (which can be used e.g. to identify that species: Chapter 16). The complete sequence of nucleotides in chromosomal DNA has been ascertained and published for various species of bacteria – for example:

Bacillus subtilis	[Nature (1997) *390* 249–256]
Bacteroides thetaiotaomicron	[Science (2003) *299* 2074–2076]
Borrelia burgdorferi	[Nature (1997) *390* 580–586]
Buchnera sp	[Nature (2000) *407* 81–86]
Chlamydia trachomatis	[Science (1998) *282* 754–759]
Escherichia coli	[Science (1997) *277* 1453–1474]
E. coli O157:H7	[Nature (2001) *409* 529–533; erratum *410* 240]
Helicobacter pylori	[Nature (1997) *388* 539–547]
Lactobacillus plantarum	[PNAS (2003) *100* 1900–1995]
Mycobacterium leprae	[Nature (2001) *409* 1007–1011]
Mycobacterium tuberculosis	[Nature (1998) *393* 537–544]
Neisseria meningitidis	[Nature (2000) *404* 502–506]
Pasteurella multocida	[PNAS (2001) *98* 3460–3465]
Prochlorococcus marinus SS120	[PNAS (2003) *100* 10020–10025]
Streptococcus pneumoniae	[Science (2001) *293* 498–506]
Streptococcus pyogenes	[PNAS (2001) *98* 4656-4663]
Synechococcus sp	[Nature (2003) *424* 1037–1042]
Tropheryma whipplei	[Lancet (2003) *361* 637–644]
Vibrio cholerae	[Nature (2000) *406* 477–483]
Vibrio parahaemolyticus	[Lancet (2003) *361* 743–750]

(Details of human DNA [the *human genome*: Nature (2001) *409* 813–958; Science (2001) *291* 1304–1351] provide valuable assistance in the task of combatting various infectious and inheritable diseases.)

7.1.1.1 Genomics

Genomics is the systematic study of an organism's genome; it includes determination of the entire sequence of nucleotides *and* the mapping of all genes and non-coding sequences. Whole-genome sequences on their own can be useful e.g. in taxonomy – in which differences/similarities between sequences from different organisms can help in assessing the degree of relatedness. Such data are also useful for identifying highly conserved sequences of nucleotides, permitting the construction of so-called *broad-range primers* (section 10.8). Moreover, sequence data can be used to reveal differences between related organisms; for example, such data have shown extensive variation in strains of UPEC [PNAS (2002) *99* 17020–17024] and in genes of the restriction cluster in *Escherichia coli* C and related strains [NAR (2004) *32* 522–534].

For other purposes (e.g. identifying potential targets for new drugs) it's not enough simply to know the sequence of nucleotides in a genome – we need to know the precise *in vivo* functions of gene products. Moreover, if we are looking for potential targets for drugs, an initial question is: *which* genes are likely to be important? For pathogens, one approach to this question is to identify those genes which are switched on *only* during infection of a host animal (see e.g. IVET: Fig. 11.3); the thinking here is that such genes *may* be virulence genes whose expression is required for pathogenesis. We can also begin to determine the function of particular genes by examining the different patterns of expression of an organism's genes, under differing conditions, by detecting the specific mRNAs produced under a given set of conditions; this can be done e.g. by using DNA chip technology (section 8.5.15).

Some (computer-based) studies determine the presence/functions of genes by looking for homology (similarity, relatedness) between the whole-genome sequence of a given organism and the sequences of other (usually related) organisms whose genomes contain specific genes of known function; this approach is called *comparative genomics*. For example, such studies have identified a range of glycosyl hydrolase enzymes in rumen bacteria [FEMSMR (2003) *27* 663–693] and various taxis-linked signalling proteins in different organisms [FEMSMR (2004) *28* 113–126].

The physiology and pathogenesis of *Leptospira interrogans* (see Leptospirosis in section 11.11) have been clarified by comparative genomics [JB (2004) *186* 2164–2172].

A genomic approach has revealed that the evolution of *Bordetella* spp involved a *loss* of genes [JB (2004) *186* 1484–1492].

7.1.2   Plasmids

In addition to the chromosome, many bacteria (both Gram-positive and Gram-negative species) contain one or more plasmids. A *plasmid* is an extra piece of DNA, usually much smaller than the chromosome, which can replicate independently. (Replication of plasmid DNA is discussed in section 7.3.1.)

Many or most plasmids are 'circular', i.e. closed loops of DNA. However, some are linear [see e.g. MM (1993) *10* 917–922]; for this reason it is not appropriate to define a plasmid as 'a small circle of DNA' (as is done in some textbooks and research papers). Some bacteria (e.g. *Borrelia burgdorferi*) contain circular *and* linear plasmids.

Plasmids encode various functions. Some encode enzymes which inactivate particular antibiotics; such *R plasmids* (resistance plasmids) usually make the host cell resistant to the relevant antibiotic(s). In some pathogenic bacteria important virulence factors are encoded by plasmids; these include the anthrax toxin in *Bacillus anthracis* and an adhesin in enterotoxigenic strains of *E. coli* (ETEC). [The large virulence plasmid of *Escherichia coli* O157:H7 (complete DNA sequence and analysis): NAR (1998) *26* 4196–4204.] Some plasmids encode the ability to metabolize particular substrates. For example, the Cit plasmid encodes a transport system (section 5.4) for the uptake of citrate; strains of *E. coli* which contain this plasmid can use citrate as the sole source of carbon and energy – something which cannot be done by common ('wild-type') strains of the organism. Again, the TOL plasmid confers on strains of *Pseudomonas* the ability to metabolize toluene and xylene [see e.g. EMBO Journal (2001) *20* 1–11]. In the archaean *Halobacterium*, structural elements of the gas vacuoles (section 2.2.5) are encoded by a plasmid.

Other plasmid-mediated functions are mentioned elsewhere in the book.

Although plasmids often encode functions that are useful to the host cell, they are typically not indispensable to their host cells.

Plasmids are normally *absent* in certain obligate pathogens of the alpha subclass of Proteobacteria (Fig. 16.5) – but present in other (plant-associated and phototrophic) bacteria within the same subclass. It has been suggested that species of *Anaplasma*, *Bartonella*, *Brucella* and *Rickettsia* evolved without plasmids because their relatively constant (intracellular) environment does not select for the genetic diversity needed by species in more challenging environments [FEMSMR (1998) *22* 255–275]; however, plasmids occur e.g. in the (intracellular) human pathogen *Chlamydia* [see e.g. Microbiology (1997) *143* 1847–1854].

Many bacterial plasmids contain transposable element(s) (section 8.3).

Some (not all) plasmids encode the means to transfer themselves from one cell to another by the process of *conjugation* (section 8.4.2). Such *conjugative* plasmids may persist in the absence of a selective pressure [Genetics (2003) *165* 1641–1649].

Plasmids are widely used in recombinant DNA technology (section 8.5).

7.2   NUCLEIC ACIDS: STRUCTURE

A nucleic acid is a polymer made of subunits called *nucleotides*; each nucleotide has three parts: (i) a sugar molecule, (ii) a nitrogen-containing base, and (iii) a phosphate group (Fig. 7.1). Nucleotides containing the sugar D-ribose are *ribo*nucleotides – which form *ribo*nucleic acid (RNA); those containing 2'-deoxy-D-ribose are *deoxyribo*nucleotides – which form *deoxyribo*nucleic acid (DNA) (Fig. 7.2). The nitrogen base in any given nucleotide is either a substituted *purine* or a substituted *pyrimidine* (Fig. 7.3); a ribonucleotide may contain adenine, guanine, cytosine or uracil, while a deoxyribonucleotide may contain adenine, guanine, cytosine or thymine. The names of the various nucleotides, and their corresponding nucleosides, are given in Table 7.1.

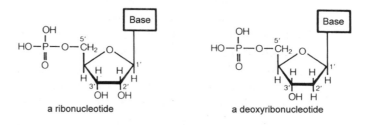

Figure 7.1   A generalized nucleotide. Note the difference between a nucleotide and a nucleoside.

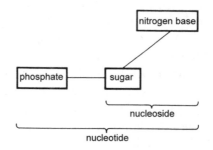

a ribonucleotide   a deoxyribonucleotide

Figure 7.2   Nucleotides: a ribonucleoside monophosphate and a *deoxy*ribonucleoside monophosphate; these generalized molecules differ only in their sugar (ribose) residues: the deoxyribonucleotide lacks oxygen at the 2' position. (The numbers 1', 2' etc. refer to the positions of the carbon atoms in the ribose molecule.) The base can be adenine, guanine or cytosine in either molecule, but uracil occurs only in ribonucleotides, while thymine is found only in deoxyribonucleotides. (In a *di*deoxyribonucleotide (not shown here) oxygen is missing at both the 2' and 3' positions; dideoxyribonucleotides are used e.g. in DNA sequencing – section 8.5.6.)

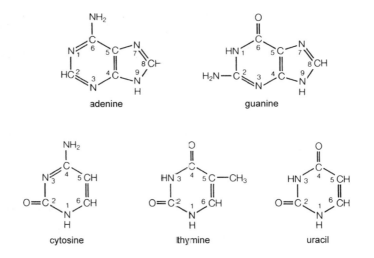

Figure 7.3 Nitrogen bases found in DNA and RNA. Adenine and guanine are substituted *purines*; the others are substituted *pyrimidines*. Thymine occurs only in DNA, uracil only in RNA; adenine, guanine and cytosine occur in both DNA and RNA. Within a nucleotide (Fig. 7.2) a purine is linked via its 9-position to the sugar molecule, while a pyrimidine is linked via its 1-position.

Table 7.1 Nucleosides and nucleotides found in RNA and DNA

Sugar	Nitrogen base	Nucleoside (base + sugar)	Nucleotide[1] (base + sugar + phosphate)
Ribose	Adenine	Adenosine	Adenosine 5'-monophosphate (AMP) (or adenylic acid)
Ribose	Guanine	Guanosine	Guanosine 5'-monophosphate (GMP) (or guanylic acid)
Ribose	Cytosine	Cytidine	Cytidine 5'-monophosphate (CMP) (or cytidylic acid)
Ribose	Uracil	Uridine	Uridine 5'-monophosphate (UMP) (or uridylic acid)
Deoxyribose	Adenine	Deoxyadenosine	Deoxyadenosine 5'-monophosphate (dAMP) (or deoxyadenylic acid)
Deoxyribose	Guanine	Deoxyguanosine	Deoxyguanosine 5'-monophosphate (dGMP) (or deoxyguanylic acid)
Deoxyribose	Cytosine	Deoxycytidine	Deoxycytidine 5'-monophosphate (dCMP) (or deoxycytidylic acid)
Deoxyribose	Thymine	Deoxythymidine	Deoxythymidine 5'-monophosphate (dTMP) (or deoxythymidylic acid)

[1] Only the monophosphate is shown in each case. Diphosphates and triphosphates are named in a similar fashion, e.g. adenosine 5'-diphosphate (ADP) and adenosine 5'-triphosphate (ATP).

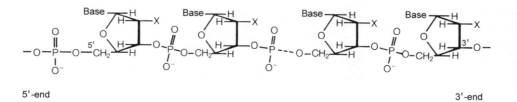

5'-end   3'-end

Figure 7.4 The structure of a single strand of nucleic acid; note the *polarity* of the molecule. 'X' is a hydrogen atom (H) in DNA but a hydroxyl group (OH) in RNA. The sugar residues are linked together by *phosphodiester* bonds.

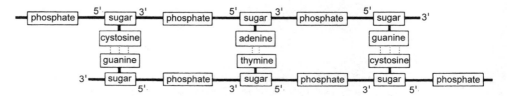

Figure 7.5 A DNA duplex (diagrammatic). The dotted lines represent hydrogen bonds. Note that the strands are *antiparallel* (section 7.2.1).

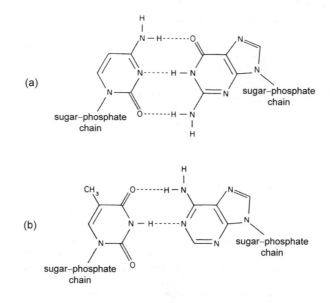

Figure 7.6 Base-pairing between nucleotides. (a) Cytosine (left) pairing with guanine (right). (b) Thymine (left) pairing with adenine (right). (*Note.* In RNA synthesis (section 7.5) adenine pairs with uracil.) The dotted lines represent hydrogen bonds.

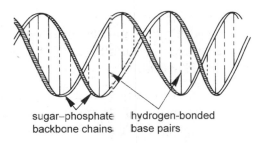

sugar–phosphate   hydrogen-bonded
backbone chains   base pairs

Figure 7.7 The DNA 'double helix' (diagrammatic). Sugar residues are linked by phosphodiester bonds (Fig. 7.4). The (planar) nitrogen bases are stacked roughly perpendicular to the axis of the helix; chemical groups on these bases protrude into the grooves of the helical molecule – thus allowing e.g. recognition by enzymes.

7.2.1 Deoxyribonucleic acid (DNA)

In DNA the nucleotides form an unbranched chain (a *strand*) in which sugar–base and phosphate residues are alternate links (Fig. 7.4). Note that a strand has *polarity*: there is a 5'-end and a 3'-end. DNA commonly consists of two strands which are held together by hydrogen bonding between their nitrogen bases; this double-stranded structure is called a DNA *duplex* (Fig. 7.5). Note that the two strands in a duplex are *antiparallel*, i.e. when read left to right (Fig. 7.5), one strand is 5'-to-3' while the other is 3'-to-5'.

Hydrogen bonding between the nitrogen bases is quite specific: adenine pairs with thymine, and guanine pairs with cytosine (Fig. 7.6); this specificity in *base-pairing* is referred to by saying that each of the two bases in a *base-pair* (bp) is *complementary* to its partner, and that (therefore) each strand in a DNA duplex is complementary to the other strand.

The ladder-like DNA duplex forms a *helix* (Fig. 7.7) – each turn of which occupies about 10 base-pairs (10 bp) of distance along the duplex; this is the *double helix* worked out by Watson and Crick in the 1950s.

The length of a DNA duplex can be described in terms of the number of nucleotides in *one* of its strands; alternatively – as each nucleotide is one-half of a single *base-pair* in the duplex – we can give the length of a duplex as the number of base-pairs it contains. In *Escherichia coli*, the chromosomal DNA is about 4.7 million base-pairs (bp) in length, in *H. pylori* about 1.7 million bp, and in *M. tuberculosis* about 4.4 million bp. The (linear) DNA in *B. burgdorferi* is about 900000 bp in length.

In most bacterial chromosomes the DNA duplex forms a closed *loop* (i.e. there are no free 5' and 3' ends); this is *covalently closed circular* DNA (cccDNA). A circular duplex can exist in various states. A *relaxed* loop of DNA can theoretically lie in a plane, like a rubber band on a table, but suppose that the loop is cut, and that one end is twisted (the other held still) before the ends are rejoined; this will either increase or decrease the 'pitch' of the helix (forming an *underwound* or *overwound* helix, respectively) – depending on the direction of twisting. Such a molecule is under strain: there is a tendency to restore the pitch to that of a relaxed molecule. To relieve the strain, the molecule contorts: the axis of the helix itself becomes coiled; a molecule in this state is said to be *supercoiled*.

Supercoiled DNA and topoisomerases. Naturally occurring DNA is usually underwound (i.e. *negatively* supercoiled). In the nucleoid (section 2.2.1), supercoiling is segregated into many topologically distinct *domains*; a nick in one domain relaxes the DNA in that domain only.

An appropriate degree of supercoiling is generally necessary *in vivo* for the normal expression of genes.

Supercoiling can be affected by those environmental factors which affect the cell's energy status; this may involve environmental influence on the intracellular level of ATP because the activity of *gyrase* (see below) is known to be sensitive to the ratio ATP:ADP. During the transition from aerobic to anaerobic growth there is a fall in the cell's energy charge and a corresponding, transient fall in overall superhelical density; this alters the expression of a range of genes (e.g. by affecting the activity of their promoters).

Increased levels of supercoiling generated during *divergent* transcription (section 7.5) may also affect the expression of genes [MM (2001) *39* 1109–1115].

Within a living cell, supercoiling is generated and controlled by enzymes called *topoisomerases*. The name of these enzymes derives from their ability to convert one topological isomer (= topoisomer) of cccDNA to another. *Topoisomers* of DNA are molecules which differ from one another only in their topological properties; thus, e.g. a given circular double-stranded (ds) DNA molecule may exist in a relaxed form or in a supercoiled form – each of these forms being a topoisomer. Moreover, a DNA molecule may be given different *degrees* of supercoiling – producing correspondingly different topoisomers of the supercoiled molecule.

Topoisomerases are also required e.g. to disentangle two or more molecules of DNA – as in the *decatenation* of chromosomes (section 3.2.1).

During activity, a given topoisomerase (transiently) breaks either one or both strands of a DNA duplex:

- type I topoisomerases (also called untwisting, relaxing or nick-closing enzymes, or swivelases) break only one strand
- type II topoisomerases break both strands

Somewhat confusingly, individual topoisomerases are referred to as topoisomerase (topo) I, topo II, topo III etc.; thus, for example, topo III is a type I topoisomerase.

Type I topoisomerases include the ω (omega) protein of *Escherichia coli* (= topo I) and the *E. coli* topo III. Both enzymes can partially relax negatively supercoiled DNA; in a supercoiled ds cccDNA molecule, the unbroken strand is passed through the break before the latter is repaired. In *E. coli*, topo I appears to prevent excessive supercoiling.

Type II topoisomerases pass a T-segment ('transfer segment') of *duplex* DNA through the double-stranded gap in a G-segment ('gate segment') of DNA. In one model, the enzyme initially binds to the G-segment, producing a sharp bend in the DNA and adopting a specific orientation in relation to the bend. Subsequently, ATP-dependent capture of the T-segment, by the enzyme, is followed by transient cleavage of the G-segment to permit unidirectional passage of the T-segment [PNAS (2001) *98* 3045–3049].

Type II topoisomerases include gyrase and topo IV. Gyrase is a tetrameric enzyme that introduces negative supercoiling in an ATP-dependent manner; it is apparently essential *in vivo* for DNA replication (section 7.3) and for some other processes. Gyrase is the target of e.g. quinolone antibiotics and novobiocin (section 15.4.6) [anti-gyrase antibiotics: TIM (1997) *5* 102–109, TIM (1998) *6* 269–275]. (The CcdB toxin (section 7.3.1) is reported to inhibit bacterial gyrase.)

Topoisomerase IV can relax both positive and negative supercoiling, but relaxes the former much more efficiently. Chirality sensing by the *E. coli* topo IV (i.e. its discrimination between positively and negatively supercoiled DNA) may depend on the enzyme's ability to recognize the spatial relationship between the G-segment and T-segment (see above) [PNAS (2003) *100* 8654–8659].

Topo IV appears to be the sole enzyme responsible for decatenation of chromosomes in *E. coli* [GD (2001) *15* 748–761].

The maintenance of appropriate levels of negative supercoiling in *E. coli* may involve the joint activities of gyrase, topo I, and topo IV.

Gyrase and topo IV may both oppose the positive supercoiling introduced by DNA replication (gyrase by generating negative supercoiling and topo IV by relaxing the positive supercoiling).

7.2.1.1 *PNA (peptide nucleic acid)*

PNA is an analogue of DNA in which the backbone chain is a modified polypeptide rather than a chain of sugar–phosphate units; two strands of PNA can hybridize to form a helical duplex resembling a DNA duplex [Nature (1994) *368* 561–563]. PNA may have evolutionary significance: it has been suggested that pre-biotic nucleic acids may not necessarily have had a sugar–phosphate backbone.

In biotechnology, certain properties of PNA have been exploited in the construction of probes (section 8.5.3). For example, PNA binds to a strand of DNA more stably than does the complementary strand of DNA (primarily because the electrostatic repulsion between two strands of DNA is absent in a PNA–DNA duplex); moreover, PNA is unaffected by the concentration of electrolyte (hybridization between two strands of DNA is sensitive to the concentration of electrolyte). In this context, PNA has been used e.g. for the preparation of so-called *light-up* probes for the real-time detection of PCR products [BioTechniques (2001) *31* 766–771].

(See also section 8.5.15.)

7.2.2 Ribonucleic acid (RNA)

RNA is a polymer of ribonucleotides (Fig. 7.4); although each ribonucleotide generally contains adenine, guanine, cytosine or uracil, modified bases occur in some molecules of RNA. Bacterial RNA (unlike DNA) is typically single-stranded. Double-stranded regions can be formed by base-pairing between complementary regions within the same RNA molecule – so that many three-dimensional structures are possible; metal ions (particularly magnesium ions) are usually important for the stability and activity of such structures.

Some of the roles of RNA are considered in section 7.9.

7.3 DNA REPLICATION

Before cell division, chromosomal DNA must be duplicated precisely so that each daughter cell can receive an exact copy (replica). DNA synthesis (*replication*) is a complex process, still incompletely understood; the following account is an outline of the replication of an *E. coli* chromosome during the normal cell cycle.

What starts (= *initiates*) a round of DNA replication? We don't know (see section 3.2.1), but replication is known to begin at a specific location in the double-stranded chromosomal DNA: the *origin* (*oriC*). [Replication origin in archaeans: TIBS (2000) *25* 521–523.] Initiation of replication is also known to require a number of different proteins, and in at least some cases it appears to involve supercoiled DNA.

One of the early events in DNA replication (in all bacteria: Gram-negative and Gram-positive) is the binding of a number of molecules of DnaA protein to specific sites *(DnaA boxes)* within the *oriC* sequence. The binding of DnaA proteins induces ATP-dependent *melting* (i.e. strand separation in DNA) within the *oriC* region; where strand separation has occurred, the single strands of DNA may be stabilized by the binding of so-called single-strand binding proteins (SSB proteins). Strand separation is a prerequisite for replication because one new strand of DNA is synthesized on each of the two strands of parent DNA (see later).

DnaA proteins are unique to bacteria. In the domain Eukarya the analogous function is carried out by six proteins of the *origin recognition complex* (Orc1–6), while in the Archaea this function depends on the Orc1 and CdC6 proteins. DnaA and some Orc proteins nevertheless have the common feature of an ATP-binding site; in order to function, these proteins undergo ATP-dependent conformational change. [Initiators of DNA replication: FEMSMR (2003) *26* 533–554.]

One factor that has been linked to the initiation of DNA replication is the Fis protein (*fis* gene product). Fis has several binding sites in *oriC* near DnaA boxes and it also seems to regulate a certain cluster of genes (*dnaA-dnaN-recF*) [JB (1996) *178* 6006–6012] whose products include DnaA protein and the β subunit of DNA polymerase III (Fig. 7.8(b)). Fis binds to *oriC* in a cell-cycle-specific way [EMBO Journal

(1995) *14* 5833–5841] and seems to form an 'initiation-preventive' complex [NAR (1996) *24* 3527–3532]; it also has a weak repressor effect on the *dnaA* p2 promoter.

Where replication begins, strand separation produces two *replication forks* (Fig. 7.8). Each fork is the site of a *primosome*: a complex of proteins whose functions include e.g. ongoing unwinding of duplex DNA (i.e. strand separation) and 'priming' of new strands (see later). Currently, evidence from certain bacteria (*Escherichia coli* and *Bacillus subtilis*) indicates that the primosomes occur at a static, mid-cell location which acts as a 'replication factory' or *replisome*; other evidence suggests that the two replication complexes migrate, in opposite directions, from an initial central position in the cell (see chromosome replication in section 3.2.1).

In *one* of the DNA strands, bases newly exposed by strand separation pair, sequentially, with complementary bases of individual molecules of *ribo*nucleoside triphosphate, the ribonucleotides being enzymically polymerized (covalently joined together) in the 5'-to-3' direction to form a short strand of RNA: a *primer* (Fig. 7.8(a), *top*); in this process, the two terminal phosphate groups of each ribonucleoside triphosphate are cleaved (forming pyrophosphate) in a reaction that provides energy for polymerization. Polymerization may be carried out by the DnaG protein (a *primase*) or by RNA polymerase (both enzymes can synthesize RNA).

The primer forms one strand of a short RNA/DNA hybrid duplex (Fig. 7.8(a), *top*). On further strand separation, newly exposed bases pair with molecules of *deoxy*ribonucleoside triphosphate so that the primer is extended in the 5'-to-3' direction by a new strand of DNA; this polymerization, which is mediated by the enzyme DNA polymerase III, forms the start of the so-called *leading strand* of DNA (Fig. 7.8(a), *top*). Note that the leading strand, together with one strand of the original duplex, is the beginning of a new duplex. Note also that the strand from the original duplex has determined the base sequence in the new strand (by specifying *complementary* bases); the strand from the original duplex is therefore called a *template* strand.

An RNA primer is needed to start the process because a DNA polymerase cannot initiate a strand – it can only extend an *existing* strand of nucleotides.

The DnaB protein, present in each primosome, is a *helicase*: an enzyme which, in the presence of gyrase (section 7.2.1) and ATP, *unwinds* DNA – i.e. separates the strands of the DNA duplex; thus, DnaB allows DNA replication to continue by opening up the duplex (like a zip) so that the two replication forks can progress, away from *oriC*. At 37°C, a replication fork apparently moves at about 1000 nucleotides per second.

So far we have considered only that strand of the original DNA duplex on which the leading strand is synthesized. What of the other strand? That, too, acts as a template strand. However, because the strands of the parent duplex are antiparallel (see section 7.2.1), and because DNA can be synthesized only in the 5'-to-3' direction, synthesis of DNA on *this* template strand occurs in the opposite direction (Fig. 7.8(a), *bottom*). In fact, this new strand of DNA (the *lagging strand*) is synthesized as a series of short fragments (*Okazaki* fragments), as follows. When strand separation in the parent duplex has reached a certain extent, the DnaG protein (primase) synthesizes the 1st

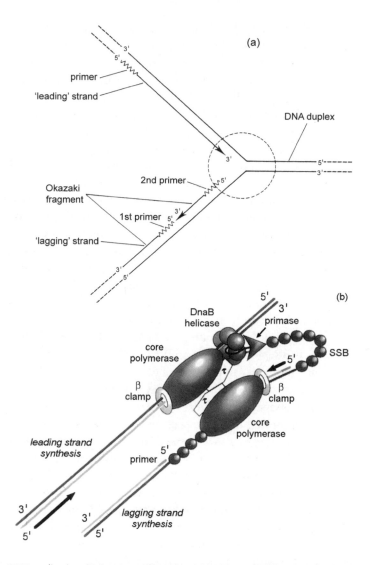

Figure 7.8 DNA replication (Cairns-type, diagrammatic). (a) In a circular DNA duplex the molecule has split, locally, into its two component strands. The diagram shows only one half of the split region; in this half, the strands continue to separate in a left → right direction. On one exposed DNA strand (top) an RNA primer is synthesized (see text); the subsequent addition of deoxyribonucleotides forms the 'leading' strand of DNA – which is synthesized continuously in the 5'-to-3' direction. On the other DNA strand (bottom) the first primer is extended by synthesis of a DNA fragment (an *Okazaki fragment*) – again in the 5'-to-3' direction (because synthesis cannot occur in the 3'-to-5' direction); only when the duplex has opened out further is the second primer formed and the next Okazaki fragment synthesized. (All primers are subsequently replaced with DNA.) The encircled region (*replication fork*) moves to the right (relative to *oriC*) as replication continues, while the other replication fork (not shown) moves to the left.

RNA primer near the replication fork; this primer is extended by the addition and polymerization of deoxyribonucleotides, in the 5'-to-3'-direction, to form the first fragment of the lagging strand. On further strand separation, the 2nd primer is formed and the second fragment is synthesized – and so on. Each Okazaki fragment is about 2000 bases long (= 2 kilobases, 2 kb). Before completion of replication all the (RNA) primers are replaced with DNA.

Co-ordination of leading/lagging strand synthesis at a replication fork, and the rapid rate of DNA synthesis, appear to involve (i) a physical link between the two DNA polymerases, and (ii) a coupling between polymerases and the helicase – loss of polymerase–helicase interaction resulting in a drastic reduction in the efficiency of the helicase [Cell (1996) 84 643–650]. One view of a replication fork is shown in Fig. 7.8(b). Note that the helicase (which unwinds the duplex) is a hexamer (a ring-shaped structure comprising six DnaB proteins) [Cell (1996) 86 177–180]; in the figure, both strands of duplex DNA are shown passing through the ring, but whether both (or only one) pass through is not known.

Termination of replication occurs at the *ter* site on the chromosomal DNA, 180° from *oriC*.

Because each new duplex consists of one strand of the original (parent) duplex and one newly synthesized (daughter) strand, replication is said to occur in a *semi-conservative* fashion.

Other modes of DNA replication. The replication described above is 'routine' replication in an *E. coli* chromosome, and it should be stressed that replication can occur in other ways. Even in *E. coli*, chromosomal replication can be mechanistically different under certain conditions; for example, in cells induced for the SOS system (section 7.8.2.2), DNA replication – inducible stable DNA replication (iSDR) – begins at a different origin (*oriMs*) and is independent of DnaA protein [JB (1994) *176* 1807–1812].

Some bacteria have *linear* chromosomes (and plasmids) in which DNA synthesis runs in *both* directions from an *internal* origin of replication; given the 5'-to-3' direction of DNA synthesis (Fig. 7.8), an unexplained aspect of this is the way in which the replication machinery avoids leaving the 3' end of the template strand (in each

---

(b) A diagrammatic view of a replication fork in *E. coli* (not drawn to scale). The two polymerases (one forming the leading strand, the other the lagging strand) work in opposite directions, and are shown physically linked via the τ subunit of each polymerase; a polymerase–helicase link is also shown (the helicase being a hexamer of DnaB proteins). The single-stranded DNA is protected by so-called *single-strand binding proteins* (SSBs) prior to serving as template for the lagging strand; the primase synthesizes primers on this template. The β-clamp (= β subunit of DNA polymerase III, *dnaN* gene product) may affect *processivity*, i.e. the coupled movement and function of the polymerase.

Drawing (b) reproduced from Cell (1996) 86 177–180 by courtesy of the author, Dr Stephen C West, Imperial Cancer Research Fund, UK, and with permission from Cell Press, Cambridge MA 02138, USA.

daughter duplex) as single-stranded DNA. Several models for completing synthesis on the template strand have been suggested [see TIG (1996) *12* 192–196]. Two models suggest that the 3' end of the template strand folds over and base-pairs with itself to provide a double-stranded starting point for replication. In another model, DNA synthesized *on the new strand* at the *fully duplexed* end of a daughter duplex displaces the 5' terminal nucleotide sequence of the template – which is then transferred to (complementary) single-stranded DNA on the other daughter duplex (thus filling the single-stranded gap).

In phage ϕ29 (Table 9.1) the *linear* dsDNA genome has a terminal protein (TP) covalently bound to each 5' end of the duplex. In the cell, replication of phage DNA starts at both ends of the duplex. A (phage-encoded) DNA polymerase binds to a *free* molecule of (phage-encoded) TP, and this complex localizes at the 3'end of each strand – perhaps by interacting with the TP at the adjacent 5' end. The polymerase covalently links dAMP (Table 7.1) to the complexed TP, and synthesis of the DNA strand (5' → 3') later continues from this initial nucleotide; the primer (TP) and the polymerase dissociate from one another after insertion of nucleotide 10 – the TP remaining covalently bound at the 5' end of the new strand, and the polymerase continuing to extend the short DNA primer. Such *protein priming* is only one of the strategies used for replicating linear genomes; these genomes cannot be replicated with RNA primers (as in Fig. 7.8) because subsequent removal of the primer would leave single-stranded DNA at the corresponding end of the new duplex. [Protein priming of DNA replication in phage ϕ29: EMBO Journal (1997) *16* 2519–2527.]

The rolling circle mechanism (Fig. 8.5) is used for DNA replication in certain phages – e.g. M13 (section 9.2.2), λ (Fig. 9.2) and ϕX174 – and in various plasmids (section 7.3.1).

7.3.1   Replication and partition of plasmid DNA

A plasmid controls its own replication – although the cell's biosynthetic machinery is needed to synthesize any plasmid-encoded protein etc. involved in replication.

The frequency of replication of a plasmid depends on the rate at which replication is *initiated*, and this is controlled in different ways by different plasmids. Some early reports indicated that, in a minority of plasmids (e.g. pSC101), initiation of replication involved the DnaA protein, but it now seems that DnaA/*oriC* binding is specific to the initiation of *chromosomal* replication (section 7.3); plasmid replication involves a range of mechanisms and a range of plasmid-encoded replication initiators [FEMSMR (2003) *26* 533–554].

In plasmid ColE1, control of initiation involves the synthesis of two, free (non-duplexed) strands of plasmid-encoded RNA, one of which (RNA II) can bind at the plasmid's origin of replication and act as a primer, thus permitting replication; the other strand (RNA I) can inhibit initiation by base-pairing with RNA II, and the outcome of RNA I–RNA II interaction determines whether or not replication will be initiated. Thus, in ColE1, initiation is controlled at the primer level. Once initiated, replication in ColE1 proceeds as in the *E. coli* chromosome (see above) – except that, in this plasmid, replication occurs *uni*directionally from the origin.

Seemingly extraneous factors may affect plasmid replication. For example, ColE1 RNA I has a 3' polyadenylate tail (i.e. -A-A-A-A......3') that facilitates degradation of the molecule; addition of this tail to RNA I is mediated by the enzyme poly(A)polymerase I (encoded by gene *pcnB*). Mutations in *pcnB* that prevent polyadenylation of RNA I tend to stabilize the molecule and thus enhance its inhibitory effect on RNA II; this results in less frequent initiation of replication of ColE1 [Microbiology (1996) *142* 3125–3133].

Many of the small circular plasmids in Gram-positive bacteria replicate by a rolling circle mechanism (Fig. 8.5). Replication is initiated by a plasmid-encoded *Rep protein* which, by nicking a specific strand at the origin, provides a 3' end for synthesis; one round of replication produces a (complete) dsDNA plasmid and a (circular) ssDNA copy, the latter then being replicated to the dsDNA state from an RNA primer. Initiation is controlled in various ways; because increased quantities of Rep protein cause increased replication of these plasmids, initiation can be controlled by regulating the synthesis and/or activity of Rep. In some cases, small (plasmid-encoded) RNA molecules (*countertranscript RNA, ctRNA*) bind to *rep* mRNA and cause either premature termination of transcription or inhibition of translation of the *rep* gene. In another mechanism, the Rep protein is inactivated *after only one use* by the binding of a small plasmid-encoded oligonucleotide. [Replication control in rolling circle plasmids: TIM (1997) *5* 440–446.]

In many plasmids of Gram-negative bacteria, the Rep initiator protein binds to 'direct repeats' of nucleotides (*iterons*) in the region of the plasmid's origin of replication, thus contributing to a nucleoprotein complex that initiates replication; in the F plasmid, RepE (E protein) carries out this function (replication then occurring by a *bi*directional Cairns-type mechanism: Fig. 7.8). For each plasmid, the iterons have a characteristic number, spacing and composition. In some plasmids, Rep can also repress transcription of its *own* gene by binding to *inverted* repeat sequences that overlap the promoter; in some cases (including F) it has been shown that, while *monomers* of Rep bind to iterons (promoting replication), *dimers* bind to the promoters of their own genes (thus inhibiting replication). Work on the RepA initiator protein of *Pseudomonas* plasmid pPS10 has suggested a model for the activation of this type of Rep protein; the model proposes that dissociation of the dimers produces monomers that are suitable, structurally, for binding to iterons – thus suggesting a structural basis for the activation of Rep initiator proteins [EMBO Journal (1998) *17* 4511–4526].

To understand replication control in a given plasmid it is necessary to know whether control molecules are synthesized continually or periodically, how their synthesis/activity is regulated, and whether other factors/molecules are involved.

In a bacterial cell, the number of copies of a given plasmid, per chromosome, is called the *copy number*. Some plasmids (e.g. the F and R6 plasmids) have a copy number of 1 or 2. Other (*multicopy*) plasmids (such as ColE1 and R6K) can have a copy number of e.g. 10–30. Copy number can be affected by mutations (see e.g. ColE1, above).

Copy number depends on the plasmid's replication control system, on the partition mechanism, if any (see later), on the bacterial strain and on growth conditions.

Some plasmids have a mechanism for ensuring their persistence within a bacterial strain. For example, the F plasmid encodes two proteins: CcdB and CcdA; together, these proteins constitute a system which results in the death of any cell which, following cell division, does *not* contain a plasmid. CcdB is a lethal toxin for the host cell and is a stable molecule. CcdA can neutralize CcdB, but CcdA is gradually degraded by the host cell's Lon protease; hence, the cell is protected by CcdA only while the latter is being synthesized from the (plasmid's) *ccdA* gene – which can occur only in those cells which actually contain the plasmid. Following cell division, both daughter cells will contain molecules of CcdB; if one cell lacks the F plasmid it will be killed by CcdB because there is no *ccdA* gene present from which the CcdA antidote can be transcribed.

Partition refers to the segregation of copies of replicated plasmids to daughter cells during cell division. In high-copy-number plasmids, partition seems likely to involve random distribution of the plasmids between daughter cells; in such cases there appear to be no special mechanisms for active partition.

Active partition is essential for low-copy-number plasmids in order to achieve a stable inheritance of plasmids in daughter cells and to avoid the segregation of plasmid-free cells at cell division. Such plasmids have special loci (designated e.g. *par*, *sop* or *sta*) which are associated with partition. Studies on the R1 plasmid in *Escherichia coli* have concluded that segregation involves a remarkable mitosis-like process that is mediated by actin-like filaments. In R1, the *par* (partitioning) locus includes: (i) *parC*, a centromere-like region, and (ii) the genes *parM* and *parR*. In a model for the partitioning process of R1, a pair of plasmids (formed at the mid-cell replication machinery of the host bacterium) become linked by the binding of the ParR protein to the *parC* region of each plasmid; *parC*-ParR-*parC* is referred to as the *partitioning complex*. The partitioning complex acts as a nucleation site on which ATP–ParM molecules polymerize to form actin-like filaments; the continual addition of ATP–ParM to the filaments causes them to lengthen, in an axial direction, concurrently forcing the plasmids towards opposite poles of the cell. Depolymerization of the filaments results from hydrolysis of the filament-bound ATP. Interestingly, ParM is related to the family of proteins that include (eukaryotic) actin and the MreB protein found in various rod-shaped bacteria (section 2.1.1); it has been suggested that MreB may be involved in the segregation of prokaryotic chromosomes. [Segregation of R1 plasmids in *E. coli*: EMBO Journal (2002) *21* 3119–3127.]

7.3.1.1 *Stringent and relaxed control of plasmid replication*

Many plasmids (including the F plasmid) fail to replicate if their host cells are treated with an antibiotic (e.g. chloramphenicol) which inhibits protein synthesis; the replication of such plasmids is said to be under *stringent control*. In order to replicate, plasmids in this category require *de novo* protein synthesis – i.e. synthesis of those proteins (e.g. Rep) which act as initiators of replication.

Other plasmids (e.g. ColE1) continue to replicate in the presence of antibiotics that inhibit protein synthesis – and can achieve higher-than-normal copy numbers; the replication of such plasmids is said to be under *relaxed control*. Plasmids in this category are able to replicate in the absence of *de novo* protein synthesis as they require only the (pre-formed) host proteins (such as RNA polymerase etc.).

7.4   DNA MODIFICATION AND RESTRICTION

In many (perhaps all) bacteria, DNA replication is followed by *DNA modification*: methylation of certain bases in *specific sequences of nucleotides* in each of the newly synthesized strands; in these sequences, the methyl group (CH_3-) is added at the N-6 position of adenine and/or the C-5 position of cytosine (see Fig. 7.3 for positions). Modification involves specific enzymes, methyltransferases (= 'methylases'), which use e.g. *S*-adenosylmethionine as a source of methyl groups.

Why is DNA methylated? Methylation protects the DNA from the cell's own *restriction endonucleases* (REs): enzymes which 'restrict' (i.e. cut) any DNA which lacks methylation in the appropriate sequence of nucleotides. A DNA duplex is not cleaved if at least one of the strands is methylated: immediately after replication, the newly synthesized (*un*methylated) daughter strand is normally protected from restriction by the (methylated) template strand until methylation occurs.

Different sequences of nucleotides are methylated in different strains of bacteria. Thus, each strain has its own strain-specific methylases and REs, both of which recognize a given sequence of nucleotides; for example, in *E. coli* the *Eco*RI methylase methylates a specific adenine residue:

$$CH_3$$
$$|$$
$$5'\text{-GAATTC-}3'$$

while the RE *Eco*RI cuts an (unmethylated) strand at:

$$5'\text{-G|AATTC-}3'$$

where '|' indicates the cut. Note that, in this example, the *complementary* strand is cut as follows: 3'-CTTAA|G-5'. This results in a *staggered* cut, i.e. each of the cut ends of the duplex has a single-stranded region (in this case 5'-AATT) (see Fig. 8.11). (See also Dam methylation: section 7.8.8.)

Many strains of bacteria contain more than one type of methylase and RE.

The sequence recognized by a given RE is called its *recognition sequence*. Most REs can be classified as type I, II or III according e.g. to the relationship between their recognition sequence and their cutting site; for example, type I REs cut at random sites distant from their recognition sequences, and type II REs cut between a specific pair of bases within the recognition sequence.

A different type of RE makes a staggered cut on *both* sides of the recognition sequence – thus cutting out a small piece of DNA from the target molecule; this small piece of DNA includes the recognition sequence. Cleavage by this type of RE typically requires magnesium ions and *S*-adenosylmethionine. An example of these REs is *Bae*I (from *Bacillus sphaericus*) [NAR (1996) *24* 3590–3592]; cleavage by *Bae*I is shown in Fig. 8.11.

REs are usually named as in the following example. *Eco*RI: an RE from *E. coli* strain R, 'I' indicating a particular RE from strain R. Other examples of REs are given in Table 8.1.

What is the purpose of restriction? Its main role seems to be to protect the cell from 'foreign' DNA – particularly phage DNA – which has entered the cell [restriction (review): MR (1993) *57* 434–450]. However, this protective mechanism is not always successful against phages: see section 9.6. Moreover, antirestriction proteins (offering protection against restriction) are known to be encoded by some plasmids [JB (1992) *174* 5079–5085]; such proteins may be needed to overcome the restriction of the recipient cell during conjugation (section 8.4.2). Antirestriction also operates in conjugative transposition (section 8.4.2.3).

'Non-classical' forms of restriction. Some REs (e.g. *Eco*RII) are active only if they bind simultaneously to *two* recognition sequences – which need not necessarily be on the same DNA molecule. (The need for two sites resembles the requirement of the enzymes involved in site-specific recombination – section 8.2.2.)

The RE *Dpn*I from *Streptococcus* (incorrect name: *Diplococcus*) *pneumoniae* is one example of an enzyme which cleaves only at a *methylated* site, i.e. it is a methyl-dependent RE (in contrast to the REs described above); *Dpn*I cleaves 5'-GATC-3' between A and T only if the adenine is methylated.

Restriction and modification in recombinant DNA technology. Type II REs (which need Mg^{2+} but not ATP) are widely used for cutting DNA at specific sites (section 8.5.1.3).

[Methods for studying DNA methylation (review) BioTechniques (2002) *33* 632–649.]

7.5 RNA SYNTHESIS: TRANSCRIPTION

In a cell, functions encoded by DNA are carried out by various types of RNA molecule. These RNA molecules are synthesized by the polymerization of ribonucleotides on a template strand of DNA – i.e. a process resembling the polymerization of deoxyribonucleotides when a new strand of DNA is synthesized (section 7.3). The process in which ribonucleotides are polymerized on a template strand of DNA is called *transcription*; an RNA molecule formed by transcription is called a *transcript*.

For a given protein to be synthesized, the corresponding *gene* – i.e. that sequence of nucleotides (in a strand of DNA) encoding the protein – must be copied in the form of a single strand of RNA; that is, the gene must be *transcribed*. The transcript of the gene is called *messenger RNA* (mRNA) because it carries the gene's message specifying the protein.

Before looking at transcription it's relevant to ask several questions. *When* is a gene transcribed? Are all genes transcribed all the time? What decides when a gene will be transcribed? These questions, which refer to the *expression* of a gene, are considered in section 7.8.

Transcription of a gene. RNA is copied from DNA by the sequential base-pairing of individual *ribo*nucleotides with the exposed bases on a single-stranded DNA template, the ribonucleotides being polymerized in the 5'-to-3' direction. Polymerization of ribonucleotides is carried out by the enzyme *RNA polymerase.* As in DNA synthesis, nucleoside *tri*phosphates base-pair with the template strand, the two terminal phosphate groups being cleaved to provide the energy for polymerization.

Note that, in RNA synthesis, adenine pairs with uracil, not with thymine.

In order to transcribe a gene the RNA polymerase binds at a specific sequence of nucleotides in (double-stranded) DNA at the beginning of the gene; this sequence of nucleotides is called a *promoter*. Within the promoter sequence there is usually the so-called *start site*, i.e. the first nucleotide at which transcription begins; this nucleotide is designated +1. Subsequent nucleotides along the gene (i.e. *downstream*) are numbered +2, +3 . . . etc. Nucleotides in the opposite direction (i.e. *upstream*) are designated −1, −2, −3 . . . etc.

Different genes can have different types (classes) of promoter, i.e. the promoter for a given gene may not have the same sequence of nucleotides as the promoter for another gene; moreover, even promoters within the same class can vary in their sequence of nucleotides. The +1 location is therefore a useful reference point for defining certain regions of a promoter sequence that are important sites of interaction with the RNA polymerase and various transcription factors.

In the most common type of promoter in *E. coli*, important regions of interaction are (i) a six-nucleotide *consensus sequence*, TATAAT, centred on, or close to, the −10 nucleotide; (ii) another consensus sequence, TTGACA, centred on/near the −35 nucleotide; (iii) the 'spacer' between these two regions; (iv) an AT-rich *UP element* located between nucleotides −40 and −60. (A *consensus sequence* is a theoretical 'representative' sequence of nucleotides in which the nucleotide at a given location occurs more often (than the other three types of nucleotide) in the various forms of that sequence which occur in nature.) Often, the closer the −10 and −35 hexamers are to their consensus sequences the stronger is the promoter, i.e. the more active or functional is the gene. UP elements enhance promoter activity; they are found in various organisms, including phages [NAR (2004 *32* 1166–1176].

Before binding to a promoter, the RNA polymerase must first bind to another protein called a *sigma factor*. The cell encodes more than one type of sigma factor, and a given type confers on the polymerase the ability to bind to promoters of a particular class. In *E. coli*, an RNA polymerase complexed with sigma factor σ^{70} can initiate transcription from most promoters in the cell. Some of the cell's sigma factors are synthesized only when it is necessary to transcribe particular genes (see section 7.8.9).

Following binding of the RNA polymerase *holoenzyme* (= RNA polymerase + sigma factor, often symbolized $E\sigma^{70}$, $E\sigma^{32}$ etc.) to the promoter, the strands of DNA separate in the region of the start site – forming the so-called 'open complex' which allows transcription to begin. The first ribonucleoside triphosphate base-pairs with the exposed +1 deoxyribonucleotide; subsequent ribonucleoside triphosphates base-pair (at +2, +3 etc.), each of these being cleaved to the monophosphate to provide energy for the phosphodiester bond. The RNA strand grows as more ribonucleotides are polymerized; shortly, the sigma factor is released and the RNA polymerase continues on its own to synthesize the remainder of the strand.

Elongation involves progressive unwinding of the DNA duplex and the sequential addition of ribonucleotides to the growing strand of RNA. As the process continues, the RNA strand peels away from the template strand and the DNA duplex re-forms.

Transcription stops at a specific sequence of nucleotides (a termination signal, or *terminator*) in the template strand. In so-called *rho-independent* termination, parts of the transcript of the terminator region base-pair with one another, and the resulting structure may cause the release of the transcript and/or polymerase. In *rho-dependent* termination, the *rho factor* may stop transcription after recognizing a particular site on the transcript.

One important aspect of transcription has not been mentioned. Following the development of an open complex, *which* of the two strands of DNA acts as a template? Clearly, it matters, because the strands are complementary (and are therefore different). Usually, one specific strand of DNA is used as template when a given gene is transcribed. In some organisms (e.g. bacteriophage T7) the same strand is used as template for all the genes. In other organisms (e.g. *Escherichia coli*) some genes are transcribed from one strand and some from the other; in such cases transcription occurs in both clockwise and anticlockwise directions in a circular chromosome.

The strand which acts as template for the synthesis of mRNA was originally called the *coding strand* or *sense strand*. This is a logical use of terminology because this strand is the one *from which* mRNA transcribes the gene's message. 'Coding strand' is used with this meaning in Fig. 8.26. (This strand is also called the *plus strand*.) Currently, a number of authors use 'coding strand' in the opposite way, i.e. to refer to the strand which is *not* used as a template for synthesis of mRNA.

Given the confusion which continues to surround this term it may be wise to define the term, in relation to mRNA, whenever it is used. Interestingly, a certain gene in the fruitfly (*Drosophila*) contains protein-encoding information in *both* strands (the term 'coding strand' thus being redundant in this case) [Nature (2001) *409* 1000].

The foregoing is only an outline of a complex sequence of events. Further information is available on polymerase–promoter interactions [JB (1998) *180* 3019–3025] and on the role of RNA polymerase during elongation [JB (1998) *180* 3265–3275].

Divergent transcription. When different genes on the same DNA molecule are transcribed in opposite directions from closely spaced promoters (= divergent transcription), the generation of supercoiling behind each advancing polymerase can cause a local increase in superhelicity which may affect the expression of genes or operons [MM (2001) *39* 1109–1115].

7.5.1 Regulation of transcription: other factors

Section 7.5 gives a simplified account of transcription. Often, transcription of gene(s) – or the inhibition of transcription – involves the activity of proteins called *transcription factors*; these factors bind to specific sequence(s) in the DNA, either close to or distant from the promoter, and exert a positive or negative regulatory influence on transcription. One example of these *DNA-binding proteins* is the repressor (regulator) protein of the *lac* operon (Fig. 7.11). In *E. coli* the homodimeric leucine-responsive regulator protein (Lrp) is involved in the regulation of a range of genes and operons, apparently by acting as an activator (or, sometimes, repressor) of transcription [Lrp in *Escherichia coli*: JBC (2002) *277* 40309–40323].

To study these (and other) protein–DNA interactions, use is made of techniques such as Southwestern blotting and footprinting (section 8.5.12).

7.6 PROTEINS: SYNTHESIS AND OTHER ASPECTS

All the cell's proteins are encoded by DNA: by studying protein synthesis we can see *how* DNA carries this information – and how the genetic code works.

A protein consists of one or more *polypeptides*, a polypeptide being a chain of amino acids covalently linked by peptide bonds (–CO.NH–). Each polypeptide is folded into a three-dimensional structure; this structure is stabilized mainly by hydrogen bonds or disulphide bonds formed between amino acids in different parts of the chain. The specific three-dimensional structure of a given polypeptide – essential for biological activity – is determined by the nature, number and sequence of its amino acids.

For any given polypeptide, the nature, number and sequence of amino acids are dictated by a particular sequence of bases in a DNA strand; this sequence of bases conforms to one definition of a *gene*.

How does a gene bring about the synthesis of a polypeptide? That is, how is the gene *expressed*? Unlike nucleotides, amino acids cannot simply 'line up' (undergo polymerization) on a DNA template strand. In fact, protein synthesis involves several stages. First, the gene is *transcribed* (section 7.5); the (RNA) transcript of the gene, which carries the message from DNA, is called *messenger RNA* (mRNA). Now, along the length of the mRNA molecule, groups of three consecutive bases each encode a particular amino acid; each of these three-base groups is called a *codon*. Thus, e.g. the codon UCA (uracil–cytosine–adenine) encodes the amino acid serine (Table 7.2). Hence, the sequence of codons in mRNA encodes the sequence of amino acids in a polypeptide. For the synthesis of a polypeptide, each amino acid must first bind to an

Table 7.2   The 'universal' genetic code: amino acids and 'stop' signals encoded by particular codons[1-5]

First base (5' end)	Second base				Third base (3' end)
	U	C	A	G	
U	Phe	Ser	Tyr	Cys	U
	Phe	Ser	Tyr	Cys	C
	Leu	Ser	*ochre*	*opal*	A
	Leu	Ser	*amber*	Trp	G
C	Leu	Pro	His	Arg	U
	Leu	Pro	His	Arg	C
	Leu	Pro	Gln	Arg	A
	Leu	Pro	Gln	Arg	G
A	Ile	Thr	Asn	Ser	U
	Ile	Thr	Asn	Ser	C
	Ile	Thr	Lys	Arg	A
	Met	Thr	Lys	Arg	G
G	Val	Ala	Asp	Gly	U
	Val	Ala	Asp	Gly	C
	Val	Ala	Glu	Gly	A
	Val	Ala	Glu	Gly	G

[1] A = adenine; C = cytosine; G = guanine; U = uracil.
[2] Standard abbreviations are used for the amino acids: Gln = glutamine, Ile = isoleucine etc.
[3] As with other nucleotide sequences, codons are conventionally written in the 5'-to-3' direction; as anticodons also are written in this way, the *first* base in a codon pairs with the *third* base in its anticodon.
[4] The codon UAA (*ochre*), UGA (*opal*) and UAG (*amber*) are normally 'stop' signals (see text).
[5] Although called 'universal', a number of exceptions to the code have been reported. In *Escherichia coli*, for example, one codon base-pairs with a tRNA carrying serine, but the serine is converted to selenocysteine before incorporation in the polypeptide (*co-translation*) [Nature (1988) *331* 723–725].

adaptor molecule which is specific for that particular amino acid and for its codon; these adaptor molecules are small molecules of RNA – *transfer RNA* (tRNA) – and each binds to its specific amino acid to form an *aminoacyl-tRNA*. (Such binding requires ATP.) As well as a binding site for the amino acid, a given tRNA molecule also contains a sequence of three bases (an *anticodon*) which is complementary to the codon of its particular amino acid; hence, a given tRNA molecule can bind its particular amino acid and can link it to the specific codon in mRNA through codon–anticodon base-pairing.

The synthesis of a polypeptide on mRNA (the *translation* process) takes place on a ribosome (section 2.2.3); a simplified version of this process is shown diagrammatically, and explained, in Fig. 7.9.

As shown in Fig. 7.9, the *initiator codon* (i.e. the first codon to be translated) is commonly AUG. The alignment of AUG with the ribosomal P site is generally promoted by a particular sequence of nucleotides (the *Shine–Dalgarno sequence*) located upstream of AUG on the mRNA (i.e. to the left of AUG in Fig. 7.9); this sequence base-pairs with part of a 16S rRNA molecule in the ribosome.

When acting as an initiator codon, AUG encodes the modified amino acid *N*-formylmethionine in most or all bacteria (including *E. coli*). The initiator codon base-pairs with a specialized *initiator tRNA* which carries *N*-formylmethionine; formylation of the methionine occurs in a reaction requiring N^{10}-formyltetrahydrofolate (Table 6.1). Note that, after completion of translation, either the formyl group or formylmethionine itself is enzymically removed from the free polypeptide. In *E. coli*, formylmethionine is cleaved from about 50% of proteins; whether or not cleavage occurs in a given protein depends apparently on the identity of the *penultimate* N-terminal amino acid – as a generalization: the longer the side-chain on the penultimate amino acid the smaller the probability of cleavage of formylmethionine. Cleavage of formylmethionine is mediated by *methionine aminopeptidase* (MAP), one of a family of physiologically important enzymes [properties and functions of aminopeptidases: FEMSMR (1996) *18* 319–344]. Cleavage of the formyl group (only) is mediated by the enzyme *methionine deformylase* (= peptide deformylase, PDF). (PDF is inhibited by some derivatives of hydroxamic acid, but these derivatives are unlikely to be useful candidate antibiotics because (i) they were bacteriostatic in the organisms tested, and (ii) resistance develops readily [AAC (2001) *45* 1058–1064].)

The initiation of protein synthesis, outlined above, requires the participation of three protein *initiation factors* – IF-1, IF-2 and IF-3 – but the timing and mode of their involvement are not yet fully understood, and the literature contains several different models for this phase of protein synthesis. It is generally agreed, however, that IF-2 and GTP are needed for the binding of tRNA–*N*-formylmethionine to AUG at the P site; that this tRNA binds to the 30S subunit *before* addition of the 50S subunit (i.e. between steps (a) and (b) in Fig. 7.9); and that all three initiation factors have been released by completion of step (b) in Fig. 7.9. During the formation of the 70S complex, IF-2 acts

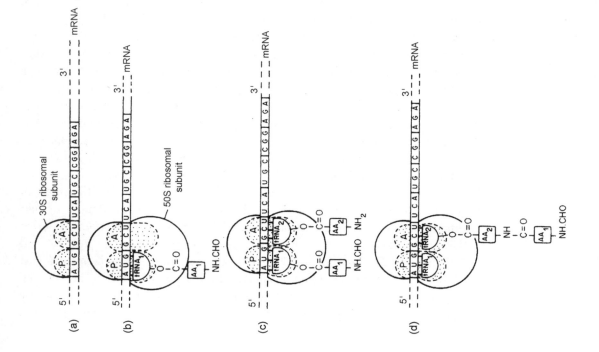

Figure 7.9 Protein synthesis in bacteria (simplified, diagrammatic) (see section 7.6). (a) The 30S ribosomal subunit binds to mRNA; the *decoding region* of 16S rRNA (in the subunit) coincides with the 'P' and 'A' sites. The *initiator codon* (AUG) of the mRNA occupies the P site.

(b) The first amino acid (AA$_1$), on its tRNA (tRNA$_1$), occupies the P site, i.e. the anticodon of tRNA$_1$ base-pairs with codon AUG. The 50S ribosomal subunit binds, completing the ribosome.

(c) The second aminoacyl-tRNA occupies the A site, i.e. the anticodon of tRNA$_2$ base-pairs with codon GCU.

(d) A peptide bond is formed between the carboxyl group of AA$_1$ and the α-amino group of AA$_2$; this is called *transpeptidation*. (Note that AA$_1$ will form the 'amino end' or 'N-terminal' of the polypeptide chain.) The enzyme which catalyses transpeptidation is *peptidyltransferase*; this enzyme is actually part of the 50S ribosomal subunit (and, incidentally, is the site of action of certain antibiotics – see section 15.4.4).

(e) The ribosome moves along the mRNA, by one codon, in the 5'-to-3' direction (*translocation*). tRNA₁ is released. The dipeptidyl tRNA now occupies the P site, and there is a vacant A ('acceptor') site opposite the third codon.

(f) A third aminoacyl-tRNA binds at the A site. Steps d–f are repeated for each codon in turn. When a stop codon (e.g. UAA) is reached, a protein *release factor* hydrolyses the ester bond between the last tRNA and the polypeptide chain – thus releasing the completed chain.

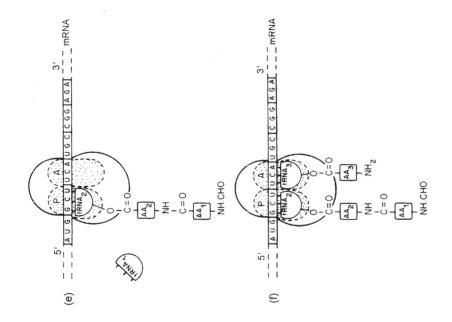

as a GTPase, the association of the 50S and 30S subunits being linked to hydrolysis of GTP. [Bacterial initiation factors in protein synthesis (review): MM (1998) *29* 409–417.]

Elongation of the polypeptide chain proceeds by sequential addition of amino acids according to the codons which follow AUG. The first step in elongation is the binding of an aminoacyl-tRNA to the vacant A site (Fig. 7.9, step c); in *E. coli* this involves GTP and the *elongation factor* EF-Tu. The binding of the aminoacyl-tRNA is followed by *transpeptidation* and *translocation* (Fig. 7.9, d–e); steps c–e are repeated throughout translation, codons generally being translated according to the *genetic code* (Table 7.2).

Base-pairing between mRNA codons and tRNA anticodons is actually not stable enough, in itself, to guarantee the degree of accuracy and efficiency needed in translation. The ribosome must therefore provide help, and this appears to be given by the rRNA [see e.g. Nature (1994) *370* 597–598].

Termination of translation (the end of polypeptide synthesis) is signalled by the presence, at the A site, of one of the *stop codons* (= *nonsense codons*): UAA, UAG and UGA (*ochre, amber* and *opal* codons, respectively). Termination involves the intervention of a protein *release factor* which must recognise the stop signal and interact directly with the mRNA; this causes hydrolysis of the polypeptide–tRNA bond at the P site, thus releasing the polypeptide. Although the stop signal is traditionally regarded as a *triplet* of bases (as above), there is evidence that the base *following* a stop codon influences the efficiency of termination, and it may be that the actual stop signal is a 4-base (or even longer) sequence [MM (1996) *21* 213–219]. In *E. coli*, the efficiency of termination at a stop codon is influenced by the 5' (upstream) nucleotide context of the codon and by the C-terminal amino acid residues in the nascent polypeptide. [Translation termination and stop codon recognition (review): Microbiology (2001) *147* 255–269.]

As one ribosome translocates along the mRNA another can fill the vacated initiation site and start translation of another molecule of the polypeptide; thus, a given mRNA molecule may carry a number of ribosomes along its length) – forming a *polyribosome* (= *polysome*) (Fig. 7.10).

Signal sequences. In many cases, a protein destined to form part of a membrane, or to pass through a membrane, is synthesized with a special N-terminal sequence of amino

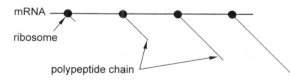

Figure 7.10 A polysome (diagrammatic). As a given ribosome travels further along the mRNA molecule its associated polypeptide chain increases in length.

acids called a *signal sequence*; this helps passage into, or through, the hydrophobic region of the membrane and is enzymically cleaved as the protein passes through the membrane (see e.g. *sec*-dependent transport: section 5.4.3). [Signal sequences (review): MM (1994) *13* 765–773.]

Experiments in which the signal sequence of a protein is changed to that of another type of protein have shown that signal sequences can determine the pathway along which a protein is exported [JB (2003) *185* 5706–5713].

Protein folding. A polypeptide chain (or chains) must be folded correctly to form a functional protein. Folding involves the development of e.g. hydrogen and/or disulphide bonds between specific amino acid residues in the polypeptide chain (these bonds holding the polypeptide in a specific three-dimensional structure).

In which part of the cell does folding occur? All proteins are synthesized (on ribosomes) in the cytoplasm, but only some of the cell's proteins remain (and are folded) within the cytoplasm; many proteins are exported through the cytoplasmic membrane or are secreted through the cell envelope. In *E. coli*, the *cytoplasmic* proteins (i.e. those remaining within the cytoplasm) generally do not contain *disulphide* bonds as structural features; this is because, in the cytoplasm, such bonds are normally reduced by the NADPH-dependent thioredoxin system [but see EMBO Journal (1998) *17* 5543–5550].

Some proteins are folded within the cytoplasm prior to export across the cytoplasmic membrane (section 5.4.7), and some are folded prior to secretion (e.g. sections 5.4.1.2, 5.4.4 and 5.4.6).

Proteins exported via the *sec* or SRP pathways (section 5.4.3) must remain *unfolded* during transport through the membrane; thus, for example, the SecB protein seems specifically to inhibit folding prior to transmembrane translocation. Within the periplasm, these proteins are commonly folded by the introduction of disulphide bonds catalysed by the Dsb proteins (products of genes *dsbA* and *dsbB*). (Mutations affecting the Dsb proteins are *pleiotropic*, i.e. they have multiple effects; this is because disulphide bonds are involved in the stabilization of different types of protein – e.g. flagellar proteins, membrane proteins and various secreted proteins. In pathogens, *dsb* mutations may affect secreted protein virulence factors, and this may reduce the virulence of the pathogen.) [Oxidative folding of proteins in the *E. coli* periplasm by DsbA: Science (2004) *303* 534–537.]

Proteins containing *proline* residues may require peptidyl-prolyl isomerases for normal folding; in *E. coli* several such isomerases (e.g. FkpA, PpiA) occur in the periplasmic region.

Synthesis of at least some protein-folding enzymes is apparently controlled by a *two-component regulatory system* (section 7.8.6) designated CpxA-CpxR. Activation of CpxA, the sensor, phosphorylates the response regulator, CpxR. CpxR~P binds upstream of the transcriptional start site in e.g. *dsbA* and *ppiA*; activation of the Cpx system correlates with increased *in vivo* transcription of both of these genes [GD (1997) *11* 1169–1182].

The folding of (usually newly synthesized) proteins, or stabilization of the unfolded state, is facilitated by so-called *molecular chaperones* (usually referred to simply as 'chaperones'); these are molecules that bind to a protein and either promote or inhibit folding. A given stage of folding, or of translocation, may require more than one type of chaperone. A chaperone may bind to a protein co-translationally, i.e. during translation [see e.g. JBC (1997) *272* 32715–32718] or post-translationally, i.e. after completion of translation.

Chaperones are produced constitutively, but under certain stress conditions (e.g. heat shock: 7.8.2.3) they are produced in greater quantities – probably for tasks such as re-folding denatured proteins.

Interestingly, the peptidyl-prolyl isomerase FkpA (see above) is reported to have an additional chaperone-like function [MM (2001) 39 199–200].

Chaperones are typically proteins; in *E. coli* they include the products of genes *groES*, *groEL* and *dnaK*. However, the membrane *lipid* phosphatidylethanolamine (PE) (Fig. 2.5) acts as a chaperone for the *E. coli* protein LacY (section 7.8.1.1); apparently, the early stages of folding of this protein (in the cytoplasmic membrane) are independent of particular lipids but later stage(s) require help specifically from PE [EMBO Journal (1998) *17* 5255–5264].

Unfolded/misfolded proteins. Unfolded/misfolded proteins in the cell's *cytoplasm* can be dealt with by a system, involving σ^{32}, which is described under heat-shock response (section 7.8.2.3).

A different mechanism operates in the *periplasmic* region in the presence of an unfolded outer membrane porin protein. In this situation, certain motifs on the C-terminal region of the porin activate a protease, DegS, located on the periplasmic side of the cytoplasmic (inner) membrane. Once activated, DegS initiates the cleavage of a complex of inner membrane proteins which, under non-stress conditions, sequester an inactive form of the sigma factor σ^E the cytoplasmic face of the inner membrane; on cleavage of these proteins, σ^E is released into the cytoplasm and activates the periplasmic stress response genes [Cell (2003) *113* 61–71].

Peptide synthesis without ribosomes. A number of very short polypeptides are synthesized by a multi-enzyme complex instead of by ribosomes and RNA molecules. This process is used e.g. for the synthesis of certain antibiotics such as *tyrocidins* (cyclic peptides containing 10 amino acid residues) and *gramicidins* (linear peptides); these antibiotics (produced by *Bacillus brevis*) act against certain Gram-positive bacteria, apparently by altering the permeability of the cytoplasmic membrane. A tripeptide *siderophore* (section 11.5.5), encoded by *Streptomyces coelicolor*, is produced by a non-ribosomal synthetase [FEMSML (2000) *187* 111–114]. [Non-ribosomal biosynthesis of peptide antibiotics (review): EJB (1990) *192* 1–15. Non-ribosomal synthesis of macrocyclic peptides: JB (2003) *185* 7036–7043.]

7.6.1   The fate of mRNA

After use, many *bacterial* mRNAs are short-lived, being rapidly degraded to re-cyclable components; this is essential: the cytoplasm would otherwise rapidly fill up with 'used' mRNA molecules. (In some cases, however, the degradation of mRNA is delayed – see section 7.8.4.) Degradation involves the action of *nucleases*: enzymes which cleave nucleic acids; the endonucleases cleave phosphodiester bonds in non-terminal positions, while exonucleases sequentially cleave terminal nucleotides.

E. coli contains membrane-associated multicomponent organelles which degrade/process RNA. These so-called *degradosomes* contain e.g. RNase E (an endoribonuclease); polynucleotide phosphorylase (= polyribonucleotide nucleotidyltransferase) which degrades mRNA from the 3' end; enolase; RhlB (RNA helicase); polynucleotide phosphate kinase; DnaK and GroEL. As well as degrading RNA, degradosomes also have essential roles in processing RNA; for example, RNase E is involved in the formation of 5S rRNA from pre-rRNA.

[RNA degradosomes: PNAS (2001) *98* 63–68.]

Degradation of mRNA can be affected by growth conditions. In *E. coli*, for example, the half-life of at least some mRNAs is greatly increased during slow anaerobic growth (doubling time: 700 minutes); under these conditions the rate of *synthesis* of mRNA is much slower, so that the increased stability of mRNA may be a mechanism for maintaining gene expression during anaerobiosis [MM (1993) *9* 375–381].

Polyadenylation may be an important factor in the stability of mRNAs. Although it's well known that most *eukaryotic* mRNAs have a 3' polyadenylate 'tail' (i.e. . . . -A-A-A-A-A-3'), the existence of polyadenylated *bacterial* mRNAs is less widely acknowledged; however, in *E. coli* there are at least two enzymes (poly(A)polymerases, PAPs) which carry out polyadenylation.

The function of polyadenylation is not known, but it may help e.g. in regulating the half-life (stability) of those RNA molecules whose 3' terminus is a stem–loop structure encoded by a rho-independent transcription terminator (section 7.5); a stem–loop tends to resist the action of certain RNA-degrading enzymes, and one suggestion is that polyadenylation at the 3' end of the stem–loop may facilitate degradation of these molecules by providing a suitable binding site for degradative enzymes such as the 3'-exonuclease polynucleotide phosphorylase.

[Polyadenylation of bacterial mRNAs (review): Microbiology (1996) *142* 3125–3133. The *E. coli* poly(A)polymerase: NAR (2000) *28* 1139–1144. Polyadenylation in mycobacteria: Microbiology (2000) *146* 633–638.]

7.6.2   Proteins within proteins: inteins

In 1990 it was discovered that certain proteins contain an internal sequence of amino acids – an *intein* – which apparently catalyses a *self-splicing* reaction in the protein: the intein is excised (forming a separate protein) and the two terminal parts of the original polypeptide (the *exteins*) are joined to form a functional protein. Thus:

$$\text{extein–intein–extein} \rightarrow \text{extein–extein} + \text{intein}$$

This phenomenon was first detected in the *VMA1* (= *TFP1*) gene of the (eukaryotic) yeast *Saccharomyces cerevisiae*, but was later found in bacteria (e.g. the *recA* gene in *Mycobacterium tuberculosis*) and in members of the Archaea (e.g. the DNA polymerase gene in *Pyrococcus* sp and *Thermococcus litoralis*).

Inteins are typically about 350–550 amino acids long. Their essential role in autocatalytic self-splicing is inferred from (i) loss of splicing following deletions in the intein-coding DNA, and (ii) 'heterologous expression' of an intein-coding gene, i.e. expression of such a gene in an organism which does not normally contain that gene; thus, expression of the *VMA1* gene in *E. coli* produced both mature protein and intein.

During self-splicing, an intermediate product is a *branched* polypeptide (with two N terminals and one C terminal); this is reminiscent of the branched mRNA formed, in eukaryotes, during the processing of pre-mRNA to mature mRNA, and there are indeed parallels between the roles of inteins and those of certain introns.

Some excised inteins have been shown to act as site-specific endonucleases (as have some translated introns). Moreover, the excised *VMA1* intein of *S. cerevisiae* will specifically cleave a *VMA1* gene lacking an intein-coding sequence, cleavage occurring at the site normally occupied by intein DNA. In a strain of *S. cerevisiae* with one intein-coding *VMA1* gene and one intein-less copy of the gene, cleavage of the latter gene was followed by insertion of intein DNA by a process (called *gene conversion*) in which intein DNA was copied from that in the intein-containing gene.

The ability of an intein to promote the insertion of its coding sequence into an intein-less copy of the given gene (as in *S. cerevisiae*, above) is called *intein homing*. Like transposons (section 8.3), inteins are referred to as *mobile genetic elements*.

[Inteins (review): TIBS (1995) *20* 351–356]. Compilation and analysis of intein sequences: NAR (1997) *25* 1087–1093.]

7.6.3   Ribosomes versus growth rate

Faster growth clearly involves an increased rate of protein synthesis, and this, in turn, requires more ribosomes per cell; we therefore find a relationship between growth rate and the rate of synthesis of ribosomes.

The production of ribosomes requires co-ordinated synthesis of their protein and rRNA components (section 2.2.3). Genes encoding the 50 or so ribosomal proteins are grouped into a number of distinct *operons* (section 7.8.1), each operon transcribed as a single molecule encoding a particular set of proteins (i.e. polycistronic mRNA). One model for the control of these operons postulates that one of the proteins encoded by a given operon acts as a regulatory molecule for that operon by inhibiting *translation* of at least some of the proteins encoded by the polycistronic mRNA; the control mechanism in one of these operons in *E. coli* is considered in section 7.8.1.3.

Some of the operon-regulatory proteins also have binding sites on either 16S or 23S rRNA. It is thought that, if the growth rate decreases, the decreased amount of 16S and 23S rRNA available in nascent ribosomes would leave these proteins free to bind to their respective transcripts and to carry out an inhibitory function. Note that, in this model, the synthesis of ribosomal proteins is linked to synthesis of rRNA.

Synthesis of rRNA is linked to the cell's translational needs and is thus susceptible to up-regulation or down-regulation according to conditions. In *E. coli*, the 16S, 23S and 5S rRNAs are co-transcribed (in that order), as a single transcript, from the *rrn* operon; the chromosome contains seven copies of this operon (A–E, G, H), each containing one copy each of the three kinds of rRNA – except the D operon, which has two copies of the 5S rRNA gene. Any reduction in the levels of 16S or 23S rRNA (due e.g. to mutation) can be compensated for by up-regulation of the *rrn* operons. However, deletion of two or more copies of the 5S rRNA gene in *E. coli* causes a sharp drop in growth rate (i.e. no compensatory effect), this reduction in growth rate being almost reversible by insertion of a plasmid-borne 5S rRNA gene [NAR (1999) *27* 637–642].

7.6.4 Genes and pseudogenes

Section 7.6 describes how a sequence of deoxyribonucleotides – referred to as a *gene* – can direct the synthesis of a protein; note that, in addition to the *coding* sequence (which specified the polypeptide), the gene necessarily included various control sequences, such as the promoter and terminator, without which it could not be expressed. This simple (yet accurate) example of the flow of information (from nucleic acid to protein) nevertheless describes only one of the ways in which genes function. For example, transcription may not occur at all unless certain regulatory factor(s) are present or absent (section 7.5.1). Even when mRNA is formed, it may not be translated directly – it may need excision of *introns* (section 7.9). Moreover, the polypeptide itself may undergo so-called post-translational modification.

In some cases two or more (contiguous) genes are transcribed jointly from a single promoter or control region; mRNA from such an *operon* (section 7.8.1) may consist of a single molecule (*polycistronic* mRNA) containing information from all the genes in the operon.

The type of gene that encodes a polypeptide (e.g. an enzyme) is called a *structural* gene. There are also genes that encode molecules of RNA (rRNA, tRNA) as end products; thus, the *rrn* operon in *E. coli* encodes the 16S, 23S and 5S rRNAs. Finally, *regulatory* genes are required for controlling/regulating/co-ordinating the expression of other genes – *parC* (section 7.3.1) is an example.

Some chromosomal sequences of nucleotides resemble known, functional genes but are nevertheless inactive (not expressed); in some cases there may be deleterious mutations in the coding and/or control regions. Such sequences are called *pseudogenes*. Pseudogenes occur in both eukaryotes and prokaryotes. In *Mycobacterium leprae*, which appears to exemplify reductive evolution, 27% of the genome is reported to consist of recognizable pseudogenes [Nature (2001) *409* 1007–1011].

7.6.5   Proteomics

Proteomics is the study of a cell's *proteome*: collectively, all the proteins (encoded by the genome) which the cell is able to express. Some researchers restrict proteomics to the study of proteins while others include, in addition, genetic data (e.g. genomics). The objective of proteomics is to gain an overall, integrated view of the cell's biology by studying the full range of a cell's proteins (instead of studying them separately). One aim is to achieve a three-dimensional picture of the locations of all the proteins in a cell, although – given the dynamics of a living cell – any one picture can give only an instantaneous view.

Studies conducted under the heading of proteomics include e.g. determination of protein localization and protein function, protein–protein interactions and post-translational modification of proteins.

Proteomics can yield information not obtainable from genomic studies, and protein-based data can complement genomic data e.g. in determining the total number of genes in a given genome.

[Proteomics (review) MMBR (2002) *66* 39–63. Initial proteome analysis of model microorganism *Haemophilus influenzae* strain Rd KW20: JB (2003) *185* 4593–4602.]

7.7   DNA MONITORING AND REPAIR

Abnormal DNA can result e.g. from the insertion of abnormal nucleotides during replication; it may be recognized and repaired immediately through *proof-reading*: DNA polymerase III can cleave a 'wrong' nucleotide from the 3' end of a growing strand, allowing replacement with a normal one.

DNA is also vulnerable to chemical change owing to the reactivity of its bases [Nature (1993) *362* 709–715]. For example, aberrant, non-enzymatic, spontaneous methylation by *S*-adenosylmethionine (normally a legitimate methyl donor: section 7.4) can produce 3-methyladenine and/or 7-methylguanine; each of these aberrantly methylated purine bases can interfere with DNA function. In *E. coli*, these aberrant bases are excised by *N*-glycosylases: enzymes which cleave the sugar–base linkage.

DNA glycosylase I (= Tag protein), which is synthesized constitutively, excises 3-methyladenine. Additionally, the inducible enzyme DNA glycosylase II (= AlkA protein) can excise not only both of these aberrantly methylated purines but also certain aberrantly methylated pyrimidines. (The AlkA enzyme is part of a DNA repair system (*adaptive response*) which is induced in *E. coli* on exposure to certain alkylating agents; the adaptive response also includes the Ada protein: a bifunctional methyltransferase which directly reverses the effects of the methylating agent.) Studies on the crystal structure of AlkA, complexed with DNA, indicate that the enzyme distorts the DNA considerably as it 'flips out' (i.e. exposes) the target site [EMBO Journal (2000) *19* 758–766].

The spontaneous, low-rate deamination of cytosine to uracil (Fig. 7.3) is repaired by the initial excision of uracil by the enzyme uracil-*N*-glycosylase (UNG). (UNG has a useful application in technology: section 8.5.4.3.)

Guanine, oxidized to 8-oxoguanine (oxoG) by reactive oxygen species, mispairs with adenine; adenine is excised specifically by the MutY glycosylase [Nature (2004) *427* 598; 652–656].

Excision of a chemically aberrant base by a glycosylase (the first stage of repair) leaves an *apurinic* (purine-less) or *apyrimidinic* (pyrimidine-less) site termed an *AP site* (= abasic site). The repair process then continues via the *base excision repair* pathway (section 7.7.1.3).

In some cases, deamination of cytosine will occur *after* it has been methylated at the 5-position by *modification* (section 7.4); if *5-methylcytosine* undergoes (spontaneous) deamination, the result is *thymine* – a normal base (Fig. 7.3) which is not removed by any enzyme. Hence, the thymine remains, and at the next round of DNA replication it will pair with adenine; thus, a *point mutation* (section 8.1.1; Fig. 8.1) is created at this site: CG in the original duplex is replaced by TA in one of the daughter duplexes. Accordingly, 5-methylcytosines are 'hotspots' of spontaneous mutation. (The replacement of one pyrimidine with another, or one purine with another, is called a *transition mutation*; the replacement of a purine with a pyrimidine, or vice versa, is called a *transversion mutation*.)

7.7.1   Excision repair systems

7.7.1.1   *Mismatch repair (Dam-directed mismatch repair, DDMR)*

During DNA replication, errors that escape proof-reading may be corrected soon afterwards – *before* modification (section 7.4) has occurred. Such repair involves the *mismatch repair system* whose main function is to repair errors that arise during replication.

In *E. coli*, a mismatched base-pair is detected by the MutS protein. (It appears that *asymmetry* of ATPase sites on the MutS dimer is crucial for the mismatch repair system and controls timing of the repair cascade [EMBO Journal (2003) *22* 746–756].) MutS activates the endonuclease MutH (an enzyme which, incidentally, exhibits sequence homology with the restriction endonuclease *Sau*3A1 (Table 8.1) [EMBO Journal (1998) *17* 1526–1534]). [Mutational analysis of the *E. coli* MutH protein: JBC (2001) *276* 12113–12119.]

MutH cleaves the new daughter strand (identified by its transient undermethylation) at a site 5' to an unmethylated GATC (Dam) sequence that may be up to 1000 nucleotides distant from the mismatch site; co-ordination between the site of mismatch and the GATC cleavage site appears to be mediated by protein MutL.

After nicking by MutH, the section of daughter strand between the nick and a site beyond the mismatch is removed. It has been assumed that removal of this section of daughter strand requires helicase II (MutU, also referred to as UvrD) and a DNA exonuclease; however, mutant strains lacking the 3'-to-5' exonuclease Exo I *and* the 5'-to-3' exonucleases Exo VII and RecJ appear to carry out mismatch repair normally [JB (1998) *180* 989–993]. The single-stranded gap is filled in by a DNA polymerase (using the parent strand as template) and is sealed by a ligase.

Strains of *E. coli* lacking *mut* products are defective in DNA repair, and they exhibit a mutator phenotype (see also sections 8.1 and 8.5.5.1). Interestingly, in humans, certain types of cancer have been linked to mutations in mismatch repair genes [DNA mismatch repair genes and colorectal cancer: Gut (2000) *47* 148–153].

7.7.1.2 *Nucleotide excision repair (UvrABC-mediated repair)*

In *E. coli*, an ATP-dependent system (UvrABC; 'ABC excinuclease') recognizes and repairs DNA which has been damaged/distorted by ultraviolet radiation or by certain other causes; UvrABC consists of three proteins encoded by genes *uvrA*, *uvrB* and *uvrC*. Initially, a UvrA dimer binds to damaged DNA. The dimer then binds UvrB, and UvrA$_2$B moves to the precise site of damage. UvrA is displaced by UvrC, and the UvrBC complex catalyses a nick on either side of the damaged site, the 3' nick being made before the 5' nick. Helicase II (UvrD) then displaces UvrC together with the short sequence of nucleotides containing the site of damage. A polymerase then fills the single-stranded gap, displacing UvrB, and a ligase completes the repair. In *E. coli* only about 10 nucleotides in and around the damaged site are involved – hence the name *short patch repair*.

Energy (from ATP hydrolysis) is apparently required e.g. for the translocation of the UvrA$_2$B complex to the damaged site; the (zinc-binding) UvrA protein is an ATPase.

7.7.1.3 *Base excision repair (repair following glycosylase activity)*

Some systems operate primarily to repair the damage caused by particular types of agent – e.g. oxidative damage [ARB (1994) *63* 915–948] or damage caused by aberrant alkylation or deamination of bases (section 7.7).

An aberrant base is initially excised, by a glycosylase, to form an AP site (section 7.7). Then, the phosphodiester bond on one side of the AP site is cleaved by a so-called AP endonuclease; in *E. coli* the major AP endonuclease is exonuclease III, a multifunctional enzyme whose structure suggests that cleavage of phosphodiester bonds involves a nucleophilic attack on the P–3'O link [Nature (1995) *374* 381–386]. A DNA polymerase can then replace the damaged nucleotide; as well as the damaged nucleotide, several nucleotides 3' of the site of damage may also be replaced [NAR (1998) *26* 1282–1287]. A ligase completes the repair.

7.7.2 Photolyase

One type of damage to DNA caused by ultraviolet radiation is the formation of *thymine dimers*; a thymine dimer is produced by chemical bonding between two adjacent thymine residues in a given strand. Some bacteria, including *E. coli*, encode an enzyme, photolyase, which can use the energy in (visible) light to repair the damage.

7.7.3 Priority repair of actively transcribed genes

Actively transcribed genes are repaired more rapidly than others. One model proposes that, during transcription, aberrant DNA blocks the movement of RNA polymerase and that a transcription-repair coupling factor (TRCF), encoded by gene *mfd* (mutation frequency decline), binds at the affected site, causing release of the polymerase and the unfinished mRNA; TRCF may then interact with the UvrABC system – which effects repair. [Minireview: JB (1993) *175* 7509–7514.]

7.7.4 Other repair systems for DNA

Certain forms of damage can be repaired through homologous recombination (section 8.2.1).

7.8 REGULATION OF GENE EXPRESSION

A cell does not express all of its genes all of the time. For example, if a particular substrate were *not* available the cell would be wasting energy if it synthesized those proteins (e.g. enzymes) needed for metabolizing that substrate. In fact, many genes can be switched on (*induced*) or switched off (*repressed*) – or their expression increased or decreased – according to the cell's requirements and/or external conditions. Such regulation often involves a mechanism which promotes or inhibits synthesis of the gene's mRNA (transcription); in other words, gene expression is often regulated 'at the level of transcription'. Transcription can be regulated in various ways – for example: by specific DNA-binding proteins (section 7.5.1); by the availability of sigma factors (section 7.8.9); and by the presence of uncharged tRNA molecules (section 7.8.10). In some cases the product of a given gene can regulate its own gene (*autoregulation*); for example, in *Rhizobium leguminosarum* the product of the *nodD* gene, NodD, competes with RNA polymerase for a binding site in the gene's promoter region – resulting in negative autoregulation [NAR (2000) *28* 2784–2793].

Other ways of regulating gene expression are also described in the following pages.

Apart from external influences, the expression of a gene is inherently regulated by the *strength* of its promoter: strong promoters facilitate transcription while weak promoters make for low-level expression. However, the strength of a promoter is not necessarily fixed: sometimes the level of transcription from a given promoter can be altered by specific regulatory factors.

Normal gene expression is dependent on a correct degree of supercoiling (section 7.2.1). Less-than-normal negative supercoiling tends to inhibit transcription from many promoters.

Other factors affecting gene expression include temperature (section 7.8.11) and circadian rhythms (section 7.8.12).

Mutation(s) (section 8.1) in the coding sequence of a gene can affect the nature and biological activity of the product (Fig. 8.1). A mutation in an *operon* (section 7.8.1) may affect the expression of other genes downstream in the same operon; this is called a *polar mutation*. In some genes expression may be lost if mutation in an *intron* (section 7.9) causes failure of the splicing mechanism, even if the coding sequence of the gene is normal.

Gene expression in the Archaea (in terms of transcription and translation) seems to be closer to the eukaryotic pattern than to the bacterial pattern [TIM (1998) *6* 222–228].

7.8.1   Operons

In some cases, two or more *contiguous* genes are transcribed as a single molecule of *polycistronic* mRNA (section 7.6.4) from a single promoter; any sequence of genes which is subject to co-ordinated expression in this way is said to form an *operon*. One example of such an operon is the *lac* operon (Fig. 7.11).

More generally, the concept of an operon can be widened to include genes that are subject to *divergent* transcription (section 7.5) from different promoters located within a common regulatory region; in this case more than one molecule of mRNA is formed.

Genes in a given operon often encode functionally related products – e.g. enzymes for a particular metabolic pathway. The organization of genes into an operon is beneficial in that it allows those genes to be jointly induced or repressed according to the cell's requirements.

Operons are controlled by various mechanisms. Some are controlled at the level of transcription (sections 7.8.1.1 and 7.8.1.2); control at the level of translation is illustrated in section 7.8.1.3.

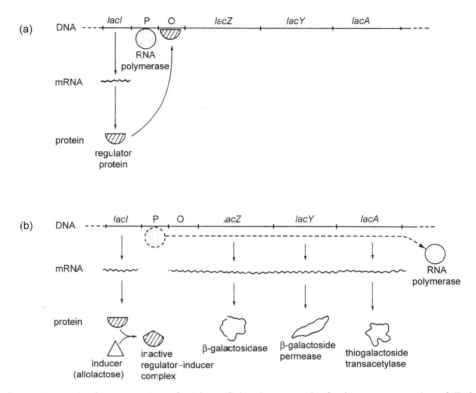

Figure 7.11 The *lac* operon in *Escherichia coli* (section 7.8.1.1). The *lac* operon consists of (i) the three regulated genes: *lacZ*, *lacY* and *lacA*; (ii) a promoter (P); (iii) an operator (O); and (iv) the regulator gene, *lacI*, encoding a repressor. The *lacI* gene is transcribed constitutively, at a low rate, from an independent promoter.

(a) In the *absence* of inducer, the regulator protein, LacI, binds to the operator, O, and minimizes transcription of the operon by the RNA polymerase (seen here bound to the promoter); there is some 'leakage': a few molecules of product are formed in each cell under these conditions.

(b) In the presence of lactose (taken up e.g. by proton–lactose symport: section 5.4), the enzyme β-galactosidase converts some of the lactose to *allolactose*; allolactose acts as an inducer of the *lac* operon by binding to – and thereby inactivating – the regulator protein. Once the regulator protein has been inactivated, the three genes can be transcribed. (For each of the three genes the single (polycistronic) mRNA transcript contains an initiator codon and a stop codon – see section 7.6.)

When the lactose has been used up, transcription of the *lac* operon is no longer needed; the inducer (allolactose) is no longer formed, and the (now active) regulator protein switches off the operon. The mRNA is rapidly degraded (section 7.6.1).

The *lac* operon is subject to catabolite repression (section 7.8.2.1).

While the above account gives the essential operation of the *lac* operon, the actual process is rather more complex. For example, the regulator protein can bind at two other sites, albeit with lower affinity; these sites are called O-2 and O-3. During repression of the operon, it seems that a tetramer of regulator protein binds at the main operator (O) and at O-2 (or O-3) – thus forming a loop of DNA – and that this stabilizes regulator–operator binding [MM (1992) *6* 2419–2422].

Isopropyl-β-D-thiogalactoside (IPTG) is a gratuitous inducer of the *lac* operon, i.e. it induces the operon but is not metabolized by the cell; IPTG is used in recombinant DNA technology (see Fig. 8.16).

Negative control and positive control of operons. If an operon is expressed unless switched off it is said to be under *negative control*. If an operon is not expressed unless switched on it is said to be under *positive control*.

7.8.1.1   Operons under promoter control

In these operons, control involves a *regulator protein* which may be formed more or less continually. The example given here, the *lac* operon, is under negative control.

The *lac* operon in *Escherichia coli* (Fig. 7.11) encodes proteins which promote the uptake and metabolism of β-galactosides such as lactose (a disaccharide composed of glucose and galactose residues); the presence of *allolactose* can 'induce' the *lac* operon (see Fig. 7.11). The *lacY* gene encodes β-galactoside permease, which promotes lactose uptake; *lacZ* encodes the enzyme β-galactosidase which can split lactose into glucose and galactose – but which also converts a small amount of lactose to allolactose. The *lacA* product (thiogalactoside transacetylase) appears not to be needed for lactose metabolism.

The term '*ara* operon' generally refers to the *araBAD* operon – genes *araB*, *araA* and *araD* which encode enzymes concerned with the metabolism of L-arabinose to D-xylulose 5-phosphate. In *E. coli* there are – additionally – genes involved with the uptake (transport: section 5.4) of arabinose, and these genes (*araE*, *araFG*) occur at two further loci on the chromosome. Genes at all three loci are controlled by a common regulator protein: AraC (encoded by *araC*). As genes at different loci are controlled by a common regulator protein, the complete set of genes (uptake + metabolism) conforms to the definition of a *regulon* (section 7.8.2).

In the absence of arabinose, AraC acts as a repressor (i.e. *negative* control, as in the *lac* operon). When present, arabinose converts AraC to an *activator* which initiates transcription; the (modified) regulator protein thus switches on the system – an example of *positive* promoter control.

7.8.1.2   Operons under attenuator control

These operons are concerned e.g. with the synthesis of amino acids. In attenuator control, the initial sequence of nucleotides in the mRNA transcript (the *leader sequence*) encodes a small peptide (*leader peptide*) which is rich in the particular amino acid whose synthesis is governed by that operon; the leader sequence also includes a rho-independent terminator (section 7.5) – the *attenuator* – located between the leader-peptide-encoding region and the first gene of the operon.

If a cell contains adequate levels of the given amino acid (i.e. no synthesis required) transcription is stopped by the attenuator. (Part of the leader peptide is synthesized because a ribosome follows closely behind the RNA polymerase, translating the newly formed transcript.) With inadequate levels of the amino acid (synthesis required) the ribosome synthesizing the leader peptide 'stalls' when it reaches codon(s) specifying the given amino acid. If this happens, it allows downstream parts of the transcript to base-pair with one another in such a way that the attenuator cannot form – so that the genes will be transcribed.

Attenuator control occurs e.g. in the *his* (histidine) operon in *E. coli*.

The *trp* (tryptophan) operon in *E. coli* is under both negative promoter control and attenuator control. With certain levels of tryptophan, the amino acid can activate a specific repressor protein that binds to the operator, blocking transcription (compare *lac* operon). With lower levels, the repressor is inactive so that the RNA polymerase can begin transcription. However, the attenuator, located just before the first gene, blocks transcription – *unless* the ribosome has stalled (owing to lack of tryptophan-bearing tRNAs); a stalled ribosome prevents the attenuator from forming, allowing transcription to proceed.

In evolutionary terms, the *trp* operon pre-dates the Bacteria and the Archaea. It may be that the finely tuned *trp* operon is inherently more stable than operons which have a less complex form of regulation [MMBR (2003) *67* 303–342].

7.8.1.3 *Operon regulation by translational control*

Translational control is exemplified by the IF3–L35–L20 operon (genes *infC–rpmI–rplT* respectively). In *E. coli*, this operon (transcribed as polycistronic mRNA) encodes two ribosomal proteins (L35, L20) and a protein factor involved in protein synthesis (IF3: translation initiation factor 3). (See also section 7.6.3.)

Translation of L35 and L20 is repressed by L20, i.e. the concentration of L20 in the cell regulates gene expression – high levels of L20 repressing *translation*. Repression by L20 appears to involve the binding of L20 to mRNA at a site upstream of *rpmI*. The mechanism of repression by L20 is unknown – L20 could simply block mRNA–ribosome binding or it could bind the ribosome to form an inactive complex. However, an alternative explanation has been suggested by *in vitro* studies which indicate that the regulation of translation may involve an RNA *pseudoknot* – a structure formed by base-pairing between mRNA sites upstream of *rpmI*; it has been suggested that L20 may stabilize the pseudoknot – which includes the Shine–Dalgarno sequence (section 7.6) and initiator codon (section 7.6) of *rpmI* – thereby inhibiting ribosome binding and preventing translation of both *rpmI* and *rplT* [EMBO Journal (1996) *15* 4402–4413].

7.8.2 Regulons and other 'global' and multi-gene control systems

A regulon is a system in which two or more typically non-contiguous genes and/or operons (each with its own promoter) are controlled by the same regulator molecule –

all the genes/operons having similar regulatory sequences that are recognized by the regulator molecule.

7.8.2.1 Catabolite repression

Diauxic growth (section 3.4) shows that the presence of lactose does not *necessarily* induce the *lac* operon (section 7.8.1.1) – i.e. the effect of allolactose is overridden in the presence of glucose. Similarly, glucose will repress the *ara* operon (section 7.8.1.1) in the presence of arabinose. These are only two examples of *catabolite repression* (the 'glucose effect'): a common phenomenon (in bacteria) in which a cell uses certain carbon/energy substrates in preference to others – even when the latter are freely available.

The molecular basis of catabolite repression has remained obscure until recent years. A current picture is as follows. There are at least two distinct mechanisms. In both mechanisms the controlling signals arise from the PTS (section 5.4.2). One of the mechanisms is characteristic of *E. coli* and other enteric bacteria; the other mechanism has been found almost exclusively in Gram-positive bacteria. Models for both mechanisms are outlined below.

Catabolite repression in enteric bacteria. In these organisms the key regulator molecule is the IIA component of glucose permease in the PTS (Fig. 5.12); note that IIA in *this* permease is a cytoplasmic (i.e. soluble, mobile) molecule.

In the *absence* of glucose, the phosphorylated form of IIA (IIA~P) is not dephosphorylated by IIB; under such conditions, IIA~P activates the enzyme adenylate cyclase – thus stimulating the synthesis of cyclic AMP (cAMP: Fig. 7.12). cAMP binds to, and activates, the so-called *cAMP-receptor protein* (CRP) (also called *catabolite activator protein*, CAP).

The cAMP–CRP complex acts as a transcriptional activator: it binds to the promoters of the *lac, ara* and other operons, allowing them to be expressed in the presence of their respective inducers. The precise way in which cAMP–CRP promotes transcription in these operons is still not clear. However, studies on e.g. the *lac* and *gal* operons

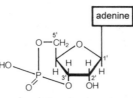

Figure 7.12 Cyclic AMP (cAMP): a molecule involved e.g. in catabolite repression (section 7.8.2.1) – and, interestingly, in the pathogenesis of cholera (section 11.3.1.1). cAMP (adenosine 3',5'-cyclic monophosphate) is synthesized from ATP by the enzyme adenylate cyclase, and is degraded to AMP by cAMP phosphodiesterase. [Various roles of cyclic AMP in prokaryotes (review): MR (1992) *56* 100–122.]

have suggested that the presence of cAMP–CRP is not needed after the establishment of a productive 'open complex' (section 7.5), and that only transient interaction is required between cAMP–CRP and the RNA polymerase [EMBO Journal (1998) *17* 1759–1767].

In the *presence* of glucose, unphosphorylated IIA binds to the membrane permeases of various sugars (e.g. lactose), inhibiting the uptake of the corresponding sugars; as these sugars induce their respective operons, this phenomenon is called *inducer exclusion*.

In some cases, a mechanism similar to that in Gram-positive bacteria (see below) has been identified in *E. coli*.

Catabolite repression in Gram-positive bacteria. The transcription of some catabolic operons is controlled by *operon-specific* regulator proteins whose activity depends on (i) whether or not they are phosphorylated, and (ii) *which* PTS protein was involved in their phosphorylation. Each regulator protein has two copies of a specific phosphorylation site; each copy is designated PRD (= PTS regulation domain). A PRD can be phosphorylated by intermediates of the PTS: IIB~P or HPr~P (Fig. 5.12). Phosphorylation by IIB~P is inhibitory, i.e. it causes the regulator protein to inhibit transcription of the corresponding operon. Phosphorylation by HPr~P tends to promote transcription.

The rationale of this system is that, for a given permease (controlling the uptake of a given substrate), the *absence* of inducer (substrate) will leave IIB phosphorylated (IIB~P) and thus able to transfer phosphate to a PRD site on the regulator protein – with consequent inhibition of the relevant operon. In the *presence* of inducer, IIB~P transfers phosphate to the inducer molecule, rather than to a PRD site, so that the regulator protein is not inhibited; *given the concomitant absence of glucose* (or other rapidly metabolizable carbon source), HPr~P will be available to phosphorylate a PRD site on the regulator protein – thus stimulating expression of the given operon.

In the presence of inducer *and* glucose, phosphate from IIB~P and HPr~P is used to phosphorylate both types of incoming molecule (i.e. inducer and glucose), so that *both* PRDs of the regulator protein remain unphosphorylated; as it lacks stimulatory phosphorylation by HPr~P, the regulator protein is inactive, i.e. the operon is not expressed.

Hence, according to this model, expression of an operon requires that its regulator protein be phosphorylated, but *only* by HPr~P (i.e. only when glucose is not available and the inducer is present); the operon is inactive if one PRD site is phosphorylated by IIB~P and the other by HPr~P (i.e. no inducer, no glucose).

Some PRD-containing regulator proteins control operons by acting as *anti-terminators*, preventing the premature termination of transcription by nucleic acid structures called *terminators*. Other regulator proteins stimulate the *initiation* of transcription.

In *E. coli*, the *bgl* operon – a β-glucoside catabolic system involved e.g. in the metabolism of salicin – is subject to a PRD-containing regulator protein, BglG (an antiterminator); in the *absence* of β-glucoside, BglG is phosphorylated (inactivated) by the IIB component of BglF (the β-glucoside permease) – so that transcription of the *bgl* operon is blocked.

[PRD regulators (review): MM (1998) *28* 865–874.]

In at least some AT-rich Gram-positive bacteria, catabolite repression may involve (additionally) a mechanism in which HPr phosphorylation can be mediated by an ATP-dependent enzyme, *HPr kinase*. When phosphorylated via this route, HPr~P can form a complex with *catabolite control protein A* (CcpA); this complex can bind to a regulatory site in target operons, inhibiting transcription. Inactivation of the gene encoding HPr kinase (*hprK*) in *Staphylococcus xylosus* has been found to abolish repression in three catabolic enzyme systems [JB (2000) *182* 1895–1902].

7.8.2.2   *The SOS system in* Escherichia coli

This system of about 30 unlinked genes is expressed when DNA is damaged and/or cannot replicate – due e.g. to the effects of ultraviolet radiation and/or certain chemicals; expression of the SOS system can (for example) stop cell division, increase DNA repair activity, affect energy metabolism and suppress restriction.

Control is exercised by the LexA protein (*lexA* gene product); under normal conditions (DNA not damaged), LexA (probably as a dimer) binds close to the promoters of the SOS genes and inhibits their transcription. For each SOS gene, the binding site of LexA (the 'SOS box') appears to be the consensus sequence 5'-CTGTN$_8$ACAG-3' (in which N$_8$ is an 8-nucleotide sequence). Damage to DNA activates the RecA protein (by an unknown mechanism). Activated RecA (designated RecA*) functions, non-enzymically, as a *co-protease* in the autocatalytic cleavage of LexA; this allows the SOS genes to be expressed. *In vitro* work suggests that cleavage of LexA (de-repression of the SOS genes) occurs more rapidly when a chi (χ) site (section 8.2.1) occurs near a double-stranded break in DNA; it may be that chi promotes the activation of RecA [Cell (1998) *95* 975–979].

The product of the SOS gene *sulA* represses formation of the septum (section 3.2.1) [SulA–FtsZ interaction: JB (1996) *178* 5080–5085], thus inhibiting cell division; cells may continue to grow as septum-less filaments. A physiological advantage of the inhibition of cell division may be that cells are given time to carry out repairs to their damaged DNA. DNA replication may continue (see end of section 7.3).

DNA repair, including that mediated by the UvrABC endonuclease (section 7.7.1.2), shows enhanced activity; genes *uvrA* and *uvrB* are included in the SOS regulon. A repair process involving genes *umuC* and *umuD* (so-called 'error-prone repair') operates only when the SOS system has been induced; as well as repairing DNA, error-prone repair results in an increased number of *mutations* (section 8.1) – this being referred to as *SOS mutagenesis*. In this process, RecA*, again acting as a co-protease, brings about the autocatalytic cleavage of UmuD to form the active fragment UmuD'. (In *E. coli*, a complex of two UmuD' proteins and one UmuC protein is called *DNA polymerase V*; it has polymerase activity and may carry out some instances of translesion synthesis – see below.) SOS mutagenesis is believed to involve DNA synthesis (repair) on a template strand containing a 'lesion' (e.g. a pyrimidine dimer) – synthesis which, owing to a lack of base-pairing specificity, is likely to introduce incorrect bases; such *translesion synthesis* may well involve specific DNA polymerase(s) induced as part of the SOS response – e.g. polymerases II (*polB* gene product), IV (*dinB* gene product) and V (see above) [TIBS (2000) *25* 74–79]. Subsequent excision repair of the lesion, and DNA replication, is therefore likely to give rise to a mutant genome. Translesion synthesis *in vitro* on a template strand containing an *abasic* site (section 7.7.1.3) has been found to require DNA polymerase III, RecA UmuD' and UmuC.

Once DNA has been repaired, RecA is inactivated and LexA again represses the SOS system.

Note that *lysogenic* bacteria may undergo lysis when the SOS system is expressed (see section 9.2.1).

The SOS system can be induced e.g. by quinolone antibiotics (such as nalidixic acid) and by the CcdB toxin encoded by the F plasmid (section 7.3.1) [TIM (1998) *6* 269–275].

The primary function of the SOS system may be seen as an adaptive genetic response to changing, unfavourable/inhibitory conditions. Under such conditions it is advantageous for bacterial populations to be able to make essential repairs – and also to adapt rapidly by making use of (fortuitous) new combinations of genetic material that may enhance their survival. This is reflected in increased repair activity, and also in SOS mutagenesis – through which advantageous (as well as lethal) mutations may arise. The suppression of restriction could also favour the exploitation of imported DNA, while inhibition of cell division may help e.g. to converse energy and preclude non-viable daughter cells.

7.8.2.3   *The heat-shock response*

Heat shock (a sudden rise in temperature) causes a characteristic adaptive response in organisms ranging from bacteria to plants and animals; the response includes increased synthesis of the so-called *heat-shock proteins* (HSPs).

In *E. coli*, the heat-shock response follows e.g. a 30° → 42°C shift in temperature; it can also be triggered by stress factors such as ultraviolet radiation, ethanol, and the intracellular accumulation of abnormal/heterologous proteins (as in overproduction: section 8.5.11.4(d)).

Some of the 17 known HSPs in *E. coli* cope with stress-induced damage; thus e.g. the molecular *chaperones* GroES, GroEL and DnaK (section 7.6) are involved e.g. in re-folding unfolded proteins and/or preventing aggregation of unfolded proteins, while the Lon protease (product of the *lon* gene) degrades damaged or abnormal proteins. Other HSPs include the products of genes *dnaJ*, *rpoD* and *lysU*.

Following heat shock, the synthesis of HSPs rises to a maximum within minutes and then decreases to a new steady state above that of the lower temperature.

In *E. coli*, control of the heat-shock regulon involves the product of gene *rpoH* (previously called *htpR*). RpoH is a sigma factor (section 7.5), σ^{32}, which is (transiently) synthesized in greater quantities following heat shock; compared with the 'routine' sigma factor (σ^{70}), σ^{32} permits more efficient transcription from the promoters of the HSP genes and so causes increased synthesis of the HSPs.

The increase in σ^{32} following heat shock is due partly to its increased translation and stabilization (rather than to increased transcription of *rpoH*). Heat-induced translation from the *rpoH* mRNA appears to involve an effect on a secondary structure in the mRNA formed by base-pairing between a region immediately downstream of the start codon (a downstream box – see section 8.5.11.2) and another region in the coding sequence; it is believed that this secondary structure represses translation under normal conditions, and that repression is relieved during heat shock [NAR (1993) *21* 5449–5455]. It appears that translation of σ^{32} is induced when the secondary structure in mRNA is de-stabilized as a result of the higher temperature; thus, the mRNA acts as a thermosensor (= RNA thermometer) for expression of σ^{32} [GD (1999) *13* 633–636].

σ^{32} is synthesized under normal conditions but it has a short half-life (about 1 minute): it is complexed by DnaK, DnaJ and GrpE but when free is degraded by the FtsH protease; during heat shock, DnaK binds preferentially to denatured proteins, and this sequestration of DnaK leaves σ^{32} free to function as a sigma factor – leading to increased synthesis of the HSPs [see e.g. EMBO Journal (1996) Journal *15* 607–617].

Heat shock in other bacteria. Different mechanisms for regulating the induction of the heat-shock response occur in some bacteria. For example, in some cases a regulatory sequence is located upstream of heat-shock genes; this inverted repeat sequence, termed CIRCE (controlling inverted repeat of chaperone expression), has been found e.g. in strains of *Bacillus* and *Clostridium* and in some Gram-negative bacteria. Control in *Bradyrhizobium japonicum* consists of at least two distinct regulatory systems which involve CIRCE and a σ^{32}-like sigma factor [JB (1996) *178* 5337–5346].

While the above account outlines an established view of the regulation of bacterial heat-shock genes, one report refers to the involvement of negative regulation in which *repressor proteins*, in conjunction with *cis*-acting DNA sequences, inhibit transcription under normal physiological conditions [MM (1999) *31* 1–8].

7.8.2.4   *The cold-shock response*

A sudden fall in temperature can bring about an altered pattern of gene expression in prokaryotes and eukaryotes. The following refers to cold shock in *E. coli*.

The cold-shock response is triggered e.g. by a 37° → 10°C down-shift (or, in general, a down-shift of at least 13°C). Growth stops, resuming (at a lower rate) after a lag of a few hours. The lag period is associated with an inhibition of translation; resumption of growth follows the resumption of normal protein synthesis.

Cold shock is characterized by e.g.:

- induction/increased synthesis of *cold-shock proteins*;
- ongoing activity of certain proteins involved in translation (despite generalized inhibition of protein synthesis);
- repression of heat-shock proteins.

These features of cold shock are believed to indicate an adaptive response.

The cold-shock proteins include CspA, a small (70 amino acid) protein whose induction (immediately on temperature down-shift) is due largely to increased stability of its mRNA in the cold. CspA appears to interact with nucleic acids; its suggested functions include (i) a low-temperature translational activator; (ii) a low-temperature 'RNA chaperone' that may prevent the formation of secondary structures in RNA and may be important e.g. for efficient translation of mRNAs at low temperatures [JBC (1997) *272* 196–202]; and (iii) an agent that inhibits translation of particular mRNAs. In fact, CspA may not function *specifically* in the cold-shock response because it is normally produced during early exponential growth at 37°C in *Escherichia coli* (i.e. under non-stress conditions) [EMBC Journal (1999) *18* 1653–1659]. In *Bacillus subtilis* at least one cold-shock protein is needed for *normal* growth; interestingly, *E. coli* translation initiation factor IF1 (section 7.6) can compensate for loss of cold-shock proteins in *B. subtilis* – suggesting e.g. that IF1 and cold-shock proteins may have at least some functional overlap [JB (2001) *183* 7381–7386].

Other cold-shock proteins include RecA; initiation factor 2 (IF-2), which mediates the binding of *N*-formylmethionine-charged tRNA at the start of translation (section 7.6); NusA, involved in termination of transcription; and the α-subunit of the topoisomerase gyrase (section 7.2.1).

The mode of regulation of the cold-shock genes is not known. Suggestions include CspA as a transcriptional activator. The small molecules ppGpp and pppGpp (in which G is guanosine, and p is phosphate) may have a regulatory role: following temperature down-shift, the concentration of these molecules decreases, and this fall in concentration has been associated with increased synthesis of some cold-shock proteins.

The cold-shock response, as described, can also be triggered by certain inhibitors of translation – e.g. the antibiotics chloramphenicol, erythromycin and tetracycline. That these agents (many of which also depress (p)ppGpp levels) have an effect similar to that of a temperature down-shift has suggested that the common factor – a decrease in translational capacity – may be important in the induction of the cold-shock response.

7.8.2.5   The stringent response

In bacteria, starvation (e.g. lack of an essential amino acid) elicits the *stringent response* – which includes decreased synthesis of proteins and e.g. rRNA; this conserves energy and material.

The response is triggered by an uncharged tRNA at the A site of a ribosome (see Fig. 7.9). This stimulates a ribosome-bound enzyme, pyrophosphotransferase (= RelA, the *relA* gene product; stringent factor), to synthesize ppGpp (in which G is guanosine, and p is phosphate) from GTP (or GDP) and ATP. ppGpp (guanosine 5'-diphosphate 3'-diphosphate) is an example of an *alarmone*: a small molecule which accumulates under certain stress conditions and which serves as a signal for re-directing the cell's metabolism.

ppGpp inhibits synthesis of both proteins and e.g. rRNA. It inhibits transcription from certain genes – but enhances transcription from others (see below). Recent studies have focused on the structural basis of ppGpp-dependent regulation of transcription [Cell (2004) *117* 299–310].

ppGpp can apparently enhance transcription from some of the operons governing the biosynthesis of particular amino acids (e.g. histidine). ppGpp may thus help to maintain a correct balance of amino acids in the cell; for example, given a level of histidine low enough to trigger the stringent response, increased synthesis of ppGpp should help to restore the balance by (i) slowing protein synthesis, and (ii) stimulating the *his* operon.

Cells having a null mutation in *relA* do not exhibit the stringent response when starved of an essential amino acid and are said to be *relaxed*.

7.8.2.6   The acid tolerance response (ATR)

Acid shock elicits a response (ATR) involving the synthesis of so-called *acid-shock proteins* (ASPs) – which are believed to deal with acid-induced damage and/or promote survival at low pH. In *Salmonella typhimurium* the response appears to be a two-stage process [JB (1991) *173* 6896–6902]; thus, synthesis and suppression of particular proteins occur at a 'pre-shock' stage (about pH 6) and again at a specific, lower pH – the cell surviving acidity as low as pH 3.3. In this organism, and in other enteric pathogens, adaptation to, or survival in, low pH is necessary for a pathogenic role because infection normally occurs via the (highly acidic) stomach.

In *S. typhimurium* at least 50 ASPs are synthesized in the ATR; expression of the ASP genes is controlled by at least two regulatory proteins: (i) the sigma factor σ^s (RpoS; *rpoS* gene product), and (ii) the Fur protein (*fur* gene product). RpoS, itself an acid-inducible ASP, regulates a subset of ASP genes; mutants lacking the σ^s function can still exhibit a transient ATR in which Fur regulates (σ^s-independent) ASP genes. Interestingly, Fur also regulates the iron-uptake function in *S. typhimurium* (section

11.5.5) – but the roles of Fur in ATR and iron-uptake are physiologically and genetically distinct [JB (1996) *178* 5683–5691].

In e.g. *E. coli*, acid stress (or anaerobiosis) induces the expression of genes encoding: (i) glutamate decarboxylase, and (ii) a putative membrane transport system for γ-aminobutyrate (the decarboxylation product of glutamate). (Expression of these genes – *gadB* and *gadC*, respectively – is regulated by the sigma factor σ^s). According to one model, glutamate is taken up and decarboxylated in a reaction that consumes a proton; γ-aminobutyrate is then exported (via the transporter), thus helping to maintain pH homeostasis. An alternative model supposes that GadB and GadC constitute an energy-generating system by effectively exporting protons – the resulting *pmf* (section 5.1.1.2) being used for synthesis of ATP at a membrane ATPase. [Possible functions of *gadCB* products: TIM (1998) *6* 214–216.]

7.8.2.7   *Assembly of the bacterial flagellum*

The structure and assembly of the flagellum (section 2.2.14.1, Fig. 2.8) involves about 50 different types of protein; assembly is highly organized – particular components being added in strict sequence – and the corresponding genes are expressed in a way which reflects this sequence. Some of the relevant genes are listed in Table 7.3.

In *E. coli*, the first genes to be expressed (class I genes) are those encoding certain transcriptional activators which seem to be necessary for expression of (class II) genes encoding components of the basal body and hook. Expression of class I and II genes is necessary for the expression of class III genes (whose products include the filament subunit protein *flagellin*).

Transcription of class III genes depends on a σ factor (section 7.5) encoded by a class II gene. This σ factor (protein FliA, *fliA* gene product) is temporarily inactivated by an anti-σ factor (FlgM, *flgM* gene product) until completion of the basal body and hook. Completion of the hook acts as a signal which leads to the secretion of FlgM (via the axial channel); this releases FliA to mediate transcription of the class III genes. Thus, genes encoding the final phase of construction are governed directly by the level of assembly of the partly completed flagellum.

In the above sequence, completion of the hook is an event without which the next stage (expression of class III genes) would not normally occur. This event represents a so-called *checkpoint* (see section 3.2.1.3). In *E. coli*, and related bacteria, flagellar assembly involves only a single checkpoint but in *Caulobacter crescentus* it involves two checkpoints (see section 4.1).

[Genetics and assembly of flagella in Gram-negative bacteria: FEMSMR (2000) *24* 21–44 (22–27).]

[Regulation cascade of flagellar expression in Gram-negative bacteria: FEMSMR (2003) *27* 505–523.]

Table 7.3 Some of the genes and gene products involved in the assembly of the flagellum in *Escherichia coli, Salmonella typhimurium* and related species[1]

Gene (product)	Function of gene product
Class I genes	
flhC (FlhC)	Transcriptional activation of class II genes
flhD (FlhD)	Transcriptional activation of class II genes
Class II genes	
flgB (FlgB)	Rod (proximal)
flgC (FlgC)	Rod (proximal)
flgE (FlgE)	Hook
flgF (FlgF)	Rod (proximal)
flgG (FlgG)	Rod (distal)
flgH (FlgH)	L ring
flgI (FlgI)	P ring
flgK (FlgK)	Hook (distal end)
flgL (FlgL)	Hook (distal end)
flgM (FlgM)	Anti-sigma factor (delays FliA activity)
flhA (FlhA)	Export apparatus
flhB (FlhB)	Export apparatus
fliA (FliA)	Sigma factor (σ^{28}) for class III genes
fliD (FliD)	Filament cap
fliF (FliF)	MS ring
fliG (FliG)	Switch/C ring?/torque generation
fliH (FliH)	Export apparatus
fliI (FliI)	Export apparatus
fliJ (FliJ)	Export apparatus/chaperone
fliK (FliK)	Regulation of hook length
fliM (FliM)	C ring/ switch
fliN (FliN)	C ring/switch
fliO (FliO)	Export apparatus
fliP (FliP)	Export apparatus
fliQ (FliQ)	Export apparatus
fliR (FliR)	Export apparatus
Class III genes	
fliC (FliC)	Flagellin (protein subunit of filament)
motA (MotA)	Torque generation
motB (MotB)	Torque generation

[1] From *Dictionary of Microbiology and Molecular Biology,* 3rd edn 2001 (ISBN 0-471-49064-4), page 310, with permission of the publisher, John Wiley & Sons Ltd, Chichester, U.K.

7.8.2.8 *The oxidative stress response*

Under certain conditions, bacteria can be damaged or killed by exogenous or en-dogenous *reactive oxygen species* (ROS) such as hydrogen peroxide (H_2O_2), hydroxyl radical (OH·) and superoxide ($\cdot O_2^-$); ROS react with e.g. nucleic acids, proteins and lipids. Pathogens may be exposed to ROS by the oxidative burst in a phagolysosome (section 11.4.1). Endogenous ROS may be formed e.g. if a cell's iron metabolism becomes deregulated; thus, excess iron (iron overload) in *E. coli* may catalyse the formation of hydroxyl radical from hydrogen peroxide.

Some bacteria respond to oxidative stress by upregulating the synthesis of anti-oxidant enzymes. These include superoxide dismutase (SOD), which degrades superoxide to H_2O_2, catalase and peroxidase, which degrade peroxides. In *E. coli*, one of the oxidative stress regulons is controlled by the OxyR protein; when activated by oxidation, OxyR promotes transcription from *katG*, *ahpC* and other genes. KatG (a catalase) catalyses the reaction: $2H_2C_2 \rightarrow 2H_2O + O_2$; AhpC is a subunit of alkyl hydroperoxide, an enzyme that degrades *organic* peroxides. In another regulon, the inducible manganese-containing SOD (*sodA* gene product) is positively regulated (in the presence of superoxide) by the *soxR*/*soxS* gene products.

Interestingly, iron *deficiency* can also lead to oxidative stress – possibly through reduced activity of haem-containing antioxidant enzymes.

Control of oxidative stress in mycobacteria is less well characterized. Some anti-oxidant enzymes, including KatG, occur in *Mycobacterium tuberculosis*, but only an inactive form of an *oxyR*-like gene is found in members of the *M. tuberculosis* complex. Protein IdeR, which negatively regulates the synthesis of *siderophores* (section 11.5.5), can apparently upregulate some antioxidant enzymes, but the mechanism is unknown.

Oxidative stress factors affect the susceptibility of *M. tuberculosis* to the important anti-tuberculosis drug *isoniazid*; isoniazid, when activated within the bacterium, apparently inhibits enzymic step(s) in the synthesis of essential cell-wall mycolic acid [Microbiology (2000) *146* 289–296]. Within the cell, activation of isoniazid relies on the peroxide-dependent action of KatG. *katG* and *ahpC* mutants of *M. tuberculosis* typically show increased resistance to isoniazid, presumably because mutations in these genes affect the activation of the drug. Unlike wild-type strains of *M. tuberculosis*, *M. leprae* (causal agent of leprosy) lacks catalase–peroxidase activity [FEMSML (1997) *149* 273–278]. (Certain bacteria which lack mycolic acid may also be susceptible to isoniazid, indicating that the drug has other target(s).)

In mycobacteria, the relationships between iron regulation, oxidative stress response and susceptibility to isoniazid remain to be elucidated [TIM (1998) *6* 354–358].

During aerobic growth, *Streptococcus pneumoniae* (catalase-negative) produces significant amounts of hydrogen peroxide, some of which can be converted to hydroxyl radical (OH·) by Fe^{2+} in the *Fenton reaction* ($H_2O_2 + Fe^{2+} \rightarrow Fe^{3+} + OH· + OH^-$). *S. pneumoniae* may escape Fenton-dependent killing (with OH·) e.g. by binding Fe^{2+} (preventing the formation of OH· near highly susceptible DNA) or by the use of an efficient system for repairing damaged DNA [JB (2003) *185* 6815–6825].

7.8.3 Recombinational regulation of gene expression

See site-specific recombination (section 8.2.2).

7.8.4 Regulation of gene expression by the rate of decay of mRNA

In some cases, the expression of a bacterial gene is regulated by the rate of decay of its mRNA – 'decay' meaning enzymatic degradation (section 7.6.1). Such regulation has been studied e.g. in the *puf* operon of a photosynthetic bacterium, *Rhodobacter capsulatus*; the *puf* genes (encoding components of the photosynthetic apparatus) are transcribed together as a single mRNA transcript (i.e. *polycistronic* mRNA). (An example of a polycistronic mRNA was seen in Fig. 7.11.) Interestingly, different parts of *puf* polycistronic mRNA decay at different rates – i.e. some sequences survive for longer than others; in this way, the correct genes are expressed at appropriate times, so that photosynthesis and cell growth occur optimally. Special decay-promoting and decay-inhibiting regions in the *puf* polycistronic mRNA may be responsible for these differential rates of decay [MM (1993) *9* 1–7].

7.8.5 Regulation of gene expression by translational attenuation

In some Gram-positive bacteria, the presence of chloramphenicol (section 15.4.4) induces the *cat* gene; *cat* encodes chloramphenicol acetyltransferase: an enzyme which inactivates chloramphenicol by acetylating it. To understand this inducible resistance to chloramphenicol we look at the regulatory process of *translational attenuation*.

The transcript of the *cat* gene consists of an initial short 'leader' sequence of nucleotides followed – in the same mRNA transcript – by the *cat* coding sequence. In the absence of chloramphenicol, the *cat* gene is transcribed but not translated because the ribosome-binding site is 'distorted' by local base-pairing between ribonucleotides. However, the leader sequence (which has its own ribosome binding site) seems to be translated continually. In the presence of chloramphenicol, the ribosome translating the leader sequence stalls; this results in a loss of the 'distortion' at the *cat* ribosome binding site – permitting translation of the *cat* coding sequence. Why is *cat* translation itself not prevented by chloramphenicol (which inhibits protein synthesis)? The *cat* induction mechanism is triggered by levels of chloramphenicol even lower than those which inhibit protein synthesis – possibly because the nascent leader peptide itself (as well as chloramphenicol) contributes to ribosome stalling [JB (1993) *175* 5309–5313].

Another example of translation attentuation is the regulation of gene *pyrC* in *E. coli*; *pyrC* encodes the enzyme dihydroorotase – involved in the synthesis of pyrimidines and, hence, nucleic acids. Transcription of *pyrC* can begin at any of several (adjacent) nucleotides on the DNA template – transcription from a given nucleotide being influenced by the availability of pyrimidines in the cell (as 'sensed' by the CTP/GTP ratio). Transcripts of *pyrC* synthesized in the presence of adequate pyrimidines are mainly of a kind in which RNA–RNA base-pairing can occur in the region of the Shine–Dalgarno sequence; such base-pairing blocks ribosome binding and gene expression. With low levels of pyrimidines, most transcripts are synthesized from a slightly different start point, and, in these transcripts, base-pairing does not occur so that the enzyme is synthesized.

7.8.6   Regulation of gene expression via signal transduction pathways

Bacteria can detect various environmental signals (such as changes in osmolality) by means of sensory systems associated with the cell envelope; within the cell, these signals act via signal transduction pathways that include so-called *two-component regulatory systems*. In these systems the first component is a *histidine kinase*: an enzyme which can undergo ATP-dependent autophosphorylation (at an active site containing a histidine residue) and transfer the phosphate to another molecule. When appropriately influenced by an environmental signal, the histidine kinase transfers phosphate to a site containing an aspartate residue in the second component – a regulator protein (= response regulator); when phosphorylated, the regulator protein modifies the expression of certain genes, e.g. by controlling their transcription, thus eliciting a response to the environmental signal.

Two-component systems are also reported to respond to signals that arise within the cell (internal signals) – e.g. during cell-cycle events in *Caulobacter crescentus* [Science (2002) *298* 1942–1946 (1942–1943)].

In some cases an environmental signal acts *directly* on the kinase, regulating its activity; that is, the kinase *is* the sensor. For example, in *E. coli*, increased osmolality in the environment stimulates the kinase activity of the sensor protein EnvZ, located in the cell envelope. EnvZ transfers phosphate to the regulator protein, OmpR, which then (i) enhances transcription of the gene encoding OmpC porin (section 2.2.9.2) and (ii) inhibits transcription of the gene encoding OmpF porin; thus, at high osmolality, the outer membrane contains a higher proportion of OmpC (which forms a slightly smaller pore) and a lower proportion of OmpF.

In *E. coli*, osmotic up-shift in the environment also induces the high-affinity Kdp transport system for the uptake of potassium ions. The sensor kinase, KdpD, apparently responds to decreased turgor pressure; when activated in this way, KdpD transfers phosphate to the regulator protein, KdpE, which promotes transcription of the *kdpABC* operon. KdpB is a membrane-associated K^+-ATPase (potassium pump) at which hydrolysis of ATP provides energy for the uptake of potassium ions.

Two-component systems regulate a wide variety of activities, including the synthesis of virulence factors by pathogens. For example, in *Clostridium perfringens* a two-component system was reported to regulate genes encoding the toxins involved in gas gangrene [RMM (1997) *8* (suppl 1) S25–S27]. In *Salmonella typhimurium*, the SsrA–SsrB system regulates transcription of genes in the SPI-2 pathogenicity island (section 11.5.7) whose products are needed for the pathogen's intracellular growth [MM (1998) 30 175–188].

Two-component systems are also involved in (for example):

- transformation competence in *Bacillus subtilis* [TIM (1998) *6* 288–294];
- adhesion, proteolysis in *Staphylococcus aureus* [JB (2000) *182* 3955–3964];
- oxygen-regulated exotoxigenesis in *S. aureus* [JB (2001) *183* 1113–1123];
- protein folding [GD (1997) *11* 1169–1182];
- membrane permeability in *Pseudomonas aeruginosa* [AAC (2003) *47* 95–101];
- cell-cycle regulation in *Caulobacter crescentus* [Science (2003) *301* 1874–1877].

Essential two-component systems. Some two-component systems are reported to be essential for viability; these systems are involved e.g. in control of the cell cycle. Examples of essential two-component systems include the *yycF–yycG* system in *Bacillus subtilis* [Microbiology (2000) *146* 1573–1583], the *mtrA–mtrB* system in *Mycobacterium tuberculosis* [JB (2000) *182* 3832–3838], and the major two-component system in *Caulobacter crescentus* which (directly or indirectly) controls 26% of the cell-cycle-regulated genes [Science (2003) *301* 1874–1877].

Variant forms of two-component systems. Sometimes the two core components (histidine kinase and regulator protein) are supplemented with other components. Thus, in some cases, the sensor (= receptor) and kinase are different molecules: signals are detected by the sensor which then activates the kinase; this occurs e.g. in the MCP-mediated pathway of chemotaxis in *E. coli* (Fig. 7.13).

In *Bacillus subtilis*, the signal transduction pathway during initiation of sporulation is often referred to as a *phosphorelay* (see section 7.8.6.1 and Fig. 7.14).

7.8.6.1 *Initiation of endospore formation in* Bacillus subtilis

When growth becomes limited by a shortage of nutrients, *B. subtilis* re-organizes its metabolism: genes are induced and repressed to reflect the new conditions – but there is no abrupt change from active growth to active sporulation. Instead, exponential (log-phase) growth is followed by a *transition state* in which both growth-related and survival-related activities occur simultaneously [transition state in *B. subtilis* (review): PNARMB (1993) *46* 121–153]; during this period the decision is made to either maintain the vegetative state, with low-level metabolism and no cell division, or to initiate sporulation.

Clearly, sporulation is a major step, and before a final commitment is made the cell must integrate and interpret signals from various sources, both environmental and intracellular. For example, one relevant signal is the level of calcium; thus, sporulation is *inhibited* if levels of calcium drop to 2 μM. The value of this particular constraint can be seen by recalling that the core of an endospore contains an accumulation of calcium dipicolinate (section 4.3.1); were sporulation to begin with insufficient calcium it may abort at an intermediate stage.

Signals that promote sporulation appear to be recognized by at least two types of sensor molecule: kinase A (KinA) and kinase B (KinB). The signals which activate these kinases are unknown. On receipt of appropriate environmental and/or internal signals, the kinases undergo ATP-dependent autophosphorylation and transfer phosphate to the protein Spo0F: the first component of a *phosphorelay* system (Fig. 7.14); Spo0F thus acts as a main junction through which are channelled signals from various external and internal sources. The phosphorelay continues with phosphorylation of proteins Spo0B and Spo0A; the phosphorylated (i.e. activated) form of Spo0A is written Spo0A~P. The intracellular level of Spo0A~P seems to be the key factor which determines the decision between sporulation and continued vegetative growth.

As shown in Fig. 7.14, Spo0A~P is negatively regulated (de-phosphorylated) by the phosphatase Spo0E. Other phosphatases, not shown in the figure, also help to regulate the levels of Spo0A~P – indirectly – by specifically de-phosphorylating Spo0F~P; these phosphatases, RapA and RapB, are themselves regulated (at the level of transcription) by the cell's physiological state. Thus, the occurrence of sporulation seems to depend on the outcome of a tug-of-war between the kinases (promoting sporulation) and the phosphatases (inhibiting sporulation) TIM (1998) *6* 366–370].

7.8.6.2 *The* agr *locus in* Staphylococcus aureus

The *agr* locus is a chromosomal sequence encoding e.g.: (i) the sensor and response regulator of a two-component regulatory system (AgrA–AgrC), and (ii) a peptide 'pheromone' which is secreted by the cell and which, at appropriate concentrations (see *quorum sensing*, section 10.1.2), activates the two-component system.

When AgrA–AgrC is activated, the effector molecule RNAIII upregulates some genes (e.g. *tst* – toxic shock toxin) and represses others (e.g. *spa* – protein A); it also affects the development of biofilms [JB (2004) *186* 1838–1850].

Another two-component system, encoded by the *srrAB* genes (staphylococcal respiratory response genes), e.g. downregulates RNAIII and the toxic shock toxin under low levels of oxygen [JB (2004) *186* 2430–2438]. [Regulation of virulence factors in *S. aureus* (complexity and applications): FEMSMR (2004) *28* 183–200.]

7.8.7 Regulation of gene expression by translational frame-shifting

In some cases, the synthesis (or composition) of a polypeptide is regulated during *translocation* (Fig. 7.9e): a specific sequence of nucleotides in the transcript causes the ribosome to 'slip' along the mRNA – commonly either a '+1 slip' (a 1-nucleotide shift downstream, i.e. in the direction of translation) or a '–1 slip' (a 1-nucleotide shift upstream, i.e. in the opposite direction). Such a shift, of course, affects all subsequent codons in the transcript (compare the frame-shift mutation, Fig. 8.1b). Frame-shifting

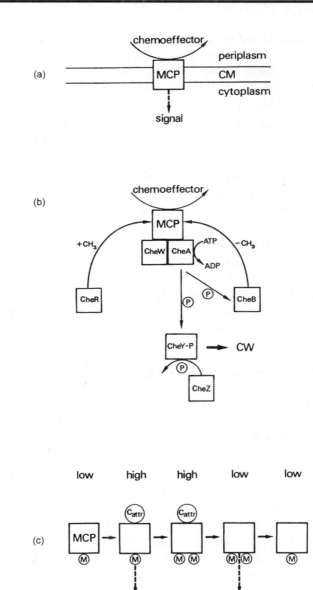

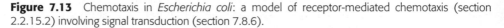

Figure 7.13 Chemotaxis in *Escherichia coli*: a model of receptor-mediated chemotaxis (section 2.2.15.2) involving signal transduction (section 7.8.6).

(a) The receptors – called *methyl-accepting chemotaxis proteins* (MCPs) – occur in the cytoplasmic membrane (CM) and have functional regions in both the periplasm and cytoplasm. There are different types of MCP, each type recognizing its own range of chemoeffectors/environmental stimuli, and there are several hundred molecules of each type of MCP in a given cell. (The MCPs are encoded by genes *tap*, *tar, trg* and *tsr*.) Some chemoeffectors (e.g. aspartate, serine) bind directly to their MCPs, but ribose, and some other sugars, first complex with certain periplasmic proteins before binding. The binding/release of chemoeffectors by an MCP regulates an intracellular signal which controls the frequency of tumbling.

(b) The intracellular signal originates at the protein CheA (encoded by gene *cheA*). CheA is a *histidine kinase* (see section 7.8.6). In the diagram, CheA is bound to the MCP; CheW is a coupling protein. CheA can transfer phosphate from ATP to a regulator protein, CheY; the phosphorylated (activated) form of CheY (CheY~P) enhances CW flagellar rotation (and, hence, increases the frequency of tumbling) by interacting with the C ring of the flagellar motor (apparently with the FliM component – see Fig. 2.8). [Role of CheY: TIM (1999) 7 16–22.] The basic signal thus consists of the transfer of phosphate from CheA to CheY; this signal is modulated (regulated) by (i) the binding/release of chemoeffector at the MCP, and (ii) the degree of methylation of the MCP. We look at each of these two factors in turn and then consider how, jointly, they regulate signal transduction and the frequency of tumbling.

(Although *phosphorylation* of CheY has been regarded as the essential mode of signalling in chemotaxis, there is some evidence that, in at least some cases, *acetylation* may be involved in the response to repellents [reported in JB (2000) *182* 1459–1471 (1466)].)

The binding of a chemoattractant *inhibits* the kinase activity of CheA, thus inhibiting the transfer of phosphate to CheY. (On binding chemoattractant it appears that subunits of the MCP move relative to one another, and that such movement is involved in the inhibition of CheA kinase activity [see e.g. JB (2000) *182* 1459–1471 (1465–1466)].) The release of chemoattractant from an MCP *stimulates* CheA, promoting the transfer of phosphate to CheY and (thus) encouraging tumbling.

The cytoplasmic side of an MCP is subject to ongoing methylation by a methyltransferase, CheR, which uses *S*-adenosylmethionine as a methyl donor. Opposing this, a methylesterase, CheB, removes the methyl groups; the activity of CheB is enhanced when it is phosphorylated by CheA. The degree of methylation of an MCP thus depends on the activities of both CheR and CheB; increased methylation tends to *enhance* the activity of CheA.

(c) In a low, uniform concentration of chemoattractant (C_{attr}) tumbling occurs at a given rate. When the cell swims *up* a concentration gradient (i.e. towards higher concentrations of the chemoattractant) there is an increase in the number of molecules of C_{attr} bound to MCPs; this results in the inhibition of CheA which, in turn, inhibits phosphorylation of CheY – and thus promotes counterclockwise (CCW) flagellar rotation, i.e. smooth swimming in the same direction. Inhibition of CheA also inhibits CheB, allowing increased methylation of the MCP (as CheR methylates continually); hence, inhibition of CheA, due to binding of C_{attr}, is subsequently offset by the stimulatory effect of increased methylation of the MCP – so that the frequency of tumbling returns to its original value. The cell has thus *adapted* to the new, higher, concentration of C_{attr} (MCP in the centre).

On swimming *down* the concentration gradient there is a decrease in the number of molecules of C_{attr} bound to the MCPs. Release of C_{attr} stimulates CheA and promotes clockwise (CW) flagellar rotation (more frequent tumbling); at the same time, CheB is stimulated, leading to de-methylation of the MCP until CheA activity has been reduced to its original level – i.e. adaptation to the new, low, uniform concentration of C_{attr} (MCP at right-hand side).

Adaptation depends on an appropriate rate of de-phosphorylation of CheY~P and CheB~P; CheB~P undergoes autodephosphorylation, while levels of CheY~P are regulated by the CheZ protein. (It appears that about 30% of the intracellular pool of CheY molecules are in the phosphorylated state in fully adapted cells [EMBO Journal (1998) *17* 4238–4248].)

Note that the degree of methylation of an MCP in an *adapted* cell reflects the extracellular concentration of the chemoattractant (compare the first, third and fifth MCPs in the diagram).

Experimental modification of the concentrations of various components (e.g. CheR) indicates that, whereas variation can occur in factors such as time taken to adapt (e.g. a > 20-fold variation with variation in CheR concentration), the *precision* of adaptation (indicated by a return to the exact pre-stimulus level of tumbling) is preserved [Nature (1999) *397* 168–171].

[Signalling components in bacterial locomotion and sensory reception: JB (2000) *182* 1459–1471. Signalling in bacterial chemotaxis (review): JB (2000) *182* 6865–6873.]

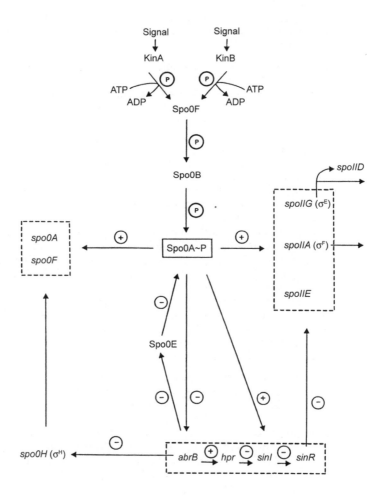

Figure 7.14 A simplified scheme for the regulation of gene expression during the initiation of sporulation in *Bacillus subtilis* (section 7.8.6.1) The scheme is based on information from various sources.

Endospore formation (section 4.3.1) involves a range of proteins that are not synthesized in the vegetative (i.e. growing) cell, and this requires the co-ordinated expression of various sporulation-specific genes. These genes must be switched on, and other genes – which have been repressing sporulation genes during vegetative growth – must be switched off.

The mechanism for initiating sporulation appears to recognize both external (environmental) and internal (intracellular) signals. These signals are believed to promote ATP-dependent autophosphorylation of certain kinases – shown in the diagram as KinA and KinB.

The kinases KinA and KinB transfer phosphate (symbolized by the encircled 'P') to a sequence of proteins: SpoOF → SpoOB → SpoOA; this is the so-called *phosphorelay* [COGD (1993) *3* 203–212; JCB (1993) *51* 55–61]. The phosphorylated form of SpoOA is written SpoOA~P. The intracellular concentration of SpoOA~P appears to be the key factor in initiating sporulation.

may result e.g. in a new stop codon, thereby ending translation prematurely (and producing a shorter polypeptide). Frame-shifting can also nullify the effect of an existing stop codon within the transcript; in one such case, frame-shifting (leading to translation of the whole transcript) occurs when levels of the polypeptide are low – but is prevented (stop codon functional) by adequate levels of the polypeptide.

Frame-shifting is a possible mechanism for the generation of new cell-surface proteins in antigenic variation (section 11.5.3).

[Translational frame-shifting (review): MM (1994) *11* 3–8.]

7.8.8 Regulation of gene expression by DNA methylation

After replication DNA is methylated (modification: section 7.4). It has been suggested that, following replication, the chromosome's origin (*oriC*) may be left initially

---

SpoOA~P regulates the expression of certain sporulation-specific genes shown in the dashed boxes. Throughout the diagram, an encircled minus sign indicates an inhibitory influence, while an encircled plus sign indicates that gene expression is enhanced

SpoOA~P promotes the expression of genes *spoOF* and *spoOA* (dashed box, left) by a positive feedback loop [MM (1993) 7 967–974]. Also, by repressing *abrB* (lower dashed box), SpoOA~P promotes the expression of *spoOH* – whose product, the sigma factor σ^H, is required e.g. for the transcription of *spoOF* and *spoOA*. σ^H is produced at low levels during vegetative growth but its production is greatly enhanced following the initiation of sporulation. During sporulation, σ^H is also needed for transcription of the *ftsZ* gene (see FtsZ in section 3.2.1) from a special promoter, p2, which is distinct from the *ftsZ* promoter used during exponential growth; FtsZ is required for the formation of the asymmetric septum (see Fig. 4.2, legend).

SpoOE is a negative regulator of the phosphorelay [PNAS (1994) *91* 1756–1760]. It is an enzyme which de-phosphorylates (i.e. inactivates) SpoOA~P and which may therefore help to prevent the initiation of sporulation until the cumulative effect of the various (extracellular and intracellular) signals dictates that sporulation is necessary.

During vegetative growth, SinR (encoded by *sinR*, lower dashed box) represses expression of the *spoII* genes (dashed box, centre right). During sporulation, SpoOA~P represses *abrB* (lower dashed box) thus (indirectly) inhibiting expression of *sinR* and (thus) helping to promote expression of the *spoII* genes. SinR is repressed at the protein–protein level, i.e. it is inhibited by the SinI protein. Note that SpoOA~P also directly promotes the transcription of *sinI*.

Sporulation can be prevented by mutations in the *spo0* genes. Thus, e.g., a null mutation (one causing total loss of a gene's product or function) in *spoOB* will block the phosphorelay. So far, no mutations have been found to affect stage I (Fig. 4.2); some workers do not regard stage I as a distinct stage.

SpoOA~P promotes the expression of *spoII* genes (dashed box, centre right).

Activation of *spoIIG* (encoding σ^E) is needed e.g. for transcription of *spoIID*. SpoIID is one of several proteins involved in degradation of peptidoglycan in the asymmetric septum – essential for subsequent engulfment of the prespore by the mother cell.

SpoIIE is needed e.g. for regulating FtsZ prior to formation of the asymmetric septum (see Fig. 4.2, legend). When the septum is formed, SpoIIE localizes on the prespore side and contributes to the activation of σ^F [GD (1998) *12* 1371–1380]. By the time septation is complete, σ^F is active in the forespore (only) and σ^E is active in the mother cell (only).

hemi-methylated (i.e. only the template strand methylated), and that the time required for further methylation (apparently needed for a functional *oriC*) may be a factor in regulating the initiation of chromosome replication in the cell cycle (section 3.2.1).

Transcription of the gene for a major regulatory factor, CtrA, in the cell cycle of *Caulobacter crescentus* (section 3.2.1.3) is timed by the transient occurrence of hemi-methylation at the P_1 promoter of *ctrA* [MM (2003) *47* 1279–1288].

In *E. coli*, the Dam methylase (encoded by the *dam* gene) methylates the N-6 position of adenine in 5'-GATC-3' sequences. Such 'Dam methylation' affects the transcription of certain genes when it occurs in their promoters. For example, Dam methylation in the promoter of the transposase gene of transposon Tn*10* (transposition: section 8.3) tends to inhibit synthesis of the transposase and, hence, to inhibit transposition of (chromosomally inserted) Tn*10* for most of the cell cycle; during DNA replication, however, this inhibition is transiently relieved immediately after the replication fork (Fig. 7.8) has passed the transposase gene but before Dam methylation has occurred – so that transposition of Tn*10* tends to be initiated during DNA replication (and, hence, to be linked to the cell cycle).

By contrast, many genes must be methylated in order to be transcribed. For example, the inability of uropathogenic strains of *E. coli* (UPEC) to transcribe genes for P fimbriae below 25°C has been attributed to the binding of a methylation-blocking factor, H-NS protein, at this temperature [MM (1998) *28* 1121–1137].

7.8.9 Regulation of gene expression by sigma factors

Specific sigma factors (section 7.5) are needed for the transcription of particular genes; this requirement gives the cell a mechanism for controlling the *timing* of certain events by synthesizing a given sigma factor only when necessary.

Some sigma factors are present at low levels during normal growth – higher levels induced by particular stress conditions leading to the expression of specific genes; these sigma factors include e.g. σ^H (*spo0H* gene product; Fig. 7.14), σ^{32} (*rpoH* gene product; section 7.8.2.3) and σ^s (*rpoS* gene product; a regulatory factor under various stress conditions). In some cases, stress conditions (heat shock for σ^{32}; e.g. osmotic shock for σ^s) promote the increased synthesis of a sigma factor by causing increased translation of the gene's mRNA (rather than increased transcription of the gene); the increased translation of σ^{32}, and of σ^s, appears to involve an effect of the inducing condition on a secondary structure in the mRNA.

σ^s, once associated solely with stationary-phase events, is now known to be involved in the regulation of a range of stress responses under stationary- and log-phase conditions [MM (1996) *21* 887–893] – often (perhaps always) in association with other forms of regulation. Thus, e.g. σ^s and the Fur protein are both regulatory elements in acid tolerance (section 7.8.2.6), while in osmotic up-shift (section 3.1.8), the resulting enhanced levels of K$^+$ and glutamate may serve to promote transcription from certain σ^s-regulated promoters [MM (1995) *16* 649–656].

Sigma factors are also involved in phage development (see e.g. section 9.1.1).

Anti-sigma factors include those molecules which can inhibit the normal activity of a given sigma factor by binding to it; clearly, such binding will inhibit the expression of those genes whose transcription depends upon the given sigma factor. One example of an anti-sigma factor is FlgM (section 7.8.2.7). Another is the AsiA protein, encoded by bacteriophage T4, which can inhibit the σ^{70} of *E. coli*. [Anti-sigma factors: ARM (1998) *52* 231–286.]

The Rsd protein ('regulator of sigma D'), encoded by *rsd*, is an *E. coli* protein which can inhibit σ^{70}. It has been proposed that Rsd is involved in the switching from σ^{70} to σ^s during the transition from exponential growth to the stationary phase [JB (1999) *181* 3768–3776].

7.8.10 Regulation of gene expression by tRNA-directed transcription antitermination

In *Bacillus subtilis* (and some other Gram-positive bacteria), certain genes whose products are involved in amino acid synthesis (and in the linkage of amino acids to tRNA molecules) are switched on when a shortage of the relevant amino acid gives rise to uncharged molecules of the corresponding tRNA; an uncharged tRNA molecule interacts with the leader region of the gene's mRNA and may cause a switch from a transcription terminator structure (formed when there are adequate amounts of the amino acid) to an antiterminator structure – so that transcription continues and the amino acid is subsequently synthesized.

[tRNA-directed transcription antitermination: MM (1994) *13* 381–387.]

7.8.11 Regulation of gene expression by temperature

Temperature can regulate a diverse range of genes. In pathogens, for example, certain virulence characteristics may be switched on at body temperature (37°C) – being absent at lower temperatures.

In *Salmonella*, the TipA *protein* (function unknown) acts as a thermosensitive regulator of its own transcription: the dimeric form of TipA represses transcription, but the monomer (promoted by temperature up-shift) does not [Cell (1997) *90* 55–64].

Repression of synthesis of *E. coli* P fimbriae (Table 2.2) at 25°C involves a small DNA-binding protein, H-NS; the binding of H-NS, which occurs at 25°C but not at 37°C, appears to inhibit transcription by acting as a methylation-blocking agent – methylation of the gene (section 7.4) being necessary for transcription [MM (1998) *28* 1121–1137]. H-NS also suppresses DNA repair in *Shigella* in a thermosensitive manner

[JB (1998) *180* 5260–5262]. Other factors which may be involved in H-NS-mediated thermoregulation of genes include: (i) temperature-induced conformational change in H-NS itself, affecting binding, and (ii) temperature-induced changes in DNA super-coiling/topology.

In some cases changes in temperature can change the structure of specific mRNA molecules such that, according to temperature, *translation* (section 7.6) is either permitted or prevented. An example is the mRNA of *IcrF* (a virulence-associated gene in *Yersinia pestis*). *Transcription* of *IcrF* is similar at 25 and 37°C, but at 25°C the mRNA forms a secondary structure that inhibits translation; at 37°C this structure 'melts', permitting translation.

In the heat-shock response (section 7.8.2.3), up-regulation of the sigma factor σ^{32} is due to increased translation of the *rpoH* mRNA; this is reported to involve de-stabilization of an inhibitory secondary structure in the mRNA as a result of the higher temperature [GD (1999) *13* 633–636].

A down-shift in temperature can also modify gene expression (section 7.8.2.4).

[Thermoregulation of genes in bacteria: MM (1998) *30* 1–6.]

7.8.12   Regulation of gene expression by circadian rhythms

Circadian rhythms are cyclical variations (approx. 24-hour cycles) observable in the characteristics of organisms kept under uniform (constant) environmental conditions; once thought to affect only eukaryotes, they are now known to affect gene expression in at least some prokaryotes.

In the unicellular cyanobacterium *Synechococcus*, gene *psbAI* (encoding a compo-nent of PSII – Fig. 5.11) is expressed least at dawn and most at dusk; some other genes are regulated in the opposite phase. Further, in constant light, the *timing* of cell division is influenced by the endogenous ~24-hour 'biological clock', even when *average* doubling time is only 10.5 hours [TIM (1998) *6* 407–410]. [Molecular bases for circadian clocks (review): Cell (1999) *96* 271–290. Circadian programmes in cyano-bacteria: ARM (1999) *53* 389–409.]

7.9   RNA

RNA molecules (section 7.2.2) have diverse forms and functions. For example:

• rRNA, mRNA, tRNA: molecules involved in protein synthesis (sections 7.5, 7.6).
• ctRNA (countertranscript RNA): any RNA molecule which, by binding to a tran-script, affects transcription, translation or replication. Examples include RNA I and other regulators of DNA replication in certain plasmids (section 7.3.1).
• SRP RNA: a molecule involved in transmembrane transport of proteins (section 5.4.3)

- pRNA: bacteriophage-encoded RNA involved in DNA packaging (section 9.1.4).
- RNase P: a *ribozyme*, i.e. an enzymic form of RNA. [Characteristics and properties of ribozymes: FEMSMR (1999) *23* 257–275.]
- siRNA: see section 7.9.2

7.9.1 Post-transcriptional modification of RNA

Many molecules of RNA are modified after their synthesis on the DNA template. For example, some bacterial mRNAs are polyadenylated (section 7.6.1). Again, the 16S, 23S and 5S rRNAs – together with tRNA molecules – are transcribed as a single unit that must be cut at specific sites by appropriate enzymes; for example, RNase E (section 7.6.1) contributes to the formation of 5S rRNA, and the ribozyme RNase P trims the 5' side of tRNA molecules.

7.9.1.1 Introns

Like typical eukaryotic genes, some bacterial, phage and archaeal genes include *introns*: sequences of nucleotides that do not encode any part of the gene product; in the transcript of such genes, any part corresponding to an intron must be removed ('spliced out') in order to produce a mature mRNA that correctly encodes the product (see Fig. 8.8).

Bacterial introns are classified into groups I and II; introns in the two groups differ in their splicing mechanisms. However, it appears that all bacterial introns are self-splicing, i.e. *autocatalytic*; this contrasts with many of the introns in eukaryotes – which are spliced out only with help from external factors such as enzymes. The group I introns seem to occur primarily within tRNA genes; they have been found in *Anabaena* and some other cyanobacteria, in *Agrobacterium tumefaciens* and in *Simkania negevensis* (a member of the Chlamydiales). [Structure–function relationships of the intron ribozymes in the purple bacterium *Azoarcus* and the cyanobacterium *Synechococcus*: NAR (2000) *28* 3269–3277.]

Bacterial group II introns have been found in e.g. *Lactococcus* spp, *Escherichia coli*, *Pseudomonas alcaligenes* and *Streptococcus pneumoniae*. Unlike group I introns they are not associated primarily with tRNA genes.

Both group I and group II introns are mobile genetic elements. For example, they can spread, replicatively, to specific sites in intron-less allelic genes by a process called *intron homing*.

[Bacterial group II introns: MM (2000) *38* 917–926. Barriers to intron promiscuity in bacteria: JB (2000) *182* 5281–5289.]

7.9.2   RNA interference (RNAi)

RNAi is primarily an antiviral response in plants and animals. In that some viral diseases (e.g. respiratory diseases) are risk factors for bacterial incursion and secondary infection, this phenomenon is relevant in bacteriology.

RNAi is triggered by the intracellular presence of double-stranded RNA (dsRNA) – an intermediate in the replication of certain viruses. Long molecules of dsRNA induce the formation of interferons (section 11.4.1.2); interferons can cause a *generalized* reaction that includes inhibition of synthesis of both viral and cellular proteins, and the degradation of both viral and cellular RNA by RNase L.

In RNAi, long molecules of dsRNA are cleaved processively by a dsRNA-specific RNase III-like endonuclease referred to as Dicer (= Dicer-RDE–1); the products are fragments of dsRNA, approximately 21–23 nucleotides in length, called *small interfering RNA* (= siRNA). Molecules of siRNA associate with a multiprotein complex (the *RNA-induced silencing complex*), and the whole can then degrade homologous mRNA in the cell – thus silencing the corresponding gene(s).

The introduction of synthetic 21-nucleotide siRNA duplexes into mammalian cells has been found to induce RNAi (independently of the Dicer enzyme).

siRNA molecules are too short to induce interferons, and their ability to inhibit gene expression in a sequence-specific way has suggested uses in medicine.

Replication of the hepatitis C virus in tissue cultures of Huh-7 cells has been specifically inhibited by siRNAs in an interferon-independent manner [PNAS (2003) *100* 2014–2018].

siRNAs specific for conserved regions of the influenza A virus genome strongly inhibited viral replication in both tissue cultures and embryonated eggs, such inhibition depending on the presence of a functional antisense sequence within the siRNA duplex [PNAS (2003) *100* 2718–2723].

The *specificity* of siRNA-mediated gene silencing has been examined in human cells using a genome-wide approach; such gene silencing was found to be highly specific for the target gene [PNAS (2003) *100* 6343–6346].

The possible therapeutic use of siRNA (including the problem of delivery to target cells) is discussed in a recent review [BMJ (2004) *328* 1245–1248].

8 Molecular biology II: changing the message

DNA can change. For example, even while replicating – and despite proof-reading and repair systems (section 7.7) – about one in 10^8–10^{10} 'wrong' nucleotides are believed to be incorporated in the new (daughter) strand. Greater changes can be brought about by chemical and physical *mutagens* (section 8.1), by recombination (section 8.2), and by *transposable elements* (section 8.3). Additionally, the cell's *genome* (its 'genetic blueprint') can be supplemented by plasmids and by other pieces of extra DNA via the processes of gene transfer (section 8.4).

Man-made changes in DNA (recombinant DNA technology) are considered in section 8.5.

8.1 MUTATION

In bacteria, a *mutation* is a stable, heritable change in the sequence of nucleotides in the DNA. (In some organisms – e.g. some bacteriophages (Chapter 9) – the genome consists of RNA, so that mutations in these organisms affect the RNA.) Note that a change in even a single *base-pair* (section 7.2.1) changes the sequence. Transcription of altered DNA produces altered RNA, and altered mRNA may specify a different polypeptide (Table 7.2) with different biological activity. Of course, mutations can affect control and recognition sequences as well as sequences encoding polypeptides.

Spontaneous mutations occur at low frequency with no obvious cause; they are mainly errors in replication/repair, but some involve *chemical* change (section 7.7). Higher rates of mutation occur when certain so-called *mutator* genes are defective; a mutator gene is one in which mutations may cause a higher rate of spontaneous mutation in *other* genes. Not surprisingly, mutator genes are found among those genes which encode the proteins involved in DNA replication and repair. For example, mutations that affect the ε subunit of DNA polymerase III (the major DNA polymerase in *E. coli*) may result in extremely high levels of spontaneous mutation. Other mutator genes in *E. coli* include genes involved in the mismatch repair system (section 7.7.1.1). (See also section 8.5.5.1.) [Mutator genes in *Escherichia coli*: TIM (1999) *7* 29–36.]

In the laboratory, mutation can be encouraged by *mutagens*: physical agents such as ultraviolet radiation and X-rays, and chemicals such as alkylating agents, bisulphite, hydroxylamine and nitrous acid as well as 'base analogues' (such as 5-bromouracil) which can be incorporated, in place of normal DNA bases, during DNA replication.

Mutagens work in various ways. Ultraviolet radiation can cause e.g. covalent cross-linking between adjacent thymines; correction of the resulting *thymine dimers* by so-called error-prone repair (section 7.8.2.2) can generate a variety of mutations. (See also section 7.7.2.) Cross-linking can also occur with some alkylating agents. Bisul-phites and nitrous acid can e.g. deaminate cytosine to uracil; although uracil is not a stable constituent of DNA (section 7.2), its different base-pairing specificity can cause the insertion of a different base (adenine instead of guanine) in the daughter strand at the next round of replication.

In a population of bacteria, mutations normally occur randomly, affecting different genes in different individuals. (Interestingly, in enterobacteria, mutations appear to occur more frequently at sites furthest from *oriC* [Science (1989) *246* 808–810].) A cell in which a mutation has occurred is called a *mutant*. Mutations are often harmful – and may be lethal if the affected sequence of nucleotides encodes a vital product or function. Beneficial mutations include e.g. those which increase the cell's resistance to antibiotic(s). For example, a mutation may result in an altered ribosome such that streptomycin (section 15.4.2) no longer binds to the ribosome and (therefore) does not inhibit protein synthesis; the (mutant) cell will thus exhibit resistance to this antibiotic.

A mutation giving increased fitness for growth under existing conditions may enable the (mutant) cell to outgrow other (non-mutant: *wild type*) individuals in the population and become numerically dominant in that population; such 'natural selection' underlies the concept of *evolution*.

8.1.1 Types of mutation

Mutations occur in various ways, and they can have various effects on the genetic message. Infrequently, a piece of DNA is lost, gained, inverted – or even *transposed* (section 8.3). A *point mutation* involves the loss, gain or substitution of a single nucleotide; even this, however, can have far-reaching consequences for the cell (Fig. 8.1).

8.1.1.1 *Nomenclature of loci, genes and mutant genes; genotype and phenotype*

On a chromosome, a given, functionally defined location (= *locus*; plural: *loci*) is designated by a group of three letters printed in italics (or underlined if handwritten). For example, *his* and *trp* are two loci concerned, respectively, with the biosynthesis of histidine and tryptophan.

When a given locus contains more than one gene, each gene is identified by a capital letter; thus, for example, the *his* locus is an operon (section 7.8.1) containing a number of genes designated *hisA*, *hisB* . . . etc. (Note that genes designated in this way do not necessarily occur in alphabetical order in the locus – see e.g. the *lac* genes in Fig. 7.11.) Genes which are *not* contiguous on the chromosome, but which have a related function, are designated in a similar way (see e.g. *ara* in section 7.8.1.1).

The above system is used when referring to a particular locus or gene – e.g. the *his* locus, the *hisA* gene.

A cell's *genotype* is its genetic make-up, a characteristic reflecting the actual sequence of nucleotides in the chromosome. To describe the genotype of a given strain we generally list the particular genes of interest, indicating whether they are wild-type or mutant. A wild-type gene is shown as e.g. *hisA*$^+$, while the corresponding mutant gene is shown as *hisA* (or sometimes *hisA*$^-$). Specific mutations (at particular sites) are designated by numbers: e.g. *hisA9* (a mutation at a particular site in the *hisA* gene).

A cell's *phenotype* is a set of *observable* characteristics. Phenotypic characteristics are symbolized as e.g. His$^-$ (inability to grow without histidine – i.e. a histidine *auxotroph*, section 8.1.2); His$^+$ (a histidine prototroph); Lac$^-$ (inability to use lactose); Met$^-$ (a methionine auxotroph); Tc^R (resistance to tetracyclines).

8.1.2   The isolation of mutants

A *particular* mutation occurs spontaneously only at very low frequency in a population of bacteria; for example, within a population of *E. coli*, the loss of ability to ferment galactose occurs (on average) once every 10^{10} cell division cycles, i.e. a *mutation rate* of 10^{-10}. Even in populations treated with a mutagen, cells with a *particular* mutation are still greatly outnumbered by wild-type cells and by those with other types of mutation. How can we isolate the one (or few) *specific* mutants from a large population of bacteria?

8.1.2.1   *Case 1: mutants which can grow under conditions that inhibit wild-type cells*

If, in a population of streptomycin-sensitive bacteria, a single cell has mutated to streptomycin resistance (section 8.1), that cell can grow on a solid medium containing streptomycin and can form a colony (section 3.3.1); all the other cells, which are inhibited by streptomycin, will not grow on such a medium. In general, this type of selective method can be used whenever the mutant is able to grow under conditions that inhibit non-mutant cells.

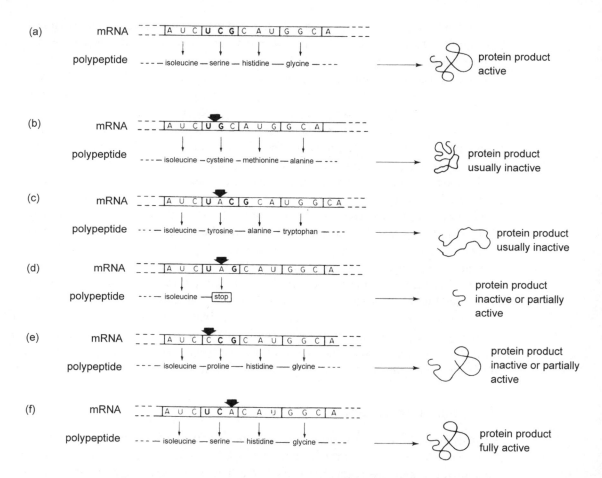

Figure 8.1 Point mutations: their effects on mRNA and on polypeptide synthesis. The effect of each point mutation is indicated by a heavy arrow (⬇); at the right is shown a possible effect of each of the mutations.

(a) mRNA and polypeptide synthesized from the normal (non-mutant, wild-type) gene.

(b) The *deletion* of a guanine nucleotide from DNA has resulted in the loss of cytosine (C) from the codon UCG in mRNA. The effect of this is that not only UCG but all subsequent codons are altered: compare the amino acids encoded in (a) and (b). Notice that, if one nucleotide is missing, the next nucleotide is read in its place, i.e., groups of three consecutive nucleotides continue to be read as codons. Because the genetic message is out-of-phase 'downstream' of the deletion, such a mutation is called a *phase-shift* or *frame-shift mutation*; if it occurs near the end of a gene, so that most of the polypeptide is normal, the product may have some biological activity. If a phase-shift mutation occurs in an operon (section 7.8.1) the effect will vary greatly according to the particular site affected.

(c) The *addition* of a thymine nucleotide to DNA has resulted in the addition of an adenine nucleotide to codon UCG in the mRNA; as in (b), above, this is a phase-shift mutation.

(d) In DNA, thymine has replaced guanine, so that the mRNA now contains UAG (a 'stop' codon) instead of UCG; this is a so-called *nonsense mutation*. Polypeptide synthesis stops at UAG; the polypeptide may have some biological activity if much or most of it has been translated prior to UAG.

8.1.2.2 Case 2: mutants whose requirements are greater than those of wild-type cells

Suppose that, as a result of mutation, a cell has lost the ability to synthesize a particular compound (e.g. amino acid) which is necessary for growth; such a metabolically dependent mutant (an *auxotroph*) can grow only if (i) it is supplied with the appropriate compound, and (ii) it has a (functional) transport (uptake) system for that compound. (The corresponding wild-type, i.e. non-mutant, cell is called a *prototroph*.)

How are auxotrophs isolated – given that any medium which allows an auxotroph to grow will also allow prototrophs to grow? One method is to use a *minimal medium*, i.e. a medium which contains the minimum range of nutrients needed by prototrophs; a given auxotroph can grow on minimal medium only if the medium has been supplemented with the auxotroph's specific growth requirement(s). If, for example, we wish to isolate a histidine-requiring auxotroph, minimal medium is supplemented with a *low* concentration of histidine – allowing *limited* growth of the auxotroph; a colony formed by an auxotrophic cell will soon exhaust the histidine in its vicinity (and therefore remain small) while the colonies of prototrophs reach a normal size. Small colonies are *presumed* to be those of auxotrophs and can be tested further.

An alternative method for isolating auxotrophs uses the antibiotic streptozotocin which kills the *growing* cells of certain types of bacteria; for susceptibility to streptozotocin, a cell must have a functional PTS (section 5.4.2) for *N*-acetylglucosamine – the transport system responsible for uptake of streptozotocin. In this method, a population of bacteria is inoculated into a liquid *minimal medium* containing streptozotocin. Prototrophs, which grow in the medium, are killed by the antibiotic; auxotrophs survive. To recover auxotrophs, the antibiotic can be rendered ineffective by diluting the bacterial population in a suitable antibiotic-free medium. Streptozotocin is useful for isolating auxotrophs because (i) cells killed by streptozotocin (unlike those killed by penicillin) do not lyse and therefore do not act as a source of nutrients (which would allow auxotrophs to grow and develop susceptibility to the antibiotic), and (ii) non-growing cells are unaffected by the antibiotic.

In yet another method for isolating auxotrophs, a low-density population containing both prototrophic and auxotrophic cells is inoculated onto a *complete medium* (on which both prototrophs and auxotrophs can grow). Following incubation, normal-sized colonies are formed by all the cells, and in order to identify the colonies of auxotrophs it is necessary to inoculate each colony onto minimal medium – on which

---

(e) In DNA, guanine has replaced adenine, so that the mRNA now contains CCG instead of UCG – the altered codon specifying proline rather than serine; this is a so-called *mis-sense mutation*. Note that the amino acids downstream of proline are not affected. The biological activity of the polypeptide will depend on the nature and position of the incorrect amino acid.

(f) In DNA, thymine has replaced cytosine, so that the mRNA now contains UCA instead of UCG; however, the altered codon still encodes serine (Table 7.2). This is a *silent mutation*; it does not, of course, affect the biological activity of the polypeptide.

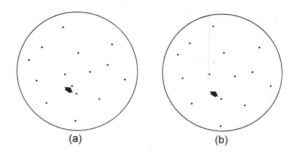

(a)   (b)

Figure 8.2 Replica plating as used for isolating an auxotrophic mutant. (a) Master plate: complete medium with colonies of both prototrophs and auxotrophs. (b) Replica plate: minimal medium with colonies of prototrophs only. On the master plate an arrow indicates a colony of a presumed auxotroph; there is no colony at the corresponding position (arrowed) on the replica plate (auxotrophs cannot grow on minimal medium).

only the prototrophs will grow. This is conveniently achieved by *replica plating*. In this method a disc of sterile velvet is pressed gently onto the surface of the complete medium (the *master plate*) containing the colonies of both prototrophs and auxotrophs; cells from each colony stick to the velvet – which is then pressed lightly onto the surface of a sterile plate of minimal medium (the *replica plate*). After incubation, the positions of colonies on the replica plate are compared with those of colonies on the master plate; any colony which occurs on the master plate but not on the replica plate is presumed to be that of an auxotroph (Fig. 8.2).

8.1.3 The Ames test for carcinogens (*Salmonella*/microsome assay)

Most known carcinogens (cancer-promoting agents) are mutagens, and this test checks for potential carcinogenicity by checking for mutagenicity (the ability to cause mutations). The mutagenicity of a chemical is checked by determining its ability to reverse a previous mutation in the test organism, *Salmonella typhimurium*. (A mutation which reverses a previous mutation is called a *back mutation*.) The (mutant) test strains of *S. typhimurium* are auxotrophic for histidine, and the Ames test checks for back-mutation to prototrophy. Essentially, prototrophs are sought in an incubated mixture containing a population of the test strain, the chemical under test, and a preparation of enzymes from rat's liver; the enzymes are included because some mutagens/carcinogens need metabolic 'activation'.

8.2 RECOMBINATION

Recombination means re-arrangement of one or more molecules of nucleic acid, i.e. a redistribution of sequence(s) of nucleotides; thus, molecules may join together, exchange strands etc., or a sequence within a molecule may be moved or inverted etc.

Categories of recombination include homologous (general) recombination (section 8.2.1), site-specific recombination (section 8.2.2) and transpositional recombination (section 8.3). Homologous recombination can play a role in the repair of damaged DNA, site-specific recombination may have a clearly regulatory function, and transpositional recombination (which can involve random insertion of a sequence of nucleotides into a gene) may have deleterious effects.

8.2.1   Homologous (general) recombination

This form of recombination can occur between two dsDNA molecules – but only if a long sequence of nucleotides in one duplex is very similar to a sequence in the other, i.e. the two duplexes must have an extensive region of *homology*. This process is involved e.g. in the repair of DNA following damage by ultraviolet radiation or X-rays. *Escherichia coli* has at least two major pathways ('recombination machines') for homologous recombination:

- RecBCD. This pathway (i) is generally responsible for the repair of double-stranded breaks in DNA, (ii) is highly effective only on *linear* substrates (e.g. circular chromosomes containing a double-stranded break), and (iii) involves a single (multisubunit) protein (330 kDa) with both helicase and nuclease functions as well as a synaptogenic function (see later);
- RecF. This pathway (i) is involved e.g. in the repair of single-stranded gaps in the chromosome, (ii) is effective on circular molecules of dsDNA, and (iii) involves the activities of a number of separate proteins.

Each of the above pathways can *initiate* recombination; recombination is initiated (by the appropriate pathway) if chromosomal DNA has a *gap* (i.e. a single-stranded region) or a double-stranded break.

In apparently all cases, homologous recombination begins with the formation of a so-called *nucleoprotein filament*; the development of this structure can be promoted by the RecBCD system, the RecF system or (see later) by a hybrid system involving components from both the RecBCD and RecF systems.

The (helical) nucleoprotein filament consists of a free 3'-end of a strand of DNA to which monomers of the RecA protein have bound. RecA has the ability to promote various interactions between molecules of DNA. (In members of the Archaea the RadA protein appears to be analogous to RecA [GD (1998) *12* 1248–1253].)

Given a double-stranded break in DNA, the RecBCD helicase unwinds the duplex while the (ATP-dependent) nuclease may degrade the 5'-terminal, leaving a free (single-stranded) 3'-terminal. RecBCD also mediates the exclusion of single-strand binding proteins, thus allowing the binding of RecA monomers to the 3'-terminal ssDNA (forming the nucleoprotein filament). In mutants defective in RecBCD, unwinding of the duplex and degradation of the 5'-terminal may be conducted by the RecQ and RecJ proteins (respectively) of the RecF system. The nucleoprotein filament may then function in a way analogous to that in the Holliday model.

The Holliday model considers nicked duplexes (*nick* = a break in the sugar–phosphate backbone). The nucleoprotein filament from one nicked duplex 'searches' the other duplex (target DNA) for a homologous sequence of nucleotides. Contrary to an earlier view, the searching is unlikely to involve a triplex DNA intermediate. It probably involves unwinding of target DNA, and local strand exchange, leading to juxtaposition of homologous sequences (*synapsis*) and the formation of a hybrid duplex (*heteroduplex*); moreover, it appears that the *polarity* of the nucleoprotein filament does not affect its ability to interact with a circular target duplex [NAR (2001) *29* 1389–1398]. The free end displaced from the target duplex may then pair with a sequence in the first duplex, again forming a heteroduplex. The resulting structure is called a *heteroduplex joint.* If the free 3'-end from each duplex is now ligated to the broken end in the other duplex, the result is a (covalently) branched structure (a *Holliday junction* or *chi structure*): a region where two duplexes are held together by two single strands crossing between the duplexes. Each strand may continue to separate from its *parent* duplex, and pair with the strand from the other duplex, so that the branched region will migrate, between the duplexes, in the corresponding direction (= *branch migration*); the extent of branch migration determines the final length of the heteroduplex.

In *Escherichia coli*, branch migration may be driven by two ring-shaped helicase complexes (one on either side of the Holliday junction) through which the DNA passes; each complex is a hexamer of the RuvB protein. (Compare this with the DnaB helicase in Fig. 7.8.) Resolution of the Holliday junction (i.e. separation of duplexes) may involve an endonuclease, RuvC, which cuts each strand of the Holliday junction, thus allowing final ligation to occur in both duplexes. In *E. coli*, RuvB and RuvC appear to interact co-operatively [EMBO Journal (1998) *17* 1838–1845].

In at least some organisms the chromosome contains certain sequences of DNA (referred to as recombinational *hotspots*) which enhance homologous recombination in their vicinity. For example, the chromosome of *E. coli* contains about 1000 copies of the so-called chi (χ) site:

5'-GCTGGTGG-3'

which enhances recombination by up to 10-fold; the influence of χ extends (with decreasing magnitude) for about 10 kb on one side of the site. RecA was reported to bind preferentially to GT-rich sequences (including the chi site); this suggested that the activity of such sites may depend, at least in part, on their ability to bind RecA and thus to promote pairing of DNA strands in their vicinity [GD (1996) *10* 1890–1903]. Subsequent work has indicated that RecBCD interacts with chi sites and that this leads e.g. to the presence of the chi site at or near the 3'-terminus of DNA in the nucleo-protein filament [see e.g. Cell (2003) *112* 741–744].

Interestingly, in *E. coli*, parts of the RecBCD and RecF systems are interchangeable; for example, a null mutation in the *recD* (nuclease) function of RecBCD can be compensated for by the RecJ nuclease of the RecF system. Hence, if part(s) of a given system are inactive (through mutation) a functional system for homologous recombination can still be operative. [Interchangeable parts of the *E. coli* recombination machinery: Cell (2003) *112* 741–744.]

8.2.2   Site-specific recombination

In the basic form of site-specific recombination (SSR), two *specific* sequences of duplex DNA are brought together, a protein catalyst (a *recombinase*) binding in the region of juxtaposition; cuts are made at staggered sites in each duplex, and the cut ends of one duplex are ligated (joined) to the cut ends of the other (see Fig. 8.3). [Action of site-specific recombinases: TIG (1992) *8* 432–439.]

SSR can control gene expression; in such *recombinational regulation*, an SSR event controls the on/off switching of gene(s) or a switch from one gene to another (Fig. 8.3c). SSR is also involved e.g. in the integration of bacteriophage λ DNA (section 9.2.1) with the *E. coli* chromosome (Fig. 8.3b), in some cases of plasmid–chromosome interaction [JB (1992) *174* 7495–7499] and in the replicative form of transposition (Fig. 8.4).

8.3   TRANSPOSITION

Transposition is the transfer of a small, specialized piece of DNA – or a 'copy' of it – from one site to another in the same duplex, to a site in a different duplex in the same cell, or (see conjugative transposons, section 8.4.2.3) to a duplex in another cell. The 'small, specialized piece of DNA' is called a *transposable element* (TE; jumping gene); TEs occur e.g. in bacterial chromosomes, in the DNA of bacteriophages (Chapter 9), and in plasmids. The following account refers to 'classical' transposable elements (i.e. excluding conjugative transposons).

The two main types of TE are the *insertion sequence* (IS) and the *transposon*. Each encodes protein(s) – including a *transposase* (see later) – needed for transposition; additionally, a transposon encodes other functions – e.g. enzyme(s) which inactive particular antibiotic(s).

In all TEs, the nucleotide sequence includes at least one pair of *inverted repeats*. An example of a pair of terminal inverted repeats:

                    5'-CTGACTA.................TAGTCAG-3'
                    3'-GACTGAT.................ATCAGTC-5'

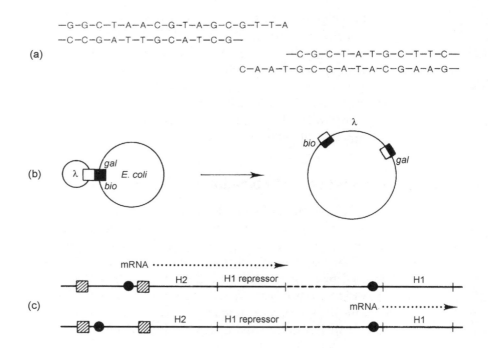

—G–G–C–T–A–A–C–G–T–A–G–C–G–T–T–A
—C–C–G–A–T–T–G–C–A–T–C–G—

(a)

—C–G–C–T–A–T–G–C–T–T–C—
C–A–A–T–G–C–G–A–T–A–C–G–A–A–G—

(b)

(c)

mRNA

H2   H1 repressor   H1

H2   H1 repressor   H1

mRNA

Figure 8.3   Site-specific recombination (diagrammatic): the principle (a) and some examples (b, c).
(a) A staggered break at one specific site in a DNA duplex; a staggered break, typically involving the same
nucleotide sequence, is made at another site in the same duplex or in another DNA molecule. Each
single-stranded region at a given breakage site can base-pair with a complementary region at the other
breakage site. (The single-stranded regions are sometimes called 'sticky ends'.) The process is mediated
by a protein (a *recombinase*). In the diagram, a staggered break is shown with two 4-nucleotide-long
single-stranded regions. However, each single-stranded region is normally either 2 nucleotides long or
6–8 nucleotides long, the length of this 'overhang' (and the actual sequence in the sticky ends)
depending on the particular recombinase involved.

(b) Integration of the (circular, double-stranded) DNA of bacteriophage λ (section 9.2.1) with the *E. coli*
chromosome. In λ DNA, the specific recombinational site (i.e. the specific sequence of nucleotides in the
duplex) is shown as a white square; in the *E. coli* chromosome it is shown as a black square flanked on either
side by the *gal* operon (galactose utilization) and the *bio* genes (biotin synthesis). *Left:* the two circular DNA
duplexes are shown with their recombinational sites juxtaposed. Initially, with the recombinase bound at the
juxtaposed sites, a staggered break is made across each duplex; sticky ends in the chromosome then base-pair
with those in the λ duplex, and the strands are ligated. *Right:* λ DNA incorporated in the *E. coli* chromosome.

(c) An example of recombinational regulation. In most strains of *Salmonella* the flagellar filament (section
2.2.14.1) can contain either of two distinct types of protein – encoded by genes H1 and H2; normally, only one
of these genes is expressed at any given time, so that the flagellar filament contains either the H1 gene product
or the H2 gene product. The promoter (●) of the H2 operon is flanked by two specific sites (▨) that are
recognized by a recombinase. *Top:* the H2 operon is transcribed; the H2 gene product forms the flagellar
filament, and the H1 repressor stops transcription of the H1 gene. *Bottom:* site-specific recombination has
occurred between the sites recognized by the recombinase, and the sequence containing the H2 promoter has

Notice that the left-hand end of the duplex has the same *polarity* as the right-hand end, i.e. the sequence is the same when read from the 5' (or 3') end in both directions – but the two ends of the duplex have opposite *orientations*. Inverted repeats seem to be necessary for transposition, probably being recognized by the appropriate transposase.

A TE may transpose by simple or replicative transposition (see Fig. 8.4).

For some TEs the target site is highly specific. Others TEs seemingly insert almost at random, although the target sites for these are likely to involve some kind of selective process [ARB (1997) *66* 437–474]. Studies on transposition of the insertion sequence IS*903* into a large (55 kb) plasmid have shown that, although insertion can occur at many different sites, there are certain *preferred* regions into which IS*903* will insert more than once on different occasions [JB (1998) *180* 3039–3048].

Transposition is usually a rare event It can generate a wide variety of re-arrangements.

8.3.1   Insertion sequences

An insertion sequence is a transposable element that encodes only those functions involved in transposition (cf. transposons, section 8.3.2). It consists of a pair of inverted repeats (see above), each about 10–40 nucleotides long, bracketting the transposition genes.

Insertion sequences are designated e.g. IS*1*, IS*2*, IS*3* etc. (note number in *italics*).

Insertion sequences occur e.g. in chromosomes and plasmids – in which their presence may affect specific properties. For example, the F plasmid is constitutively derepressed for conjugative transfer because an insertion sequence (IS*3*) occurs within (and inactivates) the plasmid's *finO* sequence (which is involved in inhibition of transfer); most plasmids related to the F plasmid (with a functional *finO* locus) are repressed for transfer.

Different insertion sequences transpose with different frequencies that depend on e.g. the structure of the donor and target sequences and the physiological state of the host cell. Transposition frequences may range from $\sim 10^{-9}$ to $\sim 10^{-5}$ per insertion sequence per cell division.

---

been inverted; without a functional promoter, the H2 operon cannot be transcribed, but the loss of the H1 repressor permits transcription from H1 – so that the filament is now made from the H1 gene product.

Variation in the composition of the *Salmonella* flagellar filament is only one example of *phase variation*: a more general phenomenon in which the composition of certain cell surface components or structures undergoes spontaneous change. For example, in individual cells of *E. coli*, the so-called type 1 fimbriae are subject to on/off switching – resulting in spontaneous changes from a fimbriate to an afimbriate state, and vice versa; this is also due to DNA re-arrangement.

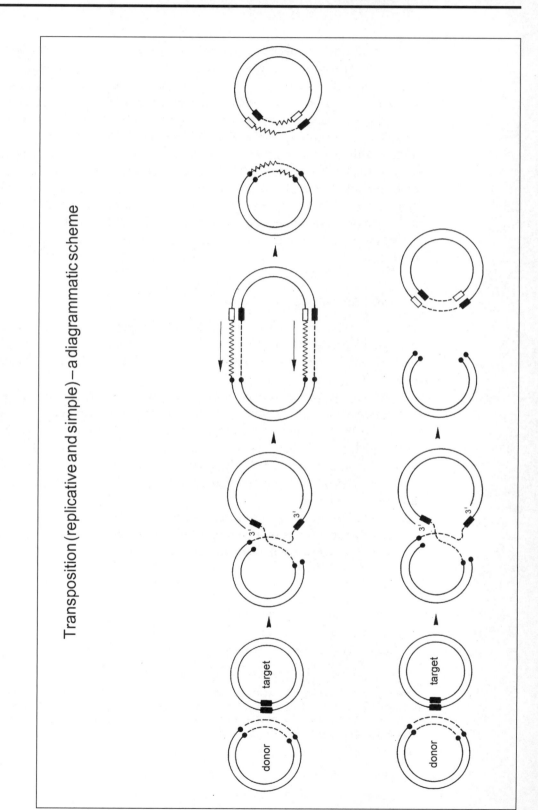

Transposition (replicative and simple) – a diagrammatic scheme

Some examples:

IS*1*.   A 768-bp insertion sequence found e.g. in R plasmids (which encode resistance to antibiotics(s)) and in the chromosome of *Escherichia coli* K12 strains. IS*1* seems to transfer preferentially to AT-rich target sites. Transposition of IS*1* is inhibited in the absence of H-NS protein (which seems to be needed for synthesis of transposase) [JB (2004) *186* 2091–2098].

IS*10*.   An insertion sequence apparently associated only with a certain transposon (Tn*10*).

IS*900*.   An insertion sequence reported to be specific to *Mycobacterium paratuberculosis*. It has been used as a target for PCR-based detection of *M. paratuberculosis* in milk [AEM (1998) *64* 3153–3158].

IS*6110*. An insertion sequence regarded as specific to members of the *Mycobacterium tuberculosis* complex; this specificity has been disputed by some [JCM (1997) *35* 799–800] and affirmed by others [JCM (1997) *35* 800–801]. The number of copies of IS*6110* per genome varies with strain (some strains of *M. tuberculosis* apparently lack the sequence); 16 copies of IS*6110* are found in the genome of

---

Figure 8.4   Transposition: a (diagrammatic) scheme for a transposon undergoing *replicative* transposition (top) and *simple* (cut-and-paste) transposition (below).

Top. Two circular, double-stranded DNA molecules are shown at the left-hand side. The donor molecule includes a transposon (dashed lines), either side of which is an old target site (●); the donor's target site was duplicated when the transposon was originally inserted into that molecule (see later). The target molecule has a single target site (■) where the transposon will be inserted.

Next, an enzyme (*transposase* – not shown) mediates at least the initial stages of transposition. In the target molecule, a staggered break has been made at the target site. In the donor molecule, a nick has been made in each strand of the transposon (at opposite ends), and the free ends have been ligated to the target molecule, as shown.

In the next stage, DNA synthesis (zigzag line) has occurred from each 3' end in the target molecule (arrows). Such synthesis first copies the target site (▢) and then continues beyond the target site – using each strand of the transposon as template; that is, the transposon has been replicated. The end of each new strand has been ligated (joined) to a free strand-end in the donor molecule. The resulting structure is called a *cointegrate*.

The final stage involves a *resolvase*: an enzyme, encoded by the transposon, which resolves the cointegrate by promoting site-specific recombination (section 8.2.2) at a site in each transposon – forming the two molecules shown.

The donor and target molecules now both contain a copy of the transposon. Notice that each copy of the transposon contains parts of the original transposon (dashed lines) as well as newly synthesized DNA (zigzag lines). Notice also that the target site in the target molecule has been duplicated.

Replicative transposition occurs e.g. in transposon Tn*3*.

Below. In simple transposition, the initial stages are similar to those shown above. However, DNA synthesis (from the 3' ends of the target molecule) is required simply to duplicate the target site. The remaining strand-ends of the transposon have been cut and ligated to the freshly duplicated target sequences, as shown. The rest of the donor molecule may be non-viable ('donor suicide').

Simple transposition occurs e.g. in transposon Tn*10*.

the H37rv reference strain of *M. tuberculosis* [Nature (1998) *393* 537–544]. IS*6110* has been used for *typing* (section 16.1.5) isolates of *M. tuberculosis*. In some strains of this pathogen the reliability of IS*6110*-based RFLP typing (section 16.2.2.6) may be compromised by the high rate of transposition of IS*6110* within the genome [JCM (1999) *37* 788–791].

8.3.2 Transposons

A transposon is a transposable element that encodes one or more functions in addition to those functions required for transposition (cf. insertion sequences, section 8.3.1).

Transposons are designated e.g. Tn*1*, Tn*2*, Tn*3* etc. (note number in *italics*).

Functions encoded by transposons include resistance to antibiotics (e.g. Tn*3*, Tn*5*, Tn*10*, Tn*1721*), resistance to heavy metals (e.g. Tn*501*), toxin production (e.g. Tn*1681* encodes the heat-stable toxin of ETEC), and lactose metabolism (e.g. Tn*951*).

Transposons have been classified according to structure:

- a *class I* transposon consists of a pair of insertion sequences (which are not necessarily identical) bracketting the gene sequence; example: transposon Tn*10*;
- a *class II* transposon consists of a pair of inverted repeats bracketting the gene sequence; example: Tn*3*.

Class I transposons are also called *composite* or *compound* transposons. Class II transposons are also called *simple* – or sometimes *complex* (!) – transposons.

Some examples:

Tn*3*.   A 4957-bp class II transposon found e.g. in plasmid R1*drd-19*. It has identical 38-bp terminal inverted repeats, carries a gene for a β-lactamase (section 15.4.1.2), and inserts preferentially into AT-rich sites. Replicative transposition (Fig. 8.4) involves the *tnpA* gene (encoding a transposase) and the *tnpR* gene (encoding a resolvase). (The Tn*3* cointegrate can also be resolved by general (homologous) recombination involving the *recA* system.)

Tn*5*.   A 5818-bp class I transposon containing genes for resistance to kanamycin and other aminoglycoside antibiotics. The three contiguous antibiotic-resistance genes are flanked by the insertion sequences IS*50*L and IS*50*R (which differ slightly from one another). Each insertion sequence is flanked by a pair of 19-bp sequences termed the outer end (OE) and inner end (IE) – the two OE sequences thus forming the two ends of the transposon. IS*50*R encodes Tnp, the transposase, and Inh, an inhibitor of Tnp; the relative levels of Tnp and Inh influence the frequency of transposition. Regulatory factors *in vivo* include Dam methylation (section 7.8.8) which negatively affects the promoter of the Tnp gene; as in Tn*10*, transposition of Tn*5* is coupled to DNA replication. Tn*5* has *in vitro* uses – e.g. in mutagenesis (section 8.5.5.3). Also, random insertion into a population of (e.g.) plasmids can generate templates (with known primer-binding sites) for complete sequencing.

Tn7. This transposon can insert into the chromosome of *E. coli* at a highly specific target site, designated *att*Tn7, apparently without disrupting cellular function; several Tn7-encoded proteins are involved, one of which (TnsD) distorts the binding site in a way that may recruit other proteins [EMBO Journal (2001) *20* 924–932]. Alternatively, Tn7 can transpose via a different mechanism, inserting into plasmids and into the *E. coli* chromosome at sites proximal to double-stranded breaks and at locations where DNA replication terminates [GD (2001) *15* 737–747].

8.3.3 Nomenclature for inserted TEs

The presence of a TE is symbolized by a double colon. For example, the presence of Tn3 in the plasmid R1*drd-19* is written R1*drd-19*::Tn3.

8.4 GENE TRANSFER

New DNA may enter a bacterium naturally through *transformation*, *conjugation* or *transduction*; transduction (requiring bacteriophages) is described in Chapter 9.

8.4.1 Transformation

In transformation, a bacterium takes up from its environment a piece of DNA; one strand of this *donor* DNA is internalized and may genetically transform the recipient cell by recombining with a homologous region of the chromosome. Transformation occurs naturally in various Gram-positive and Gram-negative bacteria (though not e.g. in *E. coli* – see section 8.4.1.1); it may e.g. contribute to the spread of antibiotic resistance in pathogenic bacteria.

Haemophilus influenzae and *Neisseria gonorrhoeae* bind donor DNA only when it includes certain *uptake-signal sequences* (which occur repeatedly in their chromosomes). This requirement seems to be lacking in *Bacillus subtilis* and *Streptococcus pneumoniae*.

The *competence* to bind and take up DNA is constitutive in *N. gonorrhoeae*. In *H. influenzae* it is induced by growth-inhibiting conditions – being promoted by high levels of intracellular cAMP. In both *B. subtilis* and *S. pneumoniae* competence is affected by nutritional status; in these species it is also affected by the population density of the bacteria, an example of *quorum sensing* (section 10.1.2). (In *B. subtilis* quorum sensing involves a pheromone, ComX, which promotes competence by activating a two-component regulatory system (section 7.8.6); this, in turn, leads to activation of the *comS* gene which is required for competence.) In *S. pneumoniae* competence has also been associated with the transmembrane transport of calcium [JB (1994) *176* 1992–1996].

[Competence in transformation: TIG (1996) *12* 150–155. Competence in *Bacillus subtilis* (pheromones): TIM (1998) *6* 288–294.]

Transformation was first observed by Griffith in the 1920s: a live, non-pathogenic strain of *S. pneumoniae* was found to become virulent when mixed with a dead, virulent strain; it was later found that DNA, released by the dead cells, had transformed the living ones – an early indication of the role of DNA as the carrier of genetic information.

8.4.1.1   *Laboratory-induced competence in transformation*

Competence can be induced (e.g. in *E. coli*) by certain procedures which increase the permeability of the cell envelope. For example, calcium chloride solution (approx. 50 mm, 0.2 ml) containing 10^8–10^9 washed, mid-log phase *E. coli* cells, is chilled on ice, and a DNA suspension (10 μl) is added to give a final DNA concentration of approx. 0.2 μg/ml; after further chilling at 0°C (15–30 min) the suspension is heat-shocked (42°C/1.5–2.0 min) and allowed to recover – e.g. returned to ice, then incubated in Luria–Bertani broth (1 ml) at 37°C for 1 hour. (LB broth contains (per litre): 10 g tryptone, 5 g yeast extract and 10 g NaCl; pH 7.5, adjusted with NaOH.)

The acquisition of competence by *E. coli* in ice-cold calcium solutions is associated with the presence of a high concentration of poly-β-hydroxybutyrate/calcium poly-phosphate complexes in the cytoplasmic membrane. It has been suggested that these complexes may form transmembrane channels which facilitate DNA transport, and that divalent cations may act as links between DNA and phosphate at the mouth of such channels [JB (1995) *177* 486–490].

Small, circular plasmids tend to transform more readily than do larger ones. In *E. coli*, linear dsDNA transforms poorly (if at all) because it is degraded by the RecBCD enzyme (section 8.2); however, it can transform some *recBCD* mutants.

8.4.1.2   *Electroporation*

High-efficiency transformation of *E. coli* with plasmids can be achieved by *electroporation*: a mixture of cells and plasmids is exposed to an electrical field of up to 16 kV/cm for a fraction of a second. The mechanism of the electrically induced uptake of DNA is not understood; important factors include field strength and pulse length. The frequency of transformation varies linearly with DNA concentration over a wide range of values, and the efficiency of transformation varies with cell concentration. 'Electrocompetent' (electroporation-competent) cells with high-level transformation efficiency are available commercially – e.g. the ElectroTen-Blue™ marketed by Stratagene.

[Electroporation in *Helicobacter pylori*: AEM (1997) *63* 4866–4871. High-efficiency electroporation: NAR (1999) *27* 910–9̄1.]

8.4.2 Conjugation

Certain (conjugative) plasmids, and conjugative transposons (section 8.4.2.3), confer on their host cells the ability to transfer DNA to other cells by *conjugation*; in this process, a *donor* (male) cell transfers DNA to a *recipient* (female) cell while the cells are in physical contact. A recipient which has received DNA from a donor is called a *transconjugant*.

Like transformation, conjugation can result in important phenotypic changes in the recipient – e.g. the acquisition of plasmid-encoded resistance to antibiotic(s).

In many cases, conjugation involves a specialized structure (a *pilus*: section 2.2.14.3) which is typically not synthesized constitutively. In general, conjugation-specific genes are expressed only under appropriate conditions or on receipt of specific signals. Thus, some systems are induced by particular antibiotics while others are triggered only when bacterial numbers reach certain levels (an example of quorum sensing: section 10.1.2). [Control of conjugative gene expression: FEMSMR (1998) *21* 291–319.]

We look first at plasmid-mediated conjugation (sections 8.4.2.1 and 8.4.2.2) and then at conjugative transposition (8.4.2.3).

8.4.2.1 *Conjugation in Gram-positive bacteria*

There are two main types of plasmid-mediated conjugation in Gram-positive bacteria: pheromone-mediated and pheromone-independent. Pili have not been demonstrated in either category.

In strains of *Enterococcus* (formerly *Streptococcus) faecalis*, potential recipients secrete small amounts of a short peptide signalling molecule (a *pheromone*) which causes the plasmid-containing donor cells to synthesize an adhesive cell-surface component. Plasmid DNA is subsequently transferred from the donor to the adherent recipient. Such matings can occur in liquid (broth) cultures. Plasmids involved in this type of conjugation typically contain genes for antibiotic resistance and/or haemolysin production.

Recipient strains often secrete several different types of pheromone, each specific for a given type of plasmid; secretion stops if the corresponding plasmid is acquired.

Some of the plasmids encode a peptide that antagonizes the action of the corresponding pheromone. The purpose of such an inhibitor may be to ensure that a mating response in the donor cell is not triggered unless the pheromone is in a sufficiently high concentration, i.e. conditions under which there is a good chance of a random collision between donor and recipient cells. [Review: JB (1995) *177* 871–876.]

Pheromone-independent conjugation requires that donor and recipient cells be present on a solid surface (e.g. a nitrocellulose filter). Such conjugation occurs e.g. in strains of *Streptococcus* and *Staphylococcus*. The plasmids, which are typically >15 kb in size, encode e.g. resistance to antibiotics.

Some non-conjugative plasmids can be *mobilized*, i.e. transferred to another cell as a result of some kind of association with a conjugative plasmid. Interestingly, pMV158, a non-conjugative plasmid of *Streptococcus* (Gram-positive) has been mobilized in *E. coli* (Gram-negative) using plasmid R388 (or RP4) as the auxiliary plasmid [Microbiology (2000) *146* 2259–2265].

8.4.2.2　*Conjugation in Gram-negative bacteria*

In Gram-negative donors the plasmid encodes (among other proteins) the protein subunits of the *pilus* (section 2.2.14.3); pili seem to be essential for conjugation in Gram-negative bacteria. Different types of pilus promote conjugation under different physical conditions, and they seem to function in different ways.

Much of the information on conjugation in Gram-negative bacteria comes from studies on the transfer of the F plasmid (and related plasmids) between strains of *Escherichia coli*; the following is based largely on this information. The features described here are not common to all transfer systems in Gram-negative bacteria.

Within each donor cell the F plasmid usually exists as an independent circular molecule. Donors, which bear pili, are designated F$^+$ cells. Recipients (which lack the F plasmid) are designated F$^-$ cells. On mixing donors and recipients, the tips of the pili bind to recipient cells (see Plate 8.1: *bottom, right*).

One report [JB (1990) *172* 7263–7264] concluded that, following pilus–recipient binding, DNA passes *through* the pilus. Other workers observed that, following initial contact, donor and recipient are quickly drawn together in wall-to-wall contact (presumably by pilus retraction) and that they maintain such contact for the next 80 minutes; specific *conjugational junctions* at juxtaposed cell envelopes were seen by electron microscopy (Plate 8.1: *top*). These latter observations, and the kinetics of DNA transfer, suggested that '. . . DNA is transferred at the state of close wall-to-wall contact rather than . . . via extended pili' [JSB (1991) *107* 146–156].

The presence of conjugational junctions has also been reported by other workers [JB (2000) *182* 2709–2715].

In RP4-mediated conjugation, the transfer apparatus in the donor may include a structure that bridges the cytoplasmic and outer membranes [JB (2000) *182* 1564–1574], although this was not detected by other workers [JB (2000) *182* 2709–2715].

At some stage of contact an (unknown) mating signal triggers the transfer of DNA from donor to recipient. In at least some cases (including the F plasmid) a particular strand of plasmid DNA (the *T-strand*) is prepared for transfer to the recipient. In the F plasmid this involves the formation of a *nick* (a break in the sugar–phosphate backbone) at a specific site (designated *nic*), in a specific strand, within the origin of transfer (*oriT*). Only the nicked strand will enter the recipient.

Nicking of the T-strand is mediated by an endonuclease (a *relaxase*) which is part of a so-called *relaxosome* (previously called a *relaxation complex*). The relaxosome associated with the F plasmid is a nucleoprotein complex consisting of the transfer origin (*oriT*), the relaxase, and certain other proteins. Relaxases are encoded by the *traI* gene of the F plasmid, the *traI* gene of plasmid RP4, and the *nikA* gene of *Salmonella* plasmid R64. As well as endonuclease activity, the TraI protein of the F plasmid has (ATP-dependent) helicase activity; it is also referred to as *helicase I*.

The relaxosome is assembled in a specific order. TraI binds after integration host factor (IHF) and TraY; TraY is required for *in vivo* nicking [JB (2000) *182* 4022–4027]. The relaxase appears to mediate an ongoing cycle of nicking and ligation at *nic*, i.e. nicking does not seem to be triggered by donor–recipient contact.

In the F plasmid–*E. coli* system, the nicked strand enters the recipient in the 5'-to-3' direction. Unwinding of this strand from its complementary strand may involve helicase I (TraI); if so, and if TraI is immobilized in relation to the cell envelope, then unwinding of the transferred strand may provide the energy for strand transfer.

A recipient which receives the transferred strand of the F plasmid synthesizes a complementary strand to form a circularized dsDNA molecule; such synthesis is called *repliconation*. A recipient containing a complete, circular copy of the plasmid has become F$^+$.

In the donor, a complementary strand is synthesized on the non-transmitted strand in order to re-form the dsDNA plasmid. Such synthesis, carried out by DNA polymerase III, is called *donor conjugal DNA synthesis* (DCDS); it may proceed according to the rolling circle model (Fig. 8.5) [JB (2000) *182* 3191–3196].

Hfr donors. Infrequently, an F plasmid integrates with the bacterial chromosome. The result is an *Hfr donor*. Hfr donors form pili, and they can conjugate with F$^-$ cells. During Hfr × F$^-$ crosses, DNA transfer starts (as in F$^+$ donors) with a nick at *oriT* (see above). The transferred strand begins with plasmid DNA, but this is followed by chromosomal DNA and, finally – if the strand doesn't break – by the remainder of the plasmid strand. (This can be understood more easily if *oriT* is imagined to be in the middle of the integrated F plasmid.) If the entire plasmid and chromosomal strands are transferred the recipient becomes an Hfr donor, but usually strand breakage occurs at some point and the recipient (which does not receive *all* of the plasmid strand) remains F$^-$. However, the recipient generally receives *some* chromosomal (as well as plasmid) DNA, and if donor genes recombine with the recipient's chromosome they may alter the genetic message; the high proportion of recombinant cells resulting from Hfr × F$^-$ crosses accounts for the designation 'Hfr' – i.e. 'high frequency of recombination'.

An F plasmid can also leave the chromosome, i.e. an Hfr donor can become an F$^+$

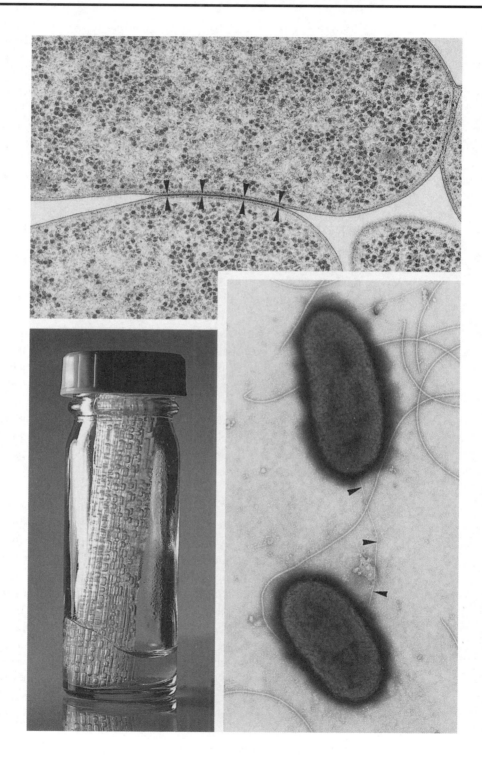

donor. Sometimes, when leaving, the plasmid takes with it an adjacent piece of the chromosome; the result is an F' (F-prime) plasmid. In F' × F⁻ crosses, the F' donor usually transfers donor ability to the recipient (as does an F⁺ donor) but it also transfers chromosomal DNA (as does an Hfr donor).

The F plasmid is only one of many types of plasmid, and it is not even typical; for example, the ability to form Hfr donors is not common. Also, in many types of plasmid the pilus-encoding and other donor genes are normally repressed, i.e. in a population of potential donors only a few cells with transiently de-repressed plasmids can actually conjugate; by contrast F⁺ cells are de-repressed, and in a population of them most or all can usually act as donors.

Universal and surface-obligatory conjugation. In Gram-negative bacteria, conjugation may be of the 'universal' or 'surface-obligatory' type, depending on the type of pilus encoded by the conjugative plasmid. Plasmids which encode long, flexible pili – as does the F plasmid (section 2.2.14.3) – promote so-called universal conjugation; this can occur equally well in a liquid medium (e.g. a broth culture) or on a moist solid surface (e.g an agar plate). Experiments in liquid media indicate that universal conjugation can be enhanced by raising the concentration of electrolyte in the medium [FEMSML (1983) *20* 151–153].

Plasmids which encode short, rigid, nail-like pili promote so-called surface-obligatory conjugation – which occurs only on moist, *non-submerged* solid surfaces or in foams. This type of conjugation seems to require that conjugating cells be present beneath or within thin films of liquid – films which are thinner than the cells themselves; in nature, cells experience such conditions, for example, when present on soil particles that are drying by evaporation. Experiments were carried out to see whether suitable conditions exist beneath the thin end of a liquid meniscus on chemically clean glass (Plate 8.1: *bottom left*); the results suggested that surface-obligatory conjugation requires, or is assisted by, surface tension [FEMSML (1983) *19* 179–182].

Plate 8.1 *Top.* Conjugating cells of *Escherichia coli* in wall-to-wall contact (scale: 6 cm = 1 μm). The electronmicrograph shows a 'conjugational junction': the electron-dense (dark) line (between the arrowheads) which marks the region of contact between donor and recipient. As well as conjugating, the cell at the top is about to divide; compare the conjugational junction with the site of cell division (far right). The masses of small, darkly stained 'dots' are ribosomes.

Bottom right. Conjugating cells of *E. coli*; this preparation has been stained to show an F pilus (arrowheads) going from one cell to the other. (Fragments of flagella are also visible.)

Bottom left. An apparatus used in the author's experiment to investigate 'surface-obligatory' conjugation (see text): a bundle of 80 chemically clean 73-mm glass capillary tubes (internal diameter approx. 0.8 mm) within a universal bottle. When shaken up-and-down, the mating medium forms a large number of small 'threads' of liquid in the capillary tubes, each thread having a meniscus (at both ends) beneath which donor and recipient cells may be held in close contact by surface tension.

Photographs of *E. coli* courtesy of Dr Markus B. Dürrenberger, University of Zürich, Switzerland.

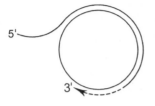

Figure 8.5   The rolling circle model of DNA synthesis in a circular, double-stranded molecule. First, a nick is made in one strand. Then, using the un-nicked strand as template, a DNA polymerase extends the 3' end of the nicked strand by adding nucleotides in the 5'-to-3' direction – the 5' end being progressively displaced. The displaced strand may itself be used as a template for the formation of Okazaki fragments (Fig. 7.8).

8.4.2.3   *Conjugative transposition (plasmid-independent conjugation)*

Conjugative transposition is conjugation mediated by a *conjugative transposon*; it occurs independently of a conjugative plasmid. First reported in Gram-positive bacteria, it is now well established in Gram-negative bacteria (e.g. *Bacteroides*). There are indications that gene transfer (e.g. the transfer of antibiotic-resistance genes) can occur between Gram-negative and Gram-positive bacteria [e.g. AEM (2001) *67* 561–568; AEM (2003) *69* 4595–4603].

Conjugative transposons, like 'classical' transposons (section 8.3), are *mobile genetic elements* which can move from one DNA duplex to another. They occur e.g. in chromosomes and plasmids, and range in size from 18 kb (Tn*916*) to >50 kb; some of the larger ones (e.g. Tn*5253*) seem to consist of a Tn*916*-like entity inserted into another transposon (both being able to transpose independently). The conjugative transposons, each of which carries at least one antibiotic-resistance gene, are readily transferred among a broad range of host species and genera, and are often responsible for transmissible resistance to antibiotics in plasmid-less Gram-positive pathogens; they have been identified in strains of e.g. *Enterococcus faecalis*, *E. faecium* and *Streptococcus pneumoniae*. Antibiotic-resistance genes carried by these elements include those conferring resistance to tetracycline, chloramphenicol, erythromycin and kanamycin.

In Gram-positive bacteria the model proposed for conjugative transposition was as follows. Contact between donor and recipient bacteria triggers excision of the transposon in the donor, this involving a transposon-encoded recombinase (product of the *int* gene). The excised transposon carries, at each end, a short sequence of nucleotides of the host molecule; these single-stranded regions are called *coupling sequences*.

The excised transposon then circularizes as a result of base-pairing (albeit mismatched) between the coupling sequences. It seems likely that a *single* strand is transferred to the recipient, and that a complementary strand is synthesized, in the recipient, prior to insertion. Insertion of the (double-stranded) transposon into a target site in the recipient duplex is not site-specific, but neither is it random: each target site apparently contains an A-rich sequence of nucleotides and a T-rich sequence – the two sequences being separated by about six nucleotides.

In Gram-positive bacteria, the frequency of transposition of Tn*916* is not limited by the frequency of excision but may be limited by the availability of transfer functions [JB (1999) *181* 5414–5418].

Conjugative transposons are characteristically resistant to restriction (section 7.4) in a new host cell. A reason for this was suggested by the finding, in Tn*916*, of a gene (*orf18*) encoding a product similar to the antirestriction proteins encoded by some plasmids.

In *Bacteroides*, a translational attenuation mechanism (section 7.8.5) may regulate the operon responsible for excision and transfer of the conjugative transposon CTnDOT [JB (2004) *186* 2548–2557].

Conjugative transposons differ from 'classical' transposons (section 8.3) in that: (i) they mediate conjugation; (ii) they do not duplicate the target site in the target molecule; (iii) the transposon is excised to form a cccDNA intermediate; (iv) the recipient duplex is characteristically in a different cell.

8.5 GENETIC ENGINEERING/RECOMBINANT DNA TECHNOLOGY AND RELATED NUCLEIC-ACID-BASED METHODOLOGY

Nucleic acids can be manipulated and altered *in vitro* and may then be inserted into cells for replication or expression. Such technology enables us e.g. to modify cells in highly specific ways; for example, bacteria can be made to synthesize mammalian proteins such as insulin. Clearly, cells can be made to synthesize proteins of a *different* species because recombinant techniques can introduce *new* genetic material; this illustrates why such technology is superior to the 'mutation and selection' approach to innovation (which is limited to modification of *existing* genes). A further advantage of the recombinant DNA approach is that genetic change can be controlled and directed (see e.g. site-specific mutagenesis: section 8.5.5.2).

Recombinant DNA technology frequently employs bacteria and/or their enzymes, plasmids or phages. Section 8.5 explains some of the methodology and introduces a number of terms, ideas and strategies. Some of the methods described – e.g. the polymerase chain reaction (PCR) – are additionally useful in other types of work, including the detection/identification and taxonomy (classification) of bacteria and other organisms, and the detection of antibiotic resistance in some pathogens (section 15.4.11.2).

The topics covered in section 8.5:
- **8.5.1** Cloning and associated general techniques – e.g. DNA restriction (cutting); screening; isolating DNA; centrifugation; blotting; vectors; linkers, tailing, ligation
- **8.5.2** Gene fusion and fusion proteins; green fluorescent protein
- **8.5.3** Probes; molecular beacon, light-up and TaqMan® probes; labelling probes (e.g. nick translation)

- **8.5.4**   PCR (polymerase chain reaction) – principle, uses; variant forms, including real-time (quantitative) PCR
- **8.5.5**   Mutagenesis (random and specific); transposon mutagenesis
- **8.5.6**   DNA sequencing – Sanger's method and Pyrosequencing™
- **8.5.7**   Expressing eukaryotic genes in bacteria – limitations
- **8.5.8**   Functional analysis of genomic DNA
- **8.5.9**   Ligase chain reaction and NASBA/TMA
- **8.5.10** Recombinant streptokinase (example of recombinant technology)
- **8.5.11** Overproduction of recombinant proteins (optimization, problems)
- **8.5.12** DNA-binding proteins; DNase I footprinting
- **8.5.13** Displaying heterologous proteins; phage display
- **8.5.14** Differential display
- **8.5.15** DNA chip technology
- **8.5.16** FRET

We look first at cloning. An overview (section 8.5.1) is followed by notes on the use of bacteria for cloning bacterial genes (8.5.1.1) and eukaryotic genes (8.5.1.2). This is followed by a detailed account of various techniques associated with cloning; many of these techniques are also used frequently in other areas of recombinant DNA technology.

8.5.1   Cloning (gene cloning, molecular cloning): an overview

Cloning is a method for obtaining many copies of a given gene (or other piece of DNA) – e.g. for sequence analysis or for making large amounts of the gene product (gp, i.e. the product encoded by the gene).

Starting with a small number of copies of a given gene we may wish to *amplify* this to many millions of copies. First, we insert copies of the gene (*in vitro*) into so-called *vector* molecules; these are often plasmids (section 7.1.2), i.e. molecules which can carry the gene into, and replicate within, bacteria. Insertion of DNA into a plasmid vector is shown in Fig. 8.6; during this process, copies of the gene are mixed with plasmids (and appropriate enzymes) so that, in many cases, a gene will be incorporated into a plasmid molecule. Hybrid plasmids, containing the given gene, can then be inserted into bacteria by transformation (section 8.4.1) and will continue to replicate as the host cells grow and divide; a bacterial population can reach very high numbers, so that we can obtain many copies of the hybrid plasmid, each incorporating a copy of the (cloned) gene.

In the above scheme we *assumed* that, during transformation, each bacterium received a copy of the plasmid. We really need a way of *selecting* those cells which have actually taken up a plasmid – because only these cells will contribute to the cloning process. One common method is to use plasmid vectors that incorporate an antibiotic-resistance gene; after transformation with such vectors, bacteria are plated on media containing the given antibiotic, thus selecting for plasmid-containing cells. However, note that a cell may grow on such a medium if (i) it is a mutant (section 8.1) that is resistant to the antibiotic, or (ii) it has taken up a plasmid which *lacks* the gene of interest, i.e. a plasmid which has failed to incorporate that gene during the insertion process (Fig. 8.6); we return to this second possibility in section 8.5.1.5.

In a different selective procedure (*repressor titration*) use is made of (i) a specially constructed strain of *E. coli* whose chromosome contains a kanamycin-resistance gene controlled by the *lacI*–promoter–operator sequence of the *lac* operon (Fig. 7.11), and (ii) multicopy plasmid vectors (section 7.3.1) that incorporate the *lac* operator sequence. Cells lacking plasmid vector molecules will fail to grow on a kanamycin-containing medium because (in the absence of lactose or IPTG) transcription from the *lac* operator is blocked by the LacI repressor protein (Fig. 7.11). A cell containing the (multicopy) plasmid will have many extra copies of the (plasmid-borne) *lac* operator sequence, and these will compete with the chromosomal *lac* operator for LacI; under these conditions, LacI fails to repress the kanamycin-resistance gene, and the cell will grow on kanamycin-containing media. As this vector does not carry an antibiotic-resistance gene (compare previous method), it can be smaller and therefore more efficiently transformable (section 8.4.1.1). [Repressor titration: NAR (1998) *26* 2120–2124.]

To harvest a cloned gene, the cells are lysed and the hybrid plasmids are isolated e.g. by isopycnic centrifugation (Fig. 8.12); alternatively, the Qiagen protocol (section 8.5.1.4; Plate 8.2) may be used. Using the original restriction endonuclease (Fig. 8.6), the gene can be cut from each hybrid plasmid and separated from plasmid DNA e.g. by gel electrophoresis.

8.5.1.1   Cloning a bacterial gene

First, the required gene must be separated from other genes in the genome. We can start by constructing a *genomic library* of the species (Fig. 8.7) and then screen the library to find the required gene.

Screening is a process of selection. Suppose, for example, that the required gene encodes an enzyme needed for the synthesis of histidine. Recall that this particular gene will occur in only a small proportion of recombinant molecules in the library (Fig. 8.7); how can we isolate the particular recombinant molecules which carry this gene? First, we use transformation (section 8.4.1) to insert the whole library of recombinant molecules into a population of mutant bacteria which are defective for the given gene (i.e. which cannot synthesize histidine). These *auxotrophs* (section 8.1.2) will not grow on media which lack histidine; however, a mutant transformed with a vector molecule carrying the functional gene will be able to grow and form a colony on media lacking histidine. The transformed mutants are therefore inoculated

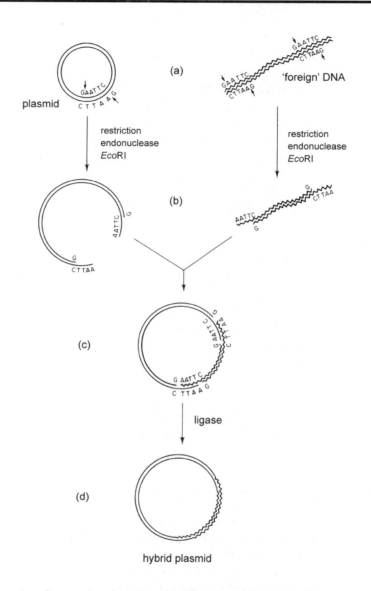

Figure 8.6   Insertion of a gene (or other piece of 'foreign' DNA) into a plasmid vector, prior to cloning (diagrammatic).

(a) A plasmid, and 'foreign' DNA. Both molecules contain the nucleotide sequence GAATTC which is recognized by the restriction endonuclease *Eco*RI (section 7.4; Table 8.1); *Eco*RI cuts between the guanosine (G) and adenosine (A) nucleotides, as shown by the arrows.

(b) As a result of *Eco*RI activity, both molecules now have 'sticky ends' (terminal, single-stranded complementary sequences of nucleotides).

(c) The 'foreign' DNA has integrated with the plasmid vector molecule through base-pairing between complementary nucleotides in the sticky ends.

onto a *minimal medium* lacking histidine so that any colony which develops is likely to consist of cells containing the vector and the given gene; we need to check this (see e.g. section 8.5.1.5) because a colony could also be formed by an auxotroph after back-mutation to prototrophy (section 8.1.3). A colony containing the given gene can be subcultured to produce large numbers of the cells – containing many copies of the (cloned) gene; then, the fragments containing the gene can be isolated as indicated in section 8.5.1.

An alternative method of screening is *colony hybridization* (see later).

8.5.1.2   *Cloning eukaryotic genes in bacteria*

The eukaryotic genome is much larger so that, for a genomic library, it has to be cut into larger restriction fragments – otherwise there would be too many pieces for cloning; these larger pieces require a different type of vector (section 8.5.1.5).

A further problem is that a eukaryotic gene typically contains one or more nucleotide sequences (called *introns*) which do not encode any part of the gene product but which interrupt the coding sequence. Within the eukaryotic cell, any part of the mRNA which has been transcribed from an intron is later removed, leaving the mature mRNA (Fig. 8.8), i.e. the uninterrupted coding sequence of the gene; this molecule is the one translated. Note that cloning the gene *with* introns would be fine if our purpose was subsequently to determine the gene's *in vivo* sequence, but, if expressed in bacteria, such molecules (unlike the corresponding mature mRNA) would not yield the normal gene product. However, we can make a DNA copy of the *mature* mRNA sequence of a given gene; this copy (= *copy DNA complementary DNA*, *cDNA*) can then be cloned.

cDNA libraries. To clone a eukaryotic gene we can start by constructing a cDNA library of the given species. Initially, all the RNA ('total RNA') is isolated from cells which are expressing the gene of interest; total RNA includes the mature mRNAs of all the *active* (i.e. expressed) genes. This may seem a lot, but actually it limits the number of clones which later have to be screened: only about 1% of the cell's genes may be active at any given time.

The isolation of mRNAs from total RNA is made easier by the polyadenylate 'tail' (A-A-A-A. . . .) which occurs at the 3' end of most eukaryotic mRNAs (see Fig. 8.9); this tail can bind, by base-pairing, to a synthetic oligo(dT)-cellulose, i.e. an oligonu-

---

(d) An enzyme, DNA ligase, has catalysed a phosphodiester bond (Fig. 7.4) between the sugar residues of each G and A nucleotide, forming a hybrid plasmid.

A population of hybrid plasmids can be inserted into bacteria by transformation (section 8.4.1) for cloning (section 8.5.1). Note that, if required, the fragment can be later cut from the vector by using the same restriction enzyme.

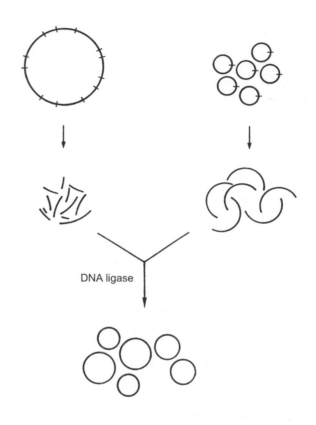

Figure 8.7   Making a bacterial genomic library (diagrammatic). Chromosomes are first isolated from a *population* of bacteria of a given strain and exposed to a particular restriction endonuclease (section 7.4). The enzyme cuts at specific sites in each chromosome (*top, left*) forming many fragments of different sizes (*centre, left*). A *population* of a given plasmid (*top, right*) provides vector molecules. The plasmid is one which can be cut by the same enzyme as that used to cut the chromosomes; however, the plasmid has only one cutting site for the given enzyme – so that cutting simply linearizes the molecule (*centre, right*).

   The chromosomal fragments and linearized plasmids are then mixed in the presence of ATP and *DNA ligase*, an enzyme which catalyses phosphodiester bonds (Fig. 7.4). Some plasmids (and fragments) may simply re-circularize via their sticky ends, but concentrations are chosen such that, in many cases, plasmids and fragments will join together – randomly – via their sticky ends (Fig. 8.6) to form a large number of recombinant molecules, many consisting of a plasmid circularized with one chromosomal fragment. (Plasmid–fragment binding can be promoted by the method described under *ligation* in section 8.5.1.6.) Because the fragments are of different sizes, the recombinant plasmids will also be of different sizes (*bottom of diagram*). *Collectively*, the recombinant plasmid molecules carry all the DNA in the chromosome, and this collection of molecules is therefore called a *genomic library*. By using transformation (section 8.4.1), the library can be taken up by a population of bacteria – *so that different cells will receive different fragments of the chromosome*. These cells can then, if required, be allowed to form individual colonies.

(a)
(b)
(c)

Figure 8.8 The formation of mRNA in a eukaryotic cell (diagrammatic). (a) A gene containing a sequence (an *intron*: dashed line) which does not encode any part of the gene product. (An example of a gene containing a single intron is the human insulin gene.) (b) mRNA transcribed from the gene: the so-called primary transcript or *pre-mRNA*. (c) After removal of the intron: the final (mature) mRNA which contains the complete, uninterrupted coding sequence of the gene. The two coding sequences, initially separated by the intron, are called *exons*. Introns also occur in some bacterial genes (see section 7.9.1.1).

Although eukaryotic genes *typically* contain intron(s) some do not; an example of an intron-less gene is the human gene encoding interferon-β.

cleotide (T-T-T-T. . . .) linked to cellulose, so that mRNAs can be isolated by affinity chromotography (section 8.5.1.4). mRNAs can also be isolated by a magnetic method (section 14.6.1).

Once isolated, the mRNAs can be converted, *in vitro*, to their corresponding cDNAs. In one method, a short synthetic oligo(dT) molecule is used as a primer; this binds to the poly(A) tail of an mRNA molecule, allowing the enzyme *reverse transcriptase* to synthesize a cDNA strand on the mRNA template. (DNA polymerase I is not used because it is relatively inefficient on an RNA template [EMBO Journal (1993) *12* 387–396].) mRNA is then removed (e.g. with RNase H) and replaced by complementary DNA, making ds cDNA. Sticky ends can be added to each molecule (section 8.5.1.6) and the library of cDNA molecules is then ready for insertion into vectors prior to cloning in bacteria.

cDNA made by the above method is sometimes found to be an incomplete copy of the mRNA. The Okayama–Berg method (Fig. 8.9) produces full-length cDNA, and is particularly useful when the cloned cDNA is subsequently to be expressed.

Screening libraries for specific genes. The colonies which form from cells containing a genomic or cDNA library can be screened for a specific gene. Screening is made easier if the gene can be expressed in a bacterial host cell. Clearly, bacterial genes can be expressed (section 8.5.1.1). However, so, too, can some cDNAs if the *vector* provides certain control functions – e.g. a promoter recognizable by the bacterial RNA polymerase (so that mRNA, and then protein, can be synthesized). Vectors which provide such functions are called *expression vectors* (see later). A *cDNA expression library* is a library of cDNA molecules inserted into expression vectors; when cells containing these vectors form colonies, it is sometimes possible to identify a colony expressing a given cDNA by the method shown in Fig. 8.10.

If a cloned gene or cDNA is *not* expressed in the bacterial host we can screen by other methods. *Colony hybridization* starts (as in Fig. 8.10) with a replica of colonies on a nitrocellulose filter; the cells on the filter are lysed, their DNA is denatured (i.e. the

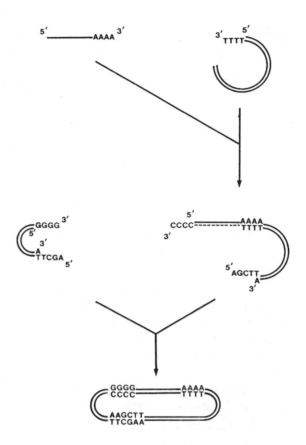

Figure 8.9 Making cDNA from a mature mRNA template by the Okayama–Berg method (diagrammatic). A mature, eukaryotic mRNA (*top*, *left*) is shown with the typical polyadenylate 'tail' at the 3' end. A specific plasmid (*top*, *right*) has been cut (i.e. linearized), and an oligo(dT) tail has been added (*in vitro*) to each 3' end (see *tailing*, section 8.5.1.6); only one tail is shown. mRNA and plasmid interact (*centre*, *right*) by base-pairing between the A and T tails. The T tail acts as a primer, allowing synthesis, by the enzyme *reverse transcriptase*, of a DNA strand (*dashed line*) on the mRNA template. An oligo(dC) tail is added to the 3' end of the new DNA strand; because only a *completed* new strand can be extended in this way, this tail allows subsequent selection for complete, full-length molecules of cDNA. Next, a sequence at the other end of the plasmid is cut by *Hind*III, leaving one 'sticky end' – the overhang 5'-AGCT. A specially constructed fragment (*centre*, *left*), together with a ligase, can now circularize the plasmid; an enzyme, *RNase H*, then removes the mRNA strand, and DNA polymerase synthesizes a DNA strand in its place – thus (with ligase action) completing the dsDNA recombinant plasmid (*bottom*) ready for insertion into a bacterial host for cloning.

Note that this method ensures that the cDNA is inserted into the plasmid in a known orientation (compare Fig. 8.6, where DNA can insert in either direction); this is important when the recombinant plasmid encodes control functions for the *expression* of the gene (see vectors, section 8.5.1.5).

strands separated), and the single-stranded DNA is bound to the filter. A labelled *probe* (section 8.5.3), which can bind to a specific sequence in the required gene or cDNA, is then added to the filter; when unbound probe is washed away, the label on the probe identifies the required DNA – and, hence, the corresponding colony on the original plate. A radioactive label, for example, can be detected as in Fig. 8.10.

How do we choose the probe's sequence of nucleotides? If the required protein has a known amino acid sequence we can 'work backwards' – deducing possible sequences for the corresponding mature mRNA (and, hence, DNA); we would choose a short sequence containing an uncommon amino acid (e.g. methionine) and then, using the genetic code (Table 7.2), work out some corresponding mRNA sequences. More than one mRNA/DNA molecule is possible for a given amino acid sequence because most amino acids are encoded by more than one codon (Table 7.2); for example:

Amino acid sequence:	Lys -Trp -Met -Lys -Glu -His -Phe
Possible mRNA:	AAA -UGG -AUG -AAA -GAG -CAU -UUC
Possible DNA:	TTT -ACC -TAC -TTT -CTC -GTA -AAG
Possible mRNA:	AAG -UGG -AUG -AAG -GAA -CAC -UUU
Possible DNA:	TTC -ACC -TAC -TTC -CTT -GTG -AAA

Each of the possible DNA sequences can be used, in separate experiments, as a probe.

8.5.1.3 *Cutting DNA precisely*

DNA can be cut precisely and predictably by type II restriction endonucleases (REs) (section 7.4); these are enzymes which cut within (or very close to) specific sequences of nucleotides (Table 8.1).

Many type II REs make staggered cuts, forming 'sticky ends', but some (e.g. *Hpa*I) cut straight across both strands, forming 'blunt' ends (Fig. 8.11).

The 8-nucleotide recognition sequence of *Not*I, being uncommon, results in infrequent cutting; this is useful for making large fragments (e.g. >1 million bp long) for eukaryotic genomic libraries (section 8.5.1.2).

The activity of REs is affected e.g. by pH and electrolyte, and it may also be affected by the type of DNA: linear or supercoiled. The *specificity* of cutting of some REs can be upset by factors such as high enzyme concentration, the use of different buffers, and glycerol concentrations >5% w/v. So-called *star activity* refers to a change or reduction in specificity which is shown by some restriction endonucleases (e.g. *Bam*HI, *Eco*RI, *Sal*I) under non-optimal conditions.

Different REs (from different species) which recognize the same sequence are called *isoschizomers*.

At least some REs retain activity at ambient temperatures; for example, *Hind*III and *Eco*RI remained active for 12 months at ambient temperature or 4°C [BioTechniques (2000) *29* 536–542].

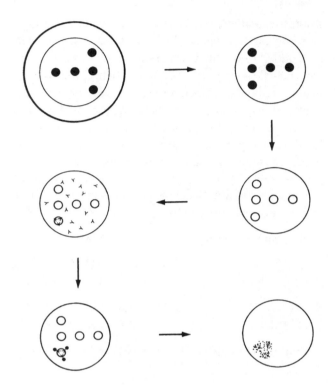

Figure 8.10 Screening a cDNA expression library (diagrammatic). The cDNA library, within expression vectors (section 8.5.1.2), has been inserted into a population of bacteria, and the latter have formed individual colonies on a plate. The object now is to find the colony whose cDNA encodes (and expresses) a particular protein. cDNA-containing colonies (*top*, *left*) are lightly overlaid with a nitrocellulose filter; when lifted off, the filter contains a (mirror image) replica of the colonies (*top*, *right*). The cells on the filter are lysed, and the proteins from these cells (including any encoded by cDNA) are bound to the filter (*centre*, *right*). The filter is then treated with *antibodies* (Y) (section 11.4.2.1) specific to the required protein (*centre*, *left*); antibodies have bound to the given protein (present in *one* of the colonies). Unbound antibodies are washed away, and the filter is treated with *protein A* – a protein (derived from *Staphylococcus aureus*) which binds to antibodies; unbound protein A is washed away, leaving some protein A (*small dots*) bound to the specific antibody–protein complex (*bottom*, *left*). Before use, the protein A is radioactively labelled so that its presence can be detected by *autoradiography*; this involves exposing the nitrocellulose filter to a photographic film and then developing the film. When the film is developed (*bottom*, *right*) it identifies that colony on the original plate which contains the cDNA of interest. This colony can then be used as an inoculum to grow more of the cells.

8.5.1.4 *Isolating nucleic acids*

For *in vitro* work, nucleic acids are obtained e.g. by lysing cells and separating the different forms of DNA and RNA. The starting point: colonies, or a pellet from a centrifuged broth culture; sedimentation of the cells in a broth culture may require centrifugation of about 5000 *g* for 10–20 minutes.

Table 8.1   Restriction endonucleases: some common examples[1]

Restriction endonuclease (source)		Recognition sequence, cutting site[2,3]
AatII	(Acetobacter aceti)	GACGT/C
AluI	(Arthrobacter luteus)	AG/CT
BamHI	(Bacillus amyloliquefaciens)	G/GATCC
BclI	(Bacillus caldolyticus)	T/GATCA
BglII	(Bacillus globigii)	A/GATCT
DpnI	See section 7.4	
EcoRI	(Escherichia coli)	G/AATTC
HindIII	(Haemophilus influenzae)	A/AGCTT
HpaI	(Haemophilus parainfluenzae)	GTT/AAC[4]
KpnI	(Klebsiella pneumoniae)	GGTAC/C
NotI	(Nocardia otitidis)	GC/GGCCGC[5]
PalI	(Providencia alcalifaciens)	GG/CC
PstI	(Providencia stuartii)	CTGCA/G
SalI	(Streptomyces albus)	G/TCGAC
Sau3AI	(Staphylococcus aureus)	/GATC
SmaI	(Serratia marcescens)	CCC/GCC[4]
SrfI	(Streptomyces sp)	GCCC/GGGC[5]
XbaI	(Xanthomonas campestris var badrii)	T/CTAGA
XhoI	(Xanthomonas campestris var holcicola)	C/TCGAG
BaeI	(Bacillus sphaericus)	/(10N)ACNNNNGTANC(12N)/[6]

[1] For further information see section 8.5.1.3.
[2] The sequence is written in the 5'-to-3' direction, and only one strand is shown. For each enzyme, compare the recognition sequence in the table with its *complementary* sequence (also read in the 5'-to-3' direction).
[3] The solidus (/) indicates the cutting site.
[4] Produces 'blunt-ended' cuts; note that the cut is in the *centre* of the recognition sequence.
[5] The 8-nucleotide recognition sequence of a 'rare-cutting' enzyme which is used e.g. for making large chromosomal fragments for eukaryotic genomic libraries.
[6] A different type of RE [NAR (1996) *24* 3590–3592] (see section 7.4 and Fig. 8.11); in this particular strand the recognition sequence is flanked by a 10-nucleotide sequence at the 5' end and a 12-nucleotide sequence at the 3' end, the cutting sites being immediately beyond the flanking sequences.

In Gram-positive bacteria the cell-wall peptidoglycan may be disrupted by buffered *lysozyme* (see Fig. 2.7). For *Staphylococcus aureus* the enzyme *lysostaphin* may be used; this enzyme cleaves the oligopeptide links between the backbone chains of peptidoglycan. For species of *Mycobacterium*, a rapid (~1 hour) method for isolating high-molecular-weight DNA involves pre-treatment of cells with chloroform/methanol (2:1 v/v), to extract lipids, followed by disruption with guanidine thiocyanate [BioTechniques (2001) *30* 272–276].

In Gram-negative bacteria the outer membrane (section 2.2.9.2) is commonly disrupted by chelating its stabilizing cations with Tris-buffered EDTA; lysozyme can then reach the peptidoglycan. (EDTA also inhibits nucleases in the cell lysate, thus protecting the nucleic acids.) Sucrose (e.g. 0.3 M) may be used to prevent violent osmotic lysis.

We often need specifically to isolate *plasmids* from bacteria – e.g. in cloning (section

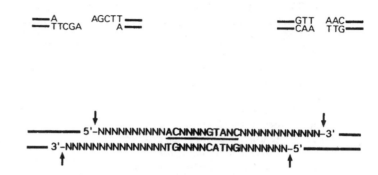

Figure 8.11   Cutting DNA by restriction endonucleases. A staggered cut (*top, left*) by *Hind*III (Table 8.1) has produced 'sticky ends' – i.e. the two 5'-AGCT overhangs; these (complementary) overhangs may bind together (thus re-forming the duplex) or each may base-pair with a complementary overhang on another molecule. (Note that some REs, e.g. *Pst*I, form 3' overhangs, while other REs form 5' overhangs.) 'Blunt' ends (*top, right*) are formed e.g. by *Hpa*I (Table 8.1); blunt ends can be fitted with *linkers* or *homopolymer tails* (section 8.5.1.6). *Below:* the cutting sites of *Bae*I [NAR (1996) **24** 3590–3592] – one of a family of REs which cut on *both* sides of the recognition sequence (see section 7.4); the recognition sequence of *Bae*I is underlined in the top strand (N = nucleotide), and the arrows indicate the cutting sites.

8.5.1). Of the various methods available, a rapid and simple procedure – which yields highly pure plasmid DNA – is described in Plate 8.2; this involves gentle, controlled lysis in which the cytoplasmic membrane is disrupted by the detergent SDS.

Isopycnic centrifugation (at appropriate **g** values) can separate macromolecules of different densities. The sample is layered on top of a solution whose concentration, and hence density, increases with depth (i.e. a *density gradient column*); centrifugation to equilibrium then brings each molecule in the sample to that part of the density gradient which corresponds to its own density. Thus, e.g. 'total RNA' (section 8.5.1.2) can be separated from DNA and from protein in a gradient of caesium trifluoroacetate; this reagent also inhibits RNase (section 7.6.1), present in cell lysate, thus helping to protect isolated RNA. Different forms of DNA can be separated in a caesium chloride gradient (see Fig. 8.12).

Chromatography. Dissimilar molecules in a sample can be separated by passing the sample over or through an appropriate stationary medium which retains or transmits molecules according to their physical properties. *Affinity chromatography* exploits the specificity of binding between certain types of molecule – e.g. the A-A-A-A. . . . tail of mRNA molecules (in the sample) and the T-T-T-T. . . . of oligo(dT)-cellulose in a stationary matrix (see e.g. section 8.5.1.2). *Spun column chromatography* uses centrifugal force (e.g. 300–400 **g**) to hasten the transit of a liquid sample through e.g. a tightly packed column of fine particles.

Electrophoresis can separate mixed, charged molecules. In *gel electrophoresis* the sample is placed in a well at one end of a gel strip; a voltage applied between the two ends of the strip causes each molecule to move, through the gel, towards the anode (+ve) or cathode (−ve), depending on the net charge (−ve or +ve, respectively) on the molecule. Small and/or highly charged molecules usually move faster than do larger and/or weakly charged molecules.

Polyacrylamide gels, which have small mesh sizes, are used for separating small fragments of DNA, or RNAs (e.g. 5–500 nucleotides in length); agarose gels are used for larger pieces of DNA (e.g. 500–50000 nucleotides in length).

Within the gel, molecules of nucleic acid are separated into discrete zones (according to size) which may be stained *in situ* (revealing spots or bands) and/or transferred for later analysis to a paper-like matrix by *Southern blotting* (Fig. 8.13) or by electro-blotting.

Pulsed-field gel electrophoresis (PFGE). Standard electrophoresis can separate DNA fragments up to about 20000 bp long. PFGE can separate molecules of over 10 million bp in length; it uses two electrical fields, applied alternately from different angles, so that the net migration of a given molecule depends on the speed with which it can change its direction of movement. [A useful source of information is *Pulsed Field Gel Electrophoresis* (ISBN 0 12 101290 5).]

8.5.1.5 *Vectors*

Vectors carry DNA into cells (section 8.5.1). Subsequently, it's often helpful if the vector can signal its presence in the cell. For example, when a population of bacteria is transformed with a library (section 8.5.1.1) some of the cells may not receive a vector; moreover, some vectors may lack an insert – the vector having re-circularized without incorporating a fragment (see Fig. 8.7). We therefore need ways of confirming the uptake of vector and insert in order to avoid the kind of uncertainty described in section 8.5.1. How can we identify those colonies which contain both vector *and* insert?

Some vectors contain antibiotic-resistance genes which can signal their presence (and that of the insert) by expression (or by lack of expression) in the host cell: see, for example, Fig. 8.14. Other vectors may contain the gene for β-galactosidase: an enzyme which cleaves lactose (section 7.8.1.1), and which can also form a blue-green product by cleaving *Xgal* (5-bromo-4-chloro-3-indolyl-β-D-galactoside); when constructing a library (Fig. 8.7) fragments can be made to insert into (and thus inactivate) this gene simply by choosing a cutting (restriction) site within the gene itself. Subsequently, colonies known to contain the vector (e.g. through expression of antibiotic resistance) can signal the presence or absence of an insert: on an Xgal-containing medium, white colonies are formed when vectors contain an insert (β-galactosidase inactivated by insert), while blue-green colonies are formed when vectors lack an insert (β-galactosidase functional). Of course, this indicator system is useful only for cells which are not themselves forming β-galactosidase.

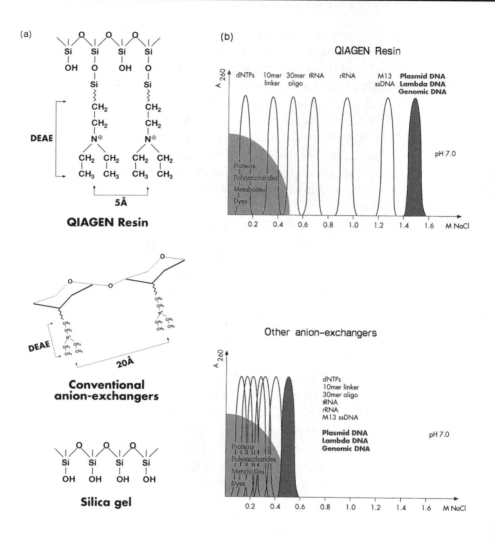

Plate 8.2 Outline of the protocol for isolating plasmids from bacteria (e.g. *E. coli*) using the Qiagen® plasmid mini kit; up to 20 μg of plasmid DNA can be isolated with this kit (larger kits can yield up to 10 mg).

The bacterial pellet is resuspended in a Tris–EDTA buffer containing RNase A; in Gram-negative bacteria Tris–EDTA disrupts the outer membrane. This is followed by controlled alkaline lysis with NaOH and the detergent sodium dodecyl sulphate (SDS). SDS disrupts the cytoplasmic membrane, releasing soluble cell contents (e.g. proteins, RNA, plasmids). NaOH denatures plasmid (and chromosomal) DNA and proteins. RNase A digests contaminating RNA. Lysis time is optimized (i) to allow maximum release of plasmids without release of chromosomal DNA, and (ii) to avoid irreversible denaturation of plasmids (denatured plasmids are resistant to restriction enzymes). The lysate is neutralized by a pre-chilled high-salt buffer (pH 5.5) which precipitates SDS; complexes of salt and SDS entrap proteins etc., but plasmids renature and stay in solution. Gentle mixing is used to ensure precipitation of all SDS; vigorous agitation is avoided as it could shear the chromosome – contaminating the supernatant with chromosomal fragments. Centrifugation (at e.g. 16000 **g**) is used to obtain a cleared lysate (containing the plasmids).

Cloning large fragments. Large fragments of DNA (section 8.5.1.2) in plasmid vectors make large recombinant molecules that do not transform efficiently (section 8.4.1.1). However, we can use certain phage vectors (phages: Chapter 9) which *inject* their nucleic acid into cells; fragments for cloning are inserted into a phage genome and are replicated when the phage genome replicates inside a cell. A *replacement vector* is made by removing a sequence of non-essential DNA (the 'stuffer') from the phage genome and replacing it with the fragment to be cloned; thus, a fragment of about 20 kb can be cloned in phage λ (compared with < 10 kb in plasmid pBR322). Even bigger fragments (up to about 40 kb) can be cloned in *cosmids* (Fig. 8.15).

Fragments of up to 300 kb can be cloned in *bacterial artificial chromosomes* (BACs): large, circular vector molecules, based on the F plasmid, which replicate in *E. coli*; the uptake of such large molecules by host cells is achieved by electroporation (section 8.4.1.2).

Yeast artificial chromosomes (YACs) can carry inserts of up to 2000 kb; these linear vector molecules, constructed *in vitro*, contain those parts of the yeast (*Saccharomyces cerevisiae*) chromosome necessary for replication and segregation in yeast cells. YACs provide a useful alternative to bacterial systems for the cloning or expression of eukaryotic genes; they can carry large eukaryotic genes, together with promoter(s) and control sequences, and are therefore suitable for studies of gene function. YACs have been used for mapping in the human genome project.

Expression vectors encode functions for transcription/translation of the insert (Fig. 8.16); not shown in the diagram are the vector's origin of replication or its marker genes (e.g. antibiotic-resistance genes). Expression can be controlled in various ways – e.g. in some systems the gene can be switched on by a specific change in temperature. If required, a gene can be cloned in bacteria and then expressed in eukaryotic cells, a single vector carrying a prokaryotic origin of replication and control functions for expression in eukaryotes.

Shuttle vectors can replicate in different types of organism. They can, for example,

---

The cleared lysate is filtered through Qiagen anion-exchange resin (a). This resin was developed specifically for isolating nucleic acids; it has a high density of (+ve) anion-exchange groups (distance between adjacent N+ only 5Å), which is ideal for binding nucleic acids. Moreover, different types of molecule elute at *widely* differing concentrations of salt – see elution profiles in (b); this permits step-wise elution and efficient separation of molecules by the use of simple salt buffers. Given the (low) electrolyte concentration and pH of the cleared lysate, only plasmid DNA binds to the resin – degraded RNA and proteins etc. are not retained. The resin is washed with buffer (which includes 1 M NaCl, pH 7) to eliminate contaminants (and also to remove any DNA-binding proteins). Plasmid DNA is then eluted with buffer containing 1.25 M NaCl, pH 8.5. The DNA is precipitated (at room temperature) with isopropanol, the tube centrifuged, and the supernatant discarded; after a wash with 70% ethanol, the DNA is air-dried (5 min) and suspended in buffer.

Line drawings courtesy of Qiagen GmbH, Hilden, Germany.

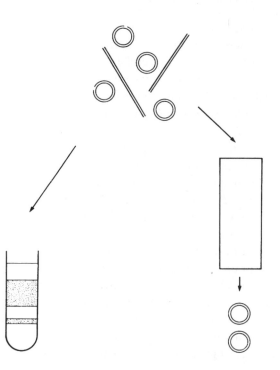

Figure 8.12   Two methods for separating supercoiled DNA from linear and 'nicked' circular DNA (diagrammatic). DNA isolated from lysed bacteria may contain (*top*): linear fragments of chromosomes; circular plasmids in which one strand has been 'nicked' (i.e. circular DNA in which a phosphodiester bond (Fig. 7.4) has been broken); and circular, supercoiled plasmids.

If DNA is treated with the dye *ethidium bromide*, molecules of dye tend to insert between adjacent base pairs – thus tending to elongate the sugar–phosphate backbone of the molecule; this distortion *decreases the density* of DNA. Linear and nicked circular DNA take up more dye than do supercoiled molecules, so that their density is decreased by a greater amount; hence, if dye-treated DNA is subjected to isopycnic centrifugation (section 8.5.1.4) in a caesium chloride gradient, supercoiled DNA forms a separate band below the less-dense linear and nicked circular molecules (*left*).

Another method, involving *spun column chromatography* (section 8.5.1.4), is quicker and easier. Alkali causes strand separation (denaturation) in DNA; on return to neutral pH, re-naturation (to the double-stranded state) is much more efficient in supercoiled molecules than in linear or nicked circular molecules – so that a column which strongly binds single-stranded DNA will tend to allow only supercoiled molecules to emerge with the effluent (*right*).

Both methods can be used to isolate supercoiled DNA in milligram quantities.

contain both prokaryotic and eukaryotic *ori* sequences, allowing replication in both prokaryotic and eukaryotic cells; such vectors have prokaryotic markers (e.g. anti-biotic-resistance genes) and eukaryotic markers (e.g. a gene for leucine synthesis, useful in auxotrophic host cells). Eukaryotic components are often from the yeast *Saccharomyces cerevisiae*.

A *gapped shuttle vector* (= *retrieval vector*) can copy e.g. a specific mutant yeast gene,

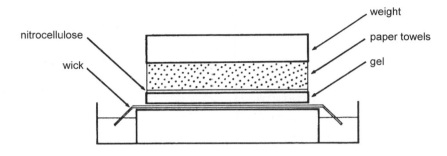

Figure 8.13 Southern blotting: the transfer of DNA fragments from a gel strip to a sheet of nitrocellulose by capillary action. Within the gel, the fragments are distributed in discrete zones, according to size, having been separated earlier by gel electrophoresis (section 8.5.1.4). The fragments are first denatured (made single-stranded) by exposing the gel to alkali; the gel is then exposed to neutral buffer and arranged as shown in the diagram. Driven by capillary action, the neutral, saline solution in the dish rises, via the wick, into and through the gel, through the (permeable) nitrocellulose, and into the stack of paper towels; this upward stream of liquid carries the DNA fragments from the gel to the sheet of nitrocellulose. (The zones of fragments on the sheet have the same relative positions as they had in the gel.) The sheet is removed and baked at 70°C under vacuum to bind the DNA. The fragments can then be examined e.g. by exposing the nitrocellulose to a specific labelled *probe* (section 8.5.3); probe–fragment binding (*Southern hybridization*), detected by a probe's label, identifies a particular sequence in a given fragment.

One alternative to nitrocellulose is a paper which incorporates 2-aminophenylthioether (*APT paper*); before use, APT is chemically modified to the reactive diazo derivative, so that single-stranded nucleic acids are bound *covalently*, i.e. no baking is required. This (more robust) preparation allows removal of probes and the subsequent use of different probes on the same sheet.

Northern blotting refers to the transfer of RNA from gel to matrix.

Western blotting refers to the transfer of proteins from gel to matrix.

Southwestern blotting: an unrelated procedure (see section 8.5.12.1).

Electroblotting is a faster method which uses an electric field (instead of capillary action) for effecting transfer.

Immunoblotting refers to the transfer of proteins to a matrix – followed by exposure of the matrix to labelled antibodies; the binding of antibodies to a particular protein may indicate the corresponding antigen.

in vivo, for subsequent analysis. The corresponding wild-type gene, together with its flanking chromosomal sequences, is inserted in the vector; the gene is then cut out (with restriction enzymes), its flanking sequences remaining in the vector. This 'gapped vector' is inserted into cells containing the mutant gene; the flanking sequences in the vector bind to complementary sequences flanking the mutant gene, and the mutant gene is copied (Fig. 8.17). The end product is a complete (circular) vector carrying a copy of the mutant gene; it can be subsequently amplified in *E. coli*.

A *phagemid* is a flexible vector (constructed *in vitro*) which includes (i) a plasmid's origin of replication; (ii) a multiple cloning site (MCS) – see Fig. 8.16; (iii) a marker gene (e.g. for kanamycin resistance); and (iv) the origin of replication of a filamentous

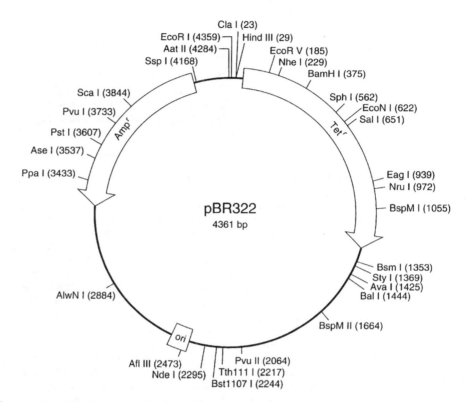

Figure 8.14 A cloning vector: the small, recombinant plasmid pBR322 which was constructed *in vitro* from other plasmids; it is 4361 bp long. Plasmid vectors are generally small (often 3–5 kb) because (i) transformation is more efficient with small plasmids, and (ii) small plasmids are more resistant to damage during isolation from cells.

Various restriction sites are shown (with their distances, in bp, clockwise from the top); each RE cuts pBR322 at one site only (linearization). The variety of restriction sites allows flexibility in the construction of recombinant molecules from various sources of DNA.

The presence of genes encoding resistance to ampicillin and tetracycline (transcribed in opposite directions, as shown by the arrows) enables us to test for (i) the presence of the plasmid in a cell, and (ii) the presence of an insert in the plasmid. Thus, if pBR322 and *Bam*HI are used when making a library (Fig. 8.7), fragments will insert into – and thus inactivate – the tetracycline-resistance gene; the ampicillin-resistance gene will remain functional. Hence, any bacterium containing the plasmid can form a colony on media containing ampicillin; if these colonies are then replica plated (section 8.1.2) onto a medium containing tetracycline, cells containing vector *and* insert should fail to grow – owing to inactivation of the tetracycline-resistance gene. (Cells containing re-circularized plasmids grow on both media.)

A cloning vector must, of course, contain an *origin* of replication (section 7.3). In pBR322, the origin (*ori*) is derived from a plasmid which replicates under relaxed control (section 7.3.1.1). Thus, pBR322, also relaxed, can continue to replicate in non-growing (chloramphenicol-inhibited) bacteria; this effect is used to obtain an *amplification* of up to several thousand copies of the vector per cell.

Line drawing of pBR322 courtesy of Pharmacia Biotech Inc., Molecular Biology Reagents Division, Milwaukee, Wisconsin, USA.

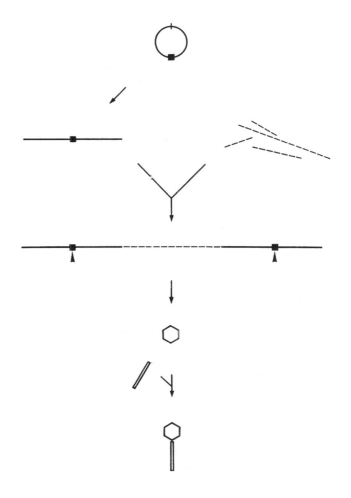

Figure 8.15 A cosmid cloning vector (diagrammatic). The cosmid (*top*) is a small plasmid into which has been inserted the *cos* site (*solid black square*) of phage λ (see Fig. 9.2); *cos* enables the plasmid, with an insert, to be packaged (*in vitro*) in a phage λ head.

Cosmids are cut at a single restriction site; the linearized molecules are then mixed, in the presence of ligase, with DNA fragments (*dashed lines*) cut by the same enzyme. Cosmids and fragments bind randomly, and in some cases a fragment will be flanked by two cosmids (*centre*); if, in such a molecule, the distance *cos* to *cos* is about 40–50 kb long, then the *cos–cos* sequence is the right size for packaging in a phage λ head. Enzymic cleavage at the *cos* sites (*arrowheads*) creates sticky ends, and packaging of the recombinant molecules occurs in the presence of phage components. After addition of the tail, the particle can inject its recombinant DNA into a suitable bacterium (just like an ordinary phage λ). The injected DNA circularizes, via the *cos* sites, and replicates as a plasmid; its presence in the cell can be detected e.g. by the expression of plasmid-encoded antibiotic-resistance genes.

Screening for specific sequences is carried out in a way similar to that used for plasmid vectors.

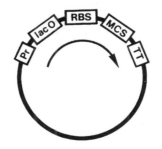

Figure 8.16   A generalized plasmid expression vector used in *Escherichia coli* (one of many possible *in vitro* constructions).

The promoter (Pr) is commonly a strong, hybrid promoter – e.g. the *tac* promoter (constructed, *in vitro*, from the promoters of the *trp* and *lac* operons of *E. coli*); the arrow indicates the direction of transcription. Promoter activity (in this case) is controlled by the host cell's *lac* regulator protein (see Fig. 7.11) which binds to the operator site (lacO) on the plasmid, inhibiting transcription. Activity can be switched on by adding isopropyl-*β*-thiogalactoside (IPTG) to the medium; IPTG binds to – and inactivates – the regulator protein, allowing transcription to proceed. For cells which do not synthesize the *lac* regulator protein (or enough regulator protein for both chromosome and plasmid) we can use a vector that incorporates a *lacI* gene (Fig. 7.11).

The ribosome-binding site (RBS) ensures that, following transcription, the mRNA product will contain the Shine–Dalgarno sequence (section 7.6) – needed for binding to the ribosome. The Shine–Dalgarno sequence is a short sequence of purine nucleotides.

MCS is a *multiple cloning site* (also called a *polylinker*): a sequence of nucleotides containing a number of different restriction sites (which often overlap); part of an MCS might be as follows:

....GGATCCCGGGAATTC....   containing the sites
   GGATCC   *Bam*HI
       CCCGGG   *Sma*I
          GAATTC   *Eco*RI

DNA to be expressed is inserted in the MCS; the range of restriction sites allows flexibility in the preparation of the insert.

Three important points. (i) The insert must be in the correct orientation relative to the direction of transcription. (ii) The insert's equivalent of the initiator codon (transcribed as AUG in mRNA – section 7.6) must be the correct distance from the RBS so that AUG can align with the 'P' site on the ribosome (see Fig. 7.9); this distance has been reported to be between 5 and 13 nucleotides – in a given case the optimal distance being necessary for maximum efficiency of translation. (iii) If the insert lacks an initiator sequence, the vector must provide one, upstream of the insert; in this case, care must be taken to ensure that, following transcription, the insert's mRNA is in the correct *reading frame*, i.e. its codons are in-phase with the initiator codon – otherwise the effect will be analogous to that of a phase-shift mutation (Fig. 8.1b).

One way of correctly orientating the insert is shown in Fig. 8.9. Alternatively, the gene to be expressed can be isolated on a fragment which has been created by cutting with two different restriction enzymes – and which has non-complementary sticky ends; each sticky end can bind to its complementary sequence in a vector molecule which has been cut by the same two enzymes. (This procedure also prevents re-circularization of the vector, and of the insert, thus encouraging vector–insert binding.)

The mRNA is in the correct reading frame when the number of nucleotides separating the initiator codon from the first base of the first complete codon is 0, 3, or a multiple of 3; if not, this can be adjusted by inserting a linker (section 8.5.1.6) of the right length between insert and initiator sequence.

TT is a sequence specifying a rho-independent transcription terminator (section 7.5); it is often included to prevent unwanted *readthrough* (continued transcription) which may otherwise occur.

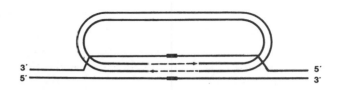

Figure 8.17 A gapped shuttle vector copying a mutant yeast gene (diagrammatic) (see text). In the yeast chromosome, strand separation has occurred in the region of the mutant gene (▬), and strands in the vector's free ends have base-paired with complementary sequences on the chromosome. The strand binding to the lower (5'-to-3') chromosomal strand acts as a *primer* (Fig. 7.8) for DNA synthesis (*dashed line*). Similarly, synthesis occurs from the opposite end of the other strand (*dashed line*) – thus, after ligation, completing the (circular) plasmid; note that the newly synthesized DNA includes a copy of the mutant gene.

phage (e.g. f1 – Table 9.1). Sample DNA can be inserted into the MCS and the phagemid then inserted into a bacterium by transformation.

A phagemid can be used for making double-stranded copies of the insert, i.e. conventional cloning. Alternatively, if we infect the phagemid-carrying bacteria with a suitable 'helper phage' (e.g. a strain of M13), replication will be initiated from the phage origin in the phagemid, thus producing single-stranded copies of the phagemid (for mechanism see section 9.2.2); these single-stranded copies will be packaged in phage coat proteins (encoded by the helper phage) and exported into the medium. Single-stranded copies of the insert (when freed from the phage coats) can be used e.g. for sequencing (section 8.5.6).

Many phagemids contain a promoter on either side of the MCS (on opposite strands); this permits transcription of the cloned fragment, if required.

8.5.1.6   *Linkers, adaptors, tailing, ligation*

To join molecules of DNA we can often use the method shown in Fig. 8.6; this method allows the molecules to be separated again, if required, by using the same restriction enzyme. However, molecules may lack restriction sites or may have incompatible ones; in such cases we can alter one or both molecules, as follows. Sticky ends (if present) are first changed to 'blunt' ends (Fig. 8.11) either by removing single-stranded overhangs (e.g. with endonuclease S1) or by synthesizing a complementary strand on each 5' overhang (thus converting overhangs to dsDNA). To each (blunt) end is now ligated (i.e. covalently bound) a *linker*: a short, synthetic dsDNA molecule containing a restriction site – for example:

5'-GGAATTCC-3'
3'-CCTTAAGG-5'

containing the *Eco*RI site (Table 8.1). Because a linker is a *palindromic* sequence (i.e.

the 5'-to-3' sequence is the same in both strands) either end can be ligated. After ligation, linkers are cleaved with the given restriction enzyme to form sticky ends. Note that it's necessary to check whether the given restriction enzyme has any cleavage sites in the fragment itself; any such cleavage site(s) in the fragment must be methylated prior to ligation of the linker – otherwise the fragment (as well as the linker) would be cleaved by the restriction enzyme.

Linkers are also used for reading-frame adjustments (Fig. 8.16).

Adaptors are similar to linkers but contain more than one type of restriction site, and may be designed to have pre-existing sticky ends. They are used e.g. to insert one or more restriction sites into a vector; for example, an adaptor with *Eco*RI sticky ends and an internal *Sma*I site can be inserted into a vector's *Eco*RI site, thus introducing the *Sma*I site.

Tailing involves adding nucleotides to the 3' end of DNA, using the enzyme *terminal deoxynucleotidyl transferase*; blunt-ended DNA can be tailed, though less efficiently than ssDNA or the 3' overhangs of sticky ends. Commonly, nucleotides of only one type are added – forming a *homopolymer tail* (e.g. a deoxythymidine (dT) tail – see Fig. 8.9). Homopolymer tailing can be used e.g. to join blunt-ended DNA molecules; for example, if one molecule is tailed with deoxycytidine (dC) the other would be tailed with deoxyguanosine (dG). One advantage of this is that base-pairing can occur only between different molecules – i.e. a given molecule cannot circularize because both ends have the same tail; however, joining molecules in this way does not usually result in the creation of restriction sites, so that it can be difficult subsequently to separate the molecules (if required).

Ligation (in this context) means covalently joining free, juxtaposed 5'-phosphoryl and 3'-hydroxyl ends of DNA (to form a phosphodiester bond – Fig. 7.4); the 5' end must be phosphorylated, and energy is required. The enzyme commonly used is an ATP-dependent ligase encoded by phage T4.

Ligation shown in Fig. 8.6 is single-strand ligation in different parts of the molecule. Blunt-ended ligation (the joining of two double-stranded ends) requires a higher concentration of enzyme, a lower temperature, and a longer time.

When making a library (Fig. 8.7) we can prevent re-circularization of vector molecules (and thus maximize vector–fragment binding) by de-phosphorylating the 5' ends of the vectors with the enzyme alkaline phosphatase; phosphoryl groups on the *fragments* allow some covalent binding, and this stabilizes the recombinant molecules during transformation. Subsequently, the cell's DNA repair enzymes can join the remaining strands.

8.5.2 Gene fusion and fusion proteins

An expression vector (section 8.5.1.5; Fig. 8.16) can carry sequences from two different genes with their reading frames in phase; both sequences can be transcribed from a single promoter to yield a single transcript – translation of which gives rise to a hybrid or *fusion protein*. Of what use is this technique? In fact, fusion can be used e.g. to (i) facilitate isolation/detection of specific gene products; (ii) determine the intracellular location of specific proteins; and (iii) study the regulation of gene expression. Examples are outlined below.

When expressed, some cloned genes yield a protein product which is difficult to detect or isolate. In such cases, a sequence from the given gene can be fused with one from a second gene (the *partner* or *carrier* gene) whose product can be easily detected or isolated; for example, if the partner gene encodes glutathione *S*-transferase (GST), the fusion protein can be readily isolated by affinity chromatography (section 8.5.1.4) using glutathione in the stationary matrix. (In this example GST is the *affinity tail* on the protein of interest.)

A gene whose product normally remains *intra*cellular can be fused with a gene whose product is normally secreted so that the fusion protein may be secreted into the medium. This may facilitate isolation/purification of the required protein and/or help to protect it from intracellular degradative enzymes – see e.g. section 8.5.11.4(b).

Fusion may also be used to obtain gene products if a given gene is normally not highly expressed – the required gene being fused downstream of a highly expressed partner gene within an expression vector; the partner gene may also confer the benefit of increased solubility and/or stability.

If a fusion protein has been isolated it may be possible to cleave it enzymatically with a site-specific protease, yielding a separate product for each of the two genes. Cleavage may also be achieved by chemical agents such as cyanogen bromide (which cleaves at methionine residues) or hydroxylamine (which cleaves between asparagine and glycine residues); problems with chemical cleavage include (i) cleavage sites *within* the required protein, and (ii) unwanted chemical modification of the required protein.

A common fusion partner is the gene encoding *green fluorescent protein* (GFP). GFP, which occurs naturally in jellyfish of the genus *Aequorea*, emits green light ($\lambda = 508$ nm) when irradiated with blue light ($\lambda = 395$ nm). Fusion with GFP can be used to reveal the intracellular location of a specific protein; thus, the gene of the target protein is fused to that of GFP, and the fusion protein (i.e. target protein + GFP) is detected by fluorescence microscopy. For example, an Smc-GFP fusion was used to monitor the Smc protein in a study on chromosome partitioning (section 3.2.1); in this particular study, the vector (containing one end of the *smc* gene fused with *gfp*) was inserted into the *smc* gene of the chromosome to create a complete *smc–gfp* sequence – insertion involving a single cross-over of the type shown in Fig. 8.21 (a) [GD (1998) *12* 1254–1259]. GFP has also been used as a partner gene to study the intracellular location of the FtsZ protein: see Plate 5.1.

Fused to a coat protein of a specific phage, GFP has been used for the rapid detection of *Escherichia coli* strain O157:H7 [AEM (2004) *70* 527–534].

In a *lacZ* fusion, the partner gene, *lacZ*, encodes β-galactosidase (see Fig. 7.11);

expression of this gene (and hence, the gene of interest) can be monitored on media containing Xgal (section 8.5.1.5).

Fusion can also be useful for studying the *regulation* of gene expression. For example, if the expression of a gene cannot be monitored (owing e.g. to difficulty in assaying the gene product), the activity of the gene's *promoter* can be studied separately by fusing it upstream of a promoter-less *reporter gene* whose product/activity can be readily monitored; reporter genes include those encoding galactokinase, chloramphenicol acetyltransferase (CAT) and green fluorescent protein. Although this approach usually works well, certain reporter genes have been found to influence the activity of some promoters [JB (1994) *176* 2128–2132]; clearly, we must bear in mind the possibility of unpredictable interactions between components of recombinant DNA.

8.5.3 Probes

Suppose that we wish to find out whether a given plasmid or genome etc. contains a specific short sequence of nucleotides. One way is to use a *probe*: a piece of single-stranded DNA or RNA (or even PNA: section 7.2.1.1) which is *complementary* to the sequence of interest and which has been 'labelled' in some way. By taking advantage of the specificity of base-pairing (section 7.2.1), the sequence of interest can be located (if present) by the binding (base-pairing) of the probe – binding being detected by the probe's label. In practice, the target DNA (i.e. the chromosome or plasmid etc.) is typically examined in denatured (single-stranded) form attached to a support such as a nitrocellulose filter. Initially, the target DNA is exposed to many molecules of the probe under conditions in which target–probe binding will occur if the specific sequence is present; when all the free (non-bound) probes are washed away, any remaining probes – detected by their label – indicate the presence of the sequence of interest.

The labelling of probes is considered in section 8.5.3.1.

Although target DNA is usually single-stranded, direct probing of *double-stranded* DNA has been reported. In this method the (ssDNA) probe is complementary to a sequence at one end of the dsDNA target fragment. The probe is used in conjunction with RecA protein (section 8.2.1); RecA mediates replacement of a strand in the terminal region of target DNA by (homologous) probe DNA. [Method: NAR (1998) *26* 5728–5733.]

Making probes. Probes can be synthesized directly according to any required nucleotide sequence; commercial sources of probes are available. Alternatively, probes can be made by cloning the particular sequence (section 8.5.1; Fig. 8.6). Before excision from the vector, the cloned probe can be labelled by so-called *nick translation*. In this process an enzyme (endonuclease) makes random, single-strand nicks in each hybrid plasmid;

then another enzyme (e.g. *E. coli* DNA polymerase I) removes nucleotides from the nicked strands (using nicks as starting points) and replaces them with labelled nucleotides which are present in excess in the reaction mixture. After ligation, each probe – labelled as part of the hybrid plasmid – is cut out by restriction endonucleases.

Probes can also be made by asymmetric PCR (section 8.5.4.2).

8.5.3.1 Labelling probes

A probe's label indicates the presence, and sometimes the location, of the sequence of interest.

Probes can be labelled with radioactive isotopes. A probe labelled with e.g. ^{32}P can be detected by autoradiography (as in Fig. 8.10). However, unless essential, the use of radioactive material is avoided owing to (i) expense and (ii) inconvenience.

Non-radioactive (= non-isotopic) labelling of probes is common. In *direct* non-isotopic labelling, the label – bound covalently to the probe – may be a fluorochrome such as fluorescein or rhodamine. Fluorescent labels can be detected by ultraviolet radiation.

In indirect non-isotopic labelling the probe is covalently bound to a *small* molecule, such as digoxigenin or biotin. After a probe has bound to its target sequence, digoxigenin is detected by adding *antidigoxigenin* – i.e. the antibody (section 11.4.2.1) to digoxigenin – which has been covalently linked to a label (e.g. alkaline phosphatase). The antibody–label complex binds to digoxigenin and can be detected by adding a substrate which the enzyme can cleave to coloured products that are assayed by colorimetry.

Biotin is detected by *streptavidin*, a protein (from the bacterium *Streptomyces avidinii*) which binds strongly to biotin and which can be linked to a fluorescent or enzymic label. Detection of biotin (and its covalently attached probe) is generally carried out by adding the streptavidin–label complex *after* the probe has been allowed to bind (*hybridize*) to the target sequence. However, it has been shown that short (e.g. 10-nucleotide) biotinylated probes which have been labelled with streptavidin–alkaline phosphatase *before* the hybridization step can give good results when used for probing target sequences on membranes [enzyme-coupled probes: NAR (1999) *27* 703–705].

An enzymic label can also be detected by *chemiluminescence*. For example, a probe labelled with alkaline phosphatase can be detected by adding a substrate which, when cleaved by this enzyme, emits light – the light being recorded either by photographic film or by an instrument. Chemiluminescent substrates include the 1,2-dioxetanes CSPD and AMPPD (both trade designations of Tropix, Bedford MA, USA). These agents can be used not only for probes but also e.g. for detecting the bands of DNA in DNA sequencing by the dideoxy method (section 8.5.6).

8.5.3.2 *Novel types of probe*

Particular sequences may be detected by two pyrene-labelled oligonucleotide probes – one labelled at the 3' end, one at the 5' end. If the probes bind *adjacently* on a complementary target sequence, in specific binding positions, pyrene–pyrene interaction modifies the fluorescence characteristics of the compound [NAR (1998) *26* 3789–3793].

Molecular beacon probes. Each probe is a single-stranded oligonucleotide whose ends base-pair to form a double-stranded sequence, i.e. the molecule has a stem–loop structure. The sequence in the loop region is complementary to the target sequence. In the (double-stranded) stem, one strand carries a fluorescent dye while the other carries a quencher of fluorescence. The *unbound* probe is non-fluorescent. When the probe's loop binds to a target sequence the two ends of the probe separate, separating dye from quencher; the probe can then fluoresce. Molecular beacon probes have been used e.g. to monitor progress in NASBA (section 8.5.9.2). These probes have greater *specificity* than linear probes of equivalent length [PNAS (1999) *96* 6171–6176].

Light-up probes. These probes carry a molecule whose ability to fluoresce increases when the probe binds to its specific target sequence; they can be used e.g. for real-time (quantitative) monitoring of PCR (section 8.5.4.7). In one test system, probes consisting of PNA (section 7.2.1.1), conjugated with the dye thiazole orange, were able successfully to monitor PCR amplification of genomic DNA from *Yersinia enterocolitica* [BioTechniques (2001) *31* 766–771].

TaqMan® probes. These probes (PE Applied Biosystems) can be used e.g. for quantitative (real-time) monitoring of PCR (section 8.5.4.7) and for estimating the pre-amplification concentration of target sequences. Each probe is a short, target-specific oligonucleotide to which is covalently bound a fluorochrome *(reporter* dye) and, *in close proximity*, a quencher of fluorescence. In PCR, these probes are added in large numbers to the initial reaction mixture; at the annealing stage they bind to an *internal* site on the templates (amplicons). During primer extension (DNA synthesis) the polymerase exerts 5'-to-3' exonuclease activity, degrading the probe and thus separating fluorochrome from quencher; fluorescence from the reporter molecules therefore increases in intensity as the number of unquenched reporter molecules (from degraded probes) rises with ongoing cycling.

    TaqMan® probes have been used e.g. for quantitative estimation of *Borrelia burgdorferi* in specimens of tissue [JCM (1999) *37* 1958–1963].

LightCycler® hybridization probes. These commercial probes (Roche) can be used in quantitative PCR (section 8.5.4.7). The reaction mixture contains two types of probe which bind to *adjacent* sites at an internal (i.e. non-terminal) region of the amplicon; binding occurs at the annealing stage of thermal cycling, and the probes are later displaced by primer extension. When bound to the target sequence, the 3' end of probe 1 (labelled with fluorescein) is juxtaposed to the 5' end of probe 2 (labelled with the dye LightCycler Red 640). When the probes bind in this way, fluorescence from probe 1 excites the dye on probe 2 by a process called FRET (section 8.5.16); the dye on probe 2 then emits radiation of wavelength 640 nm, which is monitored. During ongoing cycling the probes bind to increasing numbers of amplicons and produce a progressively increasing signal (which is measured at the annealing stage of the cycle).

The LightCycler system can also detect two targets simultaneously in the same sample by using two pairs of primers and a different pair of probes/dyes for each target – so that two independent signals are obtained from the same sample. [Duplex LightCycler assay for (i) the *mecA* gene and (ii) a *Staphylococcus aureus*-specific marker: JCM (2000) *38* 2429–2433.]

8.5.4   The polymerase chain reaction (PCR)

PCR is a method for copying specific sequences of nucleotides in DNA (or, in a modified form of the process, in RNA); repeated replication of a given sequence of nucleotides (the *amplicon*, usually <2 kb long) forms millions of copies within hours. The method depends on the ability of a (thermostable) DNA polymerase to extend a primer (section 7.3) on a template strand that includes the amplicon. The principle of PCR is shown diagrammatically in Fig. 8.18. (PCR is covered by patents owned by Hoffmann-La Roche.)

Note that PCR is one of several methods for copying sequences of nucleotides. Other methods include the ligase chain reaction (section 8.5.9.1) and NASBA (8.5.9.2).

PCR is a flexible technique with applications in many fields, both medical and non-medical. Some of the uses of PCR are given in section 8.5.4.1.

Notice that a high temperature (e.g. 72°C) is used to keep the strands of sample dsDNA apart during extension of primers; this is an essential requirement for the synthesis of DNA under these *in vitro* conditions. Not all DNA polymerases can work at this temperature, so that a *thermostable* polymerase is a key component in PCR. One such enzyme is the *Taq* polymerase from *Thermus aquaticus*, a bacterium from hot springs which has an optimum growth temperature of about 66–75°C. A modified, recombinant form of this enzyme (the *Stoffel fragment*) is rather more thermostable than the parent enzyme. However, neither of these enzymes has 3'-to-5' exonuclease activity, i.e. neither can cleave nucleotides from the 3' end of a growing strand – so that both lack *proof-reading* ability (section 7.7). This is important for some purposes (e.g. when the products of PCR are to be sequenced); for such purposes it is necessary to copy the amplicon with high fidelity.

Currently there is a wide variety of commercially available enzymes for PCR; some have e.g. very high levels of thermostability and/or proof-reading capacity. For example, the *PfuUltra*™ polymerase (Stratagene) – derived from the archaean

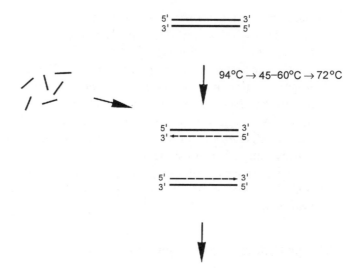

Figure 8.18 The polymerase chain reaction (PCR): principle (diagrammatic).

The double-stranded (ds) fragment of DNA (*top*) is the specific sequence of nucleotides to be amplified by PCR. This sequence of nucleotides is called the *amplicon*; *copies* of this sequence, made by PCR, are also called amplicons.

The reaction mixture includes the following, (i) The sample dsDNA. (ii) The *primers:* small pieces of ssDNA, each about 20–30 nucleotides in length. The primers are of two types: one type is complementary to the 3' end of one strand of the amplicon, and the other type is complementary to the 3' end of the other strand of the amplicon. (The reaction mixture contains many millions of copies of the primers.) (iii) Deoxyribonucleoside triphosphates of all four kinds. (v) Appropriate buffer.

The reaction mixture undergoes a series of changes in temperature. This cycle is repeated about 20–30 times, the temperature being automatically changed/controlled in a *thermocycler* device. Initial heating to about 94°C denatures the dsDNA fragment (*top*) to the single-stranded state. Transient cooling to 45–60°C allows the primers (*top, left*) to bind (= *anneal*) to their complementary sites on the amplicon. A change to 72°C then permits the (thermostable) DNA polymerase to start DNA synthesis (*dashed line*) from the 3' end of each primer. When the temperature cycle is repeated, the newly synthesized strands (as well as the pre-existing strands) act as templates.

For each target sequence (amplicon) in the sample, the number of copies made by n cycles of PCR is theoretically 2^n, but after about 25 cycles the rate of synthesis decreases owing to e.g. the increased number of template strands competing for a finite quantity of polymerase.

For clarity, the amplicon in the diagram is equal in length to the fragment of sample DNA, i.e. the whole fragment of sample DNA is the amplicon. However, it's important to note that the amplicon is usually a sequence within a much longer molecule of sample DNA; the principle, though, is the same: the amplicon is defined by the sequence of nucleotides between the 5' nucleotide in one primer and the 5' nucleotide in the other primer.

The choice of primers is important. For example, primers should not include sequences that permit base-pairing between one primer and the other. In particular, complementarity at the 3' ends of the two primers would allow each primer to act as a template for the other – leading to extension of both primers and the formation of so-called *primer-dimers*. Also, a primer should not contain sequences that permit one part of the primer to bind to another part of the same primer.

The choice of primers also influences the annealing temperature. The annealing temperature

Pyrococcus furiosus – is a highly thermostable, high-fidelity enzyme containing a factor that overcomes dUTP poisoning in archaeal polymerases (caused by dCTP deamination during PCR) [PNAS (2002) *99* 596–601].

For some purposes we require that primers bind to target sequences to which they are not fully complementary (section 16.2.2.4). In such cases we can use *low-stringency* conditions, i.e. particular values of temperature, pH and electrolyte concentration which optimize binding by helping to overcome, or minimize, the effects of mismatch(es) between the nucleotide(s) in primers and those in template strands.

Generally, however, we use *high-stringency* conditions; these allow primer–template binding to occur only when primers and target sequences are exactly complementary, or very nearly so. The greater the stringency of the conditions (e.g. the higher the temperature) the greater will be the degree of matching required between primers and target sequences for successful binding to occur. Clearly, high-stringency conditions are used when primer–target binding has to be highly specific.

Successful PCR requires close attention to detail. One important requirement is to avoid contamination by extraneous DNA – which may contain (i) site(s) able to bind the primers being used, and/or (ii) sequences able to act as primers to unwanted sites in the sample DNA; in either case, unwanted sequences (as well as the required sequence) would be amplified, forming a troublesome mixture of products and lowering efficiency. Some approaches to this problem are outlined in section 8.5.4.3.

The need for controls. PCR can give false-positive and false-negative results; control procedures are therefore mandatory.

Causes of false-positive results include: (i) contamination of sample, apparatus and/or reagents with the target sequence (derived from other specimens and/or from a previous amplification); (ii) amplification of non-specific sequence(s) due to insufficient specificity of the primers (see e.g. section 8.5.4.1) and/or a level of stringency which is too low (permitting priming of non-specific sequences).

Causes of false-negative results include: (i) absence of target sequence due to faulty extraction of DNA from sample; (ii) concentration of target in sample below the limit of detection; (iii) target's condition too poor for amplification; (iv) the presence of inhibitors of PCR in the sample (see section 8.5.4.6); (v) mutation in the primer-binding site of the target sequence, inhibiting binding; (vi) incorrect operation of the thermocycler.

Some control procedures:

A water blank may be used as a negative control [e.g. AEM (1999) *65* 2650–2653].

---

increases with increasing G+C% of the primers (for a rationale, compare the number of hydrogen bonds in GC and AT pairs in Fig. 7.6). The annealing temperature is usually between 45 and 60°C but may be outside this range. Examples of primer-binding temperatures: GC% 33: 45°C [JCM (1999) *37* 772–774]; GC% 62: 63°C [NAR (1998) *26* 3614–3615].

For variant forms of PCR see section 8.5.4.2 and Fig. 8.19.

A positive control may contain the target sequence at a known concentration [e.g. JCM (1997) *35* 566–569].

Problems with amplification, or with inhibitors, may be detected by the following method [American Review of Respiratory Diseases (1991) *144* 1160–1163]. A fragment of DNA (unrelated to the true target, and of different length) is constructed (*in vitro*) with terminal primer-binding sequences identical to those of the true target sequence. The fragment is cloned, and copies are added to the test samples. This approach avoids risking contamination with the true target sequence; moreover, amplification of the construct (but not the true target) adds confidence to a negative test result.

In assays of bacterial target sequences, so-called *broad-range primers*, which bind to highly conserved sequences in the 16S rRNA gene in *all* bacteria, have been used to detect inhibitors of PCR [e.g. JCM (1999) *37* 772–774].

Detection of PCR products. What *are* PCR products? In most cases (though not in asymmetric PCR: section 8.5.4.2) the products of a PCR assay are millions of identical copies of a specific sequence of linear dsDNA. Because these copies have a *precise* length (i.e. an exact number of base-pairs defined by the primer binding sites), they can be readily detected by gel electrophoresis (section 8.5.1.4) following by staining (e.g. with ethidium bromide) and examination by ultraviolet radiation. (SYBR Safe™ DNA gel stain (Molecular Probes Inc., Oregon, USA) is less mutagenic, i.e. safer, than ethidium bromide.) To facilitate identification of the amplicons, another lane in the same gel can be used for electrophoresis of a set of fragments of known size (100, 200, 300 . . . base-pairs); this produces a scale ('ladder') of bands with which the experimental band can be compared. This method can also be used for multiplex PCR (section 8.5.4.2), when products of more than one length are obtained.

The *identity* of products from a given reaction may also be checked e.g. by subjecting them to a particular restriction endonuclease (section 7.4, Table 8.1) and checking, by electrophoresis, to see whether fragments of the *expected* sizes have been formed.

Bands of products within a gel strip may be further examined by Southern blotting and Southern hybridization (Fig. 8.13).

Products can also be rapidly detected by a plate-based assay in which the dsDNA fragments are first denatured to ssDNA strands; these strands are bound to complementary, labelled detector probes and are then bound by immobilized capture probes on a plate. This method has been used for detecting *Bordetella pertussis* in clinical specimens [JCM (1997) *35* 117–120].

Real-time detection of products is discussed in section 8.5.4.7.

8.5.4.1   Some uses of PCR

PCR is used in fields as diverse as medicine, forensic science and palaeobiology. Some of the uses of PCR in bacteriology and molecular biology are outlined below, and some variant forms of PCR are mentioned in section 8.5.4.2.

1. In some cases PCR is used simply to determine whether or not a particular sequence of nucleotides is present in a given sample; the assumption is that, if present, the sequence will be amplified and can be detected by appropriate means. Because a given PCR product is usually of known length (in terms of nucleotides), it can be readily detected/identified by gel electrophoresis (section 8.5.1.4) as a band *in a precise location* in the gel. As a further check, PCR products may be subjected to restriction endonuclease(s) (section 8.5.1.3) and the resulting fragments examined by electrophoresis to see whether fragments of *expected* sizes have been formed. Alternatively, the products may be examined by Southern blotting and hybridization with a specific labelled probe (Fig. 8.13).

   It was soon realized that PCR could detect pathogens in clinical material by detecting specific sequence(s) of their DNA – especially useful for those pathogens that cannot be cultured (i.e. which do not grow *in vitro*) and for those that grow very slowly (e.g. *Mycobacterium tuberculosis*). Hence, PCR is useful in the diagnosis of certain diseases and in epidemiological studies. One early example was the rapid diagnosis of tuberculous meningitis [Neurology (1990) *40* 1617–1618]; the results of PCR, taken as confirmation of the disease, were available in hours (instead of the weeks required by conventional methods).

   For diagnostic purposes, primers must be highly specific to sequence(s) found *only* in the given pathogen – e.g. a sequence in a gene encoding a unique toxin. Conditions must be arranged so that extension of primers by PCR occurs only if the target sequence is present in the clinical sample. It should be noted that amplification of a pathogen-specific sequence by PCR does not necessarily indicate the presence of the *living* pathogen. It is also essential to ensure that the primers used are, in fact, unique to the given pathogen; in one (apparently rare) case, primers thought to be specific for mycoplasmas amplified a sequence from the unrelated species *Fusobacterium necrophorum* [JCM (1999) *37* 828–829].

   See also section 16.1.6.1.

2. PCR can be used to copy a particular amplicon in order to determine (or check) its actual sequence of nucleotides. If amplicons are required for sequencing (section 8.5.6) they may be made by asymmetric PCR (section 8.5.4.2); a high-fidelity DNA polymerase (section 8.5.4) should be used.

3. PCR products (made by asymmetric PCR) may be used as probes (section 8.5.3); dsDNA products may be inserted into expression vectors (section 8.5.1.5).

4. PCR is used in studies on classification/taxonomy (Chapter 16).

5. PCR, using broad-range primers (section 10.8), can be used e.g. for surveying the diversity of species in environmental samples.

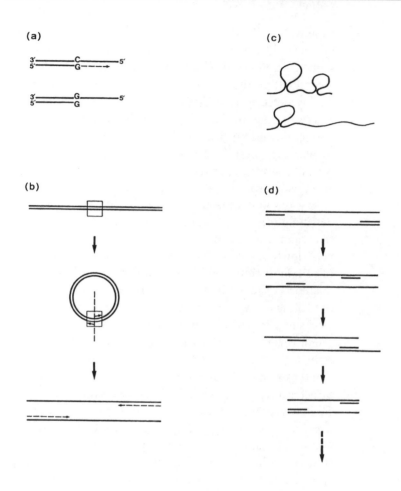

Figure 8.19 PCR: some variant forms/applications (principle, diagrammatic).

(a) Amplification-refractory mutation system (ARMS). This procedure can demonstrate/detect a specific *point mutation* (Fig. 8.1) in DNA whose wild-type sequence is known. In the diagram, a G → C (guanine to cytosine) mutation is shown as 'C' in one strand of a mutated molecule (*top*). The presence of this point mutation can be demonstrated or detected by designing a PCR primer with a 3' guanosine nucleotide that, when the primer is bound, will align with the potentially mutant base. This primer can be extended by a *Taq* DNA polymerase (*dashed line, arrow*), indicating the G → C transition in the mutant strand; however, extension by *Taq* polymerase would be inhibited on a wild-type strand (*bottom*) owing to the absence of base-pairing between two guanosines.

(b) Inverse PCR. This procedure copies DNA that *flanks* a known sequence (rather than the sequence itself); it can be used e.g. to investigate the site of insertion of a known sequence in a chromosome or plasmid. The diagram shows a linear fragment (*top*) in which the known sequence occurs within the square. The fragment is first circularized (*centre*). PCR primers are designed such that, when bound to the fragment, their 3' ends will be extended in the directions shown by the arrows. A cut at a restriction site (*dashed line*) linearizes the molecule, giving a primer–template relationship that corresponds to the standard form of PCR (*bottom*).

8.5.4.2   Some (of many) variant forms of PCR

Amplification-refractory mutation system (ARMS). This PCR-based system can be used for detecting point mutations in DNA whose wild-type sequence is known. The principle is shown in Fig. 8.19(a). The standard method may not be able to detect point mutations within stretches of identical nucleotides (e.g. TTTTTT); however, detection of such point mutations has been demonstrated by using non-repetitive bases at the 3' ends of diagnostic primers [BioTechniques (1999) *27* 662–666].

Arbitrarily primed PCR (AP-PCR). See section 16.2.2.4.

Asymmetric PCR. In this procedure, one of the two types of primer is used at a much lower concentration (e.g. 1:50), and this primer will be used up after a few cycles of PCR; however, a plentiful supply of the other primer ensures that *one* strand of the

---

(c) Single-strand conformation polymorphism (SSCP) analysis. SSCP analysis compares different (but related) samples of single-stranded DNA, detecting differences in their sequences by revealing differences in their electrophoretic speeds in polyacrylamide gel; strands which differ by even one base may be distinguished. PCR can be used to produce the samples of single-stranded DNA.

Related strands of DNA are likely to differ electrophoretically when difference(s) in their nucleotide sequences cause different levels of intra-strand base-pairing; intra-strand base-pairing confers a specific conformation which influences electrophoretic speed. The figure shows two related strands of DNA; in the lower strand a change in nucleotide sequence has resulted in a loss of local base-pairing – so that the two strands now have different conformations and are therefore likely to have different electrophoretic speeds. The gel used in SSCP analysis is a *non*-denaturing gel so as to avoid disrupting any existing intra-strand base-pairing. (See also section 15.4.11.2.)

A related method – denaturing gradient gel electrophoresis (DGGE) – compares samples of related PCR-generated (or restriction-generated) *double*-stranded DNA by two-dimensional electrophoresis. In this method [see e.g. Biotechnology (1995) *13* 137–139], the fragments are initially separated (by size) in the first phase of electrophoresis. Then, in the same gel, the fragments are moved electrophoretically at right-angles to their original path – this time through an increasing gradient of DNA-denaturing agents (e.g. urea + formamide); at given levels in the gradient, localized *sequence-dependent* 'melting' (DNA strand separation) occurs within part(s) of the different fragments (base-pairing being stronger in GC-rich regions) – this affecting electrophoretic speeds and allowing separation of fragments in the gel. DGGE has been used e.g. for characterizing organisms by comparing PCR-amplified sequences from their 16S rRNA genes [AEM (1996) *62* 340–346; AEM (1999) *65* 1251–1258], for detecting species of *Rhizobium* [LAM (1999) *28* 137–141], and for the differentiation of isolates of *Escherichia coli* by analysis of the 16S–23S intergenic spacer region [FEMSME (2001) *35* 313–321].

TTGE *(temporal temperature gradient gel electrophoresis)* is similar to DGGE in principle but uses an ongoing increase in temperature (instead of a chemical denaturing gradient) during the second phase of electrophoresis. [Comparison of DGGE with TTGE: LAM (2000) *30* 427–431.]

(d) Nested PCR (nPCR). Initially, standard PCR is carried out for about 25 cycles (*top*). Then, using part of the product as template, a further 25 cycles of PCR are carried out using a pair of primers complementary to *sub-terminal* sequences in the template. The final product is thus many copies of a sequence contained within the original sequence (*bottom*). nPCR has been reported to increase both the sensitivity and specificity of PCR; it has been used e.g. when only small amounts of target DNA are available or when the material is not in good condition.

sample dsDNA will be copied a significant number of times. This method can be used e.g. for making single-stranded DNA for sequencing (section 8.5.6) or for use as probes (section 8.5.3).

Hot-start PCR involves withholding (or blocking) an essential component of the reaction mixture until the mixture has been initially heated to above the primer-binding temperature; this enhances specificity and sensitivity by helping to avoid extension of primers from non-specific sites on the template.

In one commercial process, the polymerase is initially inactivated by its antibody during heating up to 70°C; such inactivation is lost as the temperature continues to rise for the first denaturing step, and PCR then proceeds normally.

In another system, the reaction mixture *without* polymerase is covered by a layer of wax (AmpliWax™; Perkin Elmer) which has been melted at 75–80°C and allowed to set. The polymerase, added above the wax layer, can act only after an appropriate temperature has been reached during cycling.

A different method involves a chemically modified form of DNA polymerase (AmpliTaq Gold™ DNA polymerase; Perkin Elmer Applied Biosystems) which remains inactive until (thermally) activated at the required temperature.

Inverse PCR. See Fig. 8.19(b).

Multiplex PCR. In this procedure, two or more target sequences are amplified, simultaneously, in a single assay. In most cases a different pair of primers is used for each target sequence. (In some assays one primer can be common to two target sequences [e.g. JCM (1999) *37* 1621–1624].) The use of multiple primers makes it particularly important to avoid complementarity between the 3' ends of the primers in order to prevent the formation of primer-dimers (see Fig. 8.18, legend). Examples of multiplex PCR include (i) detection of the B, C and F superantigen genes of *Staphylococcus aureus* [JMM (1998) *47* 335–340] and (ii) detection of vancomycin-resistance genes in enterococci [JCM (1999) *37* 2090–2092].

Nested PCR is carried out in two phases. In the first phase, a standard form of PCR is run for about 25 cycles. Amplicons from this phase then form the target for the second phase of amplification in which a different pair of primers is used to amplify an *internal* sequence of the first-phase amplicon. (Primers used in the two phases are often called the 'outer' and 'inner' pair of primers.) Nested PCR (nPCR) has been found to improve specificity/sensitivity, and is useful e.g. when the DNA sample is in poor condition or is of limited quantity. Examples of nPCR include (i) detection of *Mycobacterium malmoense* [JCM (1999) *37* 1454–1458] and (ii) detection of *Bordetella pertussis* and *B. parapertussis* [JCM (1999) *37* 606–610].

REP-PCR. See section 16.2.2.5.

Reverse transcriptase PCR (*rtPCR*) is used for copying an RNA target sequence. Initially, the enzyme reverse transcriptase forms a strand of DNA on the sample sequence of RNA. The RNA sequence is degraded (e.g. by RNase H) and replaced by a strand of DNA, making a dsDNA molecule; the latter is then amplified by a standard PCR procedure.

Touchdown PCR. This procedure involves repetition of a given PCR assay, the annealing temperature being decreased in a stepwise fashion in each run; the object is to find the lowest annealing temperature, for a given assay, that will permit normal primer binding in the absence of *mispriming* (i.e. the binding of primers to inappropriate sequences). [Example of use: JCM (1999) *37* 1274–1279.]

8.5.4.3   *The problem of extraneous DNA in PCR: some solutions*

It was mentioned earlier that contamination with extraneous DNA can jeopardize the success of PCR. This can be particularly important, for example, in those clinical laboratories which routinely test specimens for only one or two types of amplicon; here, new specimens may risk contamination from a build-up of DNA of the specific target sequence(s).

One approach to this problem is to carry out PCRs routinely with deoxyuridine triphosphate (dUTP) instead of the usual deoxythymidine triphosphate (dTTP) – so that *all* products contain uracil rather than thymine. This method makes use of the enzyme uracil-*N*-glycosylase (UNG) which cleaves uracil from dUMP in DNA (see also section 7.7.1.3); UNG is included in each reaction mixture. Routinely, *prior to* cycling, UNG is allowed to act on any amplicons which may have contaminated the reaction mixture from *previous* PCRs; after treatment by UNG, any such amplicons will be destroyed by double-stranded cleavage when the temperature first rises to e.g. 95°C at the start of cycling. This high temperature also inactivates UNG, so that cycling in the *new* assay can proceed in the absence of both UNG and contaminating amplicons. (Note that UNG does not affect *template* DNA – which contains dTMP.) [Effectiveness and limitations of UNG: BioTechniques (2004) *36* 44–48.]

In a different approach, *isopsoralen* is incorporated in the PCR reaction mixture. After completion of cycling the reaction mixture is exposed to ultraviolet radiation (e.g. 365 nm/4°C/15 minutes); this photo-activates the isopsoralen. Activated isopsoralen *covalently* binds together the two strands in dsDNA; thus, the final PCR products are dsDNA amplicons which cannot be denatured to single strands *and which therefore cannot serve as templates*. This method is used e.g. when PCR products are required simply to demonstrate the presence of a given sequence by gel electrophoresis; clearly, it cannot be used if products are required for sequencing. It has been reported that activation of isopsoralen is more efficient at low temperature than at room temperature [JCM (1999) *37* 261–262].

Minimization of contamination also involves so-called *amplicon containment*: division of the workplace into a number of dedicated areas – each used for only certain stage(s) of PCR; thus, two of the areas may be used e.g. for (i) cycling and (ii) analysis of products.

8.5.4.4 Cloning (blunt-ended) PCR products

A simple and convenient method for cloning PCR products – e.g. prior to sequencing (section 8.5.6) – involves the use of a 'rare-cutting' restriction endonuclease (section 8.5.1.3), *Srf*I [TIG (1996) *12* 286–287]. *Srf*I makes a blunt-ended cut (section 8.5.1.3) at its recognition sequence; because such sequences are rare, it can be assumed that, in general, PCR products will not contain an *Srf*I site.

A cloning vector was constructed by inserting an *Srf*I site into the genome of phage M13 (see section 9.2.2). To insert PCR products into the (double-stranded, circular) vector molecules, the products and vectors are incubated together in a mixture containing both *Srf*I and a ligase. A given *Srf*I site will undergo cutting (thus linearizing the vector molecule), ligation, cutting, re-ligation etc. until a cut end becomes ligated to a PCR fragment (by blunt-ended ligation – section 8.5.1.6); when this happens the vector is no longer susceptible to *Srf*I, and ligation can occur between the free ends of vector and fragment to form a circularized, fragment-containing vector. Such vectors can be inserted into bacteria by transformation.

*Srf*I cuts (/) at the following recognition sequence:

$$5'\text{-GCCC/GGGC-}3'$$
$$3'\text{-CGGG/CCCG-}5'$$

and is manufactured by Stratagene.

In another procedure (PCR cloning kit (blunt end); Boehringer Mannheim), the PCR product inserts into a blunt-ended cut that disrupts a lethal gene in the vector molecule. Vectors within which a PCR product has been ligated can be inserted into cells, by transformation, and cloning will occur when the cells grow. Vectors that re-ligate without incorporating a PCR product contain a *functional* lethal gene, so that these vectors will kill their host cells following transformation (i.e. the system has a built-in selection for insert-containing vectors).

8.5.4.5 Preparation of single-stranded DNA from PCR products

Purified single-stranded DNA – needed e.g. for sequencing (section 8.5.6) or for use as probes (section 8.5.3) – can be obtained from the products of PCR by a simple, rapid method [NAR (1996) *24* 3645–3646]. Prior to PCR, *one* of the two types of primer is labelled with biotin. Following PCR, streptavidin (a protein obtained from *Streptomyces avidinii*) is added to the reaction mixture; streptavidin binds tightly to biotin, i.e.

it binds to *one* of the two types of ssDNA product. The mixture is subjected to gel electrophoresis in a *denaturing* gel (one containing e.g. concentrated urea) which inhibits base-pairing between complementary strands; the (complementary) ssDNA products of PCR therefore cannot hybridize. Because *one* of the PCR products is bound to protein (streptavidin), its electrophoretic mobility (i.e. its speed during electrophoresis) is greatly reduced; hence the *other* product will form a well-separated band in the gel. If a *particular* strand is required the primer of the other strand is biotinylated prior to PCR.

8.5.4.6   *Inhibitors and facilitators of PCR*

Inhibitors. Samples may contain substance(s) that inhibit PCR, so that a negative result may be obtained even in the presence of the target sequence (false-negative result).

Inhibitory factors occur in various clinical specimens – e.g. bone marrow aspirates, cerebrospinal fluid, faeces [JCM (1997) *35* 995–998; NAR (1998) *26* 3309–3310] and urine. In blood cells, PCR-inhibitory components include haemoglobin (in erythrocytes) and lactoferrin (in leukocytes) [purification and characterization of PCR-inhibitory components in blood cells: JCM (2001) *39* 485–493].

Inhibitors also include the blood anticoagulants SPS (sodium polyanetholesulphonate) [JCM (1998) *36* 2810–2816] and heparin.

Agar (used e.g. in Stuart's transport medium) inhibits PCR in a concentration-dependent manner. Gellan gum may be a suitable gelling agent for microbiological media that are used in the context of PCR assays [JMM (2001) *50* 108–109].

Inhibitors may be detected by adding to the test samples a few fragments of synthetic DNA (an *internal control*) containing (i) binding sites for the primers being used and (ii) a binding site for an identification probe. Amplification of the control but not the target could indicate the absence of target sequences (or targets below the limit of detection); amplification of neither control nor target may indicate the presence of inhibitor(s). [Example of internal control: JCM (1998) *36* 191–197.]

Facilitators. These agents are able to enhance the performance of PCR in the presence of inhibitors; in some cases they can improve specificity and/or the fidelity of DNA synthesis. Facilitators of amplification include bovine serum albumin (BSA) and betaine; a particular facilitator may be effective against a range of inhibitors or against a few, specific ones. [Effects of amplification facilitators on diagnostic PCR in the presence of blood, faeces and meat: JCM (2000) *38* 4463–4470.]

8.5.4.7   *Quantitative (real-time) PCR (QPCR)*

In PCR we can follow the progress of amplification (as it is happening) and also estimate the number of target sequences *initially* present in the sample.

One method uses TaqMan® probes (section 8.5.3.2); the *unbound* probes are non-fluorescent owing to the action of the quencher. In the annealing stage of PCR, primers and TaqMan probes bind to their complementary sequences on the amplicons; a TaqMan probe binds to an *internal* sequence. On extension of the primer (i.e. DNA synthesis) the polymerase reaches the location of the probe; when this happens, the polymerase degrades the probe (liberating dye and quencher as *separate* molecules) and then continues to synthesize the remainder of the complementary strand of the amplicon. Now unquenched, the dye is able to fluoresce. Notice that one dye molecule is liberated for *each* product strand synthesized on a probe-binding amplicon; hence, as cycling continues, the level of fluorescence continues to rise as the number of amplicons continues to increase.

If, from the start, the increasing level of fluorescence is monitored, *cycle by cycle*, we can identify the first cycle in which fluorescence from the reporter molecules exceeded the level of background fluorescence; this cycle is the *threshold cycle*. From the threshold cycle we can estimate the number of target sequences that were in the sample *before* amplification began; the higher the threshold cycle the smaller the number of targets initially present. This is logical: the smaller the number of targets initially present the higher will be the number of cycles of amplification needed to reach the threshold value of fluorescence. In fact, there is an inverse linear relationship between threshold cycle and the logarithm of the number of target molecules initially present. Thus, preliminary experiments (using known numbers of target molecules) give a straight-line graph if threshold cycle (1, 2 ... 5, 6 ... 15, 16 ... 20, 21 ... etc.) is plotted against the log of initial number of target molecules; we can extrapolate from this graph when the initial number of target molecules is unknown.

Quantitative PCR can also use other types of probe: see section 8.5.3.2.

[Quantitative detection of *Borrelia burgdorferi* in specimens of tissue: JCM (1999) *37* 1958–1963. Improved titre determination of bacteriophage λ by real-time PCR: BioTechniques (2003) *35* 368–375.]

8.5.4.8 *PCR with degraded DNA*

Palaeobiological samples (and samples for forensic examination) may contain degraded DNA in which the fragments are too short to act as templates for PCR. In one approach to this problem (*reconstructive* PCR) an initial phase is carried out without primers: overlapping fragments are allowed to prime each other, i.e. the 3' end of one strand of a given fragment is extended on a template strand provided by an overlapping fragment; in this way, one can obtain longer fragments suitable for use as templates in PCR [NAR (1996) *24* 5026–5033]. An alternative approach is a modified recombinant PCR [BioTechniques (1999) *27* 480–488].

8.5.5 Mutagenesis

We have seen that nucleic acids can be altered and manipulated by methods such as restriction and ligation, gene fusion etc. Mutagenesis (involving the deliberate creation of mutations) is a further method which can be approached in various ways.

Mutations may be created for either of two main purposes: (i) to *inactivate* a gene, so that the gene no longer has a product or function, or (ii) to *modify* a gene in order to change its expression in some way. Inactivation of a given gene can be used e.g. to confirm the involvement (or otherwise) of that gene in a particular activity – cells with an active gene being compared to those with an inactivated gene. Gene modification has many applications in medical and industrial fields as well as uses in biological research.

8.5.5.1 *Making random mutations*

When bacteria are exposed to mutagens, mutations develop in different genes in different individuals (section 8.1). By using appropriate *selective* conditions we can isolate from a mutagenized population those cells which have acquired the desired type of mutation (and which display a particular altered characteristic or function); for example, to isolate those (mutant) cells which have acquired resistance to a given antibiotic, the population of mutagenized bacteria is grown on a medium containing that antibiotic – on which only the resistant cells can grow.

For general approaches to the isolation of specific types of mutant see section 8.1.2.

Mutagenesis in a cloned gene: the use of mutator strains. An effective way of introducing random mutations into a cloned gene is to insert the vector (carrying the gene) into cells of a *mutator strain* of bacteria, i.e. a strain which, through mutation, is defective in DNA replication and/or in DNA repair (see sections 7.7 and 8.1); owing to a deficiency in normal repair functions, spontaneous mutations accumulate in the bacterial genome *and* in the plasmid (or other cloning vector) containing the cloned gene during growth and multiplication of such cells. One such mutator strain is Epicurian Coli® XL1–Red (marketed by Stratagene); in this strain mutations are incorporated at a rate which is several thousand times higher than that in the corresponding wild-type strain. In this way, mutations can be introduced efficiently *without the use of mutagens*.

An error-prone enzyme. When carrying out PCR (section 8.5.4) it's usual to employ a high-fidelity DNA polymerase to ensure accurate copying from the template strand. However, if the requirement is for random mutations in the product strands one can use a commercial system such as the GeneMorph™ PCR Mutagenesis Kit (marketed by Stratagene); this kit includes Mutazyme® DNA polymerase which can introduce up to seven mutations per kb template strand in a PCR reaction.

8.5.5.2 Making specific mutations

Specific mutations can be used for studying gene expression/function and for modifying gene products in particular ways – e.g. improving and efficiency of an industrial enzyme by re-coding particular nucleotide(s) in its gene. Point mutations (section 8.1.1) and even more extensive changes can be created at pre-determined sites.

The principle of *site-specific mutagenesis* (= *site-directed* or *oligonucleotide-directed mutagenesis*) is shown in Fig. 8.20. A simpler, commercial method is described below.

QuikChange® Site-Directed Mutagenesis (Stratagene). This method does not require single-stranded templates of sample DNA (compare method shown in Fig. 8.20). Instead, it uses circular, dsDNA plasmids (of size < 8 kb), each containing the insert to be mutated.

The plasmids are mixed with a large number of mutagenic oligonucleotide primers (analogous to the primer shown in Fig. 8.20). These primers are of two types. One type binds to a sequence on one strand of the insert, while the other type binds to a sequence on the other strand of the insert (the two sequences overlapping). Both primers include the required mutation.

Thermal denaturing of the plasmids (to single-stranded copies) allows primers to bind to their respective sites on the insert. The primers are then extended by a (thermostable) DNA polymerase, forming dsDNA copies of the plasmid – each of which carries the required mutation in the strand formed from the primer. (In these copies the 3' end of each new strand remains unligated, i.e. there is a 'nick' between the end of the strand and the 5' end of the primer.)

During ongoing temperature cycling mutant strands are repeatedly synthesized on each strand of the parent plasmid.

Next, the non-mutant parent plasmids are eliminated by treatment with the restriction endonuclease *Dpn*I. Because the parent plasmids had originally replicated in *Escherichia coli*, they will have undergone *dam* methylation in the sequence 5'-GATC-3' (see section 7.8.8); hence, the parent plasmids will be susceptible to restriction by *Dpn*I (section 7.4).

Base-pairing between mutant strands yields dsDNA mutant plasmids. (*These* plasmids are not restricted by *Dpn*I: having been synthesized *in vitro* they lack *dam* methylation and are thus not susceptible to *Dpn*I.) Each mutant dsDNA plasmid is nicked in both strands; however, the nicks are in *staggered* positions (recall that the primer-binding sites overlap) so that stable ccc dsDNA plasmids are formed. When these mutant plasmids are inserted into *E. coli* the nicks are repaired *in vivo*; the plasmids can then replicate to produce many copies (and, hence, many copies of the mutant insert).

Another version of the QuikChange® system can be used with plasmids > 8 kb, and there is also a system for introducing more than one site-directed mutation.

Insertion of mutated/recombinant DNA into bacterial chromosomes. If mutated/recom-

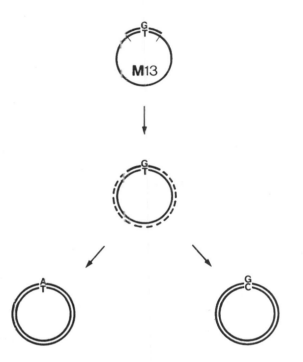

Figure 8.20 Site-specific mutagenesis (=oligonucleotide-directed mutagenesis) using a phage M13 cloning vector: creating a point mutation at a known site in a given gene (diagrammatic) (see section 8.5.5.2).

The purpose of M13 is to make single-stranded copies of the given gene. Initially, the (double-stranded) gene is inserted into a circular, double-stranded form of M13 (see section 9.2.2); replication of M13 (in a bacterium) produces a single-stranded copy of the gene *incorporated in each of the ccc ssDNA genomes of M13*. One such genome, carrying the target gene, is shown in the diagram (*top*).

A 15–20-nucleotide sequence of the target gene (its size exaggerated for clarity) is shown between two bars (*top*); within this small sequence we wish to replace deoxythymidine (T) with deoxycytidine (C). To do this we initially synthesize copies of a *complementary* sequence (an oligonucleotide, 15–20 nucleotides long) containing one mismatch at the required site: deoxyguanosine (G) opposite deoxythymidine (T); each oligonucleotide is then bound to a copy of the target gene (*top*) and used to prime *in vitro* DNA synthesis (*dashed line*) (*centre*). The resulting dsDNA molecules (each with the single mismatch) are inserted into bacteria (by transformation). Within the cells, DNA repair mechanisms (section 7.7) correct the mismatch – sometimes replacing T with C, sometimes replacing G with A (deoxyadenosine). Even without repair, replication of the mismatched duplex will produce some molecules with the required mutant sequence (*bottom, right*). Cells containing the mutant sequence can be identified e.g. by colony hybridization (section 8.5.1.2), using labelled oligonucleotides as probes.

Simpler methods, using double-stranded vectors, are available – see section 8.5.5.2.

binant DNA occurs in vector molecules (e.g. plasmids), the vectors can be inserted into bacteria by electroporation (section 8.4.1.2) or by transformation (sections 8.4.1 and 8.4.1.1). Figure 8.21 shows two methods for inserting the vector-borne DNA into a chromosome that contains a sequence homologous to the one carried by the vector.

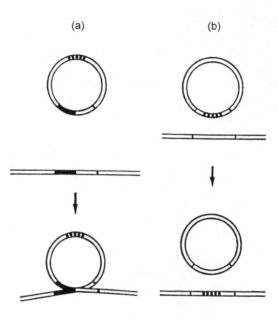

Figure 8.21 Insertion of vector-borne DNA into a specific bacterial chromosomal gene (diagrammatic).

(a) A circular plasmid, constructed *in vitro*, carrying (i) a marker gene (five dots, top of plasmid), and (ii) a fragment of DNA corresponding to *part* of the coding sequence of the (chromosomal) target gene; the left of the fragment is shown black, the right white, simply to help in depicting the fate of each part of the fragment. Copies of this plasmid are introduced, by transformation, into a population of bacteria carrying the target gene. Within the cells, recombination occurs between the fragment in the plasmid and the corresponding (*homologous*) sequence in the chromosomal target gene (also shown in black/white); this involves a single cross-over: a double-stranded break in both fragment and target gene (at the black/white junctions), and the joining of ends to form a continuous structure, as shown below. Note that (i) the entire plasmid has inserted into the target gene, and (ii) part of the coding sequence of the gene has been duplicated (black/white.....black/white); for this reason, the process is sometimes called insertion–duplication recombination (or Campbell-type integration). Note that this procedure is carried out with a *non*-replicating recombinant plasmid – one which, in dividing cells, can be replicated only as part of the chromosome (i.e. following *integration*); this ensures that subsequent selection of the marker (e.g. antibiotic resistance) will identify only those cells in which the plasmid has integrated into the gene. (If a replicating plasmid were used, copies of the plasmid would be passed to both daughter cells on cell division so that, in this case, *all* cells would be selected by the marker, regardless of whether or not the plasmid had integrated.)

If the fragment (carried by the plasmid) is a mutant form of *one end* of the target gene, insertion of the plasmid will simply replace the wild-type end of the gene, in the chromosome, with the mutant end from the plasmid; the final result would therefore be a complete, i.e. non-disrupted, chromosomal gene with a mutant end-sequence (together with DNA from the plasmid followed by the wild-type sequence of the gene).

If the plasmid carried a wild-type form of *one end* of the target gene *fused 'in frame' with a different*

Either method can be used to modify or inactivate a gene; additionally, method (a) can be used for gene fusions [e.g. an Smc–GFP fusion: GD (1998) *12* 1254–1259].

8.5.5.3   *Transposon mutagenesis*

Potential problems with plasmid-mediated mutagenesis (section 8.5.5.2) include (i) the possibility of restriction (section 7.4) of plasmid DNA within target cells, and (ii) possible failure of homologous recombination. Problem (ii) can be overcome by using transposons (section 8.3).

Transposon-mediated mutagenesis has been used e.g. to detect/identify secreted and cell-surface proteins of pathogenic bacteria; these proteins are of interest because of their possible association with *virulence*. A population of a given pathogen is mutagenized with transposons containing a (promoter-less) gene for the enzyme alkaline phosphatase (*phoA*); these transposons (Tn*phoA*) insert randomly (into different genes) within the bacterial population. On plating, the cells form individual colonies, and these are tested with a chromogenic (colour-generating) substrate for alkaline phosphatase; as only the *extracellular* enzyme can use this substrate, any colony which gives a positive (colour) reaction indicates secreted or cell-surface phosphatase. As *phoA* lacks promoter and signal sequence (section 7.6), the production of *extracellular* phosphatase by the cells of any given colony indicates that Tn*phoA* has inserted into the gene of a *secreted or cell surface* protein such that:

* *phoA* is 'in frame' and is transcribed from a genomic promoter
* the signal sequence of the gene into which *phoA* has inserted has enabled secretion of the phosphatase

The gene disrupted by Tn*phoA* is likely to non-functional, and, if it is a virulence gene, the virulence of cells in that clone may be demonstrably affected; this clone of mutants can be tested for virulence, and the relevant gene can be isolated and sequenced. Note that, despite random insertion of transposons, this method is selective in that it allows identification of specific (i.e. secreted/cell surface) proteins and their genes.

---

gene, insertion of the plasmid would result in a chromosome encoding a fusion protein (section 8.5.2).

If the plasmid carried a fragment of the *internal* coding sequence of the target gene, insertion of the plasmid would interrupt the coding sequence – thus *disrupting* (and almost certainly inactivating) the target gene.

(b) This plasmid includes a fragment of target gene (white) containing a mutant region (five dots); the chromosomal sequence corresponding to the fragment is shown as a white segment, below. In this case, a cross-over can occur on *both* sides of the mutant region; this results in an *exchange* of material between plasmid and chromosome: the mutant section is transferred from plasmid to target gene, and the corresponding wild-type sequence is transferred from chromosome to plasmid. Thus, particular mutation(s) can be inserted into a specific gene. If, instead of a mutant region, the fragment contained a fragment of foreign DNA, then insertion of this DNA would inactivate the target gene.

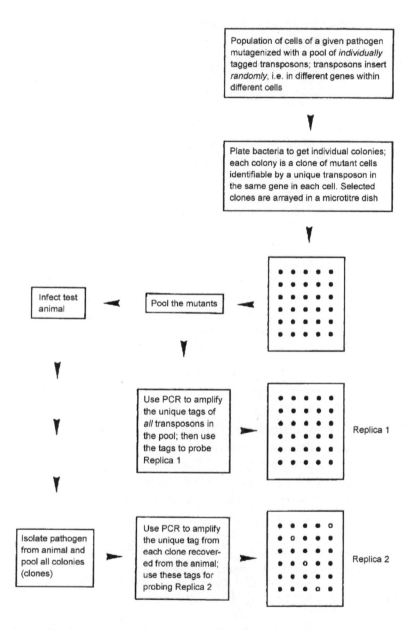

Figure 8.22 Signature-tagged mutagenesis: a method for identifying *virulence*-associated genes in a pathogen (general principle).

A *unique* sequence ('tag'), about 40 base-pairs in length, is inserted into each of a population of transposons, i.e. the tag in each transposon is different from that in all the other transposons; each tag is flanked, on both sides, by a short primer-binding site, *these* sites being the same in all transposons. The transposons are used to mutagenize cells of the pathogen, and they insert, randomly, into different genes in different cells.

A protocol for transposon-mediated mutagenesis in *Haemophilus influenzae* includes a delivery system containing the specific uptake-signal sequence (section 8.4.1) which facilitates transformation [AEM (1998) *64* 4697–4702].

Using *luxAB* genes as a reporter system, transposon mutagenesis has been used to identify genes induced in the cold-shock response (section 7.8.2.4) in *Sinorhizobium meliloti* [AEM (2000) *66* 401–405]. Transposon mutagenesis (involving gene knockouts by insertional mutation) has also been useful e.g. for genetic analysis of *Rickettsia prowazekii* [AEM (2004) *70* 2816–2822].

Signature-tagged mutagenesis (STM) can be used to detect those virulence genes whose expression, in a pathogen, is necessary for growth within the host organism; the principle of STM is outlined in Fig. 8.22. Aberrant results in STM may be due e.g. to non-random integration of transposons in the chromosome (with possible failure to detect a given virulence gene). A false-positive result may be obtained e.g. if a transposon inserts into a *non*-virulence gene in a cell which (by chance) already carries a mutation in a virulence gene. Nevertheless, STM has been useful e.g. in detecting a pathogenicity island (section 11.5.7) in *Salmonella typhimurium*.

The mutagenized cells are plated to form individual colonies. A given colony consists of a clone of mutant cells, each cell containing the same *uniquely* tagged transposon in the same gene. A number of colonies are chosen, and an inoculum from each colony is arrayed, separately, in a microtitre dish. (The dish shown has 30 wells, but larger dishes are normally used.) Two replica 'blots' of the array are made on membranes (for subsequent DNA hybridization studies); in these blots the cells are lysed and their chromosomal DNA is exposed and fixed to the membrane.

Cells are taken from each of the wells and pooled. The pool is used in two ways. First, it provides an inoculum for infection of the test animal. Second, cells from this ('input') pool are lysed, and PCR (section 8.5.4) is used – with labelled primers – to amplify the unique tags of all transposons in the pool. The amplified, labelled tags are then used as *probes* (section 8.5.3) on one of the replica blots; in this blot (Replica 1), each (unique) tag should hybridize with the DNA from cells containing the corresponding transposon. This 'pre-screening' process (on Replica 1) checks for efficient amplification of each unique tag in the pool. (The need to pre-screen may be avoidable if a set of 'dedicated' tags is used [TIM (1998) *6* 51].)

The pathogen is then recovered from the test animal by plating an appropriate specimen. The resulting colonies are pooled (forming the 'recovered' pool), and PCR is used (with labelled primers) to amplify the tag from each clone in this pool. The amplified, labelled tags are then used to probe the second replica blot (Replica 2).

Considering the original, mutagenized cells (in the 'input' pool), the cells of interest are those which, through a (transposon-mediated) mutation, are *unable* to grow within the test animal, i.e. cells whose virulence has been lowered; such cells will be absent, or few in number, in the 'recovered' pool (compared with cells which have grown normally in the animal). Hence, the signature tags of these 'virulence-attenuated' cells will be present in the input pool but absent (or rare) in the recovered pool; consequently, such cells can be identified by hybridization in the first blot but an *absence* of (or weak) hybridization in the second blot (see figure). (If 'dedicated' tags were used, virulence-attenuated cells would be identified by the absence of hybridization in a single blot.)

In a given 'virulence-attenuated' clone, the relevant gene (identifiable from the inserted transposon) can be isolated and sequenced for further study.

In vitro *transposition*. A study of transposon Tn5 [JBC (1998) *273* 7367–7374] showed that transposition could be achieved in a wholly *in vitro* system. For example, a population of plasmids can be mutagenized simply by incubating them for a few hours at 37°C with (i) DNA flanked by the 19-bp OE sequences of Tn5 (see section 8.3.2) and (ii) Tnp transposase; a hyperactive (recombinant) form of Tnp can significantly increase the rate of transposition. Insertion occurs at random sites in the plasmids.

This *in vitro* system of transposition has various practical uses. For example, because transposition occurs at *random* sites in the target molecules (plasmids etc.) it generates many different templates (with known primer-binding sites) which can be used for sequencing the entire target molecule. For this purpose, the mutagenized plasmids are inserted into a population of bacteria; following bacterial growth, and plating on an antibiotic-containing medium, individual colonies can be selected and the plasmids isolated for sequencing.

The system is also useful e.g. for the *in vitro* generation of insertion mutations and for introducing a selectable marker gene (e.g. an antibiotic-resistance gene) into vectors.

8.5.5.4   *Mutagenesis in* host *genes for studying pathogen–host interactions*

Inactivation of certain genes of the *host* has yielded valuable information on the roles of these genes in pathogen–host interactions. In so-called 'knock-out' mice, inactivation of various genes of the immune system has been found to modify susceptibility to certain bacteria (examples: Table 8.2).

8.5.6   DNA sequencing

In DNA the primary source of information is the *sequence* in which the nucleotides occur (section 7.1). From a known sequence we may, for example, be able to deduce the composition of a protein encoded by the DNA – although it must be remembered that the actual sequence of amino acids in a protein may be influenced by factors such as introns (section 7.9.1.1) and translational frame-shifting (section 7.8.7). By determining the sequence of nucleotides in a given sample of DNA (i.e. *sequencing*) we can

Table 8.2   Gene inactivation in knock-out mice: some examples

Gene encoding	Increased susceptibility to (e.g.)
T cell receptor	*Listeria, Mycobacterium bovis*
Inducible nitric oxide synthase	*Listeria monocytogenes*
Interferon-γ	*Mycobacterium tuberculosis*[1]
β_2-microglobulin	*Listeria, M. tuberculosis*

[1]See also section 11.4.1.4.

also study promoters and other control mechanisms and identify features such as REP sequences (section 16.2.2.5). Moreover, *whole-genome* sequences can be used e.g. for making overall comparisons between related species of bacteria, for predicting particular features in non-sequenced members of a given group of bacteria, and for detecting analogous structures and functions among more of less broad categories of organisms.

One method for sequencing is explained in Fig. 8.23. The single-stranded templates for sequencing may be made by asymmetric PCR (section 8.5.4.2) rather than by cloning in phage M13; in this case the primers are complementary to known sequences on either side of an unknown sequence. Single-stranded templates can also be made in phagemids (section 8.5.1.5). (See also section 8.5.4.5.)

Sequencing of about 500 nucleotides may be carried out from a single primer-binding site. Further sequencing of the given fragment can be achieved by removing those nucleotides (from the double-stranded form of the fragment) which have already been sequenced and then sequencing a new single-stranded fragment from a new primer-binding site. Repetition of this process *(nested deletions)* can be carried out until the entire fragment has been sequenced.

Once a large genome/chromosome has been fully sequenced, comparison of variant forms of it may be facilitated by probe-based methods [Science (1996) *274* 610–614].

An adaptation of the dideoxy sequencing procedure has been used for detecting antibiotic resistance in *Mycobacterium tuberculosis* (section 15.4.11.2 and Fig. 15.6).

One alternative to dideoxy sequencing is described in section 8.5.6.1.

8.5.6.1   *Pyrosequencing*™

This is a rapid, commercial method (Pyrosequencing AB, Uppsala, Sweden) which has been used for sequencing relatively short DNA templates of up to ~35 bases (although longer templates have been reported). The principle [Science (1998) *281* 363–365] is as follows.

Initially, copies of the single-stranded template – containing the unknown sequence and a contiguous, known sequence – can be produced e.g. by PCR (section 8.5.4) from the DNA under study. The use of biotinylated primers in PCR will allow the PCR-generated templates to bind to a streptavidin-coated, immobilized surface (biotin binds strongly to streptavidin). Primers then bind to the templates.

The templates (with their primers) are present in a solution that includes DNA polymerase, ATP sulphurylase, firefly luciferase, and an enzyme such as *apyrase* which degrades nucleotides.

The four types of deoxyribonucleoside triphosphate (dNTP) are added (separately) at intervals of 1 minute – the addition of e.g. [...A...T...G...C...A...T...G... C...A] taking 9 minutes. If an added nucleotide is complementary to the next free base in the template it will be incorporated *with release of pyrophosphate*. The

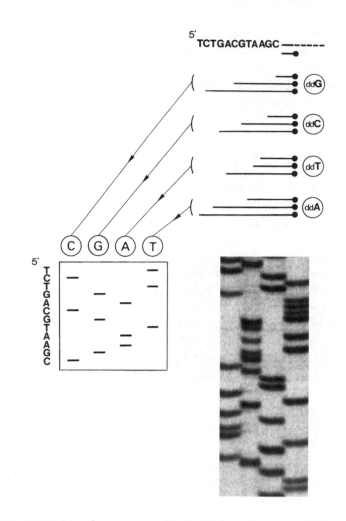

Figure 8.23 Determining the sequence of nucleotides in a fragment of DNA, i.e. *sequencing*. The method shown (in diagrammatic form) is Sanger's chain-termination method (also called the dideoxy method).

The fragment to be sequenced is first obtained in single-stranded form. This can be done e.g. by cloning the fragment in a phage M13 vector (see section 9.2.2). Alternatively, single-stranded templates can be prepared by asymmetric PCR (section 8.5.4.2).

In this example, the unknown sequence is TCT . . . . AGC *(top)*. Regardless of the method used for obtaining single-stranded templates, the unknown sequence will be flanked on its 3' side by a *known* sequence of nucleotides; this permits the design of a primer which can bind next to the unknown sequence such that the first nucleotide to be added to the primer will pair with the first 3' nucleotide of the unknown sequence. In the diagram, a primer *(short line)*, carrying a label *(black disc)*, has bound at a site flanking the 3' end of the unknown sequence *(top)*.

DNA synthesis *in vitro* is usually carried out with a reaction mixture that includes: (i) templates (in this case, single-stranded fragments that include the unknown sequence); (ii) primers; (iii) the four types of deoxyribonucleoside triphosphate (dNTP) – i.e. dATP, dCTP, dGTP and dTTP (Table 7.1); and (iv) DNA pyrophosphate is converted (by ATP sulphurylase) to ATP and, in the presence of the luciferase system, this gives rise to a burst of light. If the added nucleotide can pair with the next *two* free bases in the template, the amount of pyrophosphate released will be

---

polymerase. When base-paired to the template strand, the primer is extended (5' → 3') by the sequential addition of nucleotides dictated by the template.

For sequencing there are four separate reaction mixtures (G, C, T, A), each containing all of the constituents mentioned above (including millions of copies of the template sequence, and of the primer). In addition, each reaction mixture contains a given *dideoxyribonucleoside triphosphate* (ddNTP) (Fig. 7.2, legend); that is, the G mixture contains dideoxyguanosine triphosphate (ddG), the C mixture contains ddC, the T mixture ddT, and the A mixture ddA.

When a dideoxyribonucleotide is added to a growing strand of DNA it prevents the addition of the *subsequent* nucleotide; this is because a dideoxyribonucleotide lacks the 3'-OH group necessary for making the next *phosphodiester* bond (Fig. 7.4). Hence, extension of a primer will stop (= chain termination) at any position where a dideoxyribonucleotide has been incorporated. In a given reaction mixture, the concentration of the ddNTP is such that, in most growing strands, synthesis will be stopped – at some stage – by the incorporation of a dideoxyribonucleotide; because a given ddNTP molecule may pair with *any complementary base* in the template strand, chain termination can occur at different sites on different copies of the template strand – so that product strands of different lengths are formed in a given reaction mixture. For example, with ddG (see diagram) the three products are of different lengths because, during extension of primers, ddG has paired with cytosine residues in three different parts of the template; note that, in this case, *the length of a given product strand is related to the location of a particular cytosine residue in the unknown sequence.* Analogous comments apply to reaction mixtures containing the other ddNTPs.

At the end of the reaction new products are separated from templates by formamide. Each of the four reaction mixtures is then subjected to electrophoresis in a separate lane of a polyacrylamide gel. During electrophoresis small products move further than larger ones, in a given time; products that differ in length by only one nucleotide can be distinguished, the shorter product moving just a little further.

The gel contains urea; this inhibits base-pairing between product strands and templates; it also inhibits *intra-strand* base-pairing in the product strands. This is essential: in order to deduce the sites of a given base in the template strand it is necessary to compare the *lengths* of all the product strands, and this requires proportionality between strand length and electrophoretic mobility, i.e. between the length of a given strand and the position of its band in the gel; were intra-strand base-pairing to occur, a product's length could not be deduced, with certainty, from the position of its band in the gel.

After electrophoresis, bands in the gel may be revealed by autoradiography (if the primers had been labelled with e.g. [32]P) or by exposure to ultraviolet radiation (if the primers had a fluorescent label). In a different method, the primers are labelled with biotin before use; after electrophoresis, the gel is blotted (Fig. 8.13) and the membrane is exposed to an alkaline phosphatase–streptavidin conjugate – the products then being detected e.g. by the use of a substrate (such as CSPD) which can be split by the enzyme to produce chemiluminescence (section 8.5.3).

The locations of the bands (*bottom, left*) indicate the relative lengths of the product strands – the shorter products having moved further down the gel (from top to bottom in the diagram). Note that the first unknown 3' nucleotide (C) is identified by (i) the shortest product strand (which has moved the furthest), and (ii) the fact that this product came from the ddG mixture, indicating a base that pairs with G, i.e. C. Similarly, the next unknown (G) is the next shortest product strand – which came from the C mixture, thus indicating G in the template. The whole unknown sequence can be deduced in this way.

Bottom, right. Part of the autoradiograph of a sequencing gel (courtesy of Joop Gaken, Molecular Medicine Unit, King's College, London).

greater, and this will increase the amount of light produced – signalling the incorporation of more than one nucleotide.

Apyrase continually degrades dNTPs added to the mixture (and also degrades the ATP produced from pyrophosphate). If an added dNTP is *not* complementary to the next free base in the template it will be fully degraded by the time the next dNTP is due to be added. (Clearly, the activity of each enzyme must be carefully calculated.)

As nucleotides are sequentially added to the primer, the corresponding bursts of light are automatically recorded and presented as a time-based graph. On this graph, a given spike of light (at a particular time) corresponds to the addition of a specific, known dNTP to the test system at that time; the height of the spike of light indicates the number of consecutive bases on the template which have paired with the given nucleotide.

This method is reported to work well in most cases. Problems may arise if the template has one or more homopolymer stretches containing more than about five nucleotides (such as . . . TTTTTT . . .); this problem is due to non-linear generation of light when more than about five nucleotides of the same type are incorporated consecutively.

Using this (automated) method, DNA templates from 96 bacteriophages were sequenced in less than 1 hour with a reported failure rate of ~15% [BioTechniques (2003) *35* 317–324].

Pyrosequencing of PCR-generated products has also been used e.g. for grouping strains of *Listeria monocytogenes* [AEM (2001) *67* 5339–5342], subtyping strains of *Helicobacter pylori* [FEMSML (2001) *199* 103–107], identification and classification of clinical isolates by signature matching 16S rDNA fragments [APMIS (2002) *110* 263–272] and identification of bacterial contaminants in biotechnology reagents using broad-range primers [FEMSML (2003) *219* 87–91].

8.5.7 Expressing eukaryotic genes in bacteria: limitations

Most eukaryotic genes would not be expressed in bacteria: bacteria cannot (for example) deal with their introns – hence the use of cDNA (section 8.5.1.2). Additionally, many eukaryotic proteins undergo post-translational modification – for example, enzymic cleavage, or the addition of oligosaccharides (*glycosylation*) at particular sites; these specific modifications, which are carried out in eukaryotic cells and which are typically necessary for normal biological activity, are not normally carried out by bacteria. Hence, gene expression is often studied in the eukaryotes themselves, but this can be difficult owing to their complex control systems; in an attempt to overcome this problem, certain well-characterized bacterial gene-regulation systems have been used to control gene expression in eukaryotic cells [TIBTECH (1994) *12* 58–62]. More recently, it has been found that a *glycosylated* eukaryotic protein can be synthesized in *Escherichia coli* (thus avoiding the need for post-translational glycosylation) by introducing the relevant glycosylated amino acid into the *E. coli* cytoplasm [Science (2004) *303* 371–373].

As an alternative to the use of bacterial systems, studies on the expression of eukaryotic genes can be carried out with YACs (section 8.5.1.5). For example, wild-type genes, cloned in YACs, can be introduced into mutant eukaryotic cells to study *complementation* (i.e. the ability of a gene to compensate for defect(s) in the corresponding mutant gene when present in the same cell).

8.5.8 Functional analysis of genomic DNA

Sequencing of genomic DNA from various species has revealed many genes of unknown function. In a systematic approach to the functional analysis of genomic DNA from *Rhodobacter capsulatus*, use was made of a set of 192 cosmids (Fig. 8.15) whose overlapping inserts, collectively, cover the entire (3.7 megabase) chromosome of *R. capsulatus* [Nature (1996) *381* 653–654] Each cosmid was digested with the restriction enzyme *Eco*RV and the resulting (gene- or operon-sized) fragments were subjected to electrophoresis and Southern blotting (Fig. 8.13), producing a set of 192 blots.

Collectively, the 192 blots were used as a 'high-resolution hybridization template' (HRHT) to detect altered patterns of transcription under different physiological conditions. For example, when heat-shocked (section 7.8.2.3), the pattern of transcription differs from that under normal conditions; total RNA (section 8.5.1.2) isolated from heat-shocked cells can be suitably labelled and used as *probes* (section 8.5.3) to identify various parts of the chromosome (on particular blots) which contain sequences that are active under these conditions. In this way, specific locations on the chromosome can be associated with particular physiological state(s) of the cell.

8.5.9 Amplifying nucleic acids by methods other than PCR

Nucleic acids can be amplified by the ligase chain reaction (LCR), by nucleic acid sequence-based amplification (NASBA) and by strand-displacement amplification (SDA); like PCR, each of these methods has been developed commercially.

Figure 8.24 The ligase chain reaction (section 8.5.9.1) – binding of oligomers to the target sequence. One of the two strands of duplex DNA containing the *amplicon* (i.e. target sequence) is shown at the top: 3'-AT.....AC-5'. Two oligomers (each boxed) base-pair correctly with the target sequence and will undergo ligation; note that the 5' nucleotide of the oligomer on the *right* must be phosphorylated so that a phosphodiester bond (Fig. 7.4) can be formed. Another two oligomers will bind to the target sequence on the other strand of duplex DNA and will also undergo ligation (not shown). A more complex form of LCR is now used commercially.

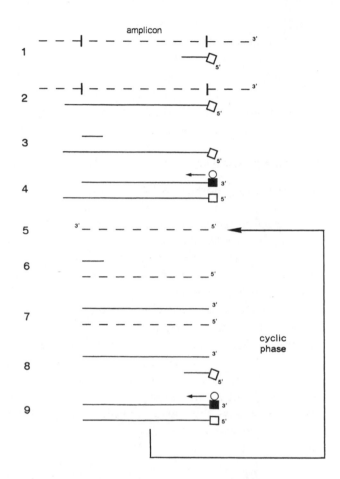

Figure 8.25 Nucleic acid sequence-based amplification (NASBA) (section 8.5.9.2) used for the amplification of an RNA sequence (diagrammatic). The dashed lines are strands of RNA, the solid lines strands of DNA. The following gives an outline of the stages involved.

1. A strand of RNA showing the target sequence (amplicon) delimited by the two, short vertical bars. A primer (primer 1) has bound to the 3' end of the amplicon. The 5' end of this primer is tagged with a short sequence (□) containing the *promoter* (section 7.5) of an RNA polymerase.

2. The enzyme reverse transcriptase (section 8.5.1.2) has extended the primer to form a strand of cDNA.

3. The enzyme RNase H has degraded (removed) the RNA strand, and a different primer (primer 2) has bound to the amplicon sequence in cDNA.

4. Reverse transcriptase (which can also synthesize DNA on a DNA template) has extended primer 2 to form double-stranded cDNA – so that a functional (double-stranded) promoter is now present. An RNA polymerase (○) has bound to the promoter and will synthesize an RNA strand in the direction of the arrow, i.e. in a 5'-to-3' direction.

5. The newly synthesized RNA strand. Note the polarity of the strand: it is *complementary* to the amplicon in the sample RNA (compare 5 with 1).

8.5.9.1 Ligase chain reaction (LCR)

Like PCR, LCR involves thermal cycling in which a heating phase is needed to separate the template strands; while the strands are separated, two oligomers bind to *each* strand of the template in such a way that oligomers on a given strand cover the entire target sequence (*amplicon*) (Fig. 8.24). If a pair of oligomers base-pairs correctly with the amplicon, the oligomers are joined by a heat-stable *ligase* (ligation: section 8.5.1.6). The cycle is repeated e.g. 25 times. Note that, for ligation to occur: (i) both juxtaposed bases in a given pair of oligomers must base-pair correctly with the corresponding nucleotides in the template strand (even if a mismatch occurs elsewhere in an oligomer), and (ii) the 5' end of one oligomer in each pair must be phosphorylated in order to permit ligation. Ligation, and amplification, argue for the presence of the given target sequence in the sample.

Commercial LCR-based assays (LCx® assays; Abbott Diagnostics) have been used to detect specific sequences of DNA in clinical specimens – e.g. in tests for *Chlamydia trachomatis*, *Neisseria gonorrhoeae* and *Mycobacterium tuberculosis*. These assays differ from the basic format described above. Thus, the two oligomers (= oligonucleotides, probes) in a bound pair are separated by a gap of one or a few nucleotides; hence, part of each cycle includes a period of extension from the 3' end of one probe by a (thermostable) polymerase present in the reaction mixture. Closure of the gap is followed by ligation. The cycle thus consists of three phases of temperature. For example, the following temperatures have been used in the LCx assay for *C. trachomatis*: 93°C for denaturing the dsDNA sample, 59°C for annealing (base-pairing) the probes, and 62°C for extension (DNA synthesis) and ligation.

An assay of 30–40 cycles can achieve e.g. a 10 million-fold amplification.

Examples of LCx assays include *C. trachomatis* [JCM (1998) *36* 94–99], *N. gonorrhoeae* [JCM (1997) *35* 239–242] and *M. tuberculosis* [JCM (1999)*37* 229–232; JCM (2002) *40* 2305–2307].

---

5. The RNA strand has bound primer 2.

7. Reverse transcriptase has synthesized cDNA on the RNA strand.

8. RNase H has removed the RNA strand, and primer 1 has bound to the amplicon sequence in cDNA.

9. Reverse transcriptase has synthesized a complementary strand of cDNA; note that the 3' end of the template cDNA has also been extended to form a functional (double-stranded) promoter for the RNA polymerase. RNA polymerase has bound to this promoter and will (repeatedly) synthesize RNA strands identical to the one shown at stage 5; these strands can participate in the cyclic phase, leading to high-level amplification of the target.

The double-stranded cDNA molecules in stage 9 are permanent products and are continually transcribed by the RNA polymerase. Operation of the cyclic phase produces many copies of the amplicon in the form of (i) *complementary* RNA (the major product) and (ii) cDNA.

NASBA involves three enzymes: reverse transcriptase, RNA polymerase and RNase H. In a similar process, TMA (see section 8.5.9.2), the role of RNase H is carried out by the reverse transcriptase, so that only two enzymes are used.

Improved reproducibility of LCx assays for *C. trachomatis* and *N. gonorrhoeae* was reported following modification of the method for reading results [JCM (2000) *38* 2416–2418]. See also section 16.1.6.1 for further information.

8.5.9.2 *Nucleic acid sequence-based amplification (NASBA) and SDA*

NASBA is more complicated than PCR but it works at e.g. 41°C – i.e. it does not need thermal cycling. This method is intended primarily for the amplification of specific sequences in RNA, and this use is outlined in Fig. 8.25.

A method that closely resembles NASBA, *transcription-mediated amplification* (TMA), is used e.g. in an assay system for *Mycobacterium tuberculosis* (section 16.1.6.1).

NASBA products may be detected, as in PCR, e.g. by gel electrophoresis followed by staining or blotting/probing. As in PCR (section 8.5.4.7), NASBA may be carried out as a quantitative (real-time) procedure – using e.g. *molecular beacon probes* (see section 8.5.3.2) [NAR (1998) *26* 2150–2155].

[NASBA in diagnostic microbiology, research and other non-commercial applications: RMM (1999) *10* 185–186.]

The *strand-displacement amplification* (SDA) method [NAR (1996) *24* 348–353] is another isothermal (non-cycling) technique. It is rather more complex than the other methods and is beyond the scope of this book; for details see e.g. *Dictionary of Microbiology and Molecular Biology* (ISBN 0-471-49064-4), pages 692–694. When SDA was tested against other nucleic-acid-amplification methods it was the only assay with 100% specificity for detecting *Chlamydia trachomatis* and *Neisseria gonorrhoeae* in endocervical specimens [JCM (2001) *39* 1751–1756].

[Isothermal nucleic acid amplification methods (NASBA, TMA, SDA): *DNA Methods in Clinical Microbiology* (ISBN 0-7923-6307-8).]

8.5.10 Recombinant streptokinase: an example of recombinant DNA technology

Streptokinase is a protein which, in nature, is produced by certain species of *Streptococcus*. It has the ability to convert human plasminogen to plasmin (fibrinolysin), an enzyme which breaks down fibrin clots; thus, during infection, streptokinase may act as an *aggressin* (section 11.3.2) – helping to spread infection by promoting the lysis of fibrin barriers which may enclose (and hence localize) streptococcal lesions.

The therapeutic use of streptokinase derives from its ability to act as a thrombolytic agent (promoting solubilization of the fibrin in blood clots) and, hence, its value e.g. for treating thrombosis. Once widely used in this capacity, streptokinase is now being replaced by other agents owing e.g. to (i) its slow and somewhat unpredictable lytic action, and (ii) its antigenicity (which limits the potential for repeated

administration). Nevertheless, the procedure described in Fig. 8.26 is a good example
of the way in which biotechnology can be used for a specific purpose.

As a commercial source of streptokinase, streptococci give low yields; expressing the
streptokinase gene in *Escherichia coli* increases the yield >10-fold [Biotechnology
(1992) *10* 1138–1142].

8.5.11 Overproduction of recombinant proteins in *Escherichia coli*: strategies for optimization of gene expression

'Overproduction' refers to the synthesis of abnormally high concentrations of specific
proteins – usually heterologous ('foreign') proteins – under experimental conditions;
this procedure has been used e.g. for the production of therapeutic agents such as
streptokinase (section 8.5.10).

Although *E. coli* is widely used for overproduction, it cannot be used for the
expression of every gene; for example, in some cases a eukaryotic gene product requires
post-translational modification which is not carried out in a prokaryote (see section
8.5.7). For genes which *can* be expressed, efficient, high-level expression requires
attention to a range of factors which can greatly affect the yield of gene product; some
of these factors are considered below. The gene of interest is assumed to be inserted in a
suitable *expression vector* (Fig. 8.16); an alternative approach (not considered here) is
to insert the gene of interest into the bacterial chromosome [see e.g. BioTechniques
(2001) *30* 252–256].

8.5.11.1 *Optimization of transcription*

The promoter of the given gene should be *strong* (section 7.8). It should also be 'tightly
regulated' – i.e. the gene should not be expressed prior to its induction or derepression.
Why? Typically, cells are grown to high density prior to expression of the given gene in
order to maximize the yield of gene product; gene expression before the appropriate
time may e.g. depress growth rate and lower the yield of recombinant protein. Tight
regulation is even more important if the gene product is toxic for *E. coli*.

The transcription regulator must also be appropriate for the purpose; thus, the
inducer IPTG (Fig. 7.11), which is toxic for man, is not an optimal regulator in the
overproduction of therapeutic proteins.

8.5.11.2 *Optimization of translation*

The efficiency of translation is affected e.g. by the precise sequence of the Shine–
Dalgarno (SD) sequence (section 7.6) and the number of nucleotides between the SD
sequence and the start codon. Moreover some genes encode a 'translational enhancer'
called the *downstream box* – a specific sequence of nucleotides downstream of, and

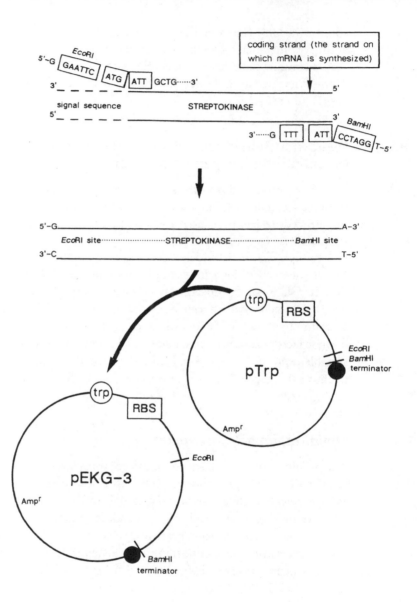

Figure 8.26 Expression of the streptokinase gene in *Escherichia coli* (section 8.5.10): a (diagrammatic) outline of the technology.

Initially, cells of *Streptococcus equisimilis* are lysed, the genomic DNA is separated, and the streptokinase gene (without its signal sequence) is copied by PCR (section 8.5.4); earlier studies had suggested that the signal sequence of this gene may cause problems with expression in *E. coli*.

The two strands of the streptokinase gene are shown at the top of the figure. The primers used for PCR bind at sites which ensure replication of the gene's coding sequence but exclude the signal sequence. The primers are:

close to, the start codon. These factors are relevant to the construction of an effective expression vector.

8.5.11.3   The problem of inclusion bodies

Inclusion bodies (in this context) are insoluble aggregates of unfolded/incorrectly folded proteins which often form in the cytoplasm during overproduction. Provided that such proteins can later be correctly folded *in vitro*, inclusion bodies can be advantageous in that they facilitate purification of the protein product (e.g. by centrifugation), and they may help to prevent degradation of the product by intra-cellular proteases.

The reasons for inclusion body formation are not fully understood so that, in some

5'-GGAATTCATGATTGCTC ······3'   (primer SK2)
5'-TGGATCCTTATTTC ······3'   (primer SK3)

Note that the template strand on which primer SK2 is extended is the *coding strand* (see note on coding strand terminology near the end of section 7.5); when primer SK2 is fully extended it will form a strand *complementary* to the coding strand. After several cycles of PCR, an increasing number of PCR-derived strands will act as templates, and the major product of PCR will be as shown in the centre of the figure: the streptokinase gene flanked on either side by a specific restriction site (Table 8.1).

Note that ATG in primer SK2, when copied in the *coding* strand, will appear as 3'····TAC····5' (see base-pairing, section 7.2.1); during transcription of the gene (section 7.5), the corresponding mRNA will be 5'-AUG-3' – i.e. the *initiator* codon (section 7.6). (See also footnote 3 in Table 7.2.) It also follows that ATT (in primer SK2) corresponds to the amino acid isoleucine. Primer SK3, when fully extended by PCR, forms the coding strand; note that 5'-TTA-3' (in SK3) corresponds, in mRNA, to the codon UAA – the *ochre* codon (a stop codon – Table 7.2).

The PCR-amplified fragment (centre of figure) is inserted into plasmid pTrp, between the *Bam*HI and *Eco*RI restriction sites, to form plasmid pEKG-3 (both plasmids shown single-stranded for simplicity); trp is the promoter of the *E. coli* tryptophan operon, RBS is a ribosome binding site (Fig. 8.16), terminator is a transcription terminator (section 7.5), and Amp^r is a gene encoding resistance to ampicillin.

A strain of *E. coli* is transformed (section 8.4.1) with plasmid pEKG-3, and the transformed cells are plated on an ampicillin-containing medium to select those cells which have taken up pEKG-3. Several colonies are chosen, and a check is carried out on the nucleotide sequence of the streptokinase gene in cells from each colony. The gene can then be cloned by culturing cells from those colonies which are known to contain the correct sequence.

In pEKG-3, trp, the promoter of the *E. coli* tryptophan operon, is used to control transcription of the streptokinase gene. In *E. coli*, regulation of the *trp* promoter involves a repressor protein (TrpR, encoded by gene *trpR*); in the presence of tryptophan TrpR represses transcription from the promoter. Some (mutant) strains of *E. coli* do not form TrpR; the strain chosen for expression of the streptokinase gene (*E. coli* W3110) is one which is *known* to form TrpR, i.e. a strain which is 'wild-type' for gene *trpR*. In strain W3110, transcription from the *trp* promoter can be induced ('switched on'), as required, by using an analogue of tryptophan, 3-β-indole acrylic acid, as inducer (compare with allolactose in the *lac* operon – Fig. 7.11).

During the growth of W3110, maximal expression of the streptokinase gene was obtained at a plasmid density of 420 copies per cell. When synthesized, the streptokinase remained *intra*cellular (reminder: the signal sequence was omitted); the yield of streptokinase was 25% of the total cell protein. After cell lysis, the recombinant streptokinase was purified by a procedure involving affinity chromatography (section 8.5.1.4), the stationary matrix containing human plasminogen.

cases, *in vitro* folding may result in little or no product with biological activity. Strategies for tackling this problem include gene fusion (section 8.5.2) to increase solubility, co-expression of chaperones (section 7.6) to promote folding, modification of pH and reduction in growth temperature. However, such approaches may be largely empirical; thus, e.g. not all fusion partners will increase the solubility of a given protein, and the particular chaperone(s) required to promote the folding of a given protein may need to be ascertained by trial and error.

8.5.11.4   *Avoiding proteolysis*

The *E. coli* cytoplasm contains many proteases (protein-degrading enzymes), and there are also some proteases in the periplasmic region; the gene product should clearly be protected from these enzymes. Strategies for avoiding – or minimizing – proteolysis include the following:

(a) The gene product may be targeted to the periplasm (fewer proteases) by incorporating a signal sequence (section 7.6) in the gene, or by *fusing* the given gene (section 8.5.2) with the gene of a periplasmic protein (e.g. DsbA).

(b) The gene product may be made secretable by fusing its gene to another gene (e.g. that of α-haemolysin – section 5.4.1.2) whose product is exported to the cell's exterior.

(c) Particular proteolytic site(s) in the protein may be eliminated by re-coding the gene in a way which does not affect the required function of the gene product.

(d) Use can be made of *E. coli* strains which are mutant in the *rpoH* gene. The accumulation of abnormal or heterologous proteins in the cytoplasm (as in overproduction) triggers the heat-shock response (section 7.8.2.3) – resulting in the production of the Lon protease (which degrades abnormal proteins). RpoH 'switches on' synthesis of the Lon protease, and *rpoH* mutants, deficient in Lon, have been found to give greatly increased yields of heterologous proteins in *E. coli*.

Proteins whose N-terminal amino acid is arginine, leucine, lysine, phenylalanine, tryptophan or tyrosine tend to be inherently unstable (i.e. they have short half-lives); this may suggest the re-coding of a gene. Note that the *penultimate* N-terminal amino acid in a bacterial protein is also important as it may potentiate removal of the terminal *N*-formylmethionine (section 7.6) and thus itself become the terminal amino acid.

8.5.11.5   *The cell's response to overproduction*

In the example of streptokinase (section 8.5.10), up to 25% of the protein synthesized by *E. coli* was – from the cell's point of view – *gratuitous* (i.e. non-functional). Commercially, this may represent a good yield, but how do the cells themselves respond to such situations? As noted above, the accumulation of abnormal proteins triggers the heat-shock response – an indication of stress. In fact, at very high levels of gratuitous protein bacteria have been found to degrade their own ribosomes, and rRNA, thus bringing about an inhibition of translation which can lead to cell death [MM (1996) *21* 1–4].

8.5.11.6 Codon bias

When a heterologous (i.e. 'foreign') gene (e.g. a mammalian gene) is expressed in *E. coli*, translation may be inefficient owing to the presence of certain codon(s) for which there are insufficient numbers of the corresponding tRNA(s). Such *codon bias* can arise in *E. coli* e.g. when a heterologous gene contains a high frequency of codons such as the proline codon (CCC) and/or the arginine codon (AGG) – both of which occur only rarely in homologous (*E. coli*) genes Attempts at high-level expression of genes containing such rare codons may lead e.g. to slowing/termination of translation and degradation of mRNA.

One solution to the problem of codon bias is to insert into the bacterial cell extra (plasmid-borne) copies of gene(s) encoding the relevant tRNA(s).

Codon bias may also arise when a same-sense mutation replaces a common codon with a synonymous but infrequent codon.

8.5.12 DNA-binding proteins: some methodology

Protein–DNA interactions are important e.g. in transcription (section 7.5.1) and in the initiation of DNA replication. Information about such interactions may be useful e.g. for engineering specific changes in the DNA in order to manipulate the binding affinity of protein(s) in studies on the regulation of gene expression etc. Outlined below are some methods for (i) detecting such interactions, and (ii) determining the binding site in the DNA.

8.5.12.1 Southwestern blotting

This technique can detect DNA-binding proteins in a cell lysate etc. Essentially, the sample is subjected to gel electrophoresis, and the *proteins* thus separated are electro-blotted onto a nitrocellulose filter. The affinity of any of the proteins for a specific DNA sequence may then be determined by using labelled DNA sequences as probes and detecting the label of any probe which has bound to a given protein.

8.5.12.2 Electrophoretic mobility-shift assay (EMSA)

EMSA can determine e.g. whether any protein (in a lysate etc.) can bind DNA of a particular sequence. Labelled DNA of the given sequence is incubated with the lysate to permit protein–DNA complexing. Electrophoresis then separates unbound from (any) bound DNA – the mobility of the latter (protein–DNA complex) being different from that of free DNA.

8.5.12.3   Footprinting

In footprinting, the sequence of nucleotides to which a protein binds can be determined by identifying that region of the DNA which is protected from enzymic (or chemical) cleavage by the shielding effect of the bound protein. Essentially, two identical populations of end-labelled DNA fragments (one population with bound protein) are subjected to enzymic (or chemical) cleavage under conditions in which (ideally) each fragment is cut at only one of a number of potential cleavage sites; under these conditions, the fragments yield labelled subfragments in a range of different lengths (produced by cleavage of the fragments at different sites). If, in the *protein-bound* fragments, the protein had obscured one or more cleavage sites, then none of the subfragments in that population will end at these particular sites; hence, when the two populations of subfragments are compared by gel electrophoresis, the *absence* of certain size(s) of subfragment from the protein-bound population will appear as a gap in the gel. This gap is the protein's *footprint*; its location in the gel, in relation to the bands of subfragments, indicates the site at which the protein binds to DNA.

DNase I footprinting (using enzymic cleavage with DNase I – see page 274) tends to give a large, clear footprint because DNase I, being quite a large protein, does not cut at sites which are very close to a DNA-bound protein. However, better resolution of the binding site may be obtained by using *in vitro*-generated hydroxyl radical, which causes sequence-independent cleavage of DNA; all unprotected phosphodiester bonds are susceptible to cleavage by this agent, so that the electrophoretic pattern contains a band for each unshielded position in the sequence.

8.5.13   Displaying heterologous ('foreign') proteins; phage display technology

Specific heterologous ('foreign') proteins can be *displayed* on a bacterial (or phage) surfaced – for various purposes. For example, when displayed, the product of an unknown gene can be characterized by affinity chromatography (section 8.5.1.4) against known, immobilized ligands (e.g. antibodies); to do this, the unknown gene is fused (section 8.5.2) to the gene of a bacterial or phage *surface* protein such that the *fusion protein* will be expressed at the cell or phage surface. Displayed proteins may also be useful e.g. for studies in vaccine development, the displayed protein contributing to the antigenic stimulus; of course, the fusion protein must give an immunologically appropriate stimulus.

In one approach, foreign DNA is fused to a bacterial flagellar gene [Biotechnology (1995) *13* 366–372]. A different approach exploits *autotransporters* (section 5.4.5) and has been termed *autodisplay*. For example, a foreign gene can be fused to DNA encoding the β-domain in a (mutant) *E. coli* lacking the OmpT protease; the foreign protein (α-domain) is therefore not cleaved but is displayed at the cell surface. Autodisplay can be used e.g. for displaying antigens of vaccine strains of *Salmonella* [INFIM (2003) *71* 1944–1952].

8.5.13.1   *Phage display technology*

This technology is used e.g. to select a natural or experimental ligand that binds with high affinity/specificity to a specified target molecule. Essentially, *immobilized* target molecules are exposed to a large population of *phages* (Chapter 9) whose coat (surface) proteins have been genetically modified, *in vitro*, so that different phages have dissimilar coat proteins; as a consequence of modification, the phage population – collectively – exhibits an enormous range of potential peptide binding sites. Such a genetically modified population of phages is termed a *phage library*. Following exposure of target molecules to the phage library, those phages which do *not* bind to target molecules are washed away and discarded. Phages which *do* bind to target molecules are subsequently eluted (i.e. detached from the target molecules) e.g. by adjusting the pH; these phages are then used to infect bacteria (e.g. *Escherichia coli*) – within which they multiply to form vast numbers. These progeny phages are thus an expanded population of those phages that were selected (from the initial phage library) for their ability to bind to the target; typically, a number of strains of phage will have been selected. The process of selecting particular phage(s) from a library in this way has been called *biopanning*, or simply *panning*.

Several consecutive stages of panning are carried out; that is, the progeny of phages selected by the initial panning are passed over target molecules under more stringent conditions of binding – so that only those phages with the *highest* affinity/selectivity for the target will be bound (the others being washed away). Panning is then repeated again.

The final batch of phages (which bind strongly/selectively to the target) can then be examined – e.g. by sequencing the genetically modified region of DNA associated with the phage's binding site.

Which phages are used? How are phage libraries prepared? In what ways can the phage's coat proteins be modified? What kinds of target molecule are studied?

Phage display exploits filamentous phages such as M13 (Table 9.1).

Phage libraries can be prepared in two main ways. In one method, a *gene library* can be inserted into the DNA of a population of unmodified phages. (Gene libraries were discussed earlier: see e.g. Fig. 8.7; in general, a library is a collection of genes, cDNAs or DNA fragments that encode a range of types of protein or peptide.) During insertion of the library, many of the genes (or cDNAs etc.) fuse with a gene that encodes a coat protein (e.g. protein pIII); the resulting *fusion protein* (see section 8.5.2) replaces the phage's normal coat protein. Each library gene that inserts in this way will form a different hybrid gene and give rise to a different fusion protein; thus, the population of phages will display a wide range of potential binding sites in their coat proteins.

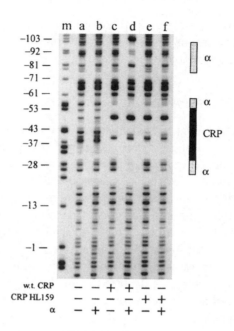

Plate 8.3. DNase I footprinting: a technique used for examining the interactions between DNA and proteins (see section 8.5.12.3 for rationale).

Here, the sample DNA is the regulatory region of the *Escherichia coli* galactose (*gal*) operon – which includes the transcription *start site* (section 7.5), the promoter P1, and a binding site for the CRP (=CAF) protein which is involved in catabolite repression (section 7.8.2.1) of the *gal* operon; the sample DNA was labelled, at one end, with ^{32}P. This DNA was examined for interactions with (i) the α-subunit of RNA polymerase, and (ii) CRP (=CAP). Specifically, the object of the work was to study the role of the α-subunit during the activation of the *gal* P1 promoter by CRP. Note that the α-subunit and CRP are only *part* of the assembly of proteins normally found at the promoter site at the start of transcription *in vivo*; the assembly normally includes other subunits of RNA polymerase and a sigma factor. Note also that CRP binds to DNA as a *dimer*, i.e. a unit consisting of two molecules of the CRP protein.

The photograph shows the gel electrophoresis pattern from six footprinting experiments (a–f) involving interactions between sample DNA and various combinations of (i) the α-subunit of RNA polymerase, (ii) wild-type (i.e. normal) CRP, and (iii) CRP HL159, a mutant (altered) form of CRP. In each experiment, sample DNA was pre-incubated with protein(s), subjected to cleavage by DNase I, and examined by gel electrophoresis.

Lane m is a form of calibration: a *Maxam–Gilbert G sequence ladder*. To prepare such a ladder, many copies of the end-labelled sample DNA are subjected to chemical action (dimethyl sulphate followed by piperidine) which cleaves phosphodiester bonds (Fig. 7.4) *specifically* between a guanine (G) base and the adjacent base. Under appropriate conditions, each of the (many) copies of sample DNA will be cleaved at only *one* of its guanine bases; the resulting fragments are in a range of sizes because cleavage occurs at different guanines in different copies of sample DNA. On electrophoresis, these fragments (of different sizes) move at different speeds, thus forming the 'ladder' seen in lane m. To the left of lane m are numbers which indicate the locations of some of the guanine bases in the sample DNA, and this provides a useful scale; −1 refers to the first base upstream of the start site (see section 7.5). The CRP-binding site is centred between bases −41 and −42. As the nucleotide sequence of the entire *gal* regulatory region is already known, and because the exact locations of

The second method of preparing a phage library does not require a gene library. It involves (*in vitro*) *randomization* of a sequence of nucleotides within a phage coat protein gene. This produces an immense number of different combinations of nucleotides within the gene – the gene in a given phage containing only one of these combinations; thus, collectively, the population of phages will display an enormous range of different potential binding sites in their coat proteins. This type of library is available commercially. One such library has been used to select ligands for use in vectors designed to deliver gene therapy to patients with cystic fibrosis [BioTechniques (2003) *35* 317–324].

The target molecules used in phage display technology are often cell-surface receptors on mammalian cells. Such receptors include e.g. those which bind cytokines (Table 11.4) and those involved in receptor-mediated endocytosis. Clearly, receptor–ligand binding is important when designing vectors for gene therapy; this technology can help to ensure that a vector's binding site has optimal affinity and specificity for a given receptor on the target cell – thus helping to avoid wasteful delivery of genes to non-target cells and reducing the required dosage.

---

the promoters and CRP-binding site are also known, this calibration ladder can be used to indicate the *location* of DNA–protein interactions detected in the footprinting experiments.

Lane a shows bands of fragments resulting from cleavage of sample DNA, by DNase I, in the absence of bound proteins. The fragments are of different sizes because DNase I can cut at any of a number of sites in the sample DNA. Only fragments carrying the end label (^{32}P) will be detectable; that is, a copy of sample DNA which is cleaved into two pieces will yield only one piece (that carrying the label) detectable in the gel.

Lane b shows the result of DNase I action on sample DNA in the presence of the α-subunit of RNA polymerase. No footprint is seen, indicating a lack of interaction between sample DNA and the subunit in the absence of other proteins.

Lane c shows the 'footprint' of CRP (between −29 and −54) (compare c with a). Not all of this region of DNA has been protected from DNase I by the presence of CRP: as shown, cleavage has occurred at e.g. −40; this region of DNA is accessible to DNase I because it presumably passes over a cleft between the two molecules of the CRP *dimer*.

Lane d shows the action of DNase I on sample DNA in the presence of both CRP *and* α-subunits. The presence of α-subunits extends the CRP footprint both upstream and downstream. This suggests that the α-subunit can bind to a site on each of the upstream and downstream components of the CRP dimer; however, while upstream binding is believed to occur – and to be important in transcription – the downstream binding was interpreted as an artefact due to the absence of the constraining effect of other polymerase subunits. The reason for the footprint in the −80 to −90 region is not known.

Lanes e and f show the action of DNase I on sample DNA in the presence of a mutant form of CRP, with and without α-subunits. The results indicate that this particular mutation in CRP suppresses binding between CRP and the subunit – because the effects of the subunit (seen in lane d) have been suppressed. As the mutation in CRP seems to inhibit binding of the α-subunit, it appears that binding between normal (non-mutant) CRP and the subunit occurs at that site in CRP which has been altered by the mutation.

Photograph reprinted from Attey *et al.* (1994) *Nucleic Acids Research 22* 4375–4380 with permission from Oxford University Press, and by courtesy of Professor Stephen Busby, University of Birmingham, UK.

8.5.14 Differential display

Differential display (DD) is a method for detecting genes that are expressed only under particular conditions; it involves comparison of the mRNAs isolated from two or more populations of cells exposed to different conditions. mRNAs from each population are first converted to cDNAs (section 8.5.1.2). To make the task manageable, only *some of* the (many) mRNAs in each population are converted to cDNAs in any given experiment. mRNAs are selected by using primers such as 5'-CCAATTTTTTTTTTGC-3'; this primer would permit conversion of those mRNAs in which CG residues occupy the first and second positions proximal to the 3'-AAAAA . . . . . . tail. cDNAs from each population are then subjected to PCR using the original primer(s) (described above) together with a second, shorter primer of *arbitrary* sequence. (See section 16.2.2.4 for other uses of arbitrary primers.) The products of PCR (amplified DNA fragments representing the 3' ends of various mRNA molecules) are subjected to gel electrophoresis. Fingerprints from the different populations of cells are compared, and any bands of interest (e.g. bands present in one fingerprint but not in others) can be removed from the gel and amplified with the same primers; the amplified products may be e.g. sequenced or used to probe a library.

[A decade of differential display (review): BioTechniques (2002) *33* 338–346.]

8.5.15 DNA chip technology

A DNA chip is *a microarray* of short, single-stranded DNA molecules – each with a different, but *known*, sequence – immobilized on a small piece ('chip') of glass, silicon or plastic; the immobilized molecules commonly function as probes (section 8.5.3).

Chips can be made by (i) immobilizing pre-existing molecules, or (ii) using 'on-chip synthesis' in which high-speed robotic devices can synthesize a vast range of combinations of nucleotides on a single chip.

A large array of immobilized probes can include all possible combinations of nucleotides for a strand of given length. Thus, for an 8-nucleotide probe, on-chip synthesis can be used to make a complete set (4^8) of combinations of nucleotides – 65536 different probes – on a 1 cm^2 glass chip.

A chip can be used e.g. to determine the sequence of an unknown, labelled, single-stranded fragment of nucleic acid; this is done by allowing the fragment to hybridize with one of the probes in the array – hybridization being detected by the fragment's label (using e.g. scanning confocal epifluorescence microscopy). As the fragment binds to a *particular* probe, its sequence will be complementary to the sequence of that probe (see *base pairing* in section 7.2.1).

Chips are useful for detecting mutations, including mutations that confer resistance to antibiotics. Thus, section(s) of the bacterial chromosome in which mutations can confer resistance to particular antibiotic(s) can be amplified by PCR (e.g. using fluorophore-labelled primers); the (fluorophore-labelled) amplicons can then be tested against a microarray that includes all possible mutations (as well as the wild-type sequence). High-density arrays have been used to identify species of *Mycobacterium* and (*simultaneously*) to test for resistance to the anti-tuberculosis drug rifampicin [JCM (1999) *37* 49–55].

Chip technology can also be used to examine the differential expression of genes. Thus, with an array of single-stranded cDNAs (which bind the mRNAs of expressed genes), fluorophore-labelled probes – with binding sites *elsewhere* on the mRNAs – can be used to detect those genes which are active under specific conditions.

One problem with microarrays is that single-stranded DNA probes have a tendency for intra-strand base-pairing, which may obscure target sequences. One solution may be to use probes made of PNA (section 7.2.1.1); PNA–DNA hybridization can occur under conditions which inhibit intra-strand base-pairing in ssDNA.

Advances in chip technology include electrically facilitated probe–target hybridization.

[DNA chips (reviews): NB (1998) *16* 27–31, 40–44; TIBTECH (1999) *17* 127–134. Biomedical discovery with DNA arrays (review): Cell (2000) *102* 9–15.]

Microarray technology has been useful for investigating the ability of furanone, an inhibitor of quorum sensing (section 10.1.2), to attenuate virulence in *Pseudomonas aeruginosa* [EMBO Journal (2003) *22* 3803–3815].

8.5.16 FRET

In the original form of FRET (fluorescence resonance energy transfer), fluorescence from one type of molecule acts as a source of excitation for a second type of fluorescent molecule *which is close to the first*; fluorescence emitted by the second type of molecule differs in wavelength from that emitted by the first.

The FRET principle has been exploited in various ways. For example, it is used for monitoring the intracellular location of specific molecules. Thus, if two types of molecule are labelled with different FRET-compatible fluorochromes it is possible to determine whether the two types of molecule bind at closely adjacent intracellular sites. This is done by using radiation that causes one of the two fluorochromes to fluoresce while simultaneously monitoring fluorescence from the other; fluorescence from the second fluorochrome is detected only if the two molecules are sufficiently close together in the cell. This approach has been used e.g. to follow the cell-cycle-dependent localization of particular proteins in *Bacillus subtilis* [EMBO Journal (2002) *21* 3108–3118].

The QSY dyes (Molecular Probes Inc., Eugene, OR, USA) are *non-fluorescent* molecules which can quench fluorescence from a range of fluorescent dyes and which can act as non-fluorescent acceptors in FRET donor–acceptor pairs. QSY dyes have been used e.g. to investigate the three-dimensional structure of proteins. They can also be used to study enzyme–substrate interactions [PNAS (2002) *99* 6603–6606].

9 Bacteriophages

Most or all bacteria can be infected by specialized viruses (*bacteriophages*, usually abbreviated to *phages*). A *virus* is an organism which does not have a cell-type structure and which cannot, by itself, metabolize or reproduce. However, when inside a suitable living cell the genome of a virus may 'take over' the synthesizing machinery and direct it to make copies of the virus; the newly-formed viruses are released and can then infect other cells.

Many phages consist simply of nucleic acid enclosed within a protein *capsid* (coat); depending on phage, the genome may be dsDNA, ssDNA, dsRNA or ssRNA. Some phages are polyhedral, others filamentous or pleomorphic, and some have a 'tail' with which they attach to the host cell (Table 9.1, Fig. 9.1, Plate 9.1). In many cases a given phage can infect the cells of only one genus, species or strain.

The effect of phage infection on a bacterium depends on the particular phage and host cell – and, to some extent, on conditions. *Virulent* phages multiply within and *lyse* their host cells, i.e. the host cells die and break open, releasing phage progeny. *Temperate* phages can establish a stable, non-lytic relationship (*lysogeny*) with their host cells. Still other phages can multiply within their hosts without destroying them, phages being released from the living cells.

The effect of phage infection on a bacterial *pathogen* may be important for the pathogen's host. In fact, in some cases a phage *confers* the pathogenicity by encoding specific virulence determinant(s); thus, phage-encoded toxins are responsible for e.g. botulism, cholera, diphtheria and haemolytic uraemic syndrome (HUS). Clearly, to understand such diseases we need to understand both the pathogens *and* their phages. This area is being actively researched. For example, attention is being focused on the regulation of toxin synthesis and on the way in which phage-encoded toxins are released from bacteria; thus, release of HUS-associated shiga-like toxins (encoded by phage H-19B) from EHEC (Table 11.2) may involve a *lysis protein* similar to that encoded by the *S* gene of phage λ (section 9.1.2).

It is also important to know whether phages multiply within the bacteria infecting a human or animal host because this may affect levels of toxin production (e.g. by

Table 9.1 Bacteriophages: some examples

Name	Genome[1]	Morphology[2]; size[3]	Main hosts(s)
λ	dsDNA (linear)	Isometric head, long non-contractile tail; ca. 200 nm	*Escherichia coli*
Mu	dsDNA (linear)	Isometric head, long contractile tail; ca. 150 nm	Enterobacteria
MV-L3	dsDNA (linear)	Isometric head, short tail; ca. 80 nm	*Acholeplasma laidlawii*
T4	dsDNA (linear)	Elongated head, long contractile tail; ca. 200 nm	*Escherichia coli*
PM2	dsDNA (ccc)	Icosahedral, with internal lipid membrane; ca. 60 nm	*Alteromonas espejiana*
f1	ssDNA (ccc)	Filamentous; >750 nm × 6 nm	Enterobacteria (only conjugative *donor* cells can be infected)
M13	ssDNA (ccc)	Filamentous; >750 nm × 6 nm	As for f1
φ29[4,5]	dsDNA (linear)	Complex (see Plate 9.1, item 5)	*Bacillus* spp
φX174	ssDNA (ccc)	Icosahedral; ca. 30 nm	*Escherichia coli*
M12	ssRNA (linear)	Icosahedral; ca. 25 nm	Enterobacteria (only conjugative *donor* cells)

[1] The form in which the nucleic acid exists within the virus.
[2] Isometric means approximately spherical, commonly icosahedral; 'icosahedral' means resembling an *icosahedron*: a solid figure bounded by 20 plane faces, all the faces being equilateral triangles of the same size.
[3] Excludes tail fibres (where present).
[4] [Assembly of φ29 studied in three dimensions: Cell (1998) *95* 431–437.]
[5] See also protein priming in section 7.3.

increasing the copy number of a toxin-encoding gene); thus, phage CTXΦ, which encodes cholera toxin, has been found to replicate within cells of *Vibrio cholerae* in the mammalian intestine [e.g. TIM (1998) *6* 295–297]. Interestingly, though, while lysogenic El Tor and O139 strains of *V. cholerae* can yield CTXΦ virions, virions are not formed by lysogenic strains of the *classical* biotype of *V. cholerae* (even when the latter form cholera toxin) [JB (2000) *182* 6992–6998].

Also important are the factors governing the transfer of toxin-encoding phages between bacteria (see section 11.5.8).

In *Shigella flexneri* (a causal agent of dysentery), phage infection can modify the O-specific chains in lipopolysaccharides (section 2.2.9.2) – affecting the serotype-determining O antigens (section 16.1.5.1). Infection by phage is therefore a factor to be remembered when developing vaccine strains of *S. flexneri* that are effective against particular serotype(s) of the pathogen. [Serotype-converting bacteriophages and O-antigen modification in *Shigella flexneri*: TIM (2000) *8* 17–23.]

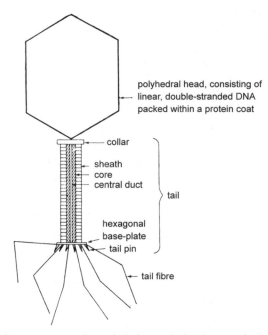

Figure 9.1 Bacteriophage T4, a complex, tailed phage which infects *Escherichia coli* (diagrammatic). The purely structural parts are made of protein; the genome (dsDNA) is enclosed within the head. Extended fibres are needed for infection. Extension of tail fibres is inhibited e.g by low pH, low temperature and low ionic strength. For size see Plate 9.1.

While certain phages confer pathogenicity, others may have beneficial effects when used as *therapy* in (human) diseases of bacterial origin (see Table 9.2).

Some phages (and/or their components) are useful in recombinant DNA technology. Phage components are used e.g. in vectors (section 8.5.1.5) such as cosmids (Fig. 8.15) and phagemids, while certain modified phages are used in mutagenesis (Fig. 8.20) and phage display technology (section 8.5.13.1).

9.1   VIRULENT PHAGES: THE LYTIC CYCLE

The lytic cycle begins when a virulent phage adsorbs to a susceptible host cell; it ends with cell lysis and the release of phage progeny.

9.1.1   The lytic cycle of phage T4 in *Escherichia coli*

Initially, a phage attaches by the tips of its tail fibres (Fig. 9.1) to specific sites on the outer membrane. The base-plate then binds, and contraction of the tail sheath causes the inner core to penetrate the cell's outer membrane so that phage DNA can pass, via the central duct, into the periplasmic region. Uptake of DNA across the cytoplasmic membrane appears to require pmf.

Within 5 minutes of infection the host cell stops making its own DNA, RNA and protein. 'Early' genes of T4 are transcribed and translated, and T4-encoded Ndd protein disrupts the host cell's chromosome – which is degraded, by phage-encoded nucleases, to nucleotides which are used later for the synthesis of T4 DNA.

Middle and then late genes are subsequently transcribed. Throughout, transcription seems to involve the host cell's RNA polymerase, though the enzyme undergoes several phage-induced changes which enable it to recognize different promoters; one such change is *ADP-ribosylation*: the transfer, from NAD (Fig. 5.1) to the polymerase, of an ADP-ribosyl group. Additionally, a phage-encoded sigma factor (section 7.5) is used for transcription of the late T4 genes.

About 5 minutes after infection, T4 DNA begins to replicate. Leading strand synthesis (Fig. 7.8) may be primed by the cell's RNA polymerase, but Okazaki fragments seem to be primed by phage-encoded proteins. New phage DNA undergoes modification (sections 7.4 and 9.6).

The T4 genome is *linear*, double-stranded DNA, about 170 kb, which contains hydroxymethylcytosine instead of cytosine; synthesis of the hydroxymethylcytosine requires phage-encoded enzymes. The DNA is *circularly permuted* and *terminally redundant* – terms which can be illustrated as follows:

5'-ABCD|EFGH...ABCDEFGH|ABCD-3'

The above shows (alphabetically for clarity) the order of nucleotides in one strand of the DNA in a given T4 genome. In another T4 genome, the sequence may be e.g.:

5'-EFGH|ABCDEFGH....ABCD|EFGH-3'

Notice that, in circularly permuted genomes, the *sequence* is the same in all molecules, but the starting points are different. 'Terminally redundant' means that the same sequence occurs at both ends of a given molecule (as shown above).

Replication of T4 DNA can be initiated at various origins within the genome; from a given origin, replication occurs bidirectionally.

During replication of T4 DNA the template of the lagging strand (Fig. 7.8) remains with a single-stranded 3' end (because the final Okazaki fragment is not synthesized on the (linear) template strand). The single-stranded ends of T4 DNA 'invade' complementary sequences in other copies of the phage genome within the cell – e.g. by displacing a homologous sequence in another duplex; such recombinational activity may lead to further DNA replication (the 3' ends acting as primers) with formation of a complex, branched network of replicating phage genomes. The result is a number of *concatemers*, each consisting of a number of phage genomes joined end-to-end. (Concatemers are also formed e.g. by phage λ – see Fig. 9.2 – but by a different mechanism.)

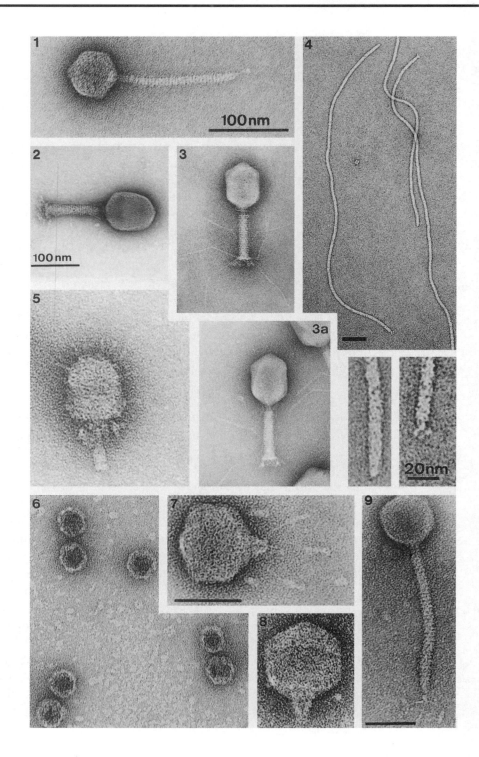

The phage head, tail etc. are encoded by late genes. Head construction, which involves about 20 genes, starts with the assembly of a *prohead* on the inner surface of the cell's cytoplasmic membrane; the prohead is built around a core of *scaffolding proteins* (which determine the shape and size of the prohead). The core is later removed, and the prohead detaches from the membrane and undergoes further conformational changes. DNA is packaged in the head by the *headful mechanism*: one end of a concatemer is inserted, and DNA is fed into the head until it is full; the DNA is then cleaved. The tail is polymerized (core first) on the base plate, and the completed tail joins spontaneously to the DNA-filled head; the fibres are then added.

Phage progeny are released by osmotic lysis: the cell envelope is weakened by T4-encoded lysozyme (Fig. 2.7) assisted by a T4-encoded *lysis protein* (section 9.1.2).

Extensive information on phage T4 is given in a detailed review of the T4 genome [MMBR (2003) *67* 86–156].

9.1.2   Lytic cycles of other phages

Lytic cycles can differ markedly from that of phage T4. The following are examples of such differences. (i) Not all phages bind to the bacterial cell wall; some bind e.g. to specific sites on a pilus: phage M12, for example, binds to the side (not the tip) of an F or F-type pilus. (ii) In some cases the host cell's chromosome survives for some time so that certain host genes can be exploited by the phage. (iii) In small ssRNA phages (e.g. M12, MS2, Qβ) the phage genome itself acts as mRNA – transcription is not required. For replication, RNA phages must themselves encode at least part of the *replicase* system needed for synthesizing RNA on an RNA template. (iv) Concatemer formation in phage λ involves the rolling circle mechanism (Fig. 9.2).

In phage T4 (and other phages with a contractile tail) it is generally believed that the phage genome is injected into the bacterial cell by a syringe-like action. However, many phages have a non-contractile tail, or no tail at all, and there has been little information on the way in which these phages transfer their genome into the host cell. Studies on phage T7 indicate that, on attaching to the host cell, the phage ejects

Plate 9.1.   Some bacteriophages (see text and Table 9.1). **1.** Bacteriophage lambda (λ). **2, 3, 3a.** Bacteriophages T6, T4 and T2, respectively (all to the same scale). These phages are morphologically very similar; they differ e.g. in their receptor (binding) sites on *E. coli*. **4.** Bacteriophage fd, a filamentous phage containing ss cccDNA; the bar is 50 nm. Below, higher magnification shows clearly that the two ends of fd are different. **5.** Bacteriophage ϕ29. The elongated head (approx. 35–40 nm), which contains linear dsDNA, bears protein fibres and is connected to a short, non-contractile tail. **6.** Bacteriophage Qβ, a small icosahedral phage (approx. 25 nm in diameter) which contains ssRNA. **7, 8.** Bacteriophages T3 and T7, respectively, both to the same scale (bar=50 nm); both phages contain linear dsDNA, and each has a short, non-contractile tail. **9.** Bacteriophage T1 (bar=50 nm); the head contains dsDNA, and the long tail is non-contractile.

Courtesy of Dr Michel Wurtz, University of Basel, Switzerland, and reprinted, with permission, from selected micrographs in 'Bacteriophage structure' by Dr M. Wurtz, *Electron Microscopy Reviews* vol. 5(2) (1992), Pergamon Press plc.

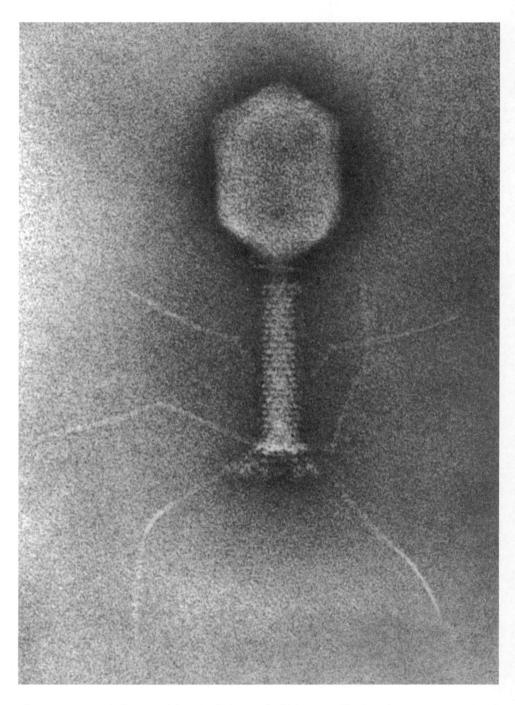

Plate 9.2.   Bacteriophage T4 (Plate 9.1, item 3 at the higher magnification of ×450000). Reprinted, with permission, from 'Bacteriophage structure' *Electron Microscopy Reviews* 5(2) (1992) Pergamon Press plc, by courtesy of the author, Dr Michel Wurtz, University of Basel, Switzerland.

proteins that may assemble to form a channel across the bacterial cell envelope; the genome then passes through this channel into the host cell. It has been speculated that some of the ejected proteins may form the components of a 'motor' that rachets phage DNA into the host cell [MM (2001) 49 1–8].

Many of the phages of Gram-negative and Gram-positive bacteria encode an enzyme which degrades peptidoglycan and which is thus able to lyse the host cell and allow release of phage progeny at the appropriate time; such enzymes, which include e.g. lysozyme and endopeptidase, have been referred to collectively as *endolysins*. The endolysins lack a 'signal sequence' (section 7.6); how, then, do they pass through the cytoplasmic membrane to reach the peptidoglycan? In addition to the endolysin, some of these phages have been shown to encode a specific type of *lysis protein* (also called a *holin*) which localizes in the cell's cytoplasmic membrane – forming a 'pore' through which the endolysin can gain access to peptidoglycan. In phage λ, for example, the holin is the product of the *S* gene; in phage T4, the *t* gene product appears to have the same role. A model for the membrane lesion ('pore') formed by the phage λ holin has been presented [JB (2003) 185 779–787].

Clearly, a holin must be active at a specific time in the phage-infected cell: it must promote lysis only after normal phage assembly has been completed; if lysis occurred earlier, fewer complete phages would be released when the cell burst (i.e. the 'burst size' would be smaller). Such early lysis has been found to occur with a strain of phage λ containing a mutation in the *S* gene [MM (1994) 13 495–504].

Interestingly, in some cases the sequence of nucleotides encoding the holin includes *two* start codons that are separated by one or two codons – i.e. two different polypeptide chains (of slightly different length) are encoded. The shorter polypeptide is the holin, while the longer polypeptide appears to be an inhibitor of holin activity; the combined effect of these two polypeptides appears to influence the timing of cell lysis [MM (1996) 21 675–682].

9.1.3 The effect of virulent phages on bacterial cultures

If virulent phages are added to a broth culture of susceptible bacteria, most or all of the cells will subsequently lyse; this can cause a dense, cloudy culture to become clear.

If a *small* number of virulent phages is added to a *confluent* layer of susceptible bacteria (section 3.3.1), each individual phage will infect and lyse a single cell, releasing many phage progeny; progeny phages can then infect and lyse neighbouring cells, and so on. In this way, a visible, usually circular, clearing – called a *plaque* – develops in the opaque layer of confluent growth at each site where a phage in the original inoculum infected a cell. Plaques are shown in Plate 16.1. A *plaque assay* assesses the concentration or *titre* of phages: a small volume of a known dilution of a phage suspension is plated to obtain a countable number of plaques from which the titre can be calculated. Alternatively, titres may be assessed by real-time PCR [BioTechniques (2003) 35 368–375].

Table 9.2   Some of the major human phage therapy studies performed in Poland and the former Soviet Union[1]

Reference(s)	Infection(s)	Etiologic agent(s)	Comments
Babalova et al. (7)	Bacterial dysentery	Shigella	Shigella phages were successfully used for prophylaxis of bacterial dysentery.
Bogovazova et al. (11)	Infections of skin and nasal mucosa	K. ozaenae, K. rhinoscleromatis, and K. pneumoniae	Adapted phages were reported to be effective in treating Klebsiella infections in all of the 109 patients.
Cislo et al. (17)	Suppurative skin infections	Pseudomonas, Staphylococcus, Klebsiella, Proteus, and E. coli	Thirty-one patients having chronically infected skin ulcers were treated orally and locally with phages. The success rate was 74%.
Ioseliani et al. (22)	Lung and pleural infections	Staphylococcus, Streptococcus, E. coli, and Proteus	Phages were successfully used together with antibiotics to treat lung and pleural infections in 45 patients.
Kochetkova et al. (25)	Postoperative wound infections in cancer patients	Staphylococcus and Pseudomonas	A total of 131 cancer patients having postsurgical wound infections participated in the study. Of these, 65 patients received phages and the rest received antibiotics. Phage treatment was successful in 82% of the cases, and antibiotic treatment was successful in 61% of the cases.
Kucharewicz-Krukowska and Slopek (27)	Various infections	Staphylococcus, Klebsiella, E. coli, Pseudomonas, and Proteus	Immunogenicity of therapeutic phages was analyzed in 57 patients. The authors concluded that the phages' immunogenicity did not impede therapy.
Kwarcinski et al. (29)	Recurrent subphrenic abscess	E. coli	Recurrent subphrenic abscess (after stomach resection) caused by an antibiotic-resistant strain of E. coli was successfully treated with phages.
Litvinova et al. (32)	Intestinal dysbacteriosis	E. coli and Proteus	Phages were successfully used together with bifidobacteria to treat antibiotic-associated dysbacteriosis in 500 low-birth-weight infants.
Meladze et al. (33)	Lung and pleural infections	Staphylococcus	Phages were used to treat 223 patients having lung and pleural infections, and the results were compared to 117 cases where antibiotics were used. Full recovery was observed in 82% of the patients in the phage-treated group, as opposed to 64% of the patients in the antibiotic-treated group.

Reference	Disease	Bacteria	Description
Miliutina and Vorotyntseva (35)	Bacterial dysentery and salmonellosis	Shigella and Salmonella	The effectiveness of treating salmonellosis using phages and a combination of phages and antibiotics was examined. The combination of phages and antibiotics was reported to be effective in treating cases where antibiotics alone were ineffective.
Perepanova et al (40)	Inflammatory urologic diseases	Staphylococcus, E. coli, and Proteus	Adapted phages were used to treat acute and chronic urogenital inflammation in 46 patients. The efficacy of phage treatment was 92% (marked clinical improvements) and 84% (bacteriological clearance).
Sakandelidze and Meipariani(45)	Peritonitis, osteomyelitis, lung abscesses, and postsurgical wound infections	Staphylococcus, Streptococcus, and Proteus	Phages administered subcutaneously or through surgical drains in 236 patients having antibiotic-resistant infections eliminated the infections in 92% of the patients.
Sakandelidze (46)	Infectious allergoses (rhinitis, pharyngitis, dermatitis, and conjunctivitis)	Staphylococcus, Streptococcus, E. coli, Proteus, enterococci, and P. aeruginosa	A total of 1380 patients having infectious allergoses were treated with phages (360 patients), antibiotics (404 patients), or a combination of phages and antibiotics (576 patients). Clinical improvement was observed in 86, 48 and 83% of the cases, respectively.
Slopek et al. (52–58)	Gastrointestinal tract, skin, head, and neck infections	Staphylococcus, Pseudomonas, E. coli, Klebsiella, and Salmonella	A total of 550 patients were treated with phages. The overall success rate of phage treatment was 92%.
Stroj et al. (67)	Cerebrospinal meningitis	K. pneumoniae	Orally administered phages were used successfully to treat meningitis in a newborn (after antibiotic therapy failed).
Tolkacheva et al. (69)	Bacterial dysentery	E. coli and Proteus	Phages were used together with bifidobacteria to treat bacterial dysentery in 59 immunosuppressed leukemia patients. The superiority ot treatment with phage-bifidobacteria over antibiotics was reported.
Weber-Dabrowska et al. (74)	Suppurative infections	Staphylococcus and various gram-negative bacteria	Orally administered phages were used to successfully treat 56 patients, and the phages were found to reach the patients' blood and urine.
Zhukov-Verezhnikov et al. (77)	Suppurative surgical infections	Staphylococcus, Streptococcus, E. coli, and Proteus	The superiority of adapted phages (phages selected against bacterial strains isolated from individual patients) over commercial phage preparations was reported in treating 60 patients having suppurative infections.

[1] Reproduced from Bacteriophage Therapy, Antimicrobial Agents and Chemotherapy (2001) 45 649–659, with permission from the authors, Drs Alexander Sulakvelidze, Zemphira Alavidze and J. Glenn Morris Jr, and from the American Society for Microbiology. (References are obtainable from the original publication.)

9.1.4   DNA packaging

In the assembly of phage T4 (section 9.1.1) it was mentioned that DNA is 'fed into' an empty phage head. The mechanism by which this occurs in T4 (and other phages) is still not understood, but recent studies give insight into the packaging of DNA in phage ϕ29 (Table 9.1) and its relatives. In ϕ29, packaging requires energy from ATP hydrolysis, and insertion of DNA into the head is also dependent on the presence of certain phage-encoded RNA molecules (designated pRNA). One early model proposed that a ring of pRNA molecules forms part of an apparatus that *rotates* (using energy from ATP hydrolysis) such that the threaded (i.e. helical) DNA molecule enters the phage head like a bolt drawn through a nut when the nut is rotated [Cell (1998) *94* 147–150]. A current model proposes that pRNA molecules form a (static) ring-shaped structure around a hole in the phage head through which the genome is inserted. Packaging is thought to require another structure called the *connector*: a dodecamer of gp10, 75 Å in length, with an axial channel; rotation of the connector, driven by ATP hydrolysis, is believed to wind DNA into the phage head like a bolt drawn through a rotating nut [Nature (2000) *408* 745–750].

9.2   TEMPERATE PHAGES: LYSOGENY

A *temperate* phage can enter into a stable, non-lytic relationship with a bacterium; the relationship is called *lysogeny*, and the host bacterium is said to be *lysogenic*. A lysogenic bacterium is immune from attack by other phages of similar type (*superinfection immunity*). One mechanism for superinfection immunity is illustrated by the lysogenic infection of *Salmonella typhimurium* with phage P22; this phage brings about the glycosylation of the host cell's lipopolysaccharides (section 2.2.9.2) – thus effectively destroying the cell-surface receptor sites for P22. In most cases of lysogeny the phage's genome (i.e. its nucleic acid, called the *prophage*) integrates with the host's chromosome; in a few cases (e.g. phage P1) the prophage does not integrate with the chromosome but exists within the bacterium as a circular molecule ('plasmid'). Either way, replication of the prophage and host cell is co-ordinated so that, at cell division, each daughter cell receives a prophage.

Maintenance of the lysogenic state involves synthesis of a phage-encoded repressor protein. Loss of active repressor protein results in *induction* of the lytic cycle – i.e. a prophage normally retains the potential for virulence. In a population of lysogenic bacteria induction may occur spontaneously in a small number of cells.

As mentioned earlier, strains of the El Tor and O139 biotypes of *Vibrio cholerae*, lysogenized by phage CTXΦ, can give rise to virions, whereas strains of the classical biotype of *V. cholerae* – lysogenized by the same phage – do not form virions. It seems that the ability of a lysogenic strain of *V. cholerae* to form virions of CTXΦ depends on the way in which phage genomes insert into the bacterial chromosome; in the El Tor and O139 biotypes multiple copies of the phage genome insert at a single site, whereas in the classical biotype either one or two (fused) genomes insert at two sites [PNAS (2000) *97* 8572–8577; JB (2000) *182* 6992–6998].

Lysogeny seems to be common in nature.

9.2.1 Lysogeny/lysis of phage λ in *Escherichia coli*

DNA enters the bacterium as linear dsDNA and immediately circularizes (Fig. 9.2). While the 'early' phage genes are being transcribed either lysogeny or lysis may follow. The lysis/lysogeny decision depends on the 'success' of one or other of two early gene products: the cI protein (which represses transcription from certain operators, and which promotes lysogeny) and the cro protein, which inhibits transcription of the *cI* gene. The cell's internal state can affect this decision. For example, lysogeny is favoured in a multiply-infected cell (i.e. one infected by a number of λ phages) under low-nutrient conditions; lysis is probable in cells infected with only one phage – regardless of nutrient levels.

In lysogeny, the λ genome integrates with the host's chromosome by site-specific recombination (Fig. 8.3b); a phage-encoded protein (product of the *int* gene) acts as the specific recombinase.

Induction (the switch from lysogeny to virulence) occurs spontaneously in a few cells in a lysogenic population. Induction in most or all of the cells can be brought about e.g. by DNA-damaging agents such as ultraviolet radiation or the antibiotic mitomycin C; under these conditions, when the SOS system is operative (section 7.8.2.2), the activated RecA protein cleaves the phage repressor protein (cI protein), allowing expression of the lytic cycle. The lytic cycle requires *excision* of the prophage – the reverse of integration; excision is mediated by the phage-encoded xis protein. The genome is then replicated, phage components are synthesized, and the assembled phages are released on cell lysis.

A study using *E. coli* with a modified (mutant) RNA polymerase found that λ lysogeny was prevented by defective transcription of certain phage-encoded factors. Factors needed for *lysis* were transcribed at sub-normal levels – but lysis could still occur in rich media, though not in poor media; this has suggested that (i) synthesis of lysis factors is influenced by levels of nutrients, and that (ii) in *non*-mutant *E. coli* growing in rich media, lysis factors may be *overproduced*, i.e. synthesized at levels in excess of those required for lysis. Overproduction in rich media could mean that (*relatively*) low levels formed in comparatively poorer media were still adequate for lysis; hence, this may be a strategy for maintaining propagation of the phage under nutritionally unfavourable conditions [Microbiology (1998) *144* 2217–2224].

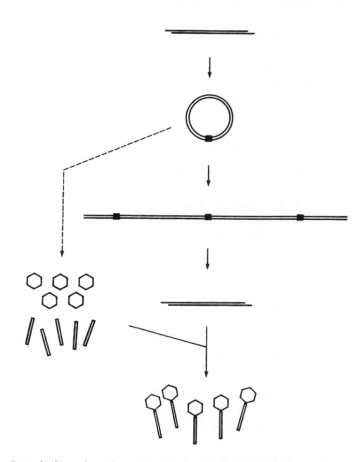

Figure 9.2 Bacteriophage λ: outline of the lytic cycle (diagrammatic). DNA which the phage injects (through its tail) into a cell is a linear, double-stranded molecule, about 49 kb long, with complementary sticky ends, each 12 bases long (*top*). Within the cell the DNA quickly circularizes via the sticky ends; the double-stranded region formed by the base-paired sticky ends is called the *cos* site (*solid black square*).

Within the host cell, transcription of the phage's 'early' genes yields regulatory proteins whose activities are mutually antagonistic; the outcome, which is influenced by conditions in the host cell (and in the cell's environment), is either lysogeny (section 9.2) or lysis, as follows.

The DNA initially replicates by a Cairns-type mechanism (Fig. 7.8), i.e. the circular DNA is replicated. Later, replication follows the rolling circle pattern (Fig. 8.5) which gives rise to a continuous chain of (double-stranded) phage genomes, joined end-to-end, in which the *cos* site is regularly repeated (*centre*); such a tandemly repeated molecule is called a *concatemer*.

Phage head and tail proteins are encoded by the late genes – transcribed from the circular DNA. In the concatemer, a *cos* site is cleaved enzymatically, and a *cos* sticky end locates in a phage head; DNA is fed into the head until the next *cos* site is reached – and cleaved. Thus, the DNA between two consecutive *cos* sites is packaged, so that the head contains linear DNA with terminal *cos* sticky ends. A tail is added to the head, and the phage – released by cell lysis – is ready to infect a new host cell.

9.2.2   DNA replication in phage M13

M13 is a filamentous phage (genome: ccc ssDNA) of the same group as phages f1 (Table 9.1) and fd (Plate 9.1); modified forms of M13 are used as cloning vectors for producing single-stranded copies of DNA for sequencing (section 8.5.6) or for site-specific mutagenesis (Fig 8.20).

Following infection of a bacterium, the host cell's enzymes synthesize a complementary (*c* or 'minus') strand on the single-stranded phage genome (the *v* or 'plus' strand), forming the ccc dsDNA *replicative form* (RF); this molecule is then super-coiled by the cell's gyrase (the supercoiled molecule is called RFI) and is transcribed from the *c* strand. The *v* strand is later nicked (by a phage-encoded enzyme), and replication occurs by the rolling circle mechanism (Fig. 8.5). Initially, genome-length pieces of *v* strand ssDNA are cleaved, circularized, and converted to daughter RFs; subsequently, a particular phage-encoded protein accumulates in the cell and coats the newly formed *v* strands, blocking their conversion to RFs. When this happens, the newly cleaved and circularized ssDNA genomes are packaged in phage coat proteins and extruded through the cell envelope of the (living) host.

9.3   ANDROPHAGES

Androphages infect only certain bacteria which contain a conjugative plasmid. For example, f1 and fd are filamentous, ssDNA phages (Table 9.1) which adsorb specifically to the *tips* of certain types of pili; penetration of the host cell may involve pilus retraction. The progeny phages of f1 and fd are released through the cell envelope; host cells remain viable, but they grow more slowly than do uninfected cells.

M12, MS2, f2 and Qβ are small, icosahedral ssRNA phages which adsorb to the *sides* of certain types of pili.

9.4   PHAGE CONVERSION

Bacteria lysogenized with phage may have certain characteristics not shown by uninfected cells; such *phage conversion* (= bacteriophage conversion; lysogenic conversion) may be due e.g. to expression of phage genes by the cells, or to inactivation of chromosomal genes through integration of the prophage. For example, those strains of *Corynebacterium diphtheriae* which cause diphtheria are lysogenized by a certain type of phage which encodes and expresses a potent toxin (section 11.3.1.7); strains which lack the phage cannot form the toxin and do not cause diphtheria. (See also cholera toxin in section 11.3.1.1.) In *Staphylococcus aureus*, integration of the genome of phage L54a causes loss of lipase activity due to inactivation of the relevant chromosomal gene. (See also *Salmonella* in the Appendix.)

As mentioned earlier, phage infection can modify the O antigens of a host bacterium, thus affecting its serotype [TIM (2000) *8* 17–23].

9.4.1   Bacteriophage Mu

The DNA of phage Mu (Table 9.1) inserts into the bacterial chromosome by simple transposition, an event requiring the phage *A* gene product (a transposase which binds to the ends of the phage DNA). Insertion occurs regardless of whether lysis or lysogeny is to follow, and it occurs at near-random sites in the chromosome; insertion of Mu DNA produces a 5 bp duplication in the target site DNA. Because phage Mu DNA often inserts into a bacterial gene, inactivating it, detectable mutations in a Mu-lysogenized bacterial population may be found in e.g. 2–3% of the cells – hence the name 'Mu' (mutator) phage.

Lysogeny involves the activity of a repressor protein (the *c* gene product) which e.g. regulates synthesis of the transposase.

Lytic infection involves about 100 *replicative* transpositions between one part of the chromosome and another, the phage genome behaving as a giant transposable element; the resulting deletions, inversions etc. in chromosomal DNA are sufficient to cause the death of the host cell. Phage genomes are excised and then packaged by the 'headful' mechanism, each genome carrying with it some bacterial DNA at each end.

9.5   TRANSDUCTION

The transfer of (typically chromosomal or plasmid) DNA from one bacterial cell to another, via a phage vector, is called *transduction.*

9.5.1   Generalized transduction

In this process, any of a variety of genes may be transferred from one cell to another. In a population of phage-infected bacteria it occasionally happens that, during phage assembly, chromosomal or plasmid DNA is incorporated in place of phage DNA; such abnormal phages can, once released, attach to other cells and donate DNA but (as it is not phage DNA) neither lysogeny nor lysis will result.

In the recipient cell (*transductant*) the transduced DNA may (i) be degraded by restriction endonucleases (section 7.4); (ii) undergo recombination with the chromosome (or plasmid) so that some donor genes can be stably inherited (*complete transduction*); (iii) persist as a stable but non-replicating molecule (*abortive transduction*). (If an abortive transductant gives rise to a colony, only one cell in the colony will contain donor DNA.)

Donor genes with functional promoters (whether in a complete or abortive transductant) may be expressed in the transductant.

In some cases, any given host gene has a similar chance of being transduced. However, in the *Salmonella*/phage P22 system, a certain region of the cell's chromosome has a greater chance of being transduced; this region resembles a sequence of the phage's genome concerned with packaging of DNA into the phage head.

The transfer of any given gene by generalized transduction is a rare event. If two or more genes are transduced together (*co-transduction*), this is taken as evidence of closeness on the chromosome; such information has been useful for the detailed mapping of donor chromosomes and plasmids – distances between genes being estimated from co-transduction frequencies.

9.5.2 Specialized (restricted) transduction

Specialized transduction can be brought about by a temperate phage in which there is a phase of integration with the host's chromosome (see e.g. section 9.2.1). On excision, a prophage will occasionally take with it some of the adjacent chromosome – an event similar (in principle) to the formation of an F' plasmid (section 8.4.2.2). In the *E. coli*/phage λ system, the prophage is normally flanked on either side by the host's *gal* and *bio* genes (Fig. 8.3b); hence, in one of the rare 'aberrant' excisions, the prophage may take with it *gal* or *bio* gene(s) – often leaving behind certain phage genes from the opposite end of the prophage. Such a recombinant phage genome may be subsequently packaged in a phage head and the rest of the phage may be assembled normally. Such a phage (a specialized transducing particle, or STP) may lack the ability to replicate (i.e. it may be *defective*); nevertheless, an STP can inject its DNA into a recipient cell and (hence) transfer specific donor genes (*gal* or *bio*). In other bacterium/phage systems, the genes flanking the prophage are those which can be specifically transduced.

In a novel type of specialized transduction, the genome of phage VGJϕ (the vector) joins covalently to that of phage CTXΦ, within cells of *Vibrio cholerae*, and a single-stranded hybrid genome is packaged in a VGJϕ capsid; this hybrid phage can infect fresh cells of *V. cholerae* via receptor sites on MSHA fimbriae. Thus, the genes for cholera toxin (in the genome of phage CTXΦ) can be transferred via phage VGJϕ, independently of the normal receptor sites for phage CTXΦ (the so-called toxin-coregulated pili) [JB (2003) *185* 7231–7240].

9.6 HOW DOES PHAGE DNA ESCAPE RESTRICTION IN THE HOST BACTERIUM?

In bacteria, the main purpose of restriction (section 7.4) seems to be to protect against 'foreign' DNA – particularly phage DNA. How, then, do phages manage to replicate in bacteria? They do so by means of various antirestriction mechanisms [MR (1993) *57* 434–450]. For example, in the DNA of some phages (e.g. phage T7 of *E. coli*) there are few (or no) cutting sites for the host cell's endonucleases, such sites having been limited or lost through the process of selection (*counterselection*).

Some phages encode specific inhibitors of restriction enzymes, and some encode methylases with the host cell's own specificity.

The T-even phages (T2, T4 and T6 of *E. coli*) are resistant to restriction because their DNA, being glycosylated, is not susceptible to *E. coli* endonucleases. Even so, on infection with T-even phages some strains of *E. coli* produce an enzyme which cleaves their own tRNA^{lys} (tRNA which carries lysine) – thus inhibiting protein synthesis in the host cell and (hence) inhibiting phage replication; however, phage T4 encodes enzymes (including an RNA ligase) which can repair the host's self-inflicted damage.

10 Bacteria in the living world

Bacteria are often thought of as pests to be destroyed, or as convenient 'bags of enzymes' – useful for experimental purposes. However, bacteria have a life of their own outside the laboratory, and many of their activities are important not only to man but also to the whole balance of nature. This aspect of bacteriology has many facets, and only a brief outline of some of them can be given here.

10.1 MICROBIAL COMMUNITIES

Most bacteria are *free-living*, i.e. they do not necessarily form specific associations with other organisms; nevertheless, they are part of the web of life, and in nature they can rarely grow without affecting – or being affected by – other organisms. Bacteria normally occur as members of mixed communities which may include fungi, algae, protozoa and other organisms. Such communities can be found in a wide variety of natural habitats – e.g. in water, in soil, on the surfaces of plants, and on and within the bodies of man and other animals. Those microorganisms which are normally present in a particular habitat are referred to, collectively, as the *microflora* of that habitat.

Microorganisms which colonize a given habitat may affect each other in various ways; for example, they may have to compete for scarce nutrients, for oxygen, or for space etc., and those organisms which cannot compete effectively are likely to be eliminated from the habitat. In some cases an organism can actively discourage at least some of its competitors by producing substances which are toxic to them – a phenomenon termed *antagonism*; for example, a microorganism that produces antibiotics (section 15.4) may have a competitive advantage. Often, such antimicrobial agents are produced only when growth is limited by nutrient levels – i.e. when the death of other organisms in the habitat may help by reducing competition for (and actually providing) nutrients.

Some bacteria secrete *bacteriocins*: antibiotics which often act specifically against related organisms. The bacteriocins include *colicins* and *microcins* (both produced by, and aimed at, Gram-negative bacteria) and *lantibiotics* (produced by, and aimed at,

Gram-positive species). Bacteriocins range from a simple modified amino acid (microcin A15), through short peptides, to high-MWt proteins; some are bacterio-static, some bactericidal. Bacteriocins inhibit/kill target cells in various ways: some form pores in the cytoplasmic membrane, some (e.g. microcin B17) inhibit DNA gyrase, some disrupt peptidoglycan, and some cleave 16S rRNA.

Colicins are high-MWt proteins produced by (colicinogenic) strains of enterobacteria (e.g. *Escherichia coli*, *Shigella boydii*) and active against (sensitive) strains of enterobac-teria; most are encoded by plasmids. Colicins bind to specific receptor sites on sensitive cells, and it appears that uptake by the target cell is usually or always an energy-dependent process. [Colicin import into *E. coli*: JB (1998) *180* 4993–5002.] Within target cells, a given colicin may e.g. act as a pore-forming agent, a non-specific DNase, or an RNase specifically against 16S rRNA, or it may inhibit the synthesis of peptido-glycan. [Colicin Y (complete coding sequence): Microbiology (2000) *146* 1671–1677.]

Microcins are low-MWt agents produced by bacteria of the family Enterobacteriaceae and effective against bacteria of that family; they are typically produced in the stationary phase of growth. In size, microcins range from one amino acid (microcin A15 is apparently a derivative of methionine) to peptides of up to ~50 amino acids in length. The type A microcins inhibit certain metabolic pathways in target cells; type B microcins inhibit DNA replication etc.

Lantibiotics are polycyclic peptide bacteriocins that contain uncommon constituents – e.g. the sulphur-containing amino acid lanthionine (hence lantibiotics); they are produced e.g. by strains of *Bacillus* and *Lactococcus*. Lantibiotics act e.g. by forming pores in the cytoplasmic membrane of a target cell. The food preservative *nisin* is one example of a lantibiotic. Lacticin 3147, a lantibiotic produced by *Lactobacillus lactis* subsp *lactis*, consists of *two* peptide components; both components are necessary for activity, and each component requires modification by a separate enzyme [Microbiol-ogy (2000) *146* 2147–2154]. [Biosynthesis and biological activities of lantibiotics: ARM (1998) *52* 41–79. Structure, biosynthesis and mode of action of lantibiotics: FEMSMR (2001) *25* 28–308.]

Clearly, bacteria have to protect themselves against their own bacteriocins. For example, the bacteriocin *lysostaphin*, secreted by *Staphylococcus simulans*, disrupts peptidoglycan (Fig. 2.7) in susceptible staphylococci; in *S. simulans*, the peptidoglycan is apparently modified by an enzyme (encoded by the *lif* gene), making it resistant to lysostaphin. [Bacteriocins (review): TIM (1998) *6* 66–71.]

There may also be relationships in a habitat in which one or both organisms benefit and neither organism is harmed. For example, an acid-producing organism can help to create favourable conditions for another organism whose growth depends on a low pH. (See also *consortia* in section 2.1.3.)

A habitat may eventually develop a stable community of organisms in which the various beneficial and antagonistic interactions have reached a delicate state of balance. An alien microorganism will often have difficulty in establishing itself in such a community – unless a disturbance in the environment upsets the balance in the community. For example, in the intestine of an animal, the natural microflora can often discourage the establishment of a pathogen because these organisms occupy space (thereby hindering access), and they are well adapted to the intestine – so that they can usually outgrow a pathogen; however, any disturbance to the microflora – due e.g. to antibiotic therapy – may enable a pathogen to become established and cause disease.

10.1.1   Transient and past communities

In contrast to the stable, mixed communities of microorganisms in many habitats, there are occasions when one, or a few, species transiently predominate.

In cholera (section 11.3.1.1), the patient's intestine becomes a living incubator for the causal organism, *Vibrio cholerae*, and the so-called 'rice-water stools' may contain up to 10^9 cells/ml of *V. cholerae*.

In lakes, reservoirs and other bodies of water, certain conditions can encourage prolific growth of particular organisms. The result is a so-called *bloom*: a visible (often conspicuous) layer of organisms at or near the surface; the organisms include (or may consist mainly of) certain cyanobacteria – particularly those which form gas vacuoles (section 2.2.5) – and/or certain eukaryotic microorganisms. Blooms can be encouraged e.g. by an excess of nutrients (such as nitrogen leached from agricultural fertilizer) and/or by thermal stratification in the water. The death/decomposition of the bloom-forming organisms (due e.g. to cessation of favourable factors) can cause a severe depletion of oxygen in the water, sometimes resulting in the asphyxiation of fish and other aquatic animals.

In some cases bloom formation can be discouraged by pumping (circulating) the water to avoid stratification and/or by using anti-cyanobacterial chemicals such as *dichlone* (dichloronaphthoquinone).

In reservoirs, a substance (*geosmin*) may be produced by *Anabaena* and some other bloom-formers, and this can impart an 'earthy' or 'musty' taste to the water (and to fish living in the water).

Some bloom-forming cyanobacteria (e.g. species of *Anabaena*, *Microcystis*, *Nodularia*) produce toxins (*cyanotoxins*) which can be lethal for fish and/or other animals. These toxins include neurotoxins, hepatotoxins and dermotoxins. [Ecological and molecular investigations of cyanotoxin production (review): FEMSME (2001) *35* 1–9.]

Stromatolites are rock-like, organosedimentary structures found in aquatic habitats; some (in Western Australia) are estimated to be over 3×10^9 years old. Stromatolites are believed to form by the incorporation of sediment in, and/or the precipitation of calcareous/siliceous material onto, microbial mats; a *microbial mat* is a complex, cohesive, benthic community of organisms often dominated by cyanobacteria and/or other photosynthetic prokaryotes. [Microstructure of cyanobacterial mats in Antarctica: AEM (2004) *70* 569–580.]

Calcified stromatolites apparently do not yield fossil microorganisms, but fossils resembling cyanobacteria (named e.g. *Gunflintia minuta, Huroniospora micro-reticulata*) have been detected in silicified stromatolites.

Stromatolites are currently in the process of formation around the Bahamas.

[Role of microbes in accretion, lamination and early lithification of modern marine stromatolites: Nature (2000) *406* 989–992.]

10.1.2   Quorum sensing

In some cases, cells in a high-density population exhibit characteristics which are absent in the *same* cells in low-density populations ('density' referring to the number of cells per unit volume). For example, *Photobacterium fischeri* (= *Vibrio fischeri*) can be either a free-living organism (occurring at low densities) or a *symbiont* (section 10.2.4) occurring at high densities in the light-emitting organ of certain fishes; in the latter role, *P. fischeri* produces blue-green light (= *bioluminescence*: light from a living organism) whereas, when free-living (low densities), this bacterium produces little or no light. (The fish use bacterial bioluminescence to signal one another; the bacteria benefit from a stable habitat in the fish.)

How do individual cells *sense* the density of the population, i.e. in the example above, how do they know *when* to 'switch on the light'? The signal actually develops as a direct consequence of growth in population. Thus, all the cells secrete signal molecules of a special kind, and if cell density reaches a certain minimum (*quorum*) these molecules accumulate to a level which activates certain genes within the cells; in this case the *lux* genes are induced, and the bacteria produce light.

The (low-molecular-weight) signalling molecule is called an *autoinducer* (because the cells themselves produce it). Different bacteria may produce different autoinducers which regulate different characteristics; in some cases different species produce the same autoinducer – but for regulating different genes. Some bacteria produce a range of autoinducers for controlling the expression of various properties. In many cases (including the *P. fischeri* example) the autoinducer is one of a group of N-acyl-L-homoserine lactones (AHLs). AHLs are common autoinducers in quorum sensing in Gram-negative bacteria. In the *P. fischeri* example, above, an AHL triggers the *lux* operon [evolution of LuxI and LuxR as regulatory molecules in quorum sensing: Microbiology (2001) *147* 2379–2387].

In some cases quorum sensing acts via two-component regulatory systems (section 7.8.6) – see e.g. sections 7.8.6.2 and 8.4.1.

Since the early examples of quorum sensing the phenomenon has been described in a wide range of situations – in eukaryotic organisms [e.g. the fungus *Candida albicans*: AEM (2001) *67* 2982–2992] as well as in both Gram-positive and Gram-negative bacteria. AHLs (0.5–50 nmol/g) have been detected in soil, suggesting the presence of quorum sensing in that environment [Microbial Ecology (2003) *45* 228–236]. Quorum sensing has been found to control the expression of virulence factors in certain pathogens; the following are a few examples.

- *Pseudomonas aeruginosa*. At least two quorum sensing systems, involving different AHLs, were reported to regulate genes encoding certain virulence factors; these two systems (the *las* and *rhl* systems) appear to act hierarchically, being activated in a definite sequence [TIM (1997) 5 132–134]. The product of gene *qscR* may be a repressor of gene *lasI* (which encodes *N*-3-(oxododecanoyl) homoserine lactone) [PNAS (2001) *98* 2752–2757].

  Interestingly, in *P. aeruginosa*, it has been reported that the stringent response (section 7.8.2.5) can activate quorum sensing *independently* of cell density [JB (2001) *183* 5376–5384].

  The existence of quorum sensing systems in pathogens may permit antimicrobial therapy based on the inhibition of autoinducers. In one study it was found that the production of virulence factors by *P. aeruginosa* strain PAO1 could be inhibited by a macrolide antibiotic (section 15.4.4), azithromycin [AAC (2001) *45* 1930–1933]. In another study, the quorum sensing inhibitor *furanone* was found to attenuate virulence in *P. aeruginosa* [EMBO Journal (2003) *22* 3803–3815].

- *Burkholderia cepacia*. Swarming, and the maturation of biofilms (aspects of this organism's physiology), appear to be regulated by a quorum sensing mechanism [Microbiology (2001) *147* 2517–2528].

- *Escherichia coli*. In EHEC and EPEC (Table 11.2), quorum sensing was reported to regulate expression of a type III protein secretion system (section 5.4.4) involving induction of genes in the LEE pathogenicity island (section 11.5.7); to account for the unusually low infectious dose of *E. coli* O157:H7 it was suggested that inducer may be supplied by non-pathogenic strains of *E. coli* in the gut [PNAS (1999) *96* 15196–15201]. Subsequent work with a mutant strain of EHEC, unable to form the autoinducer, has indicated that – within the gut – virulence genes in EHEC can be activated by the eukaryotic (host) hormone adrenalin (= epinephrine) *and/or* by a previously undescribed EHEC autoinducer (both of which may be recognized by the same receptor) [PNAS (2003) *100* 8951–8956].

  In *E. coli* O157:H7, quorum sensing has been reported to be a global regulatory mechanism – involved in basic physiological functions as well as in the formation of virulence factors [JB (2001) *183* 5187–5197].

Quorum sensing in Gram-positive bacteria. The signalling molecules used by Gram-positive bacteria are generally peptides (= *pheromones*)[peptide signalling (review):

TIM (1998) *6* 288–294]. For example, when *Bacillus subtilis* grows to a high density of cells, two types of pheromone accumulate in the extracellular environment: ComX and CSF. ComX promotes the development of competence in transformation (see section 8.4.1). CSF is taken up by an ATP-dependent oligopeptide permease (transport) system in the cytoplasmic membrane; within the cytoplasm it apparently inhibits the activity of certain phosphorylases that de-phosphorylate Spo0F~P – thereby promoting the phosphorelay (Fig. 7.14).

In *Enterococcus faecalis*, conjugation (section 8.4.2.1) involves the secretion of pheromones by potential recipients.

In *Staphylococcus aureus* quorum sensing can modulate the development of biofilms [JB (2004) *186* 1838–1850].

In *Streptococcus pneumoniae* an unusual example of convergent gene regulation involves transcriptional control of a single target by two separate quorum sensing systems [JB (2004) *186* 3078–3085].

10.2   SAPROTROPHS, PREDATORS, PARASITES, SYMBIONTS

10.2.1   Saprotrophs

Organisms which obtain nutrients from 'dead' organic matter are called *saprotrophs*. Some saprotrophs use only soluble compounds, but others can degrade cellulose (section 6.2) and other polymers, outside the cell, and assimilate soluble products. Complex substrates may be degraded in a stepwise manner, each of several species of saprotroph carrying out one (or a few) steps in the process; co-operation of this sort is important in the breakdown, and hence re-cycling, of organic matter in nature. Indeed, there are very few biological compounds which cannot be readily broken down by a community of saprotrophs working as a team. Without the (heterotrophic) saprotrophs the carbon cycle (Fig. 10.1) – and, hence, the other cycles of matter – would stop.

10.2.2   Predators

Bacteria of the order Myxobacterales are Gram-negative rods which live in soil and on dung and decaying vegetation. Most species prey on other microorganisms: they release enzymes which lyse other bacteria and fungi, and live on the soluble products. Of course, the myxobacteria are not typical predators: a predator usually ingests its prey *before* digesting it.

Another unusual predator, *Bdellovibrio* (see Appendix), attacks and lives within certain other bacteria; the name '*Bdellovibrio*' derives partly from the Greek word for 'leech'. Interestingly, *Bdellovibrio* can penetrate cells of *E. coli* even when the latter have a thick capsule [Microbiology (1997) *143* 749–753].

10.2.3   Parasites

A *parasite* is an organism which lives on or within another living organism (the *host*) and which benefits (in some way) at the expense of the host; in almost all cases a parasite obtains nutrients from its host. The host may suffer varying degrees of damage – ranging from slight inconvenience to death.

Parasitism may be adopted as an alternative way of life by certain free-living bacteria, but in some bacteria it is obligatory. Bacteria such as *Mycobacterium leprae* (which causes leprosy) can grow only within particular types of (eukaryotic) host cell; obligate parasites such as this depend heavily on their host's metabolism, and often they cannot be grown in the laboratory except in specialized preparations of living cells.

A parasite which affects its host severely enough to cause disease, or death, is called a *pathogen*. Not all parasites are pathogens, and not all pathogens are parasites; an example of a non-parasitic pathogen is *Clostridium botulinum* (section 11.3.1.2).

10.2.4   Symbionts

Originally, *symbiosis* meant any stable, physical association between different organisms (the symbionts) – regardless of the nature of their relationship. Later, the meaning of the term was restricted to cover only those instances in which the relationship was one of mutual benefit. However (despite the potential for confusion) there is now a general tendency to move back to the original meaning. Hence, symbiotic relationships may now include both mutual benefit (*mutualism*) and parasitism. Using the same terminology, a *commensal* is a symbiont which gains benefit from another symbiont such that the latter derives neither benefit nor harm from the association.

10.2.4.1   *Mutualistic symbioses*

Mutually beneficial relationships between bacteria and other organisms are quite common. For example, ruminants (e.g. sheep, cows etc.) cannot produce enzymes to digest the cellulose in their diet of plant material; however, in these animals the gut includes a specialized compartment (the *rumen*) containing vast numbers of micro-organisms (including bacteria such as *Ruminococcus*) which convert cellulose to simple products that the animal can absorb. In return, the microbes benefit from a warm, stable environment and the abundance of nutrients which the animal swallows. (See also *Wolbachia* in the Appendix.)

In certain insects, specialized cells (*mycetocytes*) in the gut lining contain (intracellular) bacteria which, in at least some cases, supply essential nutrients to the host. A *mycetome*, a distinct organelle composed of a group of mycetocytes, may be associated with the gut.

In leguminous plants (peas, beans, clover etc.) the roots have small swellings (*nodules*) typically containing bacteria of the genus *Rhizobium*; in this arrangement, the plant provides nutrients and protection, while the bacteria supply the plant with 'fixed' nitrogen from the atmosphere (section 10.3.2). Root nodules enable these plants to thrive in nitrogen-poor soils. [Legume nodulation (review): EMBO Journal (1999) *18* 281–288.] Successful symbiosis depends on bacterial quorum sensing (section 10.1.2) [MMBR (2003) *67* 574–592].

Nitrogen-fixing bacteria also form associations with non-leguminous plants. For example, nodules in the roots of the alder tree (*Alnus*) contain bacteria of the genus *Frankia*, and the small floating fern *Azolla* contains *Anabaena azollae* within specialized cavities. Strains corresponding to the cyanobacterium *Nostoc* have been detected by PCR-based methods in the corraloid roots of cycads (*Cycas, Encephalatos, Zamia*) from a botanical garden [FEMSME (1999) *28* 85–91]. (See also wheat in section 10.3.2.2.)

Nitrogen-fixing species of *Clostridium* are found e.g. in endophytic consortia in wild rice plants [AEM (2004) *70* 3096–3102].

10.3   BACTERIA AND THE CYCLES OF MATTER

The elements which make up living organisms occur on Earth in finite amounts; accordingly, for life to continue, the components of dead organisms must be re-used (re-cycled). Bacteria, together with other microorganisms, play a vital role in this process.

10.3.1   The carbon cycle

In all living organisms the major structural element is carbon (Chapter 6), so that the re-cycling of carbon is of fundamental importance. In the biological carbon cycle (Fig. 10.1) the chief contribution of bacteria is the use and degradation of 'dead' organic matter by the (heterotrophic) saprotrophs. In terrestrial systems, dead organic matter includes the remains of plants and animals.

For *aquatic* systems, dissolved organic carbon derives at least partly from *phytoplankton*, i.e. microscopic, photosynthetic (CO_2-fixing) organisms. (The growth of phytoplankton, seaweeds and terrestrial green plants – all photosynthetic organisms – is collectively referred to as *primary production*.) In lakes, carbon from phytoplankton is likely to be augmented by organic carbon of terrestrial origin (which arrives in run-off from the land). In the open ocean almost all carbon is derived from phytoplankton. Although much phytoplankton is used as food by aquatic animals, some provides carbon for bacterial growth (i.e. biomass production) and respiration. The proportion of dissolved carbon used for bacterial *respiration* is greater than once

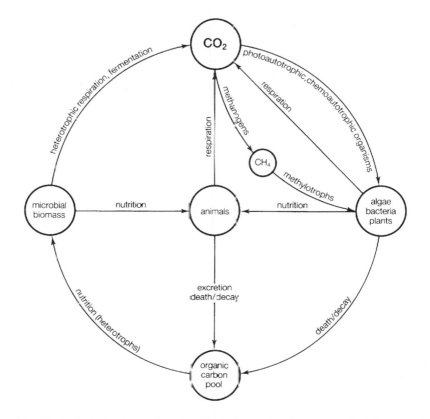

Figure 10.1 The biological carbon cycle: a simplified scheme showing some major interconversions of carbon in nature. Bacteria have significant roles as both autotrophs and heterotrophs (Chapter 6), and members of the Archaea have unique roles as methanogens (section 5.1.2.2); the methylotrophs (section 6.4) include methane-utilizing bacteria. Microorganisms (including bacteria) are responsible for the essential conversion of 'dead' organic carbon to biomass and CO_2; without this process the cycle would stop. Microbial biomass, as such, is used e.g. by filter-feeders (oysters etc.) and, via food chains, by fish and other animals. The role of *elemental* carbon seems to be minimal (compare with nitrogen and sulphur in their respective cycles).

supposed; moreover, it appears that when and where aquatic primary production is low, the production of CO_2 from bacterial respiration can actually exceed the rate of CO_2 fixation by phytoplankton. This means that, for at least certain periods of time, the cycling of carbon is unbalanced in parts of the ocean; this has suggested e.g. the possibility that periods of net autotrophy (when CO_2 fixation exceeds respiratory CO_2 output) may occur at different times of the year to compensate for the net heterotrophy [Nature (1997) *385* 148–151].

A knowledge of the carbon cycle is relevant to an understanding of the 'greenhouse effect' (section 10.6), and the evolution of the cycle is of interest in itself [past and present cycle of carbon on our planet (review): FEMSMR (1992) *103* 347–354].

10.3.1.1 *Xenobiotics*

Many *xenobiotics* (which include environmental pollutants such as pesticides, other agro-chemicals, and detergents) tend to resist biodegradation owing to the presence of certain types of bond (e.g. the ether bond: C–O–C) in their molecules. Despite this in-built resistance to biodegradation, there are nevertheless communities of bacteria which have developed the ability to degrade xenobiotics; this is fortunate because such activity helps to limit the undesirable build-up of these compounds in the environment. [Bacterial cleavage of ether bonds: MR (1996) *60* 216–232].

(See also bioremediation: section 13.8.)

10.3.2 The nitrogen cycle

Nitrogen is a component of proteins and nucleic acids and is therefore essential to all organisms. Most of the Earth's atmosphere consists of gaseous nitrogen (dinitrogen) but this form of nitrogen cannot be used by the majority of organisms; however, some organisms (all prokaryotes) are able to use gaseous nitrogen (see section 10.3.2.1).

Figure 10.2 shows some of the interconversions which occur in the nitrogen cycle when elemental and other forms of nitrogen are (i) assimilated (converted to biomass) and (ii) involved in energy-converting processes.

Ammonia assimilation. Many bacteria assimilate nitrogen as ammonia, primarily by incorporating an amino group (from ammonia) into either 2-oxoglutarate or gluta-mate – forming glutamate or glutamine, respectively; the glutamate or glutamine, in turn, acts as a nitrogen donor in various transamination reactions in the synthesis of other nitrogenous compounds. Thus, glutamate provides nitrogen for the synthesis of e.g. L-alanine and L-aspartate (Fig. 6.3) while glutamine is a nitrogen donor in the synthesis of purines and pyrimidines (bases in nucleic acids) and the amino acids histidine and tryptophan.

In *E. coli*, the pathway of ammonia assimilation depends on the concentration of ammonia; NADPH is used in both pathways.

In high concentrations of ammonia, the enzyme glutamate dehydrogenase catalyses the reductive amination of 2-oxoglutarate to glutamate.

In low concentrations of ammonia (e.g. < 1 mM) there is a two-stage reaction. First, glutamine synthetase (GS) catalyses the formation of glutamine from glutamate and ammonia; then, in the second stage, glutamine:2-oxoglutarate aminotransferase (GOGAT; glutamate synthase) catalyses a reaction in which glutamine (1 molecule) and 2-oxoglutarate (1 molecule) form glutamate (2 molecules).

In *E. coli* the pathway catalysed by glutamate dehydrogenase seems to be important under energy-limited conditions (e.g. when energy-yielding substrates are limiting); the reaction catalysed by this enzyme is independent of ATP, while the pathway involving GOGAT is ATP-dependent. [Pathway choice in glutamate synthesis in *E. coli*: JB (1998) *180* 4571–4575.]

Nitrate assimilation. Some bacteria can assimilate nitrogen as nitrate, though the nitrate is first reduced to ammonia by *assimilatory nitrate reduction* (Fig. 10.2); this differs from *nitrate respiration* (section 5.1.1.2; Fig. 10.2) e.g. in that it does not yield energy.

Nitrate, nitrite and ammonia in energy metabolism. Nitrate or nitrite can be used by some bacteria as electron acceptors in respiratory metabolism; according to species, either organic or inorganic substrates can be used as electron donors in this type of

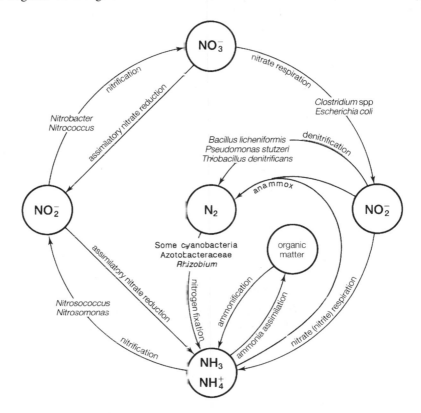

Figure 10.2 The nitrogen cycle: interconversions carried out by bacteria with some named examples (section 10.3.2). Nitrification is an aerobic energy-yielding process. Nitrate (nitrite) respiration is an anaerobic energy-yielding process. Denitrification is an energy-yielding process which has been associated with anaerobic environments but which, in some cases, can occur in the presence of oxygen. Ammonification is part of the process of mineralization (section 10.3.4). The anammox reaction, in which nitrite acts as electron acceptor in the oxidation of ammonium, occurs e.g. in anoxic marine habitats. Nitrate, or ammonia, is assimilated by many bacteria, nitrate being initially reduced intracellularly to ammonia by *assimilatory nitrate reduction*. Nitrogen fixation, which consumes much energy, occurs typically (but not always) under anaerobic conditions; in some cyanobacteria nitrogen fixation occurs in heterocysts (section 4.4.2; Plate 4.2).

metabolism. (A test for nitrate reduction is described in section 16.1.2.12.) Several pathways are known.

Denitrification is a respiratory process (section 5.1.1.2; Fig. 10.2) in which nitrate, or nitrite, is reduced to gaseous products (mainly dinitrogen and/or nitrous oxide); the process can be important in agriculture as it causes the loss of biologically useful nitrogen from the soil (section 10.3.2.2).

Dissimilatory reduction of nitrate to ammonia (DRNA) is a respiratory process carried out by certain bacteria (e.g. species of *Enterobacter* and *Vibrio*) and which has been reported to occur in habitats such as marine sediments. DRNA appears to occur more readily with low concentrations of nitrate – higher concentrations being reduced to nitrite and correspondingly smaller amounts of ammonia [FEMSME (1996) *19* 27–38].

Anaerobic ammonium oxidation. The anaerobic oxidation of ammonium, using nitrite as electron acceptor, has been found e.g. in wastewater treatment systems and in an anoxic marine habitat (Black Sea). This process (termed *anammox*) [FEMSMR (1999) *22* 421–437], in which gaseous nitrogen is produced, may be an important aspect of the marine nitrogen cycle [Nature (2003) *422* 608–611].

Nitrification (Fig. 10.2) is an oxygen-dependent, energy-yielding process in which ammonia is oxidized to nitrite, and nitrite is oxidized to nitrate, e.g. by certain chemolithotrophic bacteria (section 5.1.2).

10.3.2.1 Nitrogen fixation

The 'fixation' of atmospheric nitrogen (reduction of nitrogen to ammonia) is apparently carried out only by certain types of bacteria and by some members of the Archaea; these *diazotrophs* include some cyanobacteria (e.g. *Anabaena*, *Nostoc*), some species of *Bacillus* and *Clostridium*, *Klebsiella pneumoniae* (some strains), and members of the families Azotobacteriaceae and Rhizobiaceae and of the order Rhodospirillales.

Fixation is catalysed by the enzyme complex *nitrogenase*. Electrons from a source such as hydrogen, or NADPH, are transferred e.g. to a ferredoxin (section 5.1.1.2) and thence to nitrogenase; the reduction of nitrogen requires much energy – about 12–16 molecules of ATP for each molecule of nitrogen fixed.

Nitrogen fixation depends on the expression of the *nif* genes. In nitrogen-fixing strains of *Klebsiella pneumoniae* there are at least 17 *nif* genes, many of which encode units of the nitrogenase complex. Transcription of the *nif* genes is governed by (i) the availability of combined nitrogen and (ii) (in enterobacteria) the *ntr* genes. Among the *ntr* genes is *rpoN* (= *ntrA*), a sigma factor (section 7.5), σ^{54}, involved in controlling the initiation of transcription of all the *nif* genes. [σ^{54} (minireview): JB (2000) *182* 4129–4136.] In addition, transcription of the *nif* genes initially requires an activated NtrC protein – which is present when combined nitrogen is limiting.

The ammonia produced by nitrogen fixation may be assimilated e.g. by the amination of glutamate to glutamine (the GOGAT pathway).

Nitrogenase is highly sensitive to oxygen. Many diazotrophs (e.g. clostridia) fix nitrogen anaerobically or microaerobically, and in some cyanobacteria the process is carried out within *heterocysts* (section 4.4.2). Aerobic nitrogen-fixers have special mechanisms for protecting their nitrogenase. In unicellular, aerobic, nitrogen-fixing cyanobacteria of the genus *Cyanothece*, nitrogen fixation occurs only in the dark period during alternating 12-hour light/dark cycles, but is continuous during 24-hour illumination; nitrogen fixation during the dark period would avoid exposure to oxygen evolved during photosynthesis (section 5.2.1.1), but continuous nitrogenase activity during continuous illumination requires additional explanation – and possibly involves the 'inclusion granules' which develop under nitrogen-fixing conditions [JB (1993) *175* 1284–1292]. [Nitrogen fixation by non-heterocystous cyanobacteria (review): FEMSMR (1997) *19* 139–185.]

Diazotrophs occur as free-living organisms in soil and water, and some are involved in symbioses (section 10.2.4.1). In the open ocean, free-living and symbiotic nitrogen-fixers may contribute significantly to the nitrogen budget [nitrogen-fixing microorganisms in tropical and subtropical oceans: TIM (2000) *8* 68–73].

Finally, an intriguing question: why does nitrogen fixation occur only in certain *prokaryotes* – why not, for example, in plants? During evolution, plants are believed to have incorporated prokaryotes as organelles (e.g. mitochondria), but the ability of prokaryotes to fix nitrogen may not have developed until later, in free-living organisms; this, together with an insufficiency in selection pressure, is believed to be the reason [PTRSLB (1992) *338* 409–416].

10.3.2.2 *Agriculture and the nitrogen cycle*

Our knowledge of the roles of bacteria in the nitrogen cycle can be put to good use in improving agricultural food production. Food crops can be limited in yield by a shortage of available nitrogen in the soil; hence, by knowing how nitrogen is lost – and by exploiting nitrogen fixation – we can often take appropriate measures to increase crop yields.

Nitrogen is taken from the soil when crops are grown and harvested, and it may also be lost by denitrification and nitrification. Denitrification typically occurs maximally under anaerobic or microaerobic conditions in the presence of nitrate and organic nutrients – conditions found e.g. in waterlogged farm soils; thus, denitrification can often be reduced by improving soil structure and drainage so as to minimize the development of anaerobic conditions.

The harmful effect of denitrification is obvious, but why does *nitrification* lead to a loss of nitrogen? The answer is that, although nitrate and ammonia are both soluble, ammonium ions adsorb readily to soil particles (clay particles typically bear a net negative charge) whereas nitrate ions do not; for this reason, nitrate is much more readily washed (*leached*) from the soil by rain or flooding. Hence, if nitrogen fertilizers are required there is an advantage in choosing ammonium compounds rather than nitrates. Nitrification can often be prevented by adding a 'nitrification inhibitor' to the fertilizer; such compounds primarily block the oxidation of ammonia, and one of them, *etridiazole*, is additionally useful as a fungicide – being used e.g. for soil and seed treatment against certain 'damping off' diseases.

Nitrogenous fertilizers can replace lost nitrogen, but they are expensive and can be afforded least by countries in greatest need of them. However, there are alternatives. For example, 'nitrogen-fixing plants' such as clover and lucerne can be included in crop rotation schemes, while plants such as *Azolla* (section 10.2.4.1) can be used as 'green manure' – a practice common in South-East Asia; in rice paddies, fertility can be increased by encouraging the growth of free-living nitrogen-fixing cyanobacteria. Even better would be the creation, by genetic manipulation, of plants capable of fixing nitrogen without help from prokaryotes.

Nodulation of non-leguminous plants. Attempts to express the bacterial *nif* (nitrogen fixation) genes in plants have so far been unsuccessful, but the finding of *Rhizobium*-containing nitrogen-fixing nodules (section 10.2.4.1) in the *non*-legume *Parasponia* (a sub-tropical shrub) suggested an alternative approach. Thus, efforts have been made to promote the development of nitrogen-fixing nodules in certain non-legumes – particularly cereals such as rice and wheat – which apparently do not contain such nodules in nature; the object of this work is to reduce the need for nitrogenous fertilizer in agriculture in order to e.g. (i) lower the cost of cereal production and (ii) help minimize the use of (environmentally unfriendly) fixed-nitrogen fertilizers (such as nitrates).

In certain (tropical) legumes, 'invasion' of the plant by nitrogen-fixing bacteria occurs when the plant produces lateral rootlets; at this stage of growth the bacteria enter at sites where emergent rootlets penetrate the root cortex. This mode of bacterial invasion is called *crack entry*. The rootlets subsequently develop as nitrogen-fixing nodules. Experiments have been carried out to determine whether wheat can be 'infected' with bacteria of the nitrogen-fixing species *Azorhizobium caulinodans* (which nodulates the tropical legume *Sesbania rostrata*). *A. caulinodans* became established in the xylem and root meristem (i.e. *endophytically*), and the wheat showed significantly increased dry weight and nitrogen content compared with uninoculated control plants [PRS (1997) *264* 341–346].

10.3.3   The sulphur cycle

Sulphur is a component e.g. of the amino acids cysteine and methionine, of ferredoxins (section 5.1.1.2), and of cofactors such as coenzyme A. Green plants (and many bacteria) can assimilate sulphur in the form of sulphate, a substance commonly available in adequate amounts under natural conditions. Before incorporation, sulphate must be reduced to sulphide by *assimilatory sulphate reduction* (Fig. 10.3); this process differs from *sulphate respiration* (section 5.1.1.2; Fig. 10.3) in much the same way as assimilatory nitrate reduction differs from nitrate respiration (section 10.3.2). Some bacteria can assimilate sulphide direct from the environment.

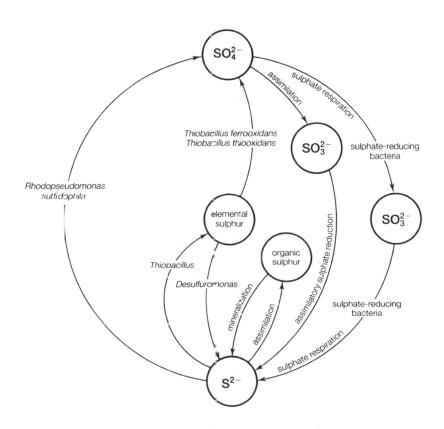

Figure 10.3 The sulphur cycle: some interconvertions carried out by bacteria. Many species can use sulphate (SO_4^{2-}) as a source of sulphur (needed e.g. for the synthesis of certain amino acids); this is shown in the figure as assimilatory sulphate reduction. In this process, sulphate is reduced, intracellularly, to sulphide, the sulphide being incorporated in different ways by different organisms; in e.g. *Escherichia coli*, sulphide is incorporated into *O*-acetylserine to form cysteine. Sulphate respiration (= dissimilatory sulphate reduction) is carried out e.g. by the 'sulphate-reducing bacteria' – organisms which use sulphate (or, e.g. sulphite) as a terminal electron acceptor in anaerobic respiration (section 5.1.1.2). *Desulfuromonas* uses elemental sulphur as a terminal electron acceptor in anaerobic respiration. *Thiobacillus* spp typically carry out aerobic respiration in which they oxidize e.g. sulphide (S^{2-}) and/or elemental sulphur. *Rhodopseudomonas sulfidophila* is one of a number of species which use sulphide as an electron donor in anaerobic phototrophic metabolism (section 5.2.1.2).

In some habitats (e.g. stagnant anaerobic ponds) much sulphide is formed by sulphate respiration; this sulphide may, in turn, be used as electron donor (i.e. oxidized to sulphite or sulphate) by anaerobic photosynthetic bacteria (section 5.2.1.2).

Elemental sulphur can be used e.g. by the archaean *Sulfolobus*, and by *Thiobacillus thiooxidans* and *T. ferrooxidans* (see Appendix). Some species of *Thiobacillus* have been shown to form a filamentous matrix with which they adhere to particles of sulphur in sewage sludge, and it has been suggested that such a matrix may play an important role in the colonization and oxidation of sulphur in the natural environment.

Thiosulphate and polythionates appear to be important intermediates in the bacterial oxidation of *pyrite* (metal sulphides), a process involved in the (environmentally harmful) formation of acid mine drainage. In this process, ferric (Fe III) ions are produced by bacterial oxidation of ferrous (Fe II) ions from the pyrite. It seems that ferric ions oxidize the sulphur in pyrite – initially to thiosulphate and then to tetrathionate; thiosulphate is regenerated, via trithionate, in a proposed cyclical reaction. A continuous supply of ferric ions is produced by the ongoing oxidative metabolism of lithotrophic bacteria such as *Thiobacillus ferrooxidans*.

10.3.4   Mineralization

Figures 10.1–10.3 show that, in the cycles of matter, complex organic substances are broken down to simple inorganic materials such as carbon dioxide, sulphide, sulphate, ammonia and nitrate; this conversion of organic to inorganic matter is called *mineralization*.

10.4   ICE-NUCLEATION BACTERIA

At temperatures just below 0°C, some bacteria promote water-to-ice transition by acting as nuclei around which ice crystals can form. Some of these 'ice-nucleation' bacteria are found on leaves [AEM (2003) *69* 1875–1883] and can promote frost damage in various agricultural crops; they include strains of *Erwinia*, *Pseudomonas* and *Xanthomonas*.

Some plant tissues can survive supercooling to several degrees below 0°C without damage – but may suffer frost injury if ice-nucleation bacteria are present. At temperatures above (approx.) – 5°C, the incidence of frost injury can be decreased by reducing the size of plant-contaminating populations of ice-nucleation bacteria – e.g. by treatment of plant surfaces with antibiotics such as streptomycin or oxytetracycline; biological control (section 13.1.2), involving treatment with appropriate non-ice-nucleating bacteria, can also be effective. A combination of antibiotic treatment and biological control acted additively in the control of frost injury in pear trees

[Phytopathology (1996) *86* 841–848]. [See also relevant text in AEM (2003) *69* 1875–1883.]

10.5   BACTERIOLOGY *in situ* – FACT OR FICTION?

Ideally, every organism should be studied *in situ*, i.e. in its natural environment; biology is, after all, about the real, living world. Of course, some aspects – e.g. intracellular structures – have to be studied *in vitro* (in the laboratory), but, where possible, a cell's normal *behaviour* is best observed under conditions which most closely resemble its normal habitat.

Clearly, any meaningful *in situ* study demands an understanding of the particular environment because, without this, the design of the experiment can be faulty; unfortunately, many studies which claim to be '*in situ*' involve obvious (sometimes extreme) distortions of nature, so that they are of little or no scientific value.

10.5.1   Membrane-filter chambers

To test for survival in rivers, suspensions of bacteria have been enclosed in 'membrane-filter chambers' – each essentially a wide plastic tube which is sealed at both ends by a membrane filter (pore size 0.22–0.45 μm); once a chamber had been immersed in a river, the bacteria inside it were considered to be essentially in 'natural' conditions, i.e. the sample was believed to be *in* the environment simply because the *chamber* had been immersed. However, the sample in such a chamber has limited contact with the environment: it generally receives less light, and it is shielded from natural turbulence (and hence from fluctuations in temperature etc.). Moreover, access or exit of molecules (including waste products) can occur only by diffusion through the minute pores of the filter; the rate of diffusion is necessarily extremely low in those membrane filters whose pores are small enough to retain the test organisms [JAB (1985) *58* 215–220].

10.5.2   Conjugation *in situ*?

In an attempt to demonstrate conjugation (section 8.4.2) in a Welsh river and canal, suspensions of donors and recipients were mixed and then filtered through a membrane filter; the filter, with its layer of bacteria, was placed (face-down) on a flat stone and held in place by a glass-fibre filter which was secured to the stone by rubber bands. After 24 hours immersion in the river or canal the whole was tested for transconjugants. Transconjugants were detected, and the overall conclusion of the experiment was that "plasmid transfer is possible between bacteria in the river epilithon" [JGM (1987) *133* 3099–3107]. Clearly – in order to draw this conclusion – the layer of bacteria on the filter must have been taken to represent *epilithon*: the sessile community of organisms which grow on the surfaces of underwater stones.

To what extent did the 'simulated epilithon' resemble real epilithon? That is, to what extent was the conclusion justified? An earlier study by different authors [Oikos (1984) *42* 10–22] had shown clearly that real epilithon consists of cells embedded in a fibrous polysaccharide matrix – electronmicrographs revealing cells and microcolonies trapped within the matrix *and separated from one another by the material of the matrix*. So, in real epilithon, would not the recipients' receptor sites for pili be hidden by the matrix? Could the pilus even reach a recipient through a matrix which can exclude macromolecules? Moreover, given the presence of the matrix, would donor and recipient cells be likely to achieve direct wall-to-wall contact (of the type shown in Plate 8.1, top)? The simulated epilithon could answer none of these awkward (but crucial) questions because it lacked the essential feature of real epilithon: the matrix!

A genuine simulation would require (in addition to a matrix) a population of cells similar to that in real epilithon – with only a realistic fraction consisting of donors and recipients. However, in this experiment, each square centimetre of the filter was packed with 10^7 donors intimately mixed with 10^8 recipients (of known compatibility). From the authors' own figures, this was over 100 times greater than the total number of viable bacteria in real epilithon, and it consisted *solely* of donors and recipients; inexplicably, this arrangement was claimed to 'mimic nature'. Nature is not like this, and it is pointless to pretend that it is.

In this experiment it seems possible that donor–recipient contacts had already been initiated during mixing and filtration – i.e. even *before* the filter had been placed in its 'in situ' test location in the river or canal.

Although seemingly concerned with 'natural habitats', this experiment in no way related to the real environment: it was little more than a laboratory-type mating experiment conducted outside the laboratory. Such experiments may be fun but they contribute nothing to science.

10.6   THE GREENHOUSE EFFECT

In the natural carbon cycle (section 10.3.1; Fig. 10.1) large amounts of CO_2 are produced by animals, plants and microorganisms during respiration or fermentation; however, large amounts of CO_2 are used by the photosynthetic autotrophs (e.g. green plants and cyanobacteria).

Globally, the important balance between biological production and uptake of CO_2 is being upset e.g. by the burning of vast amounts of fossil fuel (petroleum, gas etc.) – leading to a rising level of CO_2 and consequent warming of the planet (the 'greenhouse effect').

Is the rising level of CO_2 likely to be offset by increased photosynthesis in plants (i.e. increased sequestration of CO_2)? In those plants studied, typical responses to increased CO_2 include e.g. enhanced root growth and a decrease in tissue nitrogen (i.e. a higher carbon:nitrogen ratio). Agricultural crop-type plants, in particular, show improved growth and better yields. However, one early study [Science (1992) *257* 1672–1675] gave a less optimistic picture. This study looked at the effect of increased CO_2 on experimental tropical ecosystems; no significant increase in above-ground biomass was found, and increased root growth was associated with an increased efflux of carbon dioxide from soil to atmosphere. This efflux of carbon dioxide appeared to be due mainly to stimulated metabolism of microorganisms (including bacteria) in the *rhizosphere* (the root–soil environment). These results were interpreted to mean that raised levels of CO_2 may not necessarily result in increased sequestration of carbon but may bring about increased carbon cycling (cyclical exchange) between the atmosphere and terrestrial ecosystems.

Of course, the world's forests are a major natural 'sink' for CO_2. Unfortunately, ongoing deforestation by man continually reduces the effectiveness of this sink. More optimistically, certain deep-rooted pasture grasses, imported into the South American savannas, appear to sequester considerable amounts of carbon in their particularly massive root systems; such grasses may therefore help to offset the effect of some of the non-biological emissions of CO_2 [Nature (1994) *371* 236–238].

Another greenhouse gas, methane (section 5.1.2.2), is oxidized by methylotrophs (section 6.4), but some escapes and contributes to the greenhouse effect. Oxides of nitrogen (which are also greenhouse gases) are produced during denitrification (section 5.1.1.2).

Harte [BC (1996) *5* 1069–1083] referred to data from the Antarctic and Greenland ice cores, pointing out that global warming may itself result in increased levels of greenhouse gases, i.e. that there is a potential positive feedback effect which is not reflected in our current models of future climatic change. Moreover, this author points to the many *synergies* which exist between different forms of environmental degradation; he also indicates that the assumption of a simple, *linear* (i.e. proportional) relationship between (human) population growth and environmental problems is overly optimistic. Harte's article is well worth reading, particularly by those who believe that environmental insults can be tackled piecemeal, i.e. in isolation from the environment as a whole.

Avoidance of deforestation has been assumed to be of major importance in the context of global warming as this policy would preserve a valuable means of controlling levels of atmospheric carbon dioxide [Nature (2001) *410* 429]. However, the ability of forest systems to sequester carbon dioxide may be limited by shortages of nutrients and water and/or by the relatively rapid turnover of organic carbon in the litter layer [Nature (2001) *411* 469–472; Nature (2001) *411* 466–469; commentary: Nature (2001) *411* 431–433].

In the Amazonian region, changes have been noted in the composition of the dominant tree species, with a decline in the slower-growing subcanopy trees (whose wood is of greater density); this may adversely affect the role of these forests to act as a carbon sink [Nature (2004) *428* 171–175].

10.7 THE PROBLEM OF RECOMBINANT BACTERIA IN THE ENVIRONMENT

The use of genetically engineered bacteria – e.g. for biological control (section 13.1.2) – causes unease owing to insufficient knowledge of the way in which such organisms may behave, or transfer their genes, in the natural environment. One approach to the problem is *biological containment*: arranging for a genetic 'self-destruct' mechanism to operate automatically when the organism's function has been completed. Such mechanisms are not 100% effective: some of the recombinant cells survive owing e.g. to mutation in the killing gene; however, elimination of the majority of recombinant cells seems a worthwhile objective. ['Suicide microbes' (review): Biotechnology (1995) *13* 35–37.]

10.8 UNCULTIVABLE/UNCULTURED BACTERIA

Of the existing species of bacteria, probably only a small proportion have been grown and isolated in the laboratory. Many species may remain unknown simply because we have not yet offered them the right conditions for *in vitro* growth.

In the past, if we failed to culture and isolate a new organism (for example, one seen under the microscope) then only limited characterization would have been possible. Today, the methods of molecular biology enable us to both detect and characterize organisms which have not been cultured – or even seen! One approach depends on the fact that the 16S rRNA gene contains certain 'highly conserved' sequences of nucleotides – which occur in all members of the domain Bacteria – as well as more variable regions which differ between e.g. genera and species. Primers complementary to the conserved sequences (*broad-range primers*) can be used in PCR (section 8.5.4) to amplify *any* accessible bacterial 16S rRNA genes (e.g. in clinical or environmental samples). The PCR products can then be sequenced (section 8.5.6) and the *variable* regions of the sequence compared with the 16S rRNA database of thousands of known species from all the major bacterial groups; the amplified gene may thus indicate the presence of an organism phylogenetically related to a known species or group.

Even samples containing a *mixed* bacterial population can be examined in this way. In such cases, PCR produces mixed products (i.e. it amplifies various 16S rRNA genes), and it is necessary to clone individual products prior to sequencing and analysis. One problem in analysing a diverse community of organisms by PCR is that the 16S rRNA gene in different organisms may amplify with different degrees of efficiency – such biased PCR amplification apparently being due to interference from DNA flanking the amplified region [FEMSME (1998) *26* 141–149].

In the analysis of microbial communities by various molecular methods (including PCR), quantitatively biased results have been attributed to variation in copy number of the *rrn* operon (which includes the 16S rRNA gene) among different organisms [BioTechniques (2003) *34* 790–802].

10.8.1 'Viable but non-cultivable' bacteria

After stress (e.g. starvation) some normally cultivable bacteria fail to grow on laboratory media but still exhibit features of living cells – e.g. metabolic activity and maintenance of structure; such cells are designated *viable but non-cultivable* (VBNC, VNC). Certain VNC pathogens – e.g. *Vibrio cholerae* [Science (1996) *274* 2025–2031] and ETEC (Table 11.2) [AEM (1996) *62* 4621–4626] retain virulence.

Some have viewed the VNC state as a normal response to stress by non-spore-forming bacteria, the cells adopting a kind of dormancy [FEMSME (1998) *25* 1–9]. However, others believe that failure to culture such cells may be due, in at least some cases, to inappropriate media/conditions. For example, starved cells transferred to a nutrient-rich aerobic medium may produce toxic/lethal intracellular radicals such as superoxide anion; such radicals (which are common byproducts in aerobic metabolism) are normally detoxified by special enzymes (e.g. superoxide dismutase) – but starved, unadapted cells may lack this capability. Thus, the starved cells may be damaged, or even killed, by self-inflicted oxidative damage [Microbiology (1998) *144* 1–3].

Studies on the biochemical composition of the cell wall in VNC *Enterococcus faecalis* revealed changes in peptidoglycan and lipoteichoic acid and in the composition of the penicillin-binding proteins [AEM (2000) *66* 1953–1959]. VNC enterococci, which retain pathogenic potential, have been found to maintain resistance to vanomycin [AAC (2003) *47* 1154–1156].

11 Bacteria in medicine

Some relevant topics in other chapters:

Chapter 7   Bacterial targets for antibiotics (protein/DNA synthesis)
Chapter 8   PCR and other techniques used in diagnostic/typing procedures
Chapter 9   Phage therapy (Table 9.2)
Chapter 10 Quorum sensing (expression of virulence factors)
Chapter 12 Food poisoning, food hygiene
Chapter 14 Aseptic technique; basic practical bacteriology
Chapter 15 Sterilization, disinfection; antibiotics: mode of action; resistance to anti-
              biotics; antibiotic-sensitivity tests
Chapter 16 Identification of pathogens (traditional/DNA-based)

11.1   BACTERIA AS PATHOGENS

Some diseases are due to 'errors' in the body's chemistry, but in many diseases the symptoms result from the activities of certain microorganisms (and/or their products) on or within the body; a *pathogen* is a microorganism that causes disease. Among the many diseases of microbial origin, some are due to fungi, some to viruses, some to protozoa, and some to bacteria; a number of the latter diseases are described briefly in section 11.11.

In some cases the causal agent of a disease is highly specific: such a disease can be caused only by the appropriate species – or only by particular strains of that species. For example, anthrax is caused only by certain strains of *Bacillus anthracis*; these particular strains of the pathogen contain *plasmids* (section 7.1.2) that encode (i) anthrax toxin, and (ii) a capsule which protects the pathogen. In other cases a disease may be due to any of several (or many) different causal agents; an example is gas gangrene: a disease which can be caused by one or more of several species of *Clostridium.*

Sometimes a disease is caused by an organism which does not usually behave as a pathogen and which may actually be a member of the body's own microflora (Table 11.1). For example, species of *Bacteroides*, which are common in the intestine, can sometimes give rise to peritonitis following accidental or surgical trauma in the lower intestinal tract; in such cases the organisms are referred to as *opportunist pathogens*.

Disease does not *necessarily* follow exposure to a given causal agent. In fact, the occurrence (or otherwise) of disease typically depends on various factors – including the degree of resistance of the host and the *virulence* (capacity to cause disease) of the pathogen. This interplay of factors is illustrated by the way in which organisms that are normally weak pathogens (with low-level virulence) can cause severe disease in immunocompromised patients (such as those with AIDS). In some cases, susceptibility to a given disease, or to a severe form of the disease, has a genetic basis.

Some of the factors which can predispose to the development of disease may not be obvious. For example, infection of wounds during surgery may be promoted by the mild hypothermia induced by anaesthetics [NEJM (1996) *334* 1209–1215; Lancet (2001) *358* 876–880].

Hospitals – containing concentrations of sick people – may themselves act as sources of infection. A disease acquired in hospital is called a *nosocomial disease*. Some patients may be particularly vulnerable; infection rates of about 30% have been reported for patients in adult intensive-care units [review: Lancet (2003) *361* 2068–2077].

11.2 THE ROUTES OF INFECTION

The skin is normally an effective barrier to pathogens, but skin may be broken by wounding, surgery or the bites of insects. Wounds may admit any of a variety of potential pathogens capable of causing *systemic* disease (i.e. disease affecting the whole body) or localized disease. Bacterial pathogens which can enter the body via insect bites include the causal agents of bubonic plague and Lyme disease.

Mucous membranes of the intestinal, respiratory and genitourinary tracts tend to be more vulnerable than skin, and infections commonly begin at these sites. In some diseases (e.g. cholera) the pathogen remains external to the mucous membrane, causing disease by secreting one or more toxins; in such cases, *adhesion* of the pathogen to the membrane is likely to be an important factor (section 11.2.1). In other diseases (e.g. dysentery, typhoid) the pathogen *invades* the mucosa (section 11.2.2) and may either remain at a local site or disseminate in the body.

The intestinal tract is commonly infected via the *faecal–oral route*, i.e. by ingestion of faecally contaminated (e.g. sewage-contaminated) food or water. For any given pathogen, an important factor is the *number* of cells ingested; for EHEC (Table 11.2), the smallest number of cells able to cause disease (the *minimum infective dose*) is apparently <100.

Table 11.1   Human microflora: some of the bacteria commonly associated with the adult body

Location	Species of
Colon[1,2]	Bacteroides Clostridium Fusobacterium Peptostreptococcus
Duodenum[1]	Bacteroides Lactobacillus
Ear	Corynebacterium Mycobacterium Staphylococcus
Eye (conjunctiva)	Corynebacterium Propionibacterium Staphylococcus (coagulase-negative)
Ileum	Bacteroides Enterococcus Escherichia Lactobacillus Proteus
Mouth	Actinomyces Bacteroides Streptococcus
Nasal passages	Corynebacterium Staphylococcus
Nasopharynx	Haemophilus (e.g. H. influenzae) Streptococcus
Skin	Propionibacterium Staphyiococcus Others (according to e.g. hygiene/environment)
Stomach	Helicobacter pylori[3]
Urethra	Acinetobacter Escherichia Staphylococcus
Vagina (pre-menopausal)	Acinetobacter Corynebacterium Lactobacillus Staphylococcus

[1] The total number of bacteria (in colony-forming units/gram) increases from about 10^3 in the duodenum to about 10^{11} in the colon.
[2] Organisms resembling *Ruminococcus obeum* have been found in faecal samples [AEM (2002) *68* 4225–4232].
[3] Believed to be present in a significant part of the human population.

Having reached the gut lumen, a would-be pathogen must breach several barriers in order to initiate infection. Thus, intestinal cells secrete various antimicrobial peptides (e.g. *defensins*) which inhibit or kill invading organisms [antimicrobial peptides: Nature (2002) *415* 389–395; gut defence: Nature (2003) *422* 478–479].

A further barrier is the gut microflora: a complex population of microorganisms, including many species of bacteria, whose very presence denies space to a potential pathogen. The gut bacteria contribute significantly to the host's wellbeing; for example, (i) bacterial metabolism provides the host with short-chain fatty acids – in particular butyric acid, an important nutrient for epithelial cells in the gut; (ii) interaction between bacteria and the gut immune system promotes the maintenance of a normal immune response; (iii) the resident bacteria inhibit invading species e.g. by producing bacteriocins (section 10.1) and by competing for both attachment sites and nutrients. [Gut flora in health and disease: Lancet (2003) *360* 512–519.]

Translocation refers to the passage of viable bacteria across the epithelial barrier from the gut lumen. The translocation of resident gut bacteria may occur normally in otherwise healthy individuals, but under certain conditions – e.g. (i) bacterial overgrowth in the small intestine, (ii) increased permeability of the intestinal mucosa, and/or (iii) inadequate immune defence – these bacteria may give rise to extraintestinal disease, including sepsis, shock and death. [Translocation (review): BPRCG (2003) *17* 397–425.]

Because translocation is observed routinely, and because the experimental introduction of pathogens through the mucosal barrier of the gastrointestinal tract does not always lead to disease, the importance of translocation in pathogenesis has been disputed [Critical Care Medicine (2003) *31* 598–607].

The urinary tract may be infected upwards from the urethra or downwards from the kidneys, the former route being more common; infections of the urinary tract are generally more frequent in women than in men. A potential pathogen must have effective adhesin(s) in order to cope with the flushing action of the urine flow.

Many urinary tract infections (UTIs) are caused by UPEC (section 11.3.7). Nosocomial infections associated with catheterization often involve endogenous faecal flora (e.g. species of *Escherichia*, *Klebsiella* or *Proteus*). Infections that arise through sexual contact include gonorrhoea *(Neisseria gonorrhoeae)* and non-specific urethritis (frequently *Chlamydia trachomatis*). In some cases even a simple laboratory examination of urine sample(s) can indicate the likelihood of a UTI (section 11.8.1).

The respiratory tract is commonly infected by droplets/aerosols (see *droplet infection* in section 11.7). This type of infection occurs e.g. in diphtheria, legionnaires' disease and meningitis (section 11.11).

Iatrogenic infection. Exposure to infection during surgery is generally minimized by adherence to an aseptic technique. Further preventive measures during orthopaedic work may include e.g. incorporation of antibiotics into medical device polymers

[RMM (1996) 7 195–205] and the use of antibiotic-impregnated bone cement [AAC (1996) 40 2675–2679].

Infection may arise during catheterization of the blood vascular system or of the urinary tract; in general, the risk of infection rises with the duration of catheterization. In short-term (<3 weeks) urinary catheterization, a significant reduction in the incidence of infection can be achieved with catheters having anti-infection surface activity – e.g. nitrofurazone, minocycline + rifampicin, or a silver alloy–hydrogel [EID (2001) 7 342–347].

Infection transmitted during infusion of blood, or blood products, is particularly important in the context of certain viruses (e.g. HIV, hepatitis B and C, TT virus) but certain bacteria (e.g. *Bartonella bacilliformis*, *Treponema pallidum*) are potentially transmissible by this means.

11.2.1   Adhesion as a factor in infection

In many diseases there is an early phase in which the pathogen adheres to particular sites in the host. The need for attachment becomes clear when we consider, for example, that the common sites of infection, the mucous membranes, are continually flushed by their own secretions and may be subject to movements such as peristalsis – factors which tend to discourage the establishment of a pathogen. Adhesion may help a pathogen to compete more effectively with the host's own microflora.

Adhesion is typically mediated by specific structures or molecules at the surface of the pathogen: so-called *adhesins* (section 11.5.6). Adhesins are often fimbriae (section 2.2.14.2). Fimbria-mediated adhesion is important e.g. for virulence in *enterotoxigenic* strains of *E. coli* (ETEC: Table 11.2) which bind to intestinal mucosa. [Expression of fimbriae by enteric pathogens (review): TIM (1998) 6 282–287.] Fimbrial adhesion is also important for the initial binding of *Haemophilus influenzae* type b to the respiratory tract epithelium; once within the bloodstream this pathogen can become non-fimbriate, apparently by a random, reversible genetic switch. [Adhesins in *Haemophilus*, *Actinobacillus* and *Pasteurella* (review): FEMSMR (1998) 22 45–59.]

Non-fimbrial adhesins include streptococcal M proteins (section 2.2.13), the filamentous haemagglutinin (FHA) of *Bordetella pertussis*, and the high-molecular-weight adhesion proteins (HMW1, HMW2) of 'non-typable' strains of *Haemophilus influenzae*.

Mammalian *receptors* for adhesins include various cell-surface molecules, but binding is typically specific; for example, type 1 fimbriae of *E. coli* (Table 2.2) bind to mannose residues, while P fimbriae of uropathogenic *E. coli* bind to the α-D-galacto-pyranosyl-(1–4)-β-D-galactopyranoside receptors of glycolipids on urinary tract epithelium. Some pathogens bind to (specific) *integrins*. The integrins are a family of mammalian cell-surface glycoproteins whose normal roles include the binding of host cells to matrix components such as collagen and fibronectin; integrins occur e.g. on epithelial and endothelial cells. (Note that, in keeping with their binding role, integrins of e.g. intestinal epithelial cells occur on *basolateral* surfaces rather than on apical, i.e. lumen-facing, surfaces.) Pathogens which bind to (specific) epithelial integrins include *Bordetella pertussis*, *Borrelia burgdorferi* and *Yersinia enterocolitica*. In *Listeria monocytogenes* the *internalin* adhesin binds to E-cadherin, a member of another family of cell-surface glycoproteins (the *cadherins*) which, like the integrins, normally function in host cell adhesion.

Adhesion can involve more than simple host–pathogen binding. For example, binding of the FHA adhesin of *B. pertussis* to a monocyte integrin causes signals (within the monocyte) that upregulate the activity of a *second* type of integrin – one which binds to a different site on FHA. Thus, *B. pertussis* uses the host's internal signalling system to induce additional binding sites.

Uropathogenic strains of *E. coli* (UPEC) bind to *uroplakins*: proteins which form the innermost layer of the mammalian urinary bladder. UPEC binds via type I fimbriae (Table 2.2); such binding promotes the *uptake* (internalization) of UPEC by bladder cells.

Complex events occur when EPEC (Table 11.2) binds to the intestinal epithelium. Initial attachment may involve type IV fimbriae (also called bundle-forming pili, BFP) [Microbiology (2004) *150* 527–538]. Contact with the target (eukaryotic) cell triggers the expression (in EPEC) of the LEE pathogenicity island (section 11.5.7) and formation of a type III secretion system (section 5.4.4) that includes proteins EspA and EspB. (EspA may play a role in the attachment of EPEC [Microbiology (2004) *150* 527–538].) Using the type III system, EPEC secretes a protein into the epithelial cell; this protein is phosphorylated and then inserted into the membrane of the epithelial cell as a receptor for the *intimin* adhesin of EPEC. The (modified bacterial) protein, called Tir (translocated intimin receptor), was originally thought to be a host cell protein (Hp90).

Following the binding of EPEC to Tir (via intimin), the epithelial cell suffers effacement (loss) of microvilli; moreover, rearrangement of its actin, below the site of effacement, forms a local pedestal, protruding into the lumen, to which the bacterium is attached (Plate 11.1).

11.2.1.1 *Adhesion in tooth decay*

Tooth decay (*dental caries*) is promoted by those bacteria which adhere to tooth surfaces and gum margins and which contribute to *dental plaque*: a film composed mainly of bacteria, bacterial products, and salivary substances. *Streptococcus mutans*, a common component of plaque, forms extracellular water-insoluble glucans which assist bacterial adhesion; waste products of bacterial metabolism (e.g. lactic acid) cause localized demineralization in the teeth, permitting bacterial penetration.

Adhesion appears to be important also in the further decay of *filled* teeth, microbial colonization occurring in the small gap between the filling material and the wall of the

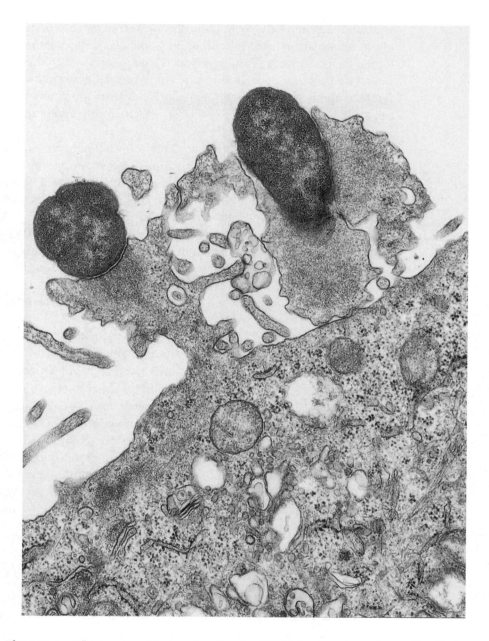

Plate 11.1   Adhesion: two (electron-dense) cells of EPEC (enteropathogenic *E. coli*) adhering via pedestals to the membrane of a host cell (section 11.2.1). Photograph courtesy of Dr Michael S. Donnenberg, University of Maryland School of Medicine, Baltimore, USA.

cavity (Plate 2.1, *centre*); from this site, microbial penetration of the dentinal tubules can bring about destruction of the dentine.

11.2.1.2 Biofilms

Some pathogens resist antibiotics (and the immune system) because they form *biofilms*: populations of cells embedded in a secreted matrix material. Biofilm development appears to be regulated by quorum sensing (section 10.1.2) in e.g. *Burkholderia cepacia* [Microbiology (2001) *147* 2517–2528] and *Staphylococcus aureus* [JB (2004) *186* 1838–1850].

11.2.2 Bacterial invasion of mammalian cells

In some diseases the pathogen normally remains on the surface of host cells; thus, pathogens *adhere* to their target cells e.g. in cholera (*Vibrio cholerae*), gonorrhoea (*Neisseria gonorrhoeae*) and whooping cough (*Bordetella pertussis*). By contrast, some pathogens *invade* host cells – invasion being a process in which bacteria induce their own uptake (internalization) by the host cells; these pathogens include e.g. species of *Brucella*, enteroinvasive *E. coli* (EIEC), *Legionella pneumophila*, *Listeria monocytogenes*, *Mycobacterium tuberculosis*, *Salmonella typhi*, *Shigella flexneri* and *Yersinia* spp.

Invasion appears to require prior adhesion to the host cell (section 11.2.1).

The mechanism of invasion depends on the pathogen and type of host cell. However, uptake generally requires a positive contribution from the *host* cell. This usually includes e.g. re-arrangement of *actin* microfilaments in the cytoplasm below the site of bacterial attachment; because of this, uptake is typically inhibited, *in vitro*, by agents such as *cytochalasins* – which inhibit re-arrangement of actin molecules in mammalian cells.

Certain strains of group A streptococci can invade (and survive within) cells of the human respiratory tract; they are thus protected from those antibiotics (e.g. β-lactams) which are generally excluded from eukaryotic cells. Moreover, these invasive strains are associated with resistance to erythromycin, an antibiotic which can penetrate eukaryotic cells [Lancet (2001) *358* 30–33].

11.2.2.1 Invasion of intestinal epithelium

There are different modes of invasion. Several are outlined below.

Salmonella. S. typhi, and other invasive salmonellae (e.g. *S. typhimurium*), invade via the apical (lumen-facing) surfaces of epithelial cells. Initial attachment of the pathogen may be mediated by type 1 fimbriae. Attachment seems to induce the formation of *invasomes*: bacterial appendages that resemble very short, thick flagella and which are believed to play an important role in invasion.

Uptake of the bacterium is preceded by *ruffling* in the host cell membrane; extrusions of host membrane, supported internally by actin microfilaments, quickly surround and engulf the bacterium – which is thus internalized in a large ('spacious') vacuole (= *phagosome*) that also encloses extracellular fluid. This so-called 'trigger' mechanism of uptake is called *macropinocytosis*. Prior to uptake, adhesion of the bacterium is associated with upregulation of calcium ions in the host cell; the significance of this is that calcium ions are needed for mobilization and deployment of stored actin during macropinocytosis. The host's GTPase (Cdc42), which regulates actin re-arrangements, seems to be necessary for the uptake of *Salmonella*. Invasive salmonellae can multiply within the phagosome. (See also section 11.3.3.1.)

The *Salmonella* type III secretory system (section 5.4.4), encoded by the *inv–spa* complex of genes in a chromosomal pathogenicity island (section 11.5.7), transports effector proteins that induce membrane ruffling and subsequent uptake of the bacterium.

Shigella. As in *Salmonella*, invasion by *Shigella* involves the trigger mechanism but, in contrast, *Shigella* enters via the *basolateral* surface of epithelial cells. The adhesin(s) are unknown. The receptor(s) probably include integrins. Invasion requires a type III secretory system (section 5.4.4), encoded by the plasmid-borne *mxi–spa* genes, that injects various proteins. One of these, IpaA, modulates the host cell's actin. Another, VirA, promotes membrane ruffling by destabilizing microtubules [EMBO Journal (2002) *21* 2923–2935]. Host cell GTPases (e.g. Rho) are also involved.

Shigella escapes from the phagosome, multiplies rapidly, and spreads from cell to cell within the epithelium (a feature of dysentery: see section 11.3.3.2).

Following invasion by *Shigella*, lipopolysaccharides (section 2.2.9.2) activate the host cell's nuclear transcription factor NFκB; this results in the expression of various genes, including the gene for interleukin-8 (IL-8) – a potent chemotactic agent for neutrophils.

In animal studies, *Shigella* is taken up initially by the (phagocytic) *M cells* of Peyer's patches, and a similar route may be taken in human dysentery. [M cells in infection (review): TIM (1998) *6* 359–365.] *Shigella* may pass through the M cells and then (i) be engulfed by macrophages (which occur beneath the M cells) and/or (ii) invade adjacent epithelial cells via their basolateral surfaces. *Shigella* can kill macrophages (section 11.5.2); this may enhance spreading within the epithelium but may subsequently contain the infection (section 11.3.3.2).

[Rupture, invasion and inflammatory destruction of the intestinal barrier by *Shigella*: FEMSMR (2001) *25* 3–14.]

Listeria monocytogenes. Invasion is promoted by the bacterial cell-surface adhesin *internalin A* (product of gene *inlA*). The receptor is *E-cadherin*, one of a family of glycoproteins (*cadherins*) normally involved in cell–cell adhesion; the intracytoplasmic part of the E-cadherin molecule interacts with the cell's actin cytoskeleton via proteins called *catenins*. E-cadherin molecules occur on the basolateral surfaces of epithelial cells.

E-cadherin–internalin A interaction *per se* seems to initiate uptake of the bacterium. (Interestingly, even *latex beads* coated with internalin A are internalized by some types of mammalian cell [INFIM (1997) *65* 5309–5319].)

Invasion by *L monocytogenes* (Plate 11.2(a)) resembles the 'zipper' uptake of *Yersinia enterocolitica* (described below). *L monocytogenes* escapes from the phagosome by secreting *listeriolysin O* – a toxin of the family of *thiol-activated cytolysins* [RMM (1996) *7* 221–229] which lyse the cholesterol-containing membranes of eukaryotic cells.

Within the host's cytoplasm, *L. monocytogenes* grows and spreads to adjacent cells by the actin-based motility described for *Shigella* (section 11.3.3.2) – the ActA protein playing the same role as IcsA in *Shigella*. When propelled into an adjacent cell, the bacterium becomes enclosed within a sac bounded by a double membrane (one membrane from each cell); escape from this sac involves secretion of the enzyme *lecithinase* (product of the *plcB* gene) which hydrolyses the lecithin (phosphatidyl-choline) component of the double membrane.

[Invasion and actin-based motility: EMBO Journal (1998) *17* 3797–3806.]

Yersinia enterocolitica. In mice, this pathogen is initially taken up by the M cells of Peyer's patches. Studies with cell cultures have suggested that the pathogen may then invade adjacent epithelial cells via their *basolateral* surfaces. *In vitro*, certain adhesins of the pathogen *(invasin* and the YadA protein) bind to host cell *integrins* (section 11.2.1); this recruits further integrins to the site – leading to the formation of multiple points of attachment between pathogen and host (so-called *zippering*), activation of host cell tyrosine kinase, and uptake of the bacterium within a close-fitting vacuole. *In vivo*, invasion in this way would occur only at the basolateral surfaces of intestinal epithelial cells because integrin receptors are found only on these surfaces.

Y. enterocolitica (as well as *Y. pestis* and *Y. pseudotuberculosis*) contains a 70-kb virulence plasmid that encodes a type III secretion system and various effector molecules (Yops), some of which are secreted into the target cell. [The Yop virulon: PNAS (2000) *97* 8778–8783.]

Following invasion, *Y. enterocolitica* apparently survives within the vacuole.

This pathogen can resist *phagocytosis* by the mechanism outlined in section 11.5.1.

11.2.2.2   *Invasion of macrophages*

Some bacteria can *invade* macrophages and may then even multiply and escape (Plate 11.2(c)). Compared with phagocytosis (section 11.4.1), invasion may involve e.g. different adhesins and receptors. The *initial* contact between pathogen and macrophage may determine the pathogen's fate; for example, of the various modes of binding, some appear to favour bacterial survival by failing to promote phagosome–lysosome fusion and/or the oxidative burst following uptake of the bacterium. After uptake:

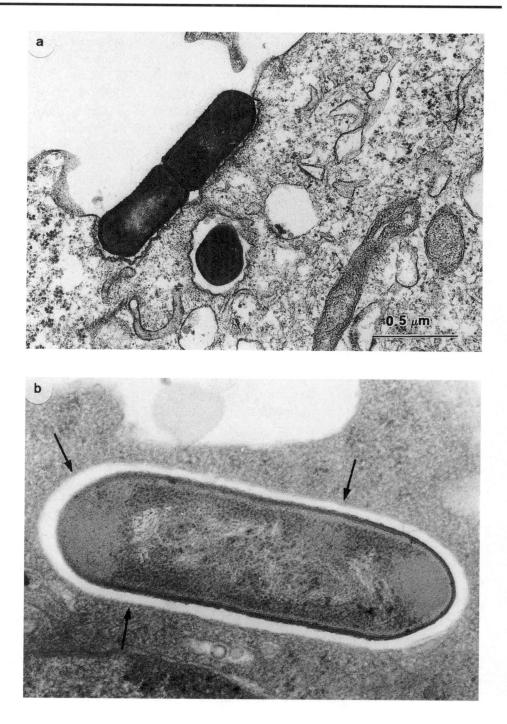

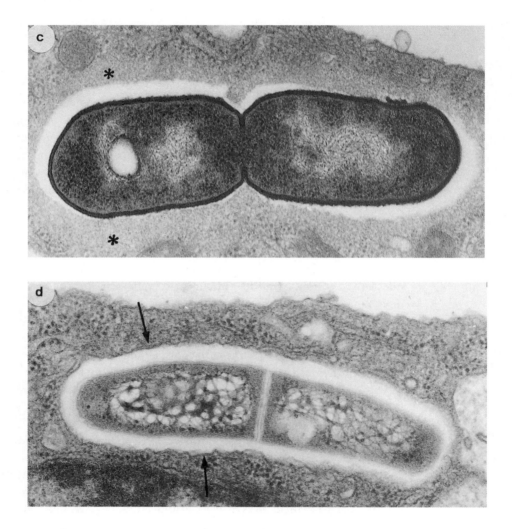

Plate 11.2 Invasive pathogens. (a) Two (electron-dense) cells of *Listeria monocytogenes* being taken up by the 'zipper' mechanism (section 11.2.2.1). (b) *L. monocytogenes* within a phagosome in a mouse macrophage (arrows indicate the phagosome's membrane). Scale: 68 mm = 1 μm. (c) *L. monocytogenes* within a mouse macrophage, escaping from the phagosome and dividing within the cytoplasm, surrounded by a network of actin filaments (asterisks). Scale: 50 mm = 1 μm. (d) *Mycobacterium avium* dividing within a phagosome in a mouse macrophage; arrows indicate the phagosome's membrane. Scale: 60 mm = 1 μm.

Photograph (a) reproduced from EMBO Journal (1998) *17* 3797–3806 with copyright permission from Nature Publishing Group and by courtesy of the authors, Drs Pascale Cossart and Marc Lecuit, Unité des Interactions Bactéries–Cellules, Institut Pasteur, Paris, France.

Photographs (b)–(d) courtesy of Dr Chantal de Chastellier, Faculté de Médecine Necker-Enfants Malades, INSERM U 411, Paris, France.

- phagosome–lysosome fusion can be absent in *Chlamydia* spp, *Legionella pneumophila* and *Mycobacterium tuberculosis*

- *Coxiella burnetii* and *Francisella tularensis* grow and multiply in the phagolysosome

Legionella pneumophila. Phagosome–lysosome fusion generally does not occur, and the vacuole is not acidified. The pathogen multiplies, killing the macrophage. The pathogen's *icm* (intracellular multiplication) genes, also called *dot* (defect in organelle trafficking) genes, are believed to encode a secretory system and effector molecules involved e.g. in blocking phagosome–lysosome fusion; mutations in these genes have been found to affect the pathogen's ability to multiply and kill the macrophage [e.g. TIM (1998) 6 253–255]. The oxidative burst in a macrophage may be inhibited by the zinc metalloprotease of *L. pneumophila* [JMM (2001) 50 517–525].

Mycobacterium tuberculosis. The outcome of contact with a macrophage may depend on e.g. (i) the cell-surface chemistry of the given strain of *M. tuberculosis*; (ii) the site of infection (e.g. lung); and (iii) the timing of pathogen–macrophage contact (at initial infection or during subsequent dissemination in the body).

The low concentration of *complement* (section 11.4.1.1) in normal bronchoalveolar fluid increases the importance of complement-independent binding/uptake of pathogens by macrophages. For some pathogens, uptake by a macrophage is induced by the binding of their cell-surface polysaccharides to the *lectin* site on the macrophage CR3 receptor. *M. tuberculosis* can bind to this site (via β-glucans), and this may allow uptake before any inflammatory increase occurs in the levels of complement; uptake in this way may favour bacterial survival (and the development of disease) during the *initial* infection of normal lung. [*M. tuberculosis*–phagocyte interactions: TIM (1998) 6 328–335.]

Species of *Mycobacterium* (including *M. tuberculosis*) can inhibit acidification of the phagosome; for example, *M. avium* seems to prevent the insertion/activity of the proton-ATPase which gives rise to acidification. A dividing cell of *M. avium* in the phagosome of a mouse macrophage is shown in Plate 11.2(d).

11.2.2.3　Paracytosis

Paracytosis is a form of invasion in which a pathogen crosses a layer of cells by passing *between* the cells. For example, *Haemophilus influenzae* can pass between the cells of lung epithelium; paracytosis in this pathogen has been studied in tissue cultures of lung epithelial cells [INFIM (2000) 68 4616–4623].

11.2.2.4　Transcytosis

Transcytosis is a process in which certain Opa[+] strains of *Neisseria* pass through a layer of epithelial cells *without disrupting mammalian cell–cell junctions*; this process involves initial engulfment of the bacteria following their binding to specific receptors on the host cells [INFIM (2000) *68* 896–905].

11.2.3   Latency

Usually, a pathogen will either cause disease or be killed by the body's defences. In *latent* infection with *Mycobacterium tuberculosis*, the pathogen remains viable but is constrained to quiescence by the immune system (section 11.4); an individual may be serologically positive for the pathogen but does not transmit the disease.

Latency occurs e.g. in ~ 60% of those infected with *Mycobacterium tuberculosis*, ~ 40% proceeding to active tuberculosis (TB); latent TB carries a lifetime risk of active disease (the risk being increased e.g. by HIV infection or the use of immunosuppressive drugs). About one in three humans is believed to be latently infected with *M. tuberculosis*.

The mechanism of latency in TB is unknown. After uptake by macrophages the pathogen can survive (section 11.2.2.2) and spread around the body. The site(s) of latent *M. tuberculosis* are not known; it is not always possible to detect AFBs (section 14.9.2) in tissues, and it is not clear whether the pathogen survives in a non-acid-fast/spore-like form or in small numbers that are not easily detectable. The possibility of a spore-like form has been suggested by the discovery of a gene (*sigF*) for a sigma factor (section 7.5) which resembles the sporulation-specific and stress response proteins of some other bacteria.

[Latency in TB: TIM (1998) *6* 107–112.]

Detection of latency has been aided by the ELISPOT assay. This assay detects those T cells (in a blood sample) which respond to the tuberculosis-associated antigen ESAT-6; ESAT-6 is a 6 kDa protein secreted by members of the 'Mycobacterium tuberculosis complex' (which includes *M. tuberculosis*, *M. bovis* and *M. africanum*). Any T cells (in the blood sample) which react to (added) ESAT-6 are detected by their secretion of the cytokine IFN-γ (section 11.4.1.2). IFN-γ is present in high concentration in the immediate vicinity of a secreting T cell. The IFN-γ is detected by adding a conjugate consisting of an enzyme linked to a molecule that binds to IFN-γ; enzymic activity (on an appropriate colour-generating substrate) results in the formation of a visible spot on the floor of the reaction chamber in the location of the secreting T cell. Hence, each spot indicates an ESAT-6-specific T cell, allowing quantification.

[Use of ELISPOT for contact tracing in the context of tuberculosis: Lancet (2001) *357* 2017–2021. Comparison of ELISPOT with the tuberculin skin test: Lancet (2003) *361* 1168–1173.]

11.3   PATHOGENESIS: THE PROCESS, OR MECHANISM, BY WHICH DISEASE DEVELOPS

How does a pathogen cause disease? There are various ways. For example, many pathogens form toxins that disrupt specific physiological activities or which disorganize intercellular signalling – leading to localized or systemic disease. Some pathogens invade and destroy particular cells or tissues (and may also form toxins). In some diseases the body's own defence mechanisms (i.e. the immune system) may contribute to the symptoms.

Molecular studies have shown that pathogenesis can involve complex interactions between pathogen and host; some idea of this interplay is given in section 11.6.

For some clinically defined diseases there are no known causal agent(s). One example is Crohn's disease: a granulomatous/ulcerative condition of the human intestine in which symptoms include malabsorption, diarrhoea and the formation of fistulae. The possibility of mycobacterial aetiology has been raised a number of times [e.g. JMM (2002) *51* 3–6], but recent work suggests the involvement of mutations in a human gene [JBC (2003) *278* 5509–5512; JBC (2003) *278* 8869–8872].

Some infectious diseases may include a systemic (generalized) response known as *septicaemia* or *sepsis*. An attempt has been made to give a precise meaning to the term 'sepsis' so that it can be usefully employed as a descriptor in clinical trials of various treatments for sepsis; for this purpose, sepsis has been defined in terms of the *systemic inflammatory response syndrome* (SIRS). SIRS involves at least two of the following:

Temperature	$>38°C$ or $<36°C$
Heart rate	>90 beats per minute
Respiration	>20 breaths per minute, or $PaCO_2$ <4.3 kPa
White blood cells	$>12000/mm^3$, $<4000/mm^3$ or $>10\%$ immature forms

SIRS can arise from various causes, including e.g. ischaemia, tissue necrosis or trauma, as well as from infection; when SIRS is a response to *infection* it is an indication of *sepsis*. [Sepsis/SIRS: JAC (1998) *41* (supplement A) 1–112. Treatment of sepsis: Drugs (1999) *57* 127–132.]

Examples of pathogenesis are given in the following subsections and in Table 11.2.

11.3.1   Toxin-mediated pathogenesis

In some diseases, most or all of the symptoms can be accounted for by the effects of a given *exotoxin*: a specific protein toxin, released by the pathogen, which affects particular site(s) within the body; such diseases include cholera, botulism, tetanus and diphtheria.

However, in some diseases a toxin may be only one of several factors needed for pathogenesis. For example, a toxin may need to modify the host's synthesis (or release) of *cytokines* (Table 11.4) as a prerequisite for the disease process; accordingly, it has been suggested that, in such diseases, cytokines may play an essential role – either as (i) the main effector molecules, or (ii) factors necessary for the activity of the toxin [TIM (1997) *5* 454–458] (see e.g. section 11.3.1.6).

11.3.1.1   Cholera

In cholera (section 11.11), the pathogen (certain strains of *Vibrio cholerae*) multiplies in the gut and forms an *enterotoxin* (cholera toxin, CT): an exotoxin which acts on mucosal cells in the small intestine, stimulating the enzyme adenylate cyclase; the resulting increased intracellular levels of cyclic AMP (cAMP, Fig. 7.12) stimulate secretion of electrolyte into the gut lumen. There is an associated massive outflow of water into the gut lumen which gives rise to the characteristic 'rice-water stools' of cholera.

Infection by strains of *V. cholerae* that do not synthesize CT may give rise to (less severe) disease as a result of other toxins (the Zot and Ace toxins) encoded by the pathogen.

Following ingestion of *V. cholerae*, colonization of the small intestine and development of disease depend on two main *virulence factors* (section 11.5): (i) appendages that mediate adhesion of *V. cholerae* to the mucosal surface – the so-called 'toxin co-regulated pili' (TCP; = type IV fimbriae: Table 2.2) – and (ii) CT.

The genes for CT and the Zot and Ace toxins occur in the genome of a filamentous phage, CTXΦ. Genes for TCP occur in a chromosomal *pathogenicity island* (section 11.5.7). The genes for CT and TCP are co-ordinately (jointly) regulated by transcriptional regulator proteins (ToxR, ToxS and ToxT) which apparently promote the expression of CT and TCP on receipt of environmental signals within the intestinal tract.

The ability of signals in the host's intestinal environment to induce synthesis of TCP has lead to insight into the transmissibility of virulence factors under *in vivo* conditions (see section 11.5.8).

O1 and O139 strains of *V. cholerae* are referred to in section 11.11.

(See also anti-cholera vaccines in section 11.4.2.1.)

11.3.1.2   Botulism

Botulism (section 11.11) involves muscle paralysis; death may result e.g. from mechanical (muscular) failure of the respiratory system. The pathogen (usually a strain of *Clostridium botulinum* – see section 11.11) forms a *neurotoxin*: an exotoxin which acts at nerve–muscle junctions, inhibiting the release of acetylcholine and (hence) inhibiting nervous stimulation of muscle.

Disease can result from the action of pre-formed toxin, usually in toxin-

Table 11.2   Pathogenesis of some diseases caused by *Escherichia coli*[1,2]

Group	Disease (and virulence factors)
EIEC[3]	Usually food-borne. EIEC adheres to, invades and destroys epithelial cells in the ileum/colon, causing dysentery; it can also cause watery diarrhoea. The dysentery resembles that caused by *Shigella flexneri*, with similar pathogenesis (section 11.3.2). At least some virulence genes occur on plasmid pInv. EIEC encodes a protein functionally analogous to *Shigella flexneri* IcsA (see section 11.3.3.2) and forms at least one plasmid-encoded enterotoxin: ShET2, named after *Shigella flexneri* enterotoxin 2 and encoded by the *sen* gene (formerly *set2* gene). At least some of the EIEC virulence genes are regulated by temperature.
EHEC/VTEC[4]	Usually food-borne; minimum infective dose apparently <100 cells. Symptoms range from mild diarrhoea to severe bloody diarrhoea (i.e. haemorrhagic colitis); haemolytic uraemic syndrome may occur in adults or (particularly) in children. Vascular damage is apparently toxin-mediated. EHEC is not an invasive, intracellular pathogen; *in vitro*, many strains form 'attaching and effacing' lesions similar to those formed by EPEC. (Antibodies against *E. coli* attaching and effacing antigens have been found in samples of human milk [EID (2003) 9 545–551].) EHEC forms one or both of two phage-encoded toxins whose activity closely resembles that of the toxin of *Shigella dysenteriae* type 1 (see section 11.3.1.6). These toxins are called shiga-like toxins I and II (= SLT-I, SLT-II) – or *verocytotoxins* (= VT1, VT2) owing to toxicity for Vero cells (kidney cells of the African green monkey); another name is 'shiga toxins' (Stx1, Stx2), and some use the abbreviation STEC (shiga toxin-producing *E. coli*). (There are variant forms of SLT-II which are designated by lower-case letters: e.g. SLT-IIf (Stx2f) [AEM (2000) 66 1205–1208].) Starved cells of EHEC (strain O157:H7) can lose antigen specificity (i.e. O157 reactivity in serological tests) but retain toxigenicity [AEM (2000) 66 5540–5543]. As well as shiga-like toxins, many strains of EHEC O157:H7 secrete EAST 1: a heat-stable toxin (an analogue of the endocrine hormone *guanylin*) which stimulates fluid outflow from the intestinal mucosa by activating guanylate cyclase. In the EHEC strain O157:H7, genes of the LEE pathogenicity island (section 11.5.7) are reported to be regulated by quorum sensing (section 10.1.2) involving a bacterial autoinducer *and* the hormone adrenalin (epinephrine) [PNAS (2003) 100 8951–8956]. More than 25 serotypes of EHEC can cause disease, but world-wide, O157:H7 is isolated most often in haemorrhagic colitis. [O157:H7 (genome sequence): Nature (2001) 409 529–533, erratum: Nature (2001) 410 240.] [Lab. diagnosis: JCM (2002) 40 2711–2715.]
ETEC[5]	Usually food-borne and/or water-borne. ETEC is a common cause of diarrhoea in children of the developing countries and a regular cause of travellers' diarrhoea. ETEC binds to the epithelium of the small intestine by means of plasmid-encoded fimbriae – e.g. the so-called *colonization factors* CFI (CFAI) and CFII (CFAII); such binding is essential for virulence. (Strains from animals have e.g. fimbrial antigens K88, K99 or 987.) ETEC forms heat-stable toxins (STa, STb/ST-I, ST-II) and heat-labile toxins (LT-I, LT-II).

STa (a peptide of 18/19 amino acids) binds to receptors in the brush border membrane, thus activating guanylate cyclase and raising the levels of intracellular cyclic GMP; it has been thought that this stimulates secretion of chloride and/or inhibits the absorption of NaCl – leading to secretion of fluid into the gut lumen. An alternative view is that the toxin interrupts the luminal acidification mechanism [JAM (2001) 90 7–26]. STb (a peptide of 48 amino acids) is formed mainly, but not solely, by porcine strains of ETEC [Microbiology (1997) 143 1783–1795].

LT-I resembles cholera toxin in structure and activity (section 11.3.1.1) but produces disease which is generally less severe than that due to cholera toxin.

LT-II causes disease e.g. in pigs, rarely in humans.

EPEC[6]

Food-borne and/or water-borne. Diarrhoea, mainly in infants. EPEC is a major cause of infant mortality in developing countries but (unlike e.g. EIEC) it is characteristically non-invasive and apparently does not form major toxins; however, it contains the LEE pathogenicity island (section 11.5.7). Following binding (see section 11.2.1), EPEC forms so-called *attaching and effacing* lesions on the intestinal mucosa; the diarrhoea may result from e.g. (i) reduced absorptive capacity due to loss of microvilli, (ii) stimulation of chloride secretion, and/or (iii) increased permeability of the epithelium, the two latter effects having been attributed to the activity of the (bacterial) EspB protein.

EAggEC[7]

Strains in this group cause persistent diarrhoea in children, especially in developing countries. The pathogen appears to form adhesins and toxins but few details of pathogenesis seem to be available. Strains of EAggEC may be characterized *in vitro* by the 'aggregative' pattern of adherence to (mammalian) HEp-2 cells [see RMM (1995) 6 196–206]. In that paper, strains of HEp-2-adherent *E. coli* were divided into two main groups: diffusely adherent *E. coli* (DAEC) and enteroaggregative *E. coli* (EAggEC); the term 'enteroadherent' *E. coli* (EAEC) was not recommended.

Studies on *E. coli* serotype O126:H27 indicated that only some strains had an enteroaggregative phenotype; some strains were reported to exhibit characteristics of both EAggEC and ETEC [EID (2003) 9 1170–1173].

---

[1] Strains of *E. coli* which cause intestinal disease have been classified into groups; five of the main groups are listed here. A strain within a given group may be referred to by a serological designation based on its O and H antigens (section 16.1.5.1).

[2] Genes responsible for pathogenicity may occur in the chromosome, in a plasmid or in a phage genome; often such genes occur in clusters within so-called *pathogenicity islands* (section 11.5.7).

[3] Enteroinvasive *E. coli.*

[4] Enterohaemorrhagic *E. coli*/verocytotoxic *E. coli.*

[5] Enterotoxigenic *E. coli.*

[6] Enteropathogenic *E. coli.*

[7] Enteroaggregative *E. coli.*

contaminated foods such as cooked meats, sausage and improperly canned vegetables; that is, the pathogen itself need not be ingested.

11.3.1.3   Tetanus

Tetanus ('lockjaw') involves uncontrollable contractions of the skeletal muscles, often leading to death by asphyxia or exhaustion. The disease develops e.g. when deep, anaerobic wounds are contaminated with the pathogen, *Clostridium tetani*. *C. tetani* produces a neurotoxin (*tetanospasmin*) which acts on certain cells (interneurones) in the central nervous system; by inhibiting the release of a neurotransmitter from these cells, tetanospasmin permits the simultaneous contraction of both muscles in a protagonist–antagonist pair, producing spastic (rigid) paralysis.

11.3.1.4   Botulinum toxin and tetanospasmin: similarities

Interestingly, although tetanospasmin and botulinum toxin give rise to very different clinical symptoms, these toxins have important similarities; thus, both toxins bind to nerve cells, both are taken up by the cells (internalized), and both produce their effects by acting as enzymes (zinc-endopeptidases) which cleave specific proteins in the nerve cells, thereby inhibiting the normal release of neurotransmitter substances [mechanism of action of tetanus and botulinum neurotoxins: MM (1994) *13* 1–8]. In that these toxins are now known to be zinc-endopeptidases, it may be possible to find specific chemical inhibitors for therapeutic use in botulism and tetanus; in an analogous case, the mammalian zinc-endopeptidase angiotensin-converting enzyme is inhibited by captopril, a drug sometimes used therapeutically for hypertension (high blood pressure).

11.3.1.5   Staphylococcal food poisoning

Staphylococcal food poisoning (Table 12.2) is due to enterotoxins produced by certain species of *Staphylococcus* (mainly *S. aureus*). The enterotoxins (types A to H) usually cause vomiting, and often diarrhoea, shortly after eating contaminated food. The mode of action of the toxins is unknown. One suggestion is that vomiting may result from stimulation of intestinal ganglion cells – causing release of neuropeptides which, in turn, induce the release of e.g. histamine and leukotrienes from mast cells. As the enterotoxins are superantigens (section 11.5.4.1), pathogenesis may involve the effects of cytokines.

11.3.1.6   Shiga toxin and shiga-like toxins (verocytotoxins)

These enterotoxins, formed by some strains of *Shigella* and by EHEC (Table 11.2), respectively, have similar mechanisms. They bind to sites in the gut and are taken up by receptor-mediated endocytosis. The toxins affect protein synthesis and e.g. inhibit the absorption of NaCl; their role in dysentery (section 11.3.3.2), haemorrhagic colitis and haemolytic uraemic syndrome is not fully understood, but they appear to be responsible for the vascular damage.

The shiga toxin of *Shigella dysenteriae* type 1, and the shiga-like toxins of *E. coli* O157:H7 and other EHEC, have glycolipid receptor sites (designated Gb_3) on vascular endothelial cells (cells lining blood vessels). *In vitro* studies indicate that Gb_3 sites, which are more numerous in *small* blood vessels, can be upregulated (i.e. their numbers further increased) by the cytokines TNF-α and IL-1. In one model of EHEC-induced vascular damage, lesions are initiated in vessel walls after toxin–Gb_3 binding; activated macrophages adhere to the lesions and are stimulated (e.g. by toxin) to secrete IL-1 and TNF-α – which cause local upregulation of Gb_3 sites and further (toxin-mediated) damage [TIM (1998) *6* 228–232].

The shiga-like toxins of *E. coli* O157:H7 can be neutralized by binding to a water-soluble, multivalent carbohydrate ligand called STARFISH – a (starfish-shaped) molecule structurally related to Gb3. The possibility of neutralizing toxin is the basis of a potential anti-EHEC therapeutic agent, Synsorb P^k, which consists of synthetic Gb3 analogues linked to insoluble particles of silica. Synsorb P^k may be able to sequester toxin in the gut – thus preventing uptake and (hence) denying the toxin access to its endothelial binding sites. [STARFISH: Nature (2000) *403* 669–673.]

11.3.1.7 *Diphtheria*

Diphtheria, caused by toxigenic strains of *Corynebacterium diphtheriae*, can involve both local and systemic effects of diphtheria toxin (DT) (see section 11.11).

DT is a heat-labile exotoxin (MWt ~60000) encoded by the *tox* gene in certain phages (Chapter 9) that infect *C. diphtheriae*. In humans, DT binds to a cell-surface receptor and is internalized within a vesicle. In the vesicle DT undergoes proteolysis and an acid-induced change of conformation; subsequently, a heat-stable fragment of the toxin *(catalytic domain*: CD) is released into the target cell's cytoplasm. CD kills the target cell by catalysing the *ADP-ribosylation* (explained in section 9.1.1) of elongation factor EF-2, thus blocking protein synthesis.

Synthesis of DT in *C. diphtheriae* is repressed if the intracellular level of iron remains above a certain level; the ions of transition metals activate a repressor protein, DtxR, which binds to the operator of the *tox* gene, blocking transcription [JB (2003) *185* 6826–6840].

11.3.2 Pathogenesis involving other bacterial products

Aggressins are products which can promote the invasiveness of a pathogen. For example, certain bacteria, including *Streptococcus pyogenes* and most coagulase-positive staphylococci, produce *hyaluronate lyase* ('spreading factor'): an enzyme which cleaves hyaluronic acid, a component of the intercellular cement in animal tissues; in at least some cases this enzyme may assist bacterial penetration of an infected site. [Structural basis of the action of hyaluronate lyase from *Streptococcus pneumoniae*: EMBO Journal (2000) *19* 1228–1240.]

Another example is the streptococcal product *streptokinase*: a secreted protein which activates the plasma component *plasminogen* to form *plasmin* (= *fibrinolysin*), an agent which degrades fibrin. This ability to promote lysis of fibrin may enable the pathogen to breach the fibrin barriers of wounds/lesions; hence, in streptococci (and some other bacteria) the production of plasminogen activators has been linked to invasiveness [see e.g. TIM (1997) *5* 466–467]. (See also section 8.5.10.)

Bacterial products can also contribute to pathogenesis in a more mechanical way – see e.g. cystic fibrosis (below).

11.3.2.1 Cystic fibrosis

Cystic fibrosis (CF) [review: Lancet (1998) *351* 277–282] is an inheritable disease in which a defective ABC transporter affects transmembrane transport of chloride. Typically, lungs are congested with thick mucus and may become infected with organisms such as mycobacteria, *Burkholderia cepacia* and *Pseudomonas aeruginosa*. High levels of salt (NaCl) on airway epithelia may allow bacterial colonization by inhibiting normal antibacterial activity [Cell (1996) *85* 229–236].

P. aeruginosa can form a viscous alginate slime that inhibits phagocytosis and promotes congestion, thus making for a poor prognosis. Many strains of *P. aeruginosa* have genes for alginate, but strains isolated from the general environment typically do not express these genes; in CF patients, conditions in the lung seem to select for mucoid (i.e. alginate-producing) strains. The conversion of non-mucoid strains to mucoidy apparently occurs if a specific sigma factor (section 7.5), termed AlgU, becomes available for transcription of the alginate genes. AlgU can become constitutively (i.e. permanently) active when a mutation inactivates the *mucA* gene (the activity of AlgU being inhibited by the binding of MucA). [Alginate, biofilms and antibiotic resistance in *P. aeruginosa*: JB (2001) *183* 5395–5401.]

It has been suggested that long-term colonization of CF lungs with *Stenotrophomonas maltophila* (formerly *Xanthomonas maltophila*) may be a significant factor in prognosis [RMM (1997) *8* 15–19]. This common, Gram-negative bacillus [description: IJSB (1993) *43* 606–609] is frequently isolated from CF patients (and is also associated with disease in e.g. immunocompromised individuals and cancer patients). [Identification/detection of *S. maltophila* by a PCR-based approach: JCM (2000) *38* 4305–4309.]

11.3.3 Pathogenesis involving destruction of host cells or tissues

11.3.3.1 *Salmonelloses*

Salmonella typhi initially invades the intestinal epithelium (section 11.2.2.1). In typhoid, the pathogen subsequently enters the bloodstream (via the lymph system) and multiplies in the liver, gall bladder and spleen; secondary infection can occur from the gall bladder. Symptoms include intestinal pain and *sepsis* (section 11.3). Inflammation of the intestine can be so intense (e.g. in Peyer's patches) that it causes local necrosis (death) of tissue with formation of typhoid ulcers and sometimes perforation and haemorrhage.

Some salmonellae (e.g. *S. typhimurium*) can invade intestinal epithelium but (unlike *S. typhi*) remain as localized pathogens in the gut. (Note that *S. typhimurium* causes typhoid in *mice*.)

Yet other salmonellae cause disease without invading epithelial cells; these pathogens typically give rise to inflammatory and secretory disease in which polymorphonuclear leukocytes (PMNs) migrate to the affected tissues. Studies on the bovine pathogen *S. dublin* indicated that inflammation and fluid secretion result from injection of an effector protein, SopB, into epithelial cells via a type III secretory system [MM (1997) *25* 903–912]. Genes for *enteropathogenicity* (as opposed to systemic salmonellosis) have been located on a distinct pathogenicity island (section 11.5.7).

11.3.3.2 *Dysentery*

Dysentery caused by *Shigella* (and EIEC: Table 11.2) involves destruction of ileal/colonic epithelium with pain, fever, watery diarrhoea and (often) bloody diarrhoea. Invasion by *Shigella* is described in section 11.2.2.1, and the role of shiga toxin is discussed in 11.3.1.6.

The plasmid-encoded outer membrane protein IcsA (= VirG) of *Shigella* (and the analogous protein in EIEC) is essential for cell-to-cell spread (and thus essential for pathogenicity). After *Shigella* escapes from the phagosome (into the host's cytoplasm) IcsA induces polymerization of actin filaments at one pole of the bacterium; this propels the bacterium through the host's cytoplasm in a direction opposite to that of the growing 'tail'. (The tail remains stationary in the host's cytoplasm: ongoing deposition of actin on the tail causes the bacterium to move.) The (motile) bacterium pushes into an adjacent cell, either forming a pocket or becoming enclosed within a vacuole (bounded by a double membrane). *Shigella* lyses the membranes, thus invading the neighbouring cell.

Some enterobacteria have an outer membrane protease which cleaves/inactivates proteins analogous to IcsA; this protease is absent in strains causing dysentery, and such absence appears to be necessary for their pathogenicity [MM (1993) *9* 459–468].

In one model for dysentery (shigellosis), the death of macrophages (section 11.5.2) causes an inflammatory response in which PMNs migrate between epithelial cells to reach the gut lumen; such migration exposes basolateral surfaces of the epithelial cells to invasion by *Shigella* – but eventually results in containment of infection [see TIM (1997) *5* 201–204].

11.3.3.3   Oroya fever

Oroya fever, an often fatal disease occurring in parts of South America, involves e.g. fever and progressive anaemia. The pathogen, *Bartonella bacilliformis*, transmitted by sandflies, grows in/on the host's erythrocytes (red blood cells) and in endothelial cells (cells lining blood vessels); bacterial growth leads e.g. to destruction of erythrocytes and associated symptoms.

11.3.4   Endotoxic shock (septic shock)

The *endotoxins* of Gram-negative bacteria have been regarded as macromolecular complexes containing certain elements of the cell envelope: (i) lipopolysaccharides (LPS – section 2.2.9.2), (ii) proteins, and (iii) phospholipids. The toxic component of LPS is lipid A.

The often-fatal condition *septic shock* involves the activity of blood-borne endotoxins (e.g. following successful chemotherapy against a Gram-negative pathogen). Endotoxins act on macrophages and other cells of the immune system, stimulating the secretion of certain physiologically potent agents (*cytokines*: Table 11.4) such as TNF-α. These cytokines can recruit others, and this may lead to symptoms of shock (e.g. a marked fall in blood pressure) and the blocking of blood vessels by white cells; death may occur as a consequence of the progressive failure of organs. Endotoxins have therefore been classified as *modulins* (section 11.5.4).

A vaccine, consisting of anti-endotoxin monoclonal antibodies, was tried unsuccessfully. Certain candidate antibiotics may reduce the risk of septic shock because they inhibit lipid A biosynthesis (section 15.4.11).

11.3.5   *Helicobacter pylori*-associated intestinal disease

Helicobacter pylori (see Appendix) has been causally connected with e.g. gastritis and peptic-ulcer disease, although the details of pathogenesis are not fully understood. However, it was found that lipopolysaccharides (section 2.2.9.2) from *H. pylori* include antigens identical to the Lewis x and Lewis y antigens which occur e.g. in human gastric mucosa, and this suggested the possibility that anti-lipopolysaccharide antibodies may promote autoimmune inflammation by binding to the mucosal Lewis antigens. [Molecular mimicry of *H. pylori* LPS: TIM (1997) *5* 70–73.] This idea was questioned because the predominant type of anti-LPS antibody found in a number of

H. pylori-infected humans corresponds to an antigen which is apparently unrelated to those structures in *H. pylori* LPS which mimic Lewis antigens [INFIM (1998) *66* 3006–3011].

Attachment of *H. pylori* to the stomach wall (as opposed to survival within the layer of gastric mucus) can apparently promote the formation of autoantibodies and lead e.g. to chronic gastritis. Whether or not attachment occurs in any given individual host may depend on particular *combinations* of factors in the (genetically diverse) populations of pathogen and host [TIM (1998) *6* 379–380].

Pathogenesis seems likely to involve the *cag* pathogenicity island (section 11.5.7) and an unlinked (but often jointly expressed) gene, *vacA*. Expression of both *cagA* and *vacA* (in type I clinical strains) seems to correlate with severe gastrointestinal disease, while type II strains, which lack *cagA* and do not express VacA activity, are generally not associated with severe disease; strains causing intermediate levels of disease are also observed.

The CagA protein is injected into host cells by *H. pylori* [Science (2000) *287* 1497–1500], resulting e.g. in proliferation of gastric epithelial cells.

VacA is a 95 kDa secreted cytotoxin. *In vitro* studies have shown that it can block the activity of a major transcription factor in T cells, leading to e.g. inhibition of transcription of the interleukin-2 (IL-2) gene; the overall effect of this toxin was to inhibit both proliferation and activation of T cells [Science (2003) *301* 1099–1102].

11.3.6 The Jarisch–Herxheimer reaction

This potentially fatal reaction may follow the first effective dose of an antimicrobial agent given to combat diseases caused by certain bacteria (particularly spirochaetes) or protozoa. Symptoms, which include an initial rise in temperature, are associated with a cascade of cytokines (e.g. TNF, IL-6, IL-8) that are presumed to be responsible for at least some of the pathophysiological events; the mechanism of the sudden release of cytokines is not understood. The Jarisch–Herxheimer reaction can be prevented by treatment with antibodies to TNF [NEJM (1996) *335* 311–315].

11.3.7 Urinary tract infection with UPEC

UPEC (uropathogenic *Escherichia coli*) refers to strains of *E. coli* responsible for a high proportion of human urinary tract infections. These strains usually encode e.g. a cytotoxin, a haemolysin and various adhesins; type I fimbriae are particularly important – see Table 2.2 for details. (Strains of UPEC exhibit considerable genomic heterogeneity [PNAS (2002) *99* 17020–17024].)

Bladder infections with UPEC are typically persistent, being refractory to antibiotic therapy. The essential first step in pathogenesis is adhesion of UPEC to the uroepithelium – necessary to counteract the flushing action of the urine flow. Type I fimbriae can bind to certain *uroplakins*: proteins found on the surface of the *facet* cells that line mammalian bladder. This bacterial colonization provokes an influx of neutrophils and also exfoliation (cell shedding) from the bladder epithelium.

Type I fimbriae also promote *uptake* (internalization) of UPEC by bladder cells. (P fimbriae appear not to [EMBO Journal (2000) *19* 2803–2812].) Within the bladder cells, UPEC are protected by a polysaccharide-rich matrix and a layer of uroplakin: factors which seem to account for the persistence of these infections [Science (2003) *301* 105–107].

11.4 THE BODY'S DEFENCES

11.4.1 Constitutive (innate) defences

Constitutive defences are non-specific defence mechanisms which can operate immediately, or very quickly, against a would-be pathogen. Given the typically rapid multiplication of microorganisms, such a quick response is essential for effective defence (compare the *adaptive* response: section 11.4.2).

To potential pathogens, the normal healthy body presents a variety of obstacles and barriers. For example, the skin is more than a simple physical barrier to infection. To most bacteria it is a hostile environment: water is scarce, and sites are occupied by the well-adapted skin microflora – some of which produce antibacterial fatty acids from lipids secreted by the sebaceous glands.

The mucous membranes, too, have their own defences: the resident microflora, with which any would-be pathogen must compete, and the secretions, bathing these tissues, which exert a mechanical flushing action and which contain various antibacterial substances. Thus, tears contain *lysozyme* (Fig. 2.7), and secretions of the intestinal tract contain *defensins* and other antimicrobial peptides [defensins and antimicrobial peptides: Nature (2002) *415* 389–395; Nature (2003) *422* 478–4791.

Additionally, among and beneath epithelial cells are various types of cell which have a protective role; these include M cells (section 11.2.2.1), macrophages, $\gamma\delta$-type T cells and dendritic cells; moreover, inflammatory signals can quickly recruit other cells – polymorphonuclear leukocytes (PMNs) and NK (natural killer) cells – to a given site.

Complement (section 11.4.1.1) has roles e.g. in inflammation and phagocytosis.

Bacterial components can activate the *acute-phase response* in which cells of the immune system are stimulated to secrete various *cytokines* (Table 11.4), including IL-1, IL-6, IL-8 and TNF-α. This e.g. promotes inflammation (see below) and promotes the formation of *acute-phase proteins* in the liver; these proteins (which include *C-reactive protein* and *serum amyloid* A) can bind to microbial surfaces (*opsonization*: see below) and facilitate phagocytosis.

Acute-phase proteins recognize *patterns* in microbial cell-surface molecules; a host molecule that recognizes a given pattern is called a *pattern recognition receptor* (PRR). PRRs are also found e.g. on the surfaces of leukocytes and epithelial cells; section 11.6.1 discusses PRRs found on, and within, cells of the intestinal epithelium.

Phagocytosis. Within the tissues, blood and lymph systems, certain cells (*phagocytes*) engulf particles of foreign material, including many types of microorganism. The phagocytes include *macrophages* and *neutrophils*, and the process by which they eliminate microorganisms etc. is called *phagocytosis*. Typically, bacterium–macrophage contact (facilitated by *opsonization*: see below) triggers uptake of the bacterium – which then becomes enclosed within an intracellular membranous sac or vacuole, the *phagosome*. The phagosome's internal pH subsequently drops and the vacuole fuses with a *lysosome* (a sac containing degradative enzymes) – forming a *phagolysosome* within which microorganisms are usually killed. (The maturation of a phagosome into a phagolysosome can be monitored e.g. by the appearance of certain proteins: a phagolysosome is characterized by the presence of *lysosome-associated membrane proteins* (LAMPs).) The antimicrobial action involves various lysosomal enzymes as well as reactive agents such as superoxide, hydrogen peroxide and singlet oxygen formed in the so-called *oxidative burst*.

Not all bacteria are killed by phagocytes. Some pathogens which *invade* macrophages survive, multiply and escape. For example, following uptake of *Legionella pneumophila* phagosome–lysosome fusion generally does not occur, and the pathogen may avoid triggering the oxidative burst – possibly by secreting a zinc metalloprotease [JMM (2001) *50* 517–525]. Other pathogens which survive/multiply in macrophages include *Brucella* spp, *Chlamydia* spp, *Coxiella burnetii*, *Francisella tularensis*, *Mycobacterium avium* and *M. tuberculosis*. (See also section 11.2.2.2.)

Opsonization. Bacteria can be *opsonized*, i.e. made more susceptible to phagocytosis, by activating *complement* (section 11.4.1.1) and binding certain components, e.g. C3b; C3b has specific receptors on some types of phagocyte and can therefore link the opsonized bacterium to such phagocytes. Opsonization is also mediated by antibodies (section 11.4.2.1); thus, Fab regions (Fig. 11.2) may bind to a bacterial cell-surface antigen while the Fc portion of the antibody binds to a specific Fc receptor on a phagocyte. Opsonization can also be mediated by acute-phase proteins (section 11.4.1).

Inflammation. Within the body, bacterial components or products may provoke an inflammatory response characterized by localized reddening, swelling, warmth, pain and loss of function. These non-specific symptoms (which can also be caused e.g. by physical or chemical injury) are associated with an increased outflow of plasma to the affected tissues; this exposes a pathogen to larger amounts of antimicrobial agents, such as *complement* (section 11.4.1.1), present in the plasma. Moreover, various phagocytes – particularly neutrophils (PMNs) and macrophages –

quickly gather at the affected site and are usually able to eliminate the pathogen.

Clearly, inflammation is a complex process which involves directional movement of cells, changes in the permeability of local blood vessels, and deployment of specific antimicrobial agents. Co-ordination of these events involves a wide range of signalling molecules, the majority of which are *cytokines* (Table 11.4) and components of the complement system (section 11.4.1.1). There is also a need for mechanisms (such as *decay-accelerating factor*: Fig. 11.1, legend) to prevent harmful over-reaction of the immune system.

Some of the events likely to occur in the response to infection by Gram-negative bacteria are listed below.

- At the site of infection, lipopolysaccharides (LPS: section 2.2.9.2) activate complement via the *alternative* pathway (Fig. 11.1). Complement components C3a and C5a (called *anaphylatoxins*) induce mast cells and basophils to release *histamine*; this increases the permeability of blood vessels and promotes outflow of plasma.

  Histamine also contributes to upregulation of certain cell-adhesion molecules (*selectins*) found e.g. on the *endothelial cells* that line blood vessels. Upregulation of these selectins is also mediated by cytokines (e.g. IL-1 and TNF-α) produced by various cells in response to bacterial components/products. The selectins bind *weakly* to leukocytes in the bloodstream, causing them to slow down and *roll* along the endothelial surface (an effect called *tethering*). This is the first stage of a process in which leukocytes leave the blood vessels and move into affected tissues.

  C5a is also chemotactic for neutrophils (PMNs); attracted to the focus of infection, PMNs are quickly tethered to endothelial selectins.

- IL-8, a *chemokine* (Table 11.4) induced by LPS (and e.g. by IL-1 and TNF-α), binds to tethered PMNs, activating another type of cell-adhesion molecule, the *integrin*; the integrins of PMNs bind *strongly* to receptors (e.g. ICAM-1, ICAM-2) on endothelial cells. The PMNs then flatten and 'crawl' along the endothelium, apparently guided by the chemokine gradient. Subsequently, PMNs pass *between* endothelial cells (a process called *diapedesis*) into the affected tissues. [Leukocyte rolling and firm adhesion (biophysical view): PNAS (2000) *97* 11262–11267.] (The genetic disorder *leukocyte adhesion deficiency* (LAD) involves inadequate expression of integrins on PMNs; individuals with LAD exhibit low resistance to infections [see e.g. Lancet (1999) *353* 341–343].)

- LPS bind to a protein, CD14, that occurs free (in plasma) and also at the surface of e.g. monocytes. The binding of LPS to CD14 on *cells* (indirectly) stimulates the formation of certain cytokines (e.g. IL-1, TNF-α); this activity of LPS is greatly increased by the acute-phase protein *lipopolysaccharide-binding protein* (LBP).

- Complement component C3b acts as a potent *opsonin*, promoting phagocytosis.

- IL-6, produced e.g. by LPS-stimulated macrophages, induces the liver to form other opsonins: the acute-phase proteins; the concentration of these proteins in plasma is said to increase up to ~1000-fold during active infection.

- PMNs, attracted to the site by C5a and IL-8, arrive quickly and are often the dominant type of cell in the early stages of acute inflammation. PMNs exhibit highly effective antibacterial activity; for this purpose they have a wide range of enzymes, including lysozyme (Fig. 2.7) and myeloperoxidase (which oxidizes halide ions to hypohalite). The azurophil granules of PMNs also contain *BPI protein* (bactericidal/permeability-increasing protein): an agent bactericidal for some Gram-negative bacteria – in which it increases the permeability of the outer membrane. Opsonized bacteria stimulate an *oxidative burst* (lasting for several minutes) characterized by increased uptake of oxygen and formation of highly reactive radicals such as superoxide.

11.4.1.1 *Complement*

Normal plasma contains *complement*: a number of different proteins which, in the presence of certain types of molecule, undergo a 'cascade' of reactions involving sequential activation; on activation, various components of the system carry out specific physiological functions (Table 11.3).

Activation of the complement system can be triggered in various ways, the type of trigger determining which of four pathways are followed (Fig. 11.1). For example, the lipopolysaccharide (LPS) of a Gram-negative bacterium (section 2.2.9.2) can trigger the *alternative* pathway. Activation by a Gram-negative bacterium can have several consequences. For example, the binding of component C3b makes the cell more susceptible to phagocytosis by macrophages. Moreover, if components C5b–9 (the *membrane attack complex*) bind to the cell surface they can form a hole in the outer membrane which, in some cases, can lead to cell lysis (*immune cytolysis*). Lysis results not from the breaching of the outer membrane but from the opportunity that this provides for lysozyme (Fig. 2.7) to reach peptidoglycan in the cell envelope (Fig. 2.6);

Table 11.3 Complement: some important antibacterial roles

Component	Role
C3a, C5a	Elicits release of histamine from mast cells/basophils, increasing the permeability of certain blood vessels to plasma/cells (part of the inflammatory response)
C3b	Opsonization (enhancement of phagocytosis)
C5a	Chemotaxis: attracts neutrophils to an infected site (part of the inflammatory process)
C5b6789	Lysis of Gram-negative bacteria under certain conditions (see section 11.4.1.1)

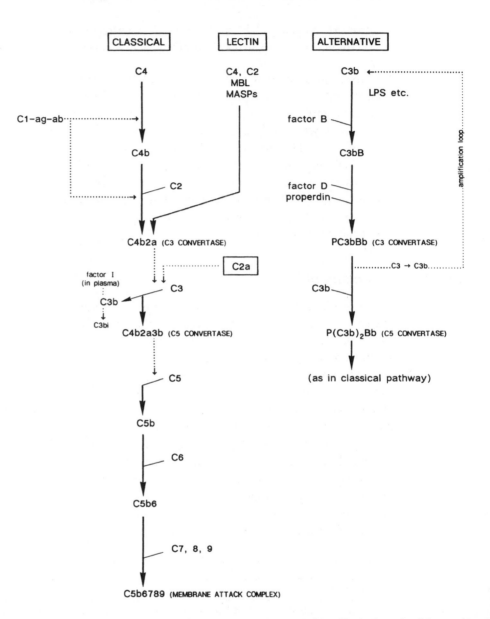

Figure 11.1 The activation of complement (section 11.4.1.1) (simplified scheme). All four pathways can lead to similar physiological effects, but they are triggered in different ways. C1, C2 etc. denote particular components of the complement system; 'a' and 'b' denote fragments of components produced by enzymic cleavage during the activation process. (For clarity, the diagram does not show *all* the cleavage products; for example, C4 is cleaved to C4a and C4b, but only the C4b fragment is considered here.) Dotted lines indicate those cases in which a given complex acts enzymically to cleave certain components of the system.

Classical pathway. Activation of this pathway can be triggered e.g. by many types of antigen–antibody complex (section 11.4.2.1); such a complex may be formed, for example, by the binding of an antibody thus, a Gram-negative pathogen (such as *Haemophilus influenzae*) on the conjunctiva would be at risk of lysis because the fluid that bathes the eye contains both complement and lysozyme.

The *classical* pathway is triggered by many types of antigen–antibody complex. (Most – but not all – antibodies are 'complement-fixing' antibodies.) The reaction cascade is shown in Fig. 11.1. In mice, the classical pathway is reported to be the dominant pathway for innate immunity to *Streptococcus pneumoniae* [PNAS (2002) 99 16969–16974].

The *lectin* pathway involves several components of normal plasma: mannose-

---

to a cell-surface bacterial antigen. Initially, C1 binds to the so-called Fc portion of the antibody (Fig. 11.2) in an antigen–antibody (ag–ab) complex. Thus activated, C1 cleaves C4. C4b binds C2, and C4b2 is cleaved, by activated C1, to C4b2a (a 'C3 convertase'). C3 convertase splits molecules of C3 to the fragments C3a and C3b.

If activation had been triggered by an antigen–antibody complex on a bacterial surface, C3b fragment(s) may bind to the Fc portion of the antibody and/or to the bacterial surface. C3b promotes phagocytosis of those cells or complexes to which it binds because specific receptor sites for C3b occur at the surface of macrophages and other phagocytic cells; C3b binds strongly to these sites. C3b may also contribute to a C5 convertase (the next stage of activation).

C3b is a short-lived molecule, but low concentrations of C3b ('tickover' C3b) are normally present (even without complement activation) owing to its spontaneous formation through low-level hydrolysis of C3. This accounts for the availability of C3b for the initiation of the alternative pathway (despite its inactivation by factor I).

The cascade of reactions continues as shown. The complex C5b67 can bind to membranes and form a *membrane attack complex* (MAC) by adding C8 and C9. If MAC develops e.g. on the outer membrane of a Gram-negative bacterium (section 2.2.9.2), it forms a pore or channel through the membrane; the wall of the channel consists of six or more molecules of C9. In some cases (see section 11.4.1.1) this can cause lysis of the bacterium.

The complex C5b67 may bind at or near the site where activation of complement was initially triggered. Alternatively, this complex may bind to *another* cell in the vicinity – so that this cell may suffer lysis if the MAC is formed; such lysis is called *reactive lysis* or *bystander lysis*.

Alternative pathway. Activation of this pathway can be triggered e.g. by lipopolysaccharides (LPS; section 2.2.9.2); a low concentration of C3b is normally available for initiation (see above). Factors B and D, and properdin, are all proteins found in normal plasma. Note that the C3 and C5 convertases found in this pathway differ from those in the classical pathway. Note also the positive feedback (amplification) loop for C3b.

Lectin pathway. See text (section 11.4.1.1).

C2a ('salvage') pathway. See text (section 11.4.1.1).

Fragments C3a and C5a. These fragments (not shown), called *anaphylatoxins*, can act on mast cells and basophils, causing release of mediators of inflammation such as *histamine*. Histamine affects the permeability of certain small blood vessels, allowing increased outflow of plasma and cells to e.g. infected tissues. Additionally, C5a acts as a chemotactic factor, attracting neutrophils to the affected site.

Regulation of the complement system. The complement system can form some highly potent physiological agents and it must be rigorously controlled. For this reason there are specific inhibitors of key stages; for example, the *decay-accelerating factor* (DAF; CD55), and certain other proteins on mammalian cells, accelerate the decay of convertases, thus helping to protect host cells from complement-mediated lysis.

binding lectin (MBL) and the MBL-associated serine proteases (MASPs). This system is triggered when MBL binds to appropriate groups on the microbial cell envelope; it effectively bypasses the C1 stage of the classical pathway. The lectin pathway may be particularly useful in infants from about 4–6 months of age (when maternal IgG antibody cover is lost) to about 18–24 months (when a more effective immune system develops); however, binding to MBL can be greatly inhibited by the presence of a bacterial capsule (e.g. in pathogens such as *Neisseria meningitidis*). [MBL binding to particular species: INFIM (2000) *68* 688–693.]

The C2a ('salvage') pathway, reported in 1997, may be significant for pathogenic mycobacteria. In this pathway, component C2a acts as a C3 convertase (Fig. 11.1) when bound to the mycobacterial surface [TIM (1998) *6* 47–49; TIM (1998) *6* 49–50]. (Note that C2a is referred to as 'C2b' by some authors.)

Interestingly, some pancreatic enzymes (e.g. trypsin) can cleave complement components to form *anaphylatoxins* (legend, Fig. 11.1), and levels of these fragments are found to be raised in experimental pancreatitis.

Complement is an early, rapid and potent form of defence. The alternative and lectin pathways belong to the body's constitutive defences, while the classical pathway is part of the adaptive response (section 11.4.2).

11.4.1.2 *Interferons*

The interferons (IFNs) are proteins secreted by certain types of animal cell in response to viruses, some types of bacteria and e.g. double-stranded RNA (dsRNA). Interferons are included within the category *cytokines* (Table 11.4); they have various roles in host defence mechanisms.

Human interferons comprise two main groups. Type 1 interferons (α, β and ω) are stable at pH 2. IFN-α, produced by leukocytes, has at least 14 distinct subtypes (IFN-α1, IFN-α2 etc.); IFNβ, produced by fibroblasts, is antigenically distinct from IFN-α. The three type 1 IFNs appear to be related; their genes are intron-less and may have developed from a common ancestral gene.

Type 1 IFNs can bind to virus-infected cells and induce the synthesis of certain proteins (e.g. MHC class I molecules) that promote CMI (section 11.4.2.2) against such cells; they can also inhibit the synthesis of both viral and cellular proteins. These IFNs can act against cells infected with any of a range of different viruses – not only the type of virus that induced the IFN; because of this generalized, non-specific potentiation of the immune system, interferons are often seen as part of the constitutive (innate) defences. However, the type 2 interferon, IFN-γ, is also a major factor in the adaptive response (section 11.4.2).

The type 2 IFN, IFN-γ, is acid-labile; in terms of amino acid sequence it is not similar to the type 1 IFNs, although its three-dimensional structure is reported to have features in common with IFN-β. The IFN-γ gene contains three introns.

Both types of human IFN have antiviral activity, but this function seems to be of major importance primarily in the type 1 IFNs.

IFN-γ is produced by T cells stimulated by (macrophage-derived) interleukin-18 (IL-18) and by antigenically and/or mitogenically stimulated T lymphocytes; it is a ~25 kDa protein, and the active form is a dimer. IFN-γ is somewhat less important as an antiviral agent but it plays an important role as a signalling molecule in various aspects of immunity. Among its known or reported roles are:

- Activation of macrophages.
- Upregulation of class I molecules.
- Upregulation of class II molecules on e.g. endothelial cells, fibroblasts, macrophages; these molecules promote the ability of a cell to present antigen to CD4$^+$ T cells (see sections 11.4.2.1 and 11.4.2.2).
- Stimulation of nitric oxide synthesis (section 11.4.1.4).
- Antagonism to class switching by interleukin-4.
- Upregulation of synthesis of complement component C3 in certain lung cells.
- Down-regulation of transferrin (iron-uptake) receptors on infected mammalian cells (the resulting reduced uptake of iron serving to inhibit intracellular pathogens).

11.4.1.3 *Sequestration of ions*

Certain metal ions are needed by pathogens, and the host's sequestration of such ions is one form of constitutive defence. Iron, for example, is sequestered by chelators in both plasma and secretions (see section 11.5.5).

Zinc occurs e.g. in certain enzymes (such as the de-acetylase involved in lipid A biosynthesis: section 2.2.9.2). Zinc is chelated by *calprotectin* – a protein released from dying neutrophils at sites of inflammation (e.g. abscesses) where concentrations of zinc can be high. Calprotectin is bacteriostatic, apparently through sequestration of zinc; when treating abscesses, this effect is likely to antagonize those antibiotics (e.g. β-lactams) which are active only on growing cells. [Zinc in antimicrobial defence: RMM (1997) *8* 217–224.]

11.4.1.4 *Nitric oxide*

When suitably activated, many types of cell (including macrophages and epithelial cells) synthesize an inducible form of the enzyme *nitric oxide synthase* (iNOS); such synthesis requires activation by certain *cytokines* (Table 11.4) (e.g. IFN-γ, TNF-α). iNOS catalyses NADPH-dependent oxidative deamination of L-arginine to L-citrulline and nitric oxide (NO).

Effect of nitric oxide on bacterial pathogens. Nitric oxide (NO) reacts with oxygen, water and/or other agents to form various reactive species that can kill pathogenic bacteria. For example, with superoxide, NO forms peroxynitrite ($ONOO^-$); this species forms a potentially toxic adduct with carbon dioxide. Other reactive species include nitrogen dioxide and nitrogen trioxide.

In a target bacterium, NO and the various reactive species can inhibit terminal oxidases and certain other iron-containing enzymes. Damage to an infecting bacterium may also include failure of DNA replication and/or loss of important metabolic processes, leading to cell death.

Inhibitors of iNOS inhibit the killing of *Mycobacterium tuberculosis* by IFN-γ-activated murine macrophages; moreover, IFN-γ *knock-out* studies (section 8.5.5.4; Table 8.2) have found that deletion of the IFN-γ gene leads to greater susceptibility to *M. tuberculosis*, correlating with deficiency in iNOS.

Bacteria can react to NO e.g. by increased activity of enzymes involved in the antioxidant response to oxidative stress (section 7.8.2.8). Resistance to NO can vary between species and strains of pathogens. [Bacterial responses to nitric oxide and nitrosative stress: MM (2000) *36* 775–783.]

In *M. tuberculosis*, resistance to NO is reported to require the proteasome (section 2.2) [Science (2003) *302* 1963–1966].

Nitric oxide in the human host. Nitric oxide is an effective vasodilator which can lead to vascular leakiness. It can stimulate the production of TNF-α and enhance the activity of NK cells. It activates cyclo-oxygenase in some types of cell but inhibits it in others. It activates the nuclear transcription factor NF-κB, inhibits degranulation of mast cells, and suppresses proliferation of T cells and the activity of antigen-presenting cells (section 11.4.2.1).

Although upregulation of NO synthesis by iNOS serves a protective role, ongoing production of high levels of NO in a persistent, unresolved infection can suppress the host's immune system and cause various forms of disease.

[Effects of NO on pathogens and host: RMM (1998) *9* 179–189. Nitric oxide in sepsis and endotoxaemia: JAC (1998) *41* (suppl A) 31–39.]

11.4.1.5   RNA interference (RNAi)

See section 7.9.2.

11.4.2   The adaptive response (acquired immunity)

In addition to constitutive defences (section 11.4.1) – which offer rapid protection against pathogens in general – the body can also respond *specifically* to a given pathogen. A given type of cell can be recognized by its chemical signature, i.e. by the particular 'shapes' of molecules (such as lipopolysaccharides) which characterize that cell.

For convenience, this topic is considered under two headings: (i) antibody-mediated immunity, and (ii) cell-mediated immunity; however, it must be remembered that the immune system operates as an integrated whole.

11.4.2.1 Antibody-mediated immunity: antibodies, antibody formation, vaccination

An *antibody* is a protein which is formed in the body in response to a particular molecule, or part of a molecule, called an *antigen*; an antibody can combine non-covalently, and reversibly, with the corresponding antigen – antigen–antibody binding exhibiting a high degree of specificity. All antibodies are classified as immuno-globulins. The basic structure of an antibody is shown in Fig. 11.2.

Antibodies are produced by a particular subgroup of leukocytes (white blood cells): the B lymphocytes (= B cells). The body's population of B cells consists of a vast number of different strains, each strain having the potential to produce antibodies of a unique type which are specific to a given antigen; in general, antibodies are produced only if a B cell is stimulated by the corresponding antigen (see later).

Antibodies elicited by a pathogen can combine with that pathogen, but of what use is this? The binding of antibodies to cell-surface antigens (e.g. lipopolysaccharides) makes a cell more susceptible to phagocytosis (i.e. the cell is *opsonized*: section 11.4.1). Moreover, most antigen–antibody complexes can activate the complement system (section 11.4.1.1), and this also promotes opsonization.

Another consequence of pathogen–antibody binding is *antibody-dependent cell-mediated cytotoxicity* (ADCC) (see section 11.4.2.2).

Antibodies may seem unimportant against *intracellular* pathogens. However, antibodies may still help – e.g. by blocking *initial* adhesion by *Chlamydia* or *Salmonella*, or by mediating ADCC in *Coxiella* [TIM (1998) 6 102–107].

Some antigens are *toxins*, and antibodies to a toxin (antitoxins) can neutralize the activity of the corresponding toxin. Toxin–antitoxin complexes are removed by certain blood cells which have receptors for the Fc portion of an antibody.

The 'down side' of antibodies is exemplified by certain *hypersensitivity* reactions which include *allergies* to antibiotics such as penicillin. In penicillin allergy, *initial* contact with the antibiotic elicits IgE antibodies; many of these antibodies bind to mast cells and basophils, leaving their *antigen*-binding sites free at the cell surface. In subsequent contact, penicillin (the antigen) binds to its IgE antibodies on the surfaces of mast cells and basophils; this causes *degranulation* of the cells, i.e. release of various physiologically active substances (e.g. histamine) with risk of potentially fatal *anaphylactic shock*.

Antibody formation. The mechanism of antibody formation varies according to the type of antigen and to the type of B cell. In the simplest case, certain large molecules containing repeated subunits – e.g. bacterial polysaccharides – can elicit antibodies

Figure 11.2   Antibodies: the basic structure of a (monomeric) immunoglobulin (Ig) molecule (diagrammatic). The Ig molecule, a glycoprotein, consists of four polypeptide chains: two identical heavy chains (*long lines*) and two identical light chains (*short lines*) – the four chains being linked by disulphide (SS) and other bonds to form the Y-shaped molecule.

The two heavy chains are adjacent for part of their length (the stem of the 'Y'), forming the so-called *Fc portion*); they diverge at the *hinge region* to contribute to the two limbs of the Y-shaped molecule. Each limb, which consists of a light chain and part of a heavy chain, is termed a *Fab region*. (Both Fab regions can be cleaved from the molecule by the enzyme *papain*. Another enzyme, *pepsin*, cleaves (as one piece) both Fab regions together with the hinge region; this (single) piece is called the *F(ab')$_2$ portion*.)

An antigen-binding site occurs at the free end of each Fab region.

Among antibodies (section 11.4.2.1), IgG, IgD and IgE have the monomeric form shown in the diagram. IgM is a pentamer, i.e. five monomers joined, radially, via their Fc portions.

IgA occurs primarily in monomeric form (molecular weight ~160000) in *plasma*, but in extravascular fluids such as tears, saliva, respiratory and gastrointestinal secretions (in which it is the predominant Ig) it occurs in dimeric form – *secretory IgA* (sIgA) – the Fc portions of the two monomers being linked by (i) a *J chain*, and (ii) a *secretory component*.

IgG (molecular weight ~150000) accounts for about 75% of the immunoglobulins in plasma, most being of the IgG1 subclass. IgG antibodies are important e.g. as opsonins and antitoxins in extravascular tissues as well as in the bloodstream; these antibodies can cross the placenta and protect the fetus and neonate. IgG antibodies generally predominate after class switching in the humoral response to antigens (section 11.4.2.1).

IgM (molecular weight ~970000) forms about 5–10% of the plasma immunoglobulins. IgM antibodies are particularly good agglutinators of LPS and other substances which display a pattern of repeated antigenic determinants; in man they do not cross the placenta or the blood–brain barrier. IgM antibodies are usually the first to be formed in the humoral response to antigens, and they are the main class of antibodies formed against thymus-independent antigens (such as pneumococcal polysaccharides) (section 11.4.2.1).

Isotype refers to a class of immunoglobulins (e.g. IgG, IgM). Note that a given isotype occurs in *all* normal individuals of the species.

Allotype refers to a variant form of Ig molecule. A given allotype does not occur in all normal individuals in the species; it occurs in only those who have the relevant allele in their genotype (an *allele* being one of several variant forms of a given gene).

Idiotype refers to a type of Ig molecule defined in terms of the specific collection of immunogenic structures in its variable region.

directly from (i) CD5$^+$ B cells (B cells found e.g. in the peritoneal cavity) and (ii) so-called *marginal zone B cells* (MZ B cells, found in the spleen); these cells rapidly produce IgM antibodies when they bind antigen. [New aspects of B cell biology: Science (2000) *290* 89–92.]

Certain antigens which activate CD5$^+$ and MZ B cells (e.g. lipopolysaccharides) can act as *mitogens*, i.e. they can stimulate many different strains of B cell (regardless of B cell specificity) to proliferate and produce antibody; these (multivalent) antigens, called *polyclonal activators*, appear to stimulate B cells by cross-linking receptor sites on the surface of the cells.

Unlike the antigens mentioned above, most antigens are smaller molecules (e.g. soluble proteins) which elicit antibodies from CD5$^-$ B cells; however, in this case, antibodies are formed only if the B cell obtains help from another type of lymphocyte, the T cell. (Because T cells mature in the thymus gland, *these* antigens are called *thymus-dependent antigens* (TD antigens). Antigens which elicit antibodies from B cells without help from T cells are called *thymus-independent antigens* (TI antigens).)

To elicit antibodies, a TD antigen must first be taken up and processed by an *antigen-presenting cell* (APC). Processing involves enzymic degradation. The processed antigen is linked to a class II MHC molecule (synthesized in the APC) and is then *presented*, at the APC's surface, to a T cell which is specifically reactive to that particular antigen. The T cell recognizes, and binds to, the combination of processed antigen and MHC class II molecule via the (antigen-specific) *T cell receptor* (TCR).

Acting as an APC, a B cell can bind specific antigen and present the (processed) antigen (with associated class II MHC molecule) to an (antigen-specific) T cell; in this context, a CD4 T cell of the Th2 subset can act as a *helper T cell*. The class II MHC molecule (on the B cell) binds to the CD4 antigen on the T cell.

Various other 'paired interactions' occur between binding sites on the B cell and T cell. Thus, e.g. receptors B7.1 and/or B7.2 on the B cell bind to CD28 on the T cell, promoting T cell development. Moreover, binding between CD40 (on the B cell) and CD40L (on the T cell) generates part of a stimulus that results e.g. in proliferation and development of the B cell.

During the above process the B cell and T cell are physically linked; this mutual (physical) co-operation is referred to as *cognate help* or *linked recognition*.

The activated, proliferating B cells form an expanded clone. Some of these cells develop as antibody-secreting *plasma cells*. Initially, antibodies of the IgM isotype are formed, but antibodies formed later are of another class, usually IgG – although these antibodies are of the same *specificity* (i.e. they still match the given antigen); this change in isotype is referred to as *class switching*. Interleukin-4 (Table 11.4) is one of the stimuli that promote class switching.

Class switching occurs in so-called *germinal centres* in spleen and lymph nodes. These sites are also involved in *affinity maturation*: a process in which certain mutant cells among the (proliferating) antigen-stimulated B cells are selected for high-affinity binding to the given antigen. This process involves *somatic hypermutation*: an extra-ordinarily high rate of mutation in those regions of the chromosome which encode the antigen-binding sites of antibodies; different B cell clones therefore produce antibodies with different degrees of affinity for the given antigen – B cells producing antibodies with the highest affinity being selected. The mechanism of somatic hypermutation may involve certain error-prone DNA polymerases. [Hypermutation in antibody genes (papers from a discussion meeting): PTRSLB (2001) *356* 1–125.]

Other activated B cells become *memory cells*. These cells do not secrete antibodies but retain the ability to produce antibodies rapidly if the body is subsequently challenged with the *same* antigen; this heightened *secondary response* appears to reflect the existence of an expanded clone of (specifically reactive) B cells resulting from antigen-stimulated proliferation.

Th1 cells and Th2 cells. The helper T cells involved in antibody formation are of the CD4$^+$ category. CD4 is a cell-surface molecule (a *co-receptor*) located close to the TCR; it helps to stabilize APC–T cell contact by binding to the MHC class II molecule and acts as a signal transducer following contact. (CD8$^+$ T cells are considered in section 11.4.2.2.)

Antigenic stimulation of a naïve CD4$^+$ T cell (i.e. one not previously exposed to its corresponding antigen) may cause the cell to develop in one of two distinct ways, its development being influenced by the presence of particular cytokines. In the presence of IL-12 it may develop characteristics of the so-called Th1 subset; Th1 cells secrete e.g. IL-2 , IFN-γ, TNF-α and TNF-β (Table 11.4). These T cells are sometimes called 'inflammatory' T cells; their cytokines can e.g. activate macrophages and may potenti-ate the activity of cytotoxic T cells (section 11.4.2.2) and NK cells.

Th2 cells may develop in the presence of IL-4. Cells of the Th2 subset secrete e.g. IL-4, IL-5, IL-6, IL-10 and TNF-α. These cells perform a major function by helping B cells in antibody formation; their cytokines can also e.g. promote the development of Th2 cells and induce class switching in B cells.

Vaccination. The body can be induced to make antibodies to a particular antigen by injecting that antigen or, in some cases, administering it orally; such a procedure is called *vaccination*. *Vaccines* are antigenic preparations of certain pathogen(s) (e.g. heat-killed cells or cell components) which are injected, or administered orally, in order to help the body defend itself against a subsequent attack by the given patho-gen(s).

Oral vaccines. In some cases oral vaccines are more efficacious than *parenteral* vaccines (which include those administered by injection). For example, good protection against bacterial *enteric* pathogens seems to require that the antigenic stimulus of a vaccine be directed at the mucosal immune system of the *gut* – access to which is apparently optimal via the alimentary canal. It is probably for this reason that the early parenteral vaccines against *cholera* (sections 11.3.1.1 and 11.11) gave poor protection (and are generally no longer recommended). Currently, *oral* anti-cholera vaccines offer a higher level of protection for a longer period of time. One of these vaccines (Dukoral® from SBL Vaccin, Sweden) consists of killed cells of *Vibrio cholerae* O1 (including Inaba and Ogawa serotypes and the classical and El Tor biotypes) together with a recombinant form of the B subunit of cholera toxin; this vaccine, designated CTB-WC (cholera toxin B-whole cell), has been approved for licencing in the E.C. It modulates expression of specific adhesion molecules in the gastrointestinal tract [INFIM (2004) *72* 1004–1009].

Anti-cholera vaccines containing live, attenuated O1 strains are also available.

Currently, efforts are being made to construct live vaccine strains of the O139 (Bengal) serotype [Vaccine (2003) *21* 1282–1291].

Caution regarding live attenuated anti-cholera vaccines is indicated by the finding that non-lysogenic (non-pathogenic) strains of *V. cholerae* can acquire genes for cholera toxin from phage VGJϕ via a novel form of specialized transduction which does not require the normal cell-surface receptors for phage CTXΦ [JB (2003) *185* 7231–7240] (see section 9.5.2).

Development of vaccines against ETEC (Table 11.2) is also based on oral, rather than parenteral, administration [RMM (1996) *7* 165–177]. Approaches have included the administration of ETEC fimbriae (section 2.2.14.2) within biodegradable microspheres and the use of live, attenuated strains which lack genes for the heat-stable and heat-labile toxins. Much of the protective action of secretory IgA (sIgA: Fig. 11.2) is directed against fimbrial antigens of ETEC.

A live, attenuated oral vaccine (Ty21a) is available for use against typhoid (*Salmonella typhi*). An older, parenteral vaccine (containing killed whole cells) was effective but gave severe side-effects in some cases. [*Salmonella* vaccines: FEMSMR (2002) *26* 339–353.]

Live, attenuated strains of *Shigella* containing null mutations in the *virG* (*icsA*) gene (see sections 11.2.2.1 and 11.3.3.2) have been tested in animals and humans [e.g. INFIM (2000) *68* 1034–1039].

Parenteral vaccines. The term *parenteral* refers to any route other than the oral route; parenteral vaccines are administered e.g. by injection into skin (intradermally) or muscle (intramuscularly). Vaccines for whooping cough (= pertussis; causal agent: *Bordetella pertussis* or *B. parapertussis*) include whole-cell vaccines (WCVs) and acellular vaccines (ACVs). The latter include inactivated pertussis toxin (a *toxoid* – see later) and (usually) fimbrial or fimbria-like components. ACVs have been widely used in Japan since 1981. Newer ACVs (used in Europe) have reduced side-effects (compared with WCVs) and are as effective as WCVs in preventing disease; however, they may give less protection against *B. parapertussis* [RMM (1996) *7* 13–21].

The vaccine against tuberculosis, BCG (bacille Calmette–Guérin), is a live attenutated strain of *Mycobacterium tuberculosis* administered by intradermal injection. The protective efficacy of BCG varies greatly geographically, apparently reflecting different degrees of exposure to environmental mycobacteria.

Anti-typhoid vaccines include not only the (oral) Ty21a vaccine but also one containing purified *Vi antigen* (a microcapsular polysaccharide antigen found e.g. on certain strains of *Salmonella*); a single subcutaneous/intramuscular dose is reported to be safe and effective for adults and children >2 years of age.

Other vaccines in this category include conjugate vaccines (see below).

Conjugate vaccines. Pathogens with a *polysaccharide* capsule or microcapsule can cause various diseases in very young children (section 11.5.1) because such children cannot respond adequately to polysaccharide antigens. [Responsiveness of infants to capsular polysaccharides: RMM (1996) *7* 3–12.] For these young children, vaccines effective against such pathogens have been made by linking a pathogen's polysaccharide to a *protein* – forming a so-called *conjugate* vaccine; a conjugate vaccine can elicit protective antibodies at e.g. 4 months of age – a timely schedule because infants are generally protected for the first 6 months or so by antibodies derived from the mother.

An example of a successful conjugate vaccine (administered subcutaneously or intramuscularly) is that used against diseases such as meningitis and epiglottitis caused by *Haemophilus influenzae* type b (Hib). [Ten years' experience with Hib conjugate vaccines in Finland: RMM (1996) *7* 231–241.]

Conjugate vaccines against menigococcal meningitis *(Neisseria meningitidis)* have so far given promising results for only certain strains of the pathogen.

Conjugate vaccines against *Streptococcus pneumoniae* (high-risk groups include young children) are useful against only a small proportion of the serotypes of this pathogen, and it would be expensive to manufacture multivalent conjugate vaccines to cover more of the serotypes. However, certain *proteins* are found in the cell envelope in nearly all clinical isolates of *S. pneumoniae*, and these proteins show minimal antigenic variation; it has been suggested that a vaccine containing several of these proteins (which seem to be involved at different stages of pathogenesis) may offer better protection than existing vaccines [TIM (1998) *6* 85–87].

A (parenteral) conjugate vaccine incorporating the *Salmonella* Vi antigen has given promising results in children 2–5 years of age. Trials are planned for children <2 years of age.

[Conjugate vaccines against group B *Streptococcus* types IV and VII (immunogenicity and efficacy in animals): JID (2002) *186* 123–126. Effectiveness of meningococcal C conjugate vaccine in teenagers in England: Lancet (2003) *361* 675–676.]

Vaccines: use of toxoids. Just as *components* of a pathogen can act as antigens, so too can many bacterial products, including toxins. When a toxin combines with its antibody the harmful properties of the toxin are neutralized, and such combination also facilitates elimination of the toxin from the body. Vaccination against a toxin (such as tetanospasmin: section 11.3.1.3) involves administration of a modified form of the toxin (a *toxoid*) which is no longer harmful but which has kept its specific antigenic characteristics; a toxoid thus induces the formation of anti-toxin antibodies. Subcutaneous or intramuscular vaccination with a toxoid is used against e.g. tetanus and diphtheria.

In the above examples a toxoid is administered *prophylactically*, i.e. to prevent disease by stimulating antibody formation. In e.g. botulism (section 11.3.1.2) an antiserum (in this case, serum containing *pre-formed* anti-toxin antibodies) can be used *therapeutically*, i.e. for treatment. The use of pre-formed antibodies is called *passive immunization*.

DNA vaccines. A DNA vaccine is a parenterally administered vaccine consisting of DNA; the DNA encodes specific antigen(s) that are synthesized in the body and which may then induce antibodies and/or a cell-mediated response. The potential of DNA vaccines against diseases such as tuberculosis (and e.g. malaria) has attracted much interest.

In mice, a DNA vaccine encoding the 65 kDa heat-shock protein of *Mycobacterium tuberculosis* has been found to act prophylactically and therapeutically in experimental infection with *M. tuberculosis*; DNA encoding interleukin-12 (IL-12) had an even greater therapeutic effect [Nature (1999) *400* 269–271].

Vaccines: other aspects. Vaccines that block the pathogen's uptake of *iron* reflect the absolute requirement for this metal. The antigens used include e.g. cell-surface receptors for siderophilins (section 11.5.5). One early study (in pigs) found that transferrin-binding protein B from *Actinobacillus pleuropneumoniae* could induce antibodies that protected the animals from challenge with a homologous strain of the pathogen. The transferrin-receptor proteins of *Neisseria gonorrhoeae* (TbpA, TbpB) have been considered as potential vaccine antigens, but they appear to lack sufficient immunogenicity in natural infections [INFIM (2004) *72* 277–283].

Monoclonal antibodies. If a B lymphocyte is 'fused' with a tumour cell, the resulting *hybridoma* can replicate as a tumour cell and secrete antibodies of the particular type specified by that B cell; we can thus obtain a large population of *identical* (monoclonal) antibodies. Monoclonal antibodies (mAbs) have many uses, including e.g. the detection of specific microorganisms by using 'labelled' monoclonal antibodies to detect their unique antigen(s). [Monoclonal antibodies specific for *E. coli* O157:H7 LPS: JCM (1997) *35* 679–684.]

11.4.2.2   *Cell-mediated immunity*

In cell-mediated immunity (CMI), cells are involved *directly* as effectors in immune responses; this contrasts with antibody-mediated responses, which are designated *humoral immunity*. CMI includes responses in which the main effector cells are antigen-specific T cells.

Many or most T cells (not all) can bind their specific antigen only when it is linked to a class I or class II MHC molecule at the surface of a cell. For example, in section 11.4.2.1, the T cell bound to a (processed) antigen linked to a class II molecule at the surface of the B cell. MHC class II molecules are usually found only on specialized *antigen-presenting cells* (APCs) such as B cells, macrophages and dendritic cells; however, they can be induced on certain other cells – e.g. IFN-γ can induce class II molecules on endothelial cells. MHC class I molecules occur on most nucleated cells of the body. T cells which bind only to surfaces containing MHC molecules are said to be subject to *MHC restriction*. In section 11.4.2.1 the T cell was MHC class II-restricted.

$\gamma\delta$ T cells (in which the T cell receptor consists of γ and δ chains, rather than the more common α and β chains), and also certain other T cells, can bind antigen in an MHC-independent manner (see later).

The antigen-binding characteristics of most T cells suggest that these cells have a major role in responding to *intracellular* pathogens whose (processed) antigens are displayed at a cell surface. When an antigen-specific, MHC-restricted T cell binds to its corresponding antigen, the outcome depends e.g. on the class of MHC molecule (I or II) linked to the (processed) antigen; some examples:

- Within a macrophage, the cells of certain pathogens, e.g. *Mycobacterium tuberculosis*, are found in vesicles or phagosomes (section 11.4.1), rather than in the cytoplasm of the macrophage. Antigens from these (intravesicular) pathogens are processed to peptide fragments that are linked to MHC class II molecules and presented at the surface of the macrophage. This combination (peptide + class II molecule) is recognized, and bound, by an antigen-specific CD4+ T cell of the Th1 subset. The bound T cell releases IFN-γ, a cytokine which promotes *activation* of the macrophage. An activated macrophage is more aggressive toward the ingested pathogen (e.g. it produces increased amounts of antimicrobial agents such as hydrogen peroxide and nitric oxide); it can kill some types of ingested pathogen that may not be killed in the absence of activation.

- Some pathogens occur in the cell's cytoplasm, rather than in a phagosome (cf. example above). It appears that (protein) antigens from this type of pathogen are processed to peptide fragments within *proteasomes* (section 2.2); these peptides are then linked to MHC class I molecules. At the cell surface, the combination (peptide + class I molecule) is recognized, and bound, by a CD8+ T cell whose T cell receptor (TCR) recognizes the peptide; these T cells, acting as *cytotoxic* cells (T_{cyt}), can kill the cells to which they bind. (This activity is apparently promoted by IL-2.) Cytotoxic T cells kill target cells e.g. by (i) releasing lethal agents (including pore-forming molecules: *perforins*) and (ii) inducing *apoptosis* (programmed cell death); as apoptosis does not involve cell lysis, this method of killing host cells may limit the pathogen's spread.

Note that activation of a T cell (through binding to specific antigen) affects not only the antigen-presenting cell but also the T cell itself. Following the binding of antigen, a T cell undergoes division, and the daughter cells continue to divide, giving rise to a clone (*clonal expansion*). Thus, in the above examples, CD4$^+$ Th1 cells and CD8$^+$ cytolytic cells undergo clonal expansion. Similarly, CD4$^+$ Th2 cells proliferate following contact with B cells (section 11.4.2.1). An expanded clone of T cells includes a number of *memory cells* involved in long-term acquired immunity to the given antigen; compared with *naïve* T cells (those not previously exposed to antigen), memory T cells can be distinguished by the presence of certain strongly expressed cell-surface molecules (e.g. CD29, CD44).

A form of cell-mediated killing which is not MHC-restricted is *antibody-dependent cell-mediated cytotoxicity* (ADCC). In this process, antibody-coated bacteria are lysed by certain white blood cells (e.g. neutrophils, NK cells, monocytes) whose CD16 receptor binds to the Fc portion of antibodies.

In another process, *natural killer* (NK) cells (related to T cells but lacking the TCR) bind to, and kill, host cells containing certain (intracellular) pathogens, including *Listeria monocytogenes*; this activity may involve the release of perforins (and/or of *granzymes* – which trigger apoptosis). In that the activity of NK cells is not antigen-specific, this aspect of CMI is sometimes considered under the heading of innate immunity.

MHC-independent binding of T cells to certain *non*-peptide antigens involves the CD1 molecule. Thus, CD1 (rather than class I or II MHC molecules) can present e.g. lipoarabinomannan (LAM, a lipoglycan) from *Mycobacterium tuberculosis*. Initially, LAM is bound by a cell-surface PRR (section 11.4.1) such as the CD14 receptor; after uptake and intracellular processing the antigen, linked to CD1, is presented at the cell surface to an antigen-specific T cell. Note the involvement of a PRR – another example of the interrelationship between innate and adaptive aspects of the immune system.

In some diseases, CMI contributes to, or may fail to prevent, tissue damage. For example, in tuberculosis, damage associated with a *tubercle* (localized, degenerate site of infection) is likely to reflect, at least partly, concentrations of activated macrophages secreting cytotoxic agents. In e.g. guinea pig and man, the mature tubercle has a caseous ('cheesy') *necrotic* centre surrounded by epithelioid macrophages and an outer mantle of monocytes and lymphocytes (Plate 11.3). During development of the lesion, the macrophages are activated e.g. by IFN-γ, secreted by lymphocytes. However, because the lymphocytes remain peripheral (i.e. do not penetrate the centre of the lesion), the concentration of IFN-γ in central parts of a large lesion may be too low to activate the macrophages (which are then likely to die); it has therefore been suggested that the peripheral distribution of lymphocytes may account for the central necrosis observed in larger tubercles [TIM (1998) 6 94–97].

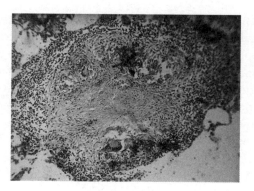

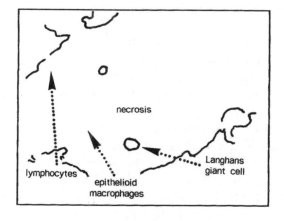

Plate 11.3   A stained section (light-micrograph) of tuberculous lung tissue showing a mature tubercle (see section 11.4.2.2); typical components of the lesion are indicated below. The centre of the tubercle consists of featureless, caseous ('cheesy') necrotic tissue. The layer of epithelioid macrophages is surrounded by an outer layer of dark-staining lymphocytes. Each Langhans giant cell is formed by the fusion of a number of macrophages; it is a multinucleate cell in which the nuclei are located at the periphery of the cytoplasm. A tubercle is classified as a granulomatous lesion.

11.5   THE PATHOGEN: VIRULENCE FACTORS

Bacterial pathogenicity (especially *in vivo* mechanisms) is being studied intensively; the reasons include increasing resistance to antibiotics and an upsurge in old problems (such as tuberculosis). It is thought that a better understanding of disease *in the host* may reveal new targets for chemotherapy or suggest alternative methods of intervention.

Research on pathogenesis has been stimulated by the development of molecular methods such as Tn*phoA* mutagenesis and signature-tagged mutagenesis (section 8.5.5.3). Another approach, *in vivo* expression technology (IVET), is outlined in Fig. 11.3. IVET detects those genes (of a pathogen) whose promoters are active only during infection of the host; such genes *may* be virulence genes, and this can be studied e.g. by animal tests using strains of the pathogen that are mutant for given gene(s). [Bacterial behaviour in the host, and the methodology for studying it: TIM (1998) 6 239–243.]

It should be appreciated that a given virulence factor is not necessarily expressed permanently. In a number of cases it has been found that virulence factors are controlled by mechanisms such as two-component regulatory systems (section 7.8.6) and quorum sensing (section 10.1.2). For example, certain virulence factors in *Staphylococcus aureus* are controlled by two-component systems that respond to environmental levels of oxygen [JB (2001) *183* 1113–1123]. Quorum sensing regulates the expression of virulence factors in *Pseudomonas aeruginosa* [TIM (1997) 5 132–134; PNAS (2001) *98* 2752–2757] and in *Escherichia coli* (EHEC) strain O157:H7 [PNAS (2003) *100* 8951–8956].

Overtly aggressive products, such as *toxins* (sections 11.3.1 to 11.3.1.7), are clearly virulence factors. However, so too are those products and strategies which help a pathogen to become established in the host and to evade the host's defences. Some of these factors are considered below.

11.5.1   Avoiding phagocytosis

In some pathogens the capsule (section 2.2.11) inhibits phagocytosis. For example, in certain strains of *Streptococcus pyogenes* a capsule of hyaluronic acid (a component of animal tissues) provides a kind of camouflage. Other inhibitors of phagocytosis include streptococcal M protein (section 2.2.13) and the poly-D-glutamic acid capsule of *Bacillus anthracis*. The virulence-promoting abilities of hyaluronic acid and M protein have been assessed [INFIM (1997) *65* 64–71.]

In *Staphylococcus aureus*, the cell wall component *protein A* binds to the Fc part of an antibody (Fig. 11.2), i.e. that part of the antibody which normally binds to Fc receptors on phagocytes. Protein A is anchored to the cell wall peptidoglycan of *S. aureus* by the enzyme *sortase* [MM (2001) *40* 1049–1057].

Polysaccharide capsules occur e.g. in *Haemophilus influenzae* type b (Hib), *Neisseria meningitidis* and *Streptococcus pneumoniae*. These capsules usually do not prevent phagocytosis in those adults, and older children, who have a normal immune system that responds with: (i) production of anti-polysaccharide antibodies, (ii) enhanced opsonization via complement (section 11.4.1.1), and (iii) phagocytosis. However, infants and young children give a poor antibody response to polysaccharides, and in this age group these pathogens can cause e.g. meningitis and respiratory tract infections; it is for this reason that the *conjugate* vaccines (section 11.4.2.1) were developed.

Yersinia spp avoid phagocytosis in a different way. On contact with a phagocyte, the

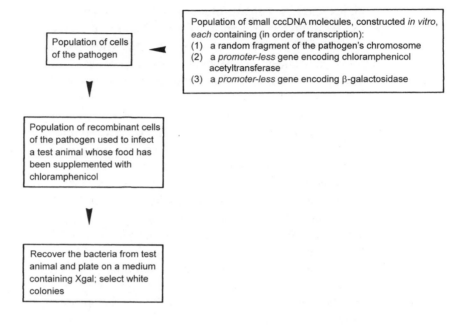

Figure 11.3 IVET (*in vivo* expression technology) (section 11.5): the principle (diagrammatic). The figure shows one of several forms of IVET (others are described at the end of the legend).

IVET detects those genes of a pathogen which are induced ('switched on') *during infection of the host animal.*

Vector molecules (see *top, right*) are inserted, by transformation (section 8.4.1), into a population of cells of the pathogen. In each cell the (random) fragment of chromosome in the vector inserts into the *corresponding* part of the pathogen's chromosome (by an insertion–duplication mechanism: Fig. 8.21a); because the vector molecules contain different fragments of the chromosome, they will insert into different chromosomal sites in different cells – forming a heterogeneous population of recombinant cells.

In some recombinant cells, the vector's two promoter-less genes will have been inserted 'in frame' with an upstream promoter: in such cells, *both* of these genes are transcribed *if the promoter is active.*

The recombinant cells are used to infect a test animal whose food contains the antibiotic chloramphenicol (section 15.4.4). Under these conditions, a recombinant cell can grow if it produces chloramphenicol acetyltransferase (CAT), i.e. if the CAT gene (in the vector) is controlled by an active promoter; thus, the fact that a given recombinant cell *grows* indicates that its CAT gene is controlled by a promoter *which is active within the test animal.* (Because synthesis of CAT points to an active promoter, the CAT gene is sometimes called a 'reporter' gene.) Cells producing CAT can form large populations which greatly outnumber cells that do not form CAT.

We need to know whether the promoter controlling a CAT gene is active *only* within the test animal or whether it is also active when the pathogen is cultured (e.g. on agar media). If active only in the test animal, this indicates that the gene *normally* controlled by the promoter is induced during infection; such a gene is of interest because of its possible association with virulence. To study promoter activity further, the recovered bacteria are plated on a medium lacking chloramphenicol but containing the agent Xgal (section 8.5.1.5); all the cells will grow. If a given promoter is active in culture, then β-galactosidase (section 8.5.1.5) will be formed, giving rise to a *blue-green* colony; hence, such a colony tells us that activation bacterial type III secretory system (section 5.4.4) injects several proteins into the cell. One of the proteins, YopH, is an enzyme that de-phosphorylates certain host proteins which, when phosphorylated, are involved in: (i) phagocytic uptake of the bacterium, and (ii) the oxidative burst. YopJ down-regulates certain kinases and, in this way, inhibits production of the pro-inflammatory cytokine TNF-α in macrophages [MM (1998) 27 953–965]. YopE has been shown to disrupt actin-based structures *in vitro*. [The *Yersinia* Yop virulon: MM (1997) 23 861–867; PNAS (2000) 97 8778–8783.]

11.5.2   Killing cells of the immune system

Cells of the immune system can be killed by certain types of molecule secreted by Gram-positive and Gram-negative bacteria. Moreover, macrophages can be killed by intracellular *Shigella flexneri*.

Killing by Gram-positive bacteria. Phagocytes can be killed by *leucocidins* secreted by some staphylococci and streptococci. For example, the staphylococcal *Panton-Valentine leucocidin* specifically lyses macrophages and polymorphonuclear leukocytes (PMNs).

Killing by Gram-negative bacteria. The so-called RTX toxins of Gram-negative bacteria appear to behave primarily as virulence factors directed against cells of the immune system. All RTX toxins can form pores in eukaryotic cell membranes, in some cases leading to osmotic lysis; moreover, at quite low concentrations, HlyA can bring about a loss of killing power in PMNs, and cyclolysin can inhibit the oxidative burst in phagocytes. (Cyclolysin has adenylate cyclase activity; increased levels of intracellular

---

of the given promoter does not occur *only* in the test animal. A *white* (*lac*⁻) colony indicates that β-galactosidase (and CAT) can be formed *only* within the test animal, i.e. the relevant promoter is active only during infection of the animal. The gene which is normally controlled by this promoter can be cloned/sequenced etc. and examined for its role in virulence.

A different version of IVET uses an *auxotrophic* (section 8.1.2) population of the pathogen which (i) can grow on agar media supplemented with the given growth requirement, but (ii) cannot grow within the test animal unless the relevant, and functioning, gene is present. Each vector molecule includes the genes encoding (i) the specific growth requirement, and (ii) β-galactosidase – both genes being promoter-less. The rationale is analogous to that given above.

In another version of IVET, each vector molecule includes (i) a tetracycline-resistance transposon, and (ii) promoter-less genes encoding a *transposase* (section 8.3) and β-galactosidase. If the promoter is active in the test animal, transposase is synthesized and excises the transposon; the excised transposon does not replicate, leading to a clone of tetracycline-*sensitive* cells. Tetracycline-sensitive cells can be detected by *replica plating* (section 8.1.2) using media which (i) lack, and (ii) contain tetracycline. As before, β-galactosidase activity is used to detect promoters which are not active in culture.

[IVET: INFIM (2002) 70 6518–6523.]

cAMP may cause a reduction in, or loss of, activity in phagocytes.) At low, sub-lytic concentrations these toxins may act as *modulins* (section 11.5.4). At least two of the toxins – the HlyA haemolysin of *Escherichia coli* and the cyclolysin of *Bordetella pertussis* – are secreted by an ABC exporter (section 5.4.1.2). [RTX toxin gene in *Vibrio cholerae*: JCM (2001) *39* 2594–2597.]

Another RTX toxin, LktA, secreted by *Pasteurella haemolytica* (a cause of pneumonia in cattle and sheep), is a potent lytic agent for the white blood cells and platelets of ruminants. The cytolytic activity of LktA is reported to be enhanced by lipopolysaccharide [MP (2001) *30* 347–357].

Killing by Shigella flexneri. Following phagocytosis by a macrophage, *Shigella* escapes from the vacuole and can kill the macrophage by a mechanism involving the (plasmid-encoded) bacterial protein IpaB; IpaB is secreted by the pathogen's type III secretory apparatus (section 5.4.4). IpaB binds to, and activates, the IL-1β-converting enzyme (ICE, also called caspase 1) of the macrophage. Activated ICE (i) converts the inactive form of IL-1β (and the inactive form of IL-18, if present) to the active form, and (ii) initiates apoptosis (programmed cell death) in the macrophage. IL-1β and IL-18 are both pro-inflammatory cytokines (IL-18 induces IFN-γ production in T cells and promotes the Th1 subset of T cells); the release of these cytokines from apoptotic *Shigella*-infected macrophages may contribute to the inflammatory reaction in dysentery – as may the release of IL-8 (Table 11.4) from epithelial cells invaded by *Shigella* [FEMSMR (2001) *25* 3–14].

11.5.3 Antigenic variation and phase variation

Certain pathogens can change their cell-surface chemistry – and, hence, their antigens; this may help them, even temporarily, to avoid the effects of specific antibodies. For example, in relapsing fever (caused by species of *Borrelia*) there are several cycles of fever and remission, and bacteria isolated from the blood during each period of fever are found to have different surface antigens.

The phrase *antigenic variation* is often used in a general sense to refer to successive changes in the surface chemistry of an organism – regardless of the way in which such changes arise. However, as explained below, changes can arise in two distinct ways.

In one scenario, each of the organism's alternative antigens is encoded by a specific, *pre-existing* allele; that is, any antigen displayed by the organism must be one of those *defined* antigens encoded by alleles in the genome. Changes involving this type of antigen are referred to as *phase variation*; an example is given in Fig. 8.3(c). Phase variation also refers e.g. to the on/off switching of type 1 fimbriae (i.e. the reversible switching between fimbriate and afimbriate cells). This involves (reversible) inversion of a 314-bp segment of DNA containing the promoter of the *fim* operon; in one orientation transcription is possible (fimbriate cells) but in the opposite orientation the *fim* operon is not transcribed (afimbriate cells). Interestingly, cross-talk between the regulatory system of type 1 fimbriae and that of P fimbriae appears to bias the on/off switching mechanism in such a way that the expression of type 1 fimbriae is inhibited [EMBO Journal (2000) *19* 1450–1457].

Phase variation in the Opa adhesins of *Neisseria* involves DNA re-arrangements that change the reading frames in *opa* genes [TIM (1998) *6* 489–495].

In the second scenario, successive new antigens are *created* by repeated recombinational events in the DNA encoding surface antigens. One example is a subunit in the fimbriae of *Neisseria gonorrhoeae*. The gene encoding this subunit, *pilE*, can undergo repeated recombination e.g. with another chromosomal gene, *pilS*, so that new versions of the subunit gene are continually being created; hence, the fimbriae of this organism can be synthesized from any of a large number of variant forms of the subunit, giving rise to extensive antigenic variation. (These fimbriae are also subject to on/off switching, i.e. phase variation.)

In *Helicobacter pylori*, antigenic variation may involve recombination between sequence-related genes encoding different cell-surface proteins. Another possibility is that variation may arise through translational frame-shifting (section 7.8.7): sequences characteristic of this mechanism have been found in the genome [Nature (1997) *388* 539–547].

11.5.4 Modulins

Some pathogens produce *modulins*: virulence factors which can damage the host by inducing inappropriate synthesis of the host's *cytokines*. Cytokines are protein signalling molecules which have a range of functions and which are involved e.g. in the normal co-ordination of the immune system (Table 11.4); however, because cytokines are potent physiological agents, their inappropriate activity may give rise to pathological effects – as e.g. in endotoxic shock and toxic shock syndrome (section 11.11).

Modulins include bacterial exotoxins, lipopolysaccharides, fragments of peptidoglycan, bacterial DNA, heat-shock proteins, superantigens (section 11.5.4.1) and protein A (a cell-surface protein found in many strains of *Staphylococcus aureus*). These molecules can induce the synthesis of cytokines in various types of mammalian cell (including epithelial cells and white blood cells). The cytokines induced by modulins frequently include e.g. IL-1 and IL-6.

11.5.4.1 *Superantigens*

A superantigen (SAg) is a protein, secreted by certain pathogens, which can bind *simultaneously* to (i) a T cell receptor (TCR) that expresses a given Vβ chain, and (ii) an MHC class II molecule on an antigen-presenting cell (APC: section 11.4.2.1); such binding causes the T cell to proliferate and secrete cytokines. Because the T cells that

Table 11.4   Cytokines: some examples[1-4]

Cytokine	Secreted by (e.g.)	Activities (e.g.)
IL-1[5] (interleukin-1)	Various types of cell, e.g. monocytes, fibroblasts, dendritic cells, endothelial cells	Known/presumed *in vivo* functions: e.g. induces macrophages to produce cytokines; induces the liver to form acute-phase proteins; induces fever; induces synthesis of IL-1 in endothelial cells; induces the synthesis of adhesion molecules in endothelial cells during inflammation
IL-2	T lymphocytes (T cells)	*In vivo*, apparently promotes the growth/differentiation of antigenically stimulated T cells and appears to potentiate CD8$^+$ cells and NK cells; it promotes the development of the Th1 subset of T cells
IL-4	Th2 subset of T cells	In activated human B cells, induces class switching to IgG1 and IgE); promotes differentiation of Th0 to Th2 cells
IL-6[6]	Th2 subset of T cells; synthesis in macrophages can be induced by lipopolysaccharides and by some cytokines (e.g. IL-1, TNF-α); the fimbriae of uropathogenic *E. coli* are reported to be potent inducers of IL-6	Various, e.g. promotes antibody formation in B cells, induces acute-phase proteins, induces fever
IL-8[7]	Macrophages, monocytes, fibroblasts, endothelial cells, neutrophils, NK cells	Chemoattractant and activator for e.g. neutrophils; recruitment of leukocytes to sites of inflammation
IL-10[8]	Th2 subset of T cells, monocytes stimulated by lipopolysaccharides	Inhibits the release of cytokines in antigen-stimulated Th1 cells, and inhibits release of pro-inflammatory cytokines from monocytes and macrophages; promotes antibody formation in activated B cells
IL-12	B lymphocytes, macrophages	Promotes development of the Th1 subset from naive T cells; in NK cells it stimulates cytotoxicity; in NK cells, T cells and macrophages it stimulates release of cytokines; it also inhibits the development of new blood vessels (anti-angiogenic activity)
TNF[9] (tumour necrosis factor)	Many types of cell when appropriately stimulated, e.g. Th1 subset of T cells, macrophages	Various, including e.g. cytolysis of certain tumours; induction of apoptosis in target cells; regulation of differentiation and proliferation in lymphocytes; upregulation of cell adhesion molecules (selectins etc.) on endothelial cells during inflammation (section 11.4.1); upregulation of toxin receptors (section 11.3.1.6); induction of chemokines (e.g. IL-8); stimulation of synthesis of cytokines (e.g. TNF-α, IL-1, IL-6) in monocytes and macrophages

γ-Interferon (IFN-γ)   T cells stimulated by antigen or by   See section 11.4.1.2.
(macrophage-derived) IL-18

[1] Cytokines are proteins which function as signalling molecules between cells; there are many different types of cytokine, and the table lists only a few of them. When activated, cells such as lymphocytes and macrophages typically secrete one or more cytokines; cells may respond to a given cytokine only if they have specific receptors for it.

[2] Note that (at least) some cytokines have (additional) functions outside the immune system; thus, for example, IL-6 has been linked to the metabolism associated with bone.

[3] Cytokines can play a role in pathogenesis (see e.g. section 11.5.4.1).

[4] Note that only some of the functions of (a few) cytokines have been listed in isolation (i.e. out of context). In reality, these cytokines work as part of a highly complex and dynamic *network* of agents. Some functions can be carried out by more than one type of cytokine, and some cytokines have a range of different functions – which can vary according to the type of cell to which they bind. Moreover, a given cytokine may be synthesized by different types of cell, under differing conditions. A further complication is that cytokines may work together (synergism) or oppose one another (antagonism).

[5] There are several forms. IL-1α and IL-1β are both pro-inflammatory cytokines with similar functions; in each case the mature protein is produced by cleavage of a precursor protein. Cells which form IL-1 generally form both the α and β types, but in some cases (e.g. a monocyte responding to lipopolysaccharide) IL-1β is produced in greater quantity. IL-1β can act synergistically with TNF-α.
IL-1ra binds to the IL-1 receptor; it acts as an antagonist to the α and β forms of IL-1.

[6] Synthesis and release of IL-6 may be induced e.g. by endotoxin. Among its many activities, IL-6 induces C-reactive protein (CRP: an acute-phase protein which is an early marker of inflammation) by stimulating transcription of the CRP gene in the liver.

[7] IL-8 is one of a subcategory of cytokines called *chemokines*. The chemokines are chemotactic for, and activate, particular subsets of leukocytes; they are important e.g. in the process by which leukocytes move out of blood vessels and into tissues during inflammation (section 11.4.1). Chemokines are divided into two types (CXC and CC) on structural/functional differences.
The leukocyte cell-surface receptors for chemokines are molecules of the rhodopsin superfamily. Some receptors bind only one type of chemokine (e.g. CXC-CKR2 is specific for the CXC-type chemokine IL-8), but others can bind more than one type. A receptor on red blood cells binds chemokines of both CXC and CC types; such binding is (for the RBC) non-functional, but it may serve as a 'sink' to prevent inappropriate activation of leukocytes.
Anti-IL-8 monoclonal antibodies exhibit anti-inflammatory activity.

[8] An anti-inflammatory cytokine; one of its effects is to down-regulate pro-inflammatory cytokines such as TNF-α and IL-1β.

[9] TNF-α is produced by many types of cell, but mainly by activated macrophages; receptors designated p55 and p75 occur on various types of cell. TNF-α and IL-1β each stimulate their own production and that of each other. TNF-β is similar to TNF-α; it is formed by lymphocytes, binds to the same receptors and has similar effects. Following endotoxaemia, the peripheral circulation contains raised levels of *soluble* (i.e. isolated) TNF receptors; this may be a mechanism for down-regulating the effects of TNF on cells. (See also section 11.10.3.)

express any given Vβ chain include many individual clones (each specific for a given, unique antigen), the effect is *polyclonal activation*, i.e. activation of many different strains of T cell as though a range of different antigens had been present. The resulting massive release of cytokines (from T cells, macrophages, or other APCs) can cause conditions such as *toxic shock syndrome* (TSS: section 11.11); superantigens are therefore classified as *modulins* (section 11.5.4).

Superantigens can activate both CD4$^+$ and CD8$^+$ cells, binding to the Vβ chain at sites *outside* the antigen-binding region. Different superantigens may bind preferentially to different Vβ chains (thus affecting different subsets of T cells).

On activation, cells of the given subset proliferate; in e.g. rodents, most cells of that subset subsequently die, and the animal exhibits (a temporary) unresponsiveness (*anergy*) to the particular superantigen. In rodents, anergy to a given superantigen can last for months, but this is apparently not the case in humans.

Possible therapeutic uses for superantigens include suppression of specific T cell subsets, or activation of anti-tumour T cells.

Superantigens include five enterotoxins of *Staphylococcus aureus* (SEA, SEB, SEC (several subtypes), SED and SEE), toxic shock syndrome toxin 1 (TSST-1, formerly exterotoxin F) of *S. aureus*, and several toxins of *Streptococcus pyogenes*. The gene for TSST-1 is carried on a mobile pathogenicity island (section 11.5.7), possibly accounting for the spread of TSST-1 toxigenicity in *S. aureus* [MM (1998) *29* 527–543].

B cell superantigens are analogous molecules that bind to the V$_H$ chain of the B cell receptor.

[T cell superantigens (clinical significance): TIM (1998) *6* 61–65.]

11.5.5   Pathogenicity and iron

Pathogens need iron, e.g. for the synthesis of certain enzymes and/or cytochromes – but iron can be scarce in the host organism; for example, in the mammalian bloodstream, the host's iron-chelating glycoprotein, *transferrin*, binds ferric iron and transports it into cells. To cope with this problem, many pathogens encode their own iron-uptake system: a *siderophore* (a low-molecular-weight ferric iron chelator, which the pathogen secretes) and associated cell-surface receptors at which the iron–siderophore complex is bound. Thus, e.g. extra-intestinal strains of pathogenic *E. coli* (such as those causing neonatal meningitis) often form *aerobactin*: a hydroxamate siderophore which binds iron and then binds to a receptor site on the outer membrane; aerobactin seems able to compete successfully with transferrin.

Theoretically, siderophores could simply deliver iron from the environment to surface receptors. However, in *E. coli*, the siderophore *enterochelin* (= *enterobactin*) first binds to a receptor (the outer membrane protein FepA) and is then transported into the cytoplasm; within the cytoplasm iron is released when the siderophore is cleaved (by an esterase). Fe^{3+} is reduced to Fe^{2+}.

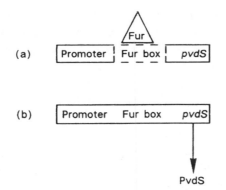

Figure 11.4 Simplified scheme for the iron-regulated synthesis of pyoverdin in *Pseudomonas aeruginosa* (section 11.5.5). (a) In the presence of adequate concentrations of iron, the transcription repressor, the Fur (ferric uptake regulation) protein (product of the *fur* gene), binds to a highly conserved region of DNA (the 'Fur box'), inhibiting transcription of gene *pvdS*. (b) When the iron concentration is insufficient, Fur no longer binds to the Fur box – thus permitting transcription of *pvdS*. The product of *pvdS* (PvdS) is a sigma factor (see sections 7.5 and 7.8.9) which is needed for the transcription of other genes involved in the biosynthesis of pyoverdin.

In *Pseudomonas aeruginosa*, a shortage of iron triggers the synthesis of a yellow-green fluorescent siderophore, *pyoverdin*, which stimulates the pathogen's growth even in the presence of transferrin. A proposed mechanism for the iron-regulated induction of pyoverdin is shown in Fig. 11.4. In *Pseudomonas fluorescens*, pyoverdin represses the synthesis of another siderophore, *quinolobactin* [AEM (2000) 66 487–492].

Salmonella typhimurium encodes a catecholamide siderophore, *enterochelin*, whose expression involves regulation by the Fur protein. In *S. typhimurium* the Fur protein is also a regulator of the acid tolerance response (section 7.8.2.6).

Mycobacterium tuberculosis has at least two strategies for obtaining iron during extracellular growth. A cell-wall-associated *mycobactin* chelates ferric ions and is believed to provide (enzymically reduced) ferrous ions for uptake. Particularly under iron-deficient conditions, iron may be transferred from the host's own iron chelators (*siderophilins*: transferrin and/or lactoferrin) to mycobactin via low-molecular-weight iron-solubilizing peptides, called *exochelins*, which are secreted by the pathogen. (Transferrin occurs in plasma, lactoferrin in secretions.)

Some pathogens obtain iron *directly* from the host's siderophilins – e.g. *Neisseria gonorrhoeae* has receptors for human transferrin, and *N. meningitidis* binds the transferrin of humans and some other primates; the (cell-surface) receptor consists of two proteins: TbpA and TbpB (Tbp = transferrin-binding protein). Strains of these pathogens which are unable to bind siderophilins (owing to lack of receptors) are not virulent; moreover, strains which *do* form receptors are virulent only in those host species in which the receptors are effective. [Iron uptake in *N. meningitidis*: MMBR (2004) 68 154–171.]

Helicobacter pylori obtains iron from lactoferrin while in the gastric *lumen*; however, when this pathogen intercalates between cells of the gastric wall its sole source of iron is apparently *haem* – which binds to specific iron-repressible protein receptors in the outer membrane.

Other strategies by which pathogens obtain iron include (i) assimilation of the haem from lysed erythrocytes (red blood cells), and (ii) use of the host's *intracellular* iron.

The method(s) used by a given pathogen to obtain iron will determine, at least partly, the species of host susceptible to infection by that pathogen and the types of tissue/cell that are affected; this is because a pathogen can multiply only in hosts/tissues/cells where it can acquire an adequate supply of iron. For example, *Actinobacillus pleuropneumoniae* has cell-surface receptors specific for *pig* transferrin, and this pathogen can cause pneumonia in pigs but not in humans. By contrast, *Listeria monocytogenes* can cause disease in cattle, sheep and humans – being able to use the siderophores of other bacteria and also various iron-chelating compounds found in the environment and in mammalian hosts; this pathogen may use a cell-surface enzyme, ferric reductase, to recognize all of these different iron chelators (rather than having individual receptors for each).

A pathogen's *absolute* need for iron has prompted the investigation of new vaccines that stimulate the production of antibodies against the pathogen's cell-surface receptors for siderophilins (see section 11.4.2.1).

[Iron metabolism in pathogenic bacteria: ARM (2000) *54* 881–941].

11.5.6 Adhesins

The ability of a pathogen to adhere to specific cells in the host's tissues is often an important virulence factor (section 11.2.1). Bacterial cell-surface structures/molecules which promote adhesion are called *adhesins*; there are different types of adhesin, and a given bacterium may have more than one type. Some adhesins have trivial names – e.g. *invasin* (*Yersinia enterocolitica*), *internalin* (*Listeria monocytogenes*), *intimin* (EHEC, EPEC).

In many cases adhesins are fimbriae (section 2.2.14.2), and they are often encoded by plasmids. Among strains of ETEC (Table 11.2) there are more than 10 different types of fimbrial adhesin, but there are also several types of non-fimbrial adhesin. Other non-fimbrial adhesins include the streptococcal M proteins (section 2.2.13) and the filamentous haemagglutinin (FHA) adhesin of *Bordetella pertussis*.

The TCP adhesin of *Vibrio cholerae* is important not only as a means of adhesion but also as a phage receptor that allows phage CTXΦ, which contains the genes for cholera toxin, to spread to non-lysogenic cells of *V. cholerae*.

Staphylococcus aureus forms protein adhesins which are typically linked to the cell-wall peptidoglycan; these adhesins bind to components of the mammalian extracellular matrix, such as collagen and fibronectin [TIM (1998) *6* 484–488].

All pathogenic strains of *Neisseria* encode adhesins designated *Opa proteins*. [Role of Opa proteins: TIM (1998) *6* 489–495.]

In UPEC (uropathogenic strains of *Escherichia coli*), type 1 fimbriae act as adhesins and *also* promote uptake of the pathogen by bladder epithelial cells; by contrast, adhesion mediated by the PapG adhesin of P fimbriae does not promote uptake [EMBO Journal (2000) *19* 2803–2812].

11.5.7   Pathogenicity islands

A pathogenicity island (PAI) is a cluster or 'cassette' of several to many genes encoding a set of virulence characteristics. The DNA of PAIs typically has a GC% (section 16.2.1) unlike that of the organism's chromosome – suggesting that PAIs have been acquired by so-called horizontal gene transfer (i.e. by processes such as transduction or conjugation); PAIs are also typically present at chromosomal sites close to genes encoding tRNAs, and many are flanked by direct repeats.

Gene clusters with the characteristics described above have been found in a number of Gram-negative pathogens. The cluster of virulence genes in the (Gram-positive) species *Listeria monocytogenes* has a GC% similar to that of the chromosome.

The 35 kb LEE (*locus of enterocyte effacement*) PAI occurs e.g. in all strains of EPEC (Table 11.2). It encodes the means by which this pathogen produces the characteristic 'attaching and effacing' lesions (section 11.2.1, Plate 11.1). LEE, which is not flanked by direct repeats, contains genes for a type III secretory system (section 5.4.4) and effector proteins which are translocated into the host cell. [Type III secretory systems and PAIs (review): JMM (2001) *50* 116–126.]

At least some of the genes in LEE are under transcriptional control from a regulatory region (*per*) in a plasmid (EPEC adherence factor plasmid: EAF plasmid) found in all strains of EPEC. This plasmid also encodes the so-called bundle-forming pili (BFP) which may be involved in the formation of microcolonies of EPEC on intestinal epithelium.

Strains of UPEC (uropathogenic *Escherichia coli*) contain a 70 kb PAI (PAI-1) encoding a haemolysin. This PAI inserts into the chromosome at a site identical to that occupied by LEE in strains of EPEC. Thus, it would seem that a non-pathogenic strain of *E. coli* receiving either LEE or PAI-1 would develop as EPEC or UPEC, respectively. However, in general, acquisition of a PAI does not necessarily mean that the recipient cell will become a pathogen because e.g. the imported genes must be able to function in the context of the chomosome as a whole.

Salmonella spp contain at least five distinct PAIs. For example, SPI-1 includes the *inv–spa* genes that encode a type III secretory system used in the invasion of epithelial cells (section 11.2.2.1). SPI-3 encodes products needed for survival in macrophages; like LEE, this PAI is located downstream of the *selC* gene (which encodes the tRNA for selenocysteine). SPI-5 apparently contributes specifically to *entero*pathogenicity [MM (1998) *29* 883–891].

In *Staphylococcus aureus*, the gene for toxic shock syndrome toxin is carried on a *mobile* PAI [MM (1998) *29* 527–543].

Other PAIs include *cag* in *Helicobacter pylori* (a major product being the CagA protein [Science (2000) *287* 1497–1500]), and, in *Vibrio cholerae*, the PAI (VPI) that encodes so-called 'toxin co-regulated pili' (TCP) which act as receptors for phage CTXΦ.

11.5.8 Influence of the *in vivo* environment on the expression of virulence factors

Signals within the mammalian intestine can promote the transfer of toxigenicity between strains of infecting bacteria. Thus, under *in vitro* (laboratory) conditions, phage CTXΦ – which encodes cholera toxin (CT) – can infect certain CT-negative strains of *V. cholerae* at only very low rates. However, within the mouse intestine, lysogenic conversion of CT-negative strains by CTXΦ (derived e.g. from a co-infecting CT-positive strain) occurs with a nearly 1 million-fold greater efficiency; this increased rate of lysogenic conversion appears to be due to the induction of toxin co-regulated pili (TCP) in response to signals in the intestinal environment – TCP acting as receptor sites for phage CTXΦ (section 11.3.1.1). This is relevant to our ideas of the general transmissibility of virulence factors: a given virulence factor which appears to be non-transmissible in the laboratory may well be readily transmissible in the appropriate *in vivo* environment.

In Lyme disease (section 11.11) the pathogen's cell-surface protein C (OspC) is induced within the tick vector [PNAS (2004) *101* 3142–3147]; OspC facilitates invasion of the tick's salivary glands – important for the infection of the humans [JCI (2004) *113* 220–230].

11.6 PATHOGEN–HOST INTERACTIONS: A NEW PERSPECTIVE

Studies at the molecular level have shown that pathogenesis can be more complex than previously supposed. Information on the activities of pathogens within their hosts has come from various methods – including e.g. STM (Fig. 8.22) and IVET (Fig. 11.3), both of which monitor the expression of a pathogen's genes *during infection* of host animals. It is now clear that pathogens frequently 'manipulate' functions of the host cell and that some mechanisms for exploiting the host may be shared by diverse pathogens. [Exploitation of mammalian cell functions by bacterial pathogens (review): Science (1997) *276* 718–725.] *Host* genes involved in pathogen–host interactions have been investigated e.g. with so-called *knock-out* mice (section 8.5.5.4).

Examples of known or proposed pathogen–host interactions include:

- Interactions between pathogens and gut epithelium (section 11.6.1).

- Upregulation of mammalian receptor sites induced by initial binding of the FHA adhesin of *Bordetella pertussis* (section 11.2.1).

- Host phosphorylation of a secreted bacterial protein, followed by insertion of this protein into the host's membrane as a receptor for the pathogen (section 11.2.1).

- Bacterial induction of cytoskeletal re-arrangement in host cells prior to invasion (e.g. *Shigella* in section 11.2.2.1).

- Induction of specific adhesins/phage receptors on *Vibrio cholerae* by signals from the host's gut environment (section 11.5.8). (Similar induction of fimbriae under *in vivo* conditions has been reported e.g. for an STb-encoding strain of ETEC.)

- Induction of apoptosis in a host cell (see *Shigella flexneri*: section 11.5.2).

- Toxin-mediated induction of cytokines, increasing the host's susceptibility (to the toxin) by upregulating toxin-binding receptors (section 11.3.1.6). (The ability of toxins to affect the host's cytokine network is common.)

- Secretion of proteins by a type III bacterial system (section 5.4.4) induced by *contact* with a (eukaryotic) target cell.

- Activation of bacterial virulence genes by the mammalian hormone adrenalin (epinephrine) [PNAS (2003) *100* 8951–8956].

11.6.1   Responses of the gut epithelium to infection

Certain molecules in the gut epithelium are able to detect, and respond to, specific *pathogen-associated molecular patterns* (PAMPs) – including those that characterize lipopolysaccharides (LPS, section 2.2.9.2) and peptidoglycan (PG, Fig. 2.7); host molecules that mediate such recognition have been called *pattern-recognition molecules* (PRMs) or *pattern-recognition receptors* (PRRs). The recognition of a PAMP by a PRR initiates intracellular signalling and e.g. production of specific cytokine(s) and recruitment of effector cells.

PRRs include the cell-membrane-associated *Toll-like receptors* (TLRs). ('Toll' derives from terminology used for *Drosophila*, the fruit fly.) Some TLRs has been reported to respond to particular PAMP(s) – for example, TLR2 to the PG (and certain lipoproteins) of Gram-positive bacteria, TLR4 to the LPS of *Escherichia coli*, and TLR5 to certain components of the flagellum of Gram-negative bacteria. In general, signals that originate from TLR–PAMP recognition appear, indirectly, to activate NF-κB, a factor that promotes e.g. the transcription of genes for certain cytokines, e.g. IL-8.

PRRs also include *intracellular* molecules that are characterized by a C-terminal leucine-rich repeat (LRR) sequence and a central nucleotide-binding site; these molecules are termed *nucleotide-binding oligomerization domain* (Nod) proteins. The intracellular location of Nod proteins permits detection of invasive pathogens. Nod1 and Nod2 are the best characterized.

Nod1–PAMP interaction leads to oligomerization of Nod1 molecules via their *caspase-activating and recruitment domain* (CARD) region; this leads to activation of NF-κB. Nod1 detects *N*-acetylglucosamine-*N*-acetylmuramic acid tripeptide in which *meso*-DAP occurs at position 3 (see Fig. 2.7) – a structure found in the PG of Gram-negative bacteria [Science (2003) *300* 1584–1587].

Nod2–PAMP interaction also leads to activation of NF-κB. Nod2 detects muramyl dipeptide (MurNAc-L-Ala-D-isoGln: Fig. 2.7) – a structure common to the PG of both Gram-negative and Gram-positive bacteria [JBC (2003) *278* 5509–5512; JBC (2003) *278* 8869–8872].

Mutations in the gene encoding Nod2 are found in individuals suffering from *Crohn's disease* (a chronic disease characterized by inflammation of the gastrointestinal tract). At least two of these mutations abolish the response of Nod2 to muramyl dipeptide, suggesting the involvement of such mutations in the aetiology of Crohn's disease [JBC (2003) *278* 5509–5512; JBC (2003) *278* 8869–8872].

11.7   THE TRANSMISSION OF DISEASE

In some diseases the pathogen does not normally spread from one individual to another. Examples include e.g. tetanus and gas gangrene; in these diseases the pathogen is usually a wound contaminant which typically originates in the soil rather than in another, infected individual.

Other diseases spread from one person (or animal) to another, either directly or indirectly. In relatively few (bacterial) diseases does transmission require direct contact between an infected individual and a healthy one, and these diseases generally involve pathogens which cannot survive for long periods of time outside the body; an example is the venereal disease syphilis, caused by *Treponema pallidum*.

Most diseases spread indirectly from person to person, usually in a way related to the normal route of infection (section 11.2). For example, pathogens which infect via the intestine are commonly transmitted in contaminated food or water; gastroenteritis, dysentery, typhoid and cholera are usually spread in this way. Such transmission generally involves some connection between the food or water and the faeces of a patient who is suffering from the disease; contamination may occur, for example, when the hands of a food-handler carry traces of faecal matter, when a housefly lands alternately on faeces and food, or when sewage has leaked into a source of drinking water (see also sections 12.3.2 and 13.5.1.3). Often the pathogen can be traced back – via the food or water – to individual(s) suffering from the disease. Sometimes, however, the source of a pathogen is a person who is *not* suffering from the disease but who is nevertheless playing host to the pathogen and acting as a *reservoir* of infection;

such an apparently healthy individual who is a source of pathogenic organisms is called a *carrier*. One notorious carrier, a cook by the name of Mary Mallon, transmitted typhoid fever to nearly thirty people before she was traced, and the name 'typhoid Mary' is sometimes used to refer to an actual or suspected carrier in outbreaks of typhoid and other diseases.

Pathogens which infect via the respiratory tract are often transmitted by so-called *droplet infection*. When a person coughs or sneezes, or even speaks loudly, minute droplets of the mucosal secretions are expelled from the mouth; these droplets can contain pathogens if such pathogens are present on the respiratory surfaces. Because the smallest droplets can remain suspended in air for some time, they can be inhaled by other individuals and can therefore act as vehicles for the transmission of pathogens. Diphtheria, whooping cough and pulmonary tuberculosis are examples of diseases which can be transmitted in this way.

A few bacterial pathogens are transmitted from one person to another by a third organism called a *vector*. For example, bubonic plague (caused by *Yersinia pestis*) is transmitted by fleas, while epidemic typhus (caused by *Rickettsia prowazekii*) is typically transmitted by lice; tularaemia (caused by *Francisella tularensis*) can be transmitted e.g. by biting flies or by ticks. (See also section 11.3.3.3). In Africa, insecticidal control of flies was associated with a lower incidence of trachoma [Lancet (2004) *363* 1093–1098].

11.8 LABORATORY DETECTION AND CHARACTERIZATION OF PATHOGENS

Rapid detection/identification/typing of a pathogen is particularly important in certain diseases, and in some cases this is possible (in hours) with nucleic-acid-based methods: see section 16.1.6. [Further details are given in *DNA Methods in Clinical Microbiology*, ISBN 0-7923-6307-8.]

Uncultivable bacteria suspected of involvement in disease may be investigated by histology and by nucleic acid amplification procedures such as PCR. This approach has been used e.g. in Whipple's disease in which the pathogen (*Tropheryma whipplei*) was uncultivable until grown in fibroblasts [NEJM (2000) *342* 620–625]. PCR is useful for confirmation.

Traditional methods are still widely used. Some of these methods are outlined below.

11.8.1 Microscopy; culture; detection of toxin

Microscopy. Light microscopy (using high-dry and oil-immersion objectives) continues to have a number of useful applications in medical bacteriology. Some examples are given below.

- Examination of Gram-stained preparations (e.g. colonies from primary isolation plates) yields basic information (Gram reaction and morphology) required in the initial stages of identification.

- Examination of sputum smears stained by the Ziehl–Neelsen method (section 14.9.2), or by the auramine–rhodamine stain, is a simple but effective initial step in the diagnosis of tuberculosis.

- Examination of a *wet mount* of (uncentrifuged) urine can give a useful indication of the abundance of polymorphonuclear leukocytes (PMNs) in the sample. PMNs occur in normal urine, but numbers greater than $\sim 10^4$/ml (*pyuria*) are generally taken to be indicative of a urinary tract infection (UTI). This assessment, together with a semi-quantitative culture of bacteria in the specimen (see *Culture* below), provides a useful indication of the probability of a UTI.

Culture. Attempts are commonly made to culture the pathogen(s) from a sample of e.g. sputum, pus, urine or faeces. In some cases this involves simply inoculating a medium, or several media, with an inoculum from a given sample (section 14.2) and incubating the media under appropriate conditions; if the primary isolation plate (i.e. the medium inoculated from the specimen) yields a *mixture* of organisms, a pure culture of the pathogen may be obtained by following the kind of procedure outlined in section 14.6. In other cases special procedures are required. For example, before attempting culture for *Mycobacterium tuberculosis*, sputum must be homogenized (liquefied) by a mucolytic agent such as *N*-acetyl-L-cysteine [JCM (1999) 37 175–178; JCM (1999) *37* 137–140]; this facilitates manipulation of the sample and exposes target cells. The sample is then *decontaminated* with e.g. 4% sodium hydroxide solution for a specified time; the purpose of decontamination is to kill a high proportion of the sample's contaminants that could otherwise overgrow a culture of *M. tuberculosis*. (Following decontamination the pH is neutralized.) Note that all procedures involving this pathogen (and some others) are carried out in a safety cabinet (section 14.3).

The choice of medium, and the conditions of incubation, will depend e.g. on the nature of the suspected pathogen; when appropriate, an enrichment medium (section 14.2.1) is used, and in some cases the protocol includes anaerobic culture (sections 11.8.1.1 and 14.7).

Semi-quantitative culture of a urine specimen is used to assess whether the sample contains bacteria (usually a single species, but sometimes two) in numbers sufficient to indicate infection, i.e. numbers higher than those likely to be present as a result of simple contamination of the specimen; a count of $\sim 10^5$ bacteria/ml is referred to as *significant bacteriuria*. In the semi-quantitative culture of urine, 0.001 ml of sample is spread onto media such as nutrient agar, blood agar, MacConkey's agar and CLED medium (Table 14.1); after appropriate incubation, colonies are counted and the number of cells per millilitre is estimated.

Significant bacteriuria is sometimes found in symptomless patients in whom disease (e.g. pyelonephritis) develops subsequently. Conversely, counts may be lower than expected in symptomatic patients, even in those whose urine contains an appreciable number of PMNs; this may occur because e.g. (i) the patient has been taking anti-biotics, (ii) the infecting organism does not grow on the media used for culture, and/or (iii) the infecting bacteria may be largely intracellular (section 11.3.7). In general, counts should be interpreted in the context of clinical findings and in conjunction with counts of PMNs in the wet mount (see above).

The procedure for culturing blood is outlined in section 11.8.1.1.

11.8.1.1 Blood culture

Blood culture is used e.g. to detect blood-borne microorganisms, mainly bacteria, in diseases such as typhoid. Essentially, 5–10 ml of the patient's blood (taken aseptically – section 14.3) is added to 50–100 ml of a medium such as trypticase soy broth; the medium usually contains a blood anticoagulant, and may also contain e.g. particular enzymes to inactivate antibiotic(s) carried over in the blood. The inoculated medium is incubated, and is examined, daily, by subculture to an appropriate solid medium.

PCR (section 8.5.4) can be used to detect/identify organisms in blood cultures by amplifying microbial DNA with species-specific primers; this can give rapid results. Note that some commercial blood culture media contain sodium polyantetholesul-phonate (SPS) as an anticoagulant and an anti-complement agent; SPS inhibits PCR, and its removal from samples used for PCR was reported to facilitate detection of bacterial DNA in blood culture media JCM (1998) 36 2810–2816].

Anaerobes (e.g. *Bacteroides fragilis*) are isolated in some cases of bacteraemia, so it is useful to attempt simultaneous detection of aerobes and anaerobes; this approach can also improve the recovery of facultative aerobes in anaerobic media. Pre-reduced, anaerobically sterilized (PRAS) media are considered to be necessary for effective isolation of anaerobes. [Anaerobic blood culture: RMM (1997) 8 (suppl 1) S87-S89.]

11.8.1.2 Detection of toxins

Samples of food suspected of containing an exotoxin may be examined by the methods referred to in section 12.3.1.1.

A culture of *Corynebacterium diphtheriae* may be examined for toxigenicity (ability to produce diphtheria toxin) by Elek's method. The test is set up as shown in Plate 11.4 *(top)*. Following incubation, toxin production by a given strain is indicated by the whitish lines of precipitate. The precipitate results from combination between toxin and antitoxin which diffuse from the lines of growth and the antitoxin-impregnated strip, respectively. Positive and negative controls, i.e. known toxigenic and non-toxigenic strains of *C. diphtheriae* respectively, must be used. The result is read after 24–48 hours' incubation; a revised form of the test (Plate 11.4, *bottom*) can be read after 16 hours [JCM (1997) 35 495–498].

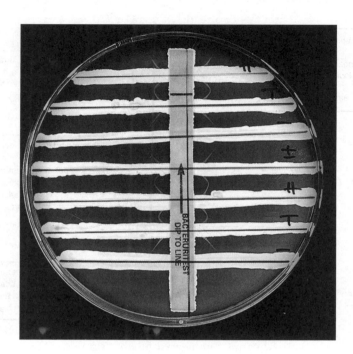

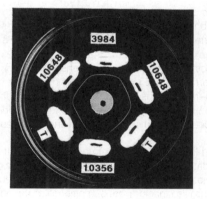

Plate 11.4   Elek plate: standard and modified forms of the test for identifying toxigenic (i.e. toxin-producing) strains of *Corynebacterium diphtheriae*.

Top: the standard method. A suitable growth medium was inoculated with seven strains of *C. diphtheriae*: five test strains, one positive (toxigenic) control and one negative (non-toxigenic) control; the inoculations were made in seven parallel lines – the positive control being at the top and the negative control at the bottom. Before incubation, the plate was overlaid with a strip of filter paper impregnated with antitoxin (i.e. serum containing antibodies to diphtheria toxin). After incubation for 24–48 hours, a toxigenic strain was indicated by fine whitish lines of specific toxin–antitoxin precipitate (section 11.8.1.2). Note that the strain next to the positive control has formed lines of precipitate that merge exactly with those of the positive control; such 'lines of identity' indicate that the test and control strains form a similar diffusible product. Note also that the last line of growth (negative control) has no lines of precipitate.

11.8.2   Immunofluorescence microscopy

This technique permits detection of a specific type of organism in a *smear* (section 14.9) containing a mixture of organisms. Antibodies (section 11.4.2.1), specific to the given organism, are first linked chemically to a fluorescent dye (e.g. fluorescein); the *conjugate* (i.e. suspension of dye-linked antibodies) is added to the smear, unbound conjugate is rinsed off, and the slide is examined by *epifluorescence microscopy*. In epifluorescence microscopy, ultraviolet radiation is beamed onto the specimen from above; visible light from any bound, fluorescent antibodies is seen, in the usual way, via the objective lens of the microscope.

11.8.3   Complement-fixation tests (CFTs)

A CFT may be used to detect specific antibodies in a sample of serum; the presence of antibodies specific to a given pathogen may indicate past or present infection with that pathogen.

Complement (section 11.4.1.1) is bound ('fixed') by most types of antigen–antibody complex (section 11.4.2.1); hence, if fixation occurs when serum is added to a mixture of specific antigen and complement then this typically indicates that the serum contains antibody which matches the antigen. Fixation in a given test mixture naturally lowers the amount of free complement remaining in the mixture; such a decrease (and its extent) – which indicates the presence (and amount) of specific antibody in the serum sample – is monitored in a CFT.

The sample of serum is first heated (56°C/30 minutes) to destroy the patient's own complement, and is then serially diluted. A known, standard amount of complement is added to each dilution. To each dilution is then added a standard amount of specific antigen, and the whole is incubated for 18 hours at 4°C. The presence/absence/quantity of specific antibody in a given dilution of serum is then determined from the amount of free (unfixed) complement remaining in that dilution. The residual complement is assayed by adding (to each dilution) a *haemolytic system*: a suspension of sensitized erythrocytes (red blood cells) which lyse in the presence of complement. If, in a given dilution of the serum, *no* erythrocytes lyse (the cells sedimenting to form a discrete button) then that dilution contains no residual complement – maximum antigen–antibody binding having occured; that is, in that dilution of serum, the concentration of antibodies was sufficient, when complexed with antigen, to fix all the complement. If, for example, *all* the erythrocytes are lysed in the highest concentration

Below: modified method. This form of the test is based on methodology used in Russia and the Ukraine. An antitoxin-impregnated disc is at the centre of the plate, and the test strain(s), together with positive and negative controls, are inoculated in a circular pattern, as shown. *C. diphtheriae* strain NCTC 10648 is a strong positive control, NCTC 3984 is a weak positive, and NCTC 10356 is a negative control; T is a (toxigenic) test strain (examined twice on the same plate).

Photographs courtesy of Dr Kathryn H. Engler, Public Health Laboratory Service, Central Public Health Laboratory, London, UK.

of serum then the test is *negative*, i.e. the serum contained maximum residual complement and (therefore) no detectable specific antibodies.

Adequate controls must be included in any CFT.

11.8.4   Enzyme-linked immunosorbent assay (ELISA)

ELISA is a highly sensitive method for detecting specific antigens or antibodies. The *principle* is explained in Fig. 11.5. The basic idea is also exploited in the ELISPOT assay (section 11.2.3).

11.8.5   *In situ* hybridization (ISH) and fluorescence *in situ* hybridization (FISH)

By using labelled *probes* (section 8.5.3) we can often detect pathogens *in situ*, i.e. actually within the infected tissue. The probes themselves contain species- or strain-

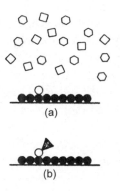

(a)

(b)

Figure 11.5   The *principle* of enzyme-linked immunosorbent assay (ELISA) (diagrammatic). Here, an attempt is being made to detect a low concentration of a particular antibody (open circle) among high concentrations of other antibodies (open squares, hexagons) in a sample of serum.

(a) Specific antigen (solid circles) is immobilized e.g. on the inner surface of a plastic test-tube, and is exposed to the serum. The single molecule of specific antibody binds to its corresponding antigen. Other antibodies, which do not bind, are subsequently removed by washing.

(b) The single, bound antibody is then detected by means of anti-immunoglobulin antibodies – i.e. antibodies whose corresponding antigens are themselves antibodies, and which thus bind specifically to other antibodies. Before use, each anti-immunoglobulin antibody is chemically linked to a particular enzyme, forming an antibody–enzyme *conjugate*. The layer of antigen is then exposed to molecules of the conjugate. One molecule of conjugate (stippled triangle) has bound to the solitary bound antibody; the remainder of the conjugate is washed away. The solitary bound antibody, now carrying a molecule of the conjugate, is thus *labelled* with the enzyme; hence, by detecting the enzyme we can detect the antibody. The enzyme is detected by adding a suitable substrate; in the presence of the enzyme, the substrate is cleaved to a measureable product, which is *amplified* by enzymic action.

specific sequences of nucleotides which hybridize with the complementary sequences in DNA or RNA targets of the pathogen.

The *direct* detection of pathogens by ISH has several advantages. ISH can detect a pathogen before the pathogen could be cultured. This is particularly useful when culture time is long (e.g. weeks for *Mycobacterium tuberculosis*) and is also important for pathogens which cannot be cultured (e.g. *M. leprae*). Species of *Mycobacterium* have been detected by ISH in both fresh-frozen and paraffin-wax sections of infected tissue.

ISH can be useful for diagnosis in immunodeficient (e.g. AIDS) patients whose poor immune response may preclude reliable serological tests for antibodies to specific pathogens. Moreover, in general, antibodies may not be detectable at all stages of an infection.

ISH can distinguish virulent from non-virulent strains of a pathogen by using probes which detect genes for specific virulence factors.

Originally, radioactive labels were used. These give good sensitivity but limited resolution owing to the length of the track of particles in the silver emulsion; moreover, hazards and cost are disadvantages of radioactive methodology. Currently, fluorescent labelling (and epifluorescence microscopy) is widely used (fluorescence *in situ* hybridization: FISH), and kits for labelling oligonucleotide probes are available commercially.

11.8.6   Characterization of the pathogen

If a pathogen is obtained in pure culture it may be examined by tests and procedures of the kind described in sections 16.1 to 16.1.2.15; this usually permits identification to the level of species (or serotype: section 16.1.5.1). Rapid methods of identification and typing are described in sections 16.1.6.1 and 16.1.6.2, respectively.

Commonly, a pathogen's pattern of resistance to a range of antibiotics (its *antibiogram*) is determined, often by a disc diffusion test (section 15.4.11.1).

11.9   PREVENTION AND CONTROL OF BACTERIAL DISEASES

Once we know how a disease spreads, it's often possible to devise methods for preventing or limiting its spread. Clearly, any disease which spreads only by direct physical contact can be prevented simply by avoiding such contact with infected persons. For other transmissible diseases, prevention or control may involve blocking the route of infection from one person to another. Thus, the spread of diseases such as typhoid and cholera can be halted by measures such as: (i) improvement in personal hygiene – e.g. washing hands after visiting the lavatory; (ii) protection of food etc. from flies and other insects likely to carry pathogens, and reduction in the numbers of such insects by the use of insecticides; (iii) protection of drinking water from contamination by sewage, and the effective treatment of communal water supplies with antimicrobial agents such as chlorine; (iv) disinfection of small quantities of untreated water before consumption – e.g. by boiling or by treatment with a disinfecting agent such as halazone; (v) detection, isolation and treatment of carrier(s) (section 11.7), if involved.

Diseases which spread by droplet infection (section 11.7) are generally more difficult to deal with. The physical exclusion of droplets (e.g. with face masks) is usually impracticable, and one of the main control measures in diseases such as diphtheria involves the protection of susceptible individuals by vaccination (section 11.4.2.1). In such diseases there may be some advantage in effectively isolating (*quarantining*) sick individuals until they are no longer able to act as sources of the pathogen.

Diseases spread by vectors (e.g. ticks, fleas, biting flies) can be controlled by eliminating the vector, by reducing its numbers and/or by preventing access of the vector to susceptible individuals.

Probiotics. A probiotic is a potentially beneficial preparation that typically consists of, or includes, a culture of bacteria (or bacterial spores) of a type normally found in the healthy gut microflora. Orally administered probiotics have been reported to restore normal gut microecology and to stimulate intestinal secretion of sIgA antibodies.

The probiotic VSL#3 (consisting of strains of *Lactobacillus*, *Bifidobacterium* and *Streptococcus*) was reported to be effective in preventing flare-up in chronic pouchitis [Gastroenterology (2000) *119* 305–309]. VSL#3 modulates the cell surface and cytokine secretion in dendritic cells [INFIM (2004) *72* 3299–3309].

Probiotic activity has been reported for various strains of *Bacillus*. An aqueous suspension of spores *(Enterogermina)* is used in Italy for preventing/treating diarrhoea of bacterial causation. [Molecular characterization of bacteria *(Bacillus clausii)* from samples of *Enterogermina*: AEM (2001) *67* 834–839.] [Characterization of some *Bacillus* probiotics: AEM (2004) *70* 2161–2171.]

11.10   NOTES ON CHEMOTHERAPY; PHAGE THERAPY; BIOLOGICALS

11.10.1   Chemotherapy

The availability of a wide range of antibiotics (section 15.4) means that many bacterial diseases can be treated by *chemotherapy*, i.e. therapy involving the use of *chem*icals – such as antibiotics. ('Keemotherapy' presumably involves the use of *keem*icals!)

Antibiotics may be wasted if used regardless of their specificity and pharmacology. Moreover, the use of inappropriate drugs may forfeit the chance of curing the disease – and may also exacerbate the general problem of resistance to antibiotic (see section 15.4.11). Factors affecting the choice of antibiotic include:

• Effectiveness against relevant pathogen(s). Ideally, the pathogen's identity and its sensitivities to antibiotics are known prior to treatment, but in some diseases (e.g. meningitis) chemotherapy is often initiated before confirmation of aetiology; in such cases chemotherapy targets *probable* pathogens. Probability reflects factors such as (local) prevalence of particular species/strains of pathogen and the patient's age.

- The (local) prevalence of antibiotic-resistant strains. Some pathogens are now resistant to antibiotic(s) to which they were once invariably sensitive (section 15.4.11); many are resistant to a *range* of drugs, and in some cases therapeutic options are almost exhausted.

- Antagonism towards other antibiotics (section 15.4.10) or to other therapeutic agents. For example, rifampicin potentiates certain mammalian enzymes that break down hormones (e.g. oestrogens); this antibiotic may therefore interfere with oral contraception.

- Possible side-effects, e.g. penicillin allergy (section 11.4.2.1).

- In pregnancy, possible teratogenicity (i.e. risk of causing problems in a fetus).

- Pharmacokinetics. The physicochemical nature of a drug affects its ability to reach particular tissues via a given route. For example, drugs with certain molecular characteristics cannot pass the blood–brain barrier – if injected intravenously they do not reach the CSF; other drugs can, and some drugs can pass the barrier only if its permeability has been increased by inflammation. Again, some penicillins (e.g. amoxycillin) are absorbed from the gut more readily than others (e.g. ampicillin). However, *lack* of absorption can also be a requirement. For example, because orally administered neomycin is poorly absorbed from the (normal) intestinal tract it has been useful e.g. for disinfecting ('sterilizing') the gut prior to abdominal surgery; in this case, poor absorption is important because significant uptake of neomycin into the bloodstream would be associated with e.g. the risk of auditory impairment (dose-dependent damage to the 8th cranial nerve). Sulphonamides (section 15.4.9), being largely excreted in the urine, have been used e.g. for the treatment of urinary-tract infections.

Clearly, to be effective at all, an antibiotic must reach adequate concentrations in appropriate tissues. In certain cases the level of antibiotic needs to be much higher than that predicted from *in vitro* tests; for example, in bacterial meningitis, concentrations in cerebrospinal fluid (CSF) may have to be more than 10 times the minimum bactericidal concentration.

In some cases a drug's activity can be extended in time by inhibiting (normal) excretion from the body; for example, secretion of penicillin at the proximal tubules of the kidney can be inhibited by another drug, *probenecid*.

In some diseases (e.g. botulism, diphtheria) treatment may also include administration of *antitoxin* – a preparation containing antibodies to the relevant toxin. In gas gangrene, treatment involves removal of dead/infected tissue and sometimes the use of hyperbaric oxygen (as well as drugs).

11.10.2 Phage therapy

The problem of bacterial resistance to antibiotics has been an important stimulus in the search for new approaches to antibacterial therapy (see section 15.4.11). In addition to novel types of chemotherapy, attention has also been focused on the use of bacteriophages (Chapter 9) for controlling disease of bacterial causation. Some of the earlier work on phage therapy, carried out in Poland and the former Soviet Union, is described in Table 9.2. [Phage therapy: AAC (2001) *45* 649–659.]

Phage can also help to reduce populations of foodborne pathogens [AEM (2003) *69* 4519–4526; (2003) *69* 5032–5036].

11.10.3 Biologicals

'Biologicals' are biologically derived therapeutic agents which are used e.g. to sequester (bind to), and thus inhibit, particular cytokines. For example, in certain cases, high levels of tumour necrosis factor-alpha (TNF-α; Table 11.4) have been linked to pathogenesis, and genetically engineered agents have been used to inhibit TNF; these agents include the chimeric monoclonal antibodies *infliximab* (Remicade®) and *etanercept* (Enbrel®). In some cases, however, tuberculosis developed soon after treatment with infliximab, apparently reflecting the role of TNF-α in the host's resistance to *Mycobacterium tuberculosis* [NEJM (2001) *345* 1098–1104].

Rituximab® is a chimeric anti-CD20 monoclonal antibody used against certain lymphomas.

11.11 SOME BACTERIAL DISEASES

The following brief, note-form descriptions give some idea of the range of types of infection which can involve bacteria. A few of the diseases (e.g. conjunctivitis) can be caused by agents other than bacteria, but most of the diseases listed are caused only by bacteria – and often only by specific pathogens.

The details given include (i) the name(s) of the causal agent(s), (ii) the characteristic symptoms, and (iii) the route(s) of infection, where known.

Anthrax *Bacillus anthracis* (virulent strains). Localized pustule on skin (anthrax boil) or, rarely, lung infection (woolsorters' disease) or intestinal anthrax; sepsis can occur in untreated cases. Infection via skin wounds, inhalation, ingestion.

[Anthrax (various aspects): JAM (1999) *87* 189–321.]

In herbivores (cattle, sheep, horses) ingestion of *B. anthracis* spores can lead to rapidly fatal sepsis (*splenic fever*). In mice, germination of *B. anthracis* spores, and expression of toxin genes, have been shown to occur within alveolar macrophages [MM (1999) *31* 9–17].

Bacterial vaginosis The vagina contains increased numbers of e.g. *Bacteroides*, *Gardnerella* and *Mobiluncus*, and fewer of the (normal) lactobacilli. Malodorous discharge. Vagina less acidic (e.g. pH > 4.5) and more reducing than is normal. 'Clue cells' (vaginal epithelial cells coated with small Gram-variable/negative rods) typical in smears. Acridine orange staining/fluorescence microscopy is one of the useful methods for diagnosis [JINF (1997) *34* 211–213].

Botulism Any strain of *Clostridium* that forms (neurotoxic) botulinum toxin (section 11.3.1.2); botulinum toxin occurs in antigenically distinct forms A, B, C, D, E, F and G. The causal agent of botulism is usually a strain of *C. botulinum*, but other species can produce the toxin – e.g. *C barati* can form type F toxin [JCM (2002) *40* 2260–2262]. (Strains of *C. botulinum* producing the G toxin were re-classified in a distinct species, *C. argentinense*.)

In humans, botulism is commonly caused by type A, B and E toxins. Classical botulism is a food-borne intoxication resulting from ingestion of foods contaminated with pre-formed toxin. Symptoms, which typically appear after 12–36 hours, may include nausea, blurred/double vision and difficulty with swallowing and speaking; a progressive flaccid paralysis may end in death from asphyxia or cardiac failure. In some cases it appears that a toxigenic strain may infect the intestine [Lancet (1987) *i* 357–360]. Rarely, a wound may be contaminated with *C. botulinum*. In infants there may be a *toxicoinfection* in which toxin is produced in the infected intestinal tract.

Among animals, botulism in cattle may be caused by the B, C or D toxin; type B or C toxin may cause the disease in horses. Birds may be affected by the A and C toxins, and in farmed salmonid fish disease has been caused by type E toxin.

Brucellosis *Brucella* spp. Headache, malaise, intermittent fever which may persist for years. Infection via mouth (e.g. unpasteurized milk), conjunctivae or wounds. In animals, the reproductive system is generally affected, often leading to abortion.

Cellulitis Usually, strains of *Staphylococcus* or *Streptococcus*. Diffuse, spreading inflammation typically affecting subcutaneous tissues. Infection via wounds etc.

Cholera *Vibrio cholerae* (virulent strains). Intestinal infection. Copious watery stools (section 11.3.1.1) and dehydration. Infection via oral–faecal route, usually via contaminated water. Until quite recently, cholera was caused only by the *V. cholerae* O1 serogroup (distinguished on the basis of O antigens) – which includes the cholerae (= 'classical') and El Tor biotypes; none of the other 137 O-serogroups of *V. cholerae* (the 'non-O1' serogroups) cause cholera. In 1993, a cholera-causing non-O1 serogroup (designated O139 Bengal) emerged in South India; immunity to O1 *V. cholerae* does not protect against O139. [*V. cholerae* O139 Bengal: RMM (1996) *7* 43–51. A rapid screening test for *V. cholerae* O139 based on monoclonal antibodies: JCM (1998) *36* 3595–3600.] In 2002, *V. cholerae* O139 caused an estimated 30000 cases of cholera in Dhaka; all of the strains involved were from one of the two ribotypes that caused the O139 outbreak in 1993 – although the organisms had a different pattern of sensitivity to antibiotics [EID (2003) *9* 1116–1122].

Conjunctivitis   Various, e.g. *Staphylococcus aureus*, other staphylococci, *Haemophilus influenzae*, *Streptococcus pneumoniae*, viruses (e.g. adenoviruses), physical or chemical irritation. Inflammation of the conjunctivae (the mucous membranes of the eye). [Ocular bacteriology, including conjunctivitis (review): RMM (1996) *7* 123–131.]

Cystitis   Various, e.g. *Escherichia coli* (see section 11.3.7), *Proteus* spp. Inflammation of the urinary bladder, typically with frequent, painful micturition and sometimes haematuria (blood in the urine) and fever. Infection downwards from the kidney or (more commonly) upwards from the urethra, with *E. coli* and *Proteus* spp being common causal agents in the latter type.

Haemorrhagic cystitis (particularly in children) may be caused by certain types of adenovirus.

Diphtheria   *Corynebacterium diphtheriae* (virulent strains). Characteristic features include the formation of a membrane (containing cell debris, cells of *C. diphtheriae*, fibrin, blood cells) on the tonsils and/or the nasopharynx, larynx or nasal passages; symptoms include difficulty in swallowing and the effects of respiratory obstruction. Diphtheria toxin (DT: section 11.3.1.7), responsible for the symptoms, is disseminated via the blood/lymph and can cause severe, often fatal, effects, including myocarditis and demyelination of nerves.

Transmission of the pathogen may involve droplet infection or ingestion of contaminated food, milk etc. A carrier state (section 11.7) is recognized.

[Microbiology and epidemiology of diphtheria: RMM (1996) *7* 31–42.]

Dysentery (bacillary)   *Shigelia* spp, EIEC (Table 11.2). Intestinal pain, watery stools (usually with fever, malaise) often followed by bloody/mucoid stools. Dehydration. Pathogenesis: see sections 11.2.2.1, 11.3.1.6 and 11.3.3.2. Infection occurs via the oral–faecal route, e.g. contaminated food, water.

Non-bacterial dysentery may be caused by the amoeba *Entamoeba histolytica* or the ciliate *Balantidium coli*.

Erysipelas   Commonly, *Streptococcus pyogenes* (group A). Usually, well-demarcated regions of erythema, often affecting the face, though other areas may be involved; tends to be more superficial than cellulitis. Fever, prostration; sepsis may occur. Infection via wounds, abrasions.

Food poisoning Various, e.g. *Bacillus cereus*, *Campylobacter jejuni*, *Clostridium perfringens*, *Escherichia coli*, *Salmonella typhimurium*, *Staphylococcus aureus*, *Vibrio parahaemolyticus*, *Yersinia enterocolitica*. Acute gastroenteritis. Abdominal discomfort/pain. Usually diarrhoea (but may be little/none in staphylococcal food poisoning and in one type caused by *B. cereus*). Nausea and vomiting are common, but not invariably present. Oral–faecal route; ingestion of pathogen and/or toxin in contaminated food. (See section 12.3.)

Gas gangrene Typically, *Clostridium perfringens* (type A), *C. septicum* and/or *C. novyi*. Spreading necrosis (death) of tissues – which contain pockets of gas formed by bacterial metabolism; in untreated cases, death is associated with severe toxaemia and shock. The *Clostridium perfringens* α-toxin (a lecithinase and sphingomyelinase) appears to act synergistically with perfringolysin [INFIM (2001) **69** 7904–7910].

Gonorrhoea *Neisseria gonorrhoeae* (the gonococcus). In both sexes infection may be initially asymptomatic or there may be a purulent discharge from the genitourinary tract. Infection may spread to adjacent tissues; in males this may lead to e.g. prostatitis/epididymitis, while in females infection may spread via the endometrium and ovaries to the pelvic peritoneum, causing *pelvic inflammatory disease*. In either sex sterility may result. Route of infection: sexual contact. (A test for detecting *N. gonorrhoeae* based on the ligase chain reaction reported results superior to those obtained by culture [JCM (1997) **35** 239–242].)

Impetigo *Streptococcus pyogenes* and/or *Staphylococcus aureus*. Highly infectious skin disease, particularly among children. Streptococcal form: spreading, inflamed pustules rupture, forming thick, brownish-yellow crusts; secondary infection with S. *aureus* may occur. Staphylococcal form (= *bullous impetigo*): lesions contain watery fluid, rather than pus, and a thin crust forms over the centre of the lesion; occurs most often in very young infants.

Kawasaki syndrome In infants and young children: a condition, occasionally fatal, characterized by fever and vasculitis (the medium-sized arteries being characteristically affected). Aetiology: unknown, but in some cases possibly associated with a superantigen (section 11.5.4.1).

Legionnaires' disease *Legionella pneumophila*. Malaise, myalgia, fever, and a consolidating pneumonia that involves mainly the alveoli and terminal bronchioles; an intra-alveolar exudate is characteristic. There may be chest/abdominal pains, vomiting, diarrhoea and mental confusion. *L. pneumophila* invades and multiplies within alveolar macrophages; the pathogen's *tissue-destructive protease* (a 38 kDa zinc metalloprotease) may inhibit the oxidative burst in phagocytes [JMM (2001) **50** 517–525].

Infection appears to occur by inhalation of contaminated aerosols associated with e.g. air-conditioning cooling towers, nebulizers etc.

Leprosy   *Mycobacterium leprae*. According to the type of leprosy, lesions in any of various tissues (usually including skin) and damage to peripheral nerves, causing loss of sensory and motor functions. *Lepromatous* leprosy: weak cell-mediated immune response, with abundant bacilli in tissues; multiple skin lesions coalesce. *Tuberculoid* leprosy: skin lesions few, bacilli rare; lesions: typical granulomas. [Epidemiology: FEMSML (1996) *136* 221–230. Global elimination of leprosy (perspective): RMM (1998) *9* 39–48. Genetic risk factors for leprosy: Nature (2004) *427* 636–640.]

Leptospirosis   *Leptospira interrogans* (a spirochaete). Mild form: fever, headache, myalgia; typically impaired kidney function. Severe form (*Weil's disease, infectious jaundice*): above symptoms with e.g. vomiting, diarrhoea, liver enlargement, jaundice, haemorrhages, meningitis. Infection via wounds or mucous membranes; the pathogen occurs e.g. in water contaminated with urine from infected animals.

Listeriosis   *Listeria monocytogenes* (virulent strains – only a proportion of food-borne strains may be pathogenic for humans [JCM (1999) *37* 103–109]). This is one of the few bacterial pathogens which can cross the blood–brain barrier and the placenta; this is reflected in the range of diseases caused by *L. monocytogenes*: meningitis (particularly in the immunocompromised); sepsis and stillbirth/abortion; sometimes gastroenteritis. The mortality rate may be >20%. Infection occurs via contaminated foods – e.g. Mexican-style cheese, certain other cheeses, coleslaw and milk. [*Listeria*: pathogenesis and molecular virulence determinants: CMR (2001) *14* 584–640.]

In animals, infection via silage (section 13.1.1.1) may lead to abortion/stillbirth.

Lyme disease   Species of *Borrelia* (e.g. *B. burgdorferi, B. afzelii, B. garinii* [IJSB (1997) *47* 1112–1117]) transmitted by the bites of (mainly ixodid) ticks. In humans: an initial rash (*erythema migrans*), spreading from the site of the tick bite, may be followed (after days/weeks) by neurological/musculo-skeletal/cardiac signs and later by joint/neurological symptoms; symptoms vary in nature and timing.

The disease has also been reported in canines [Veterinary Record (1995) *136* 244–247].

[Review: RMM (1998) *9* 99–107. Quantitative assay of *B. burgdorferi* in tissues: JCM (1999) *37* 1958–1963. Treatment of early Lyme disease (review): Drugs (1999) *57* 157–173.]

Melioidosis   *Burkholderia pseudomallei* (often strains unable to use L-arabinose). Primarily a tropical disease (e.g. S. E. Asia). Symptoms variable, ranging from a local suppurative lesion to fulminating septicaemia; a flare-up may occur after long intervals. Infection via wounds or (infrequently) by inhalation. [Person-to-person transmission: Lancet (1991) *337* 1290–1291.] In Taiwan, all isolates of the pathogen were susceptible to amoxycillin–clavulanate, piperacillin–tazobactam, imipenem and meropenem [EID (2001) *7* 428–433].

Meningitis   Various, e.g. *Neisseria meningitidis* (the 'meningococcus'), *Haemophilus* *influenzae* type b (a common cause in infants and children), *Escherichia coli* (especially in neonates), *Listeria monocytogenes, Streptococcus pneumoniae*; a case of *Streptococcus salivarius* meningitis was reported following spinal anaesthesia [Anesthesiology (1998) *89* 1579–1580].

Inflammation of the meninges (membranes covering the brain and spinal cord). The symptoms typically include severe headache, stiff neck, fever, delirium, coma. Bacterial meningitis can be rapidly fatal.

In meningococcal meningitis, caused by *N. meningitidis*, the meningitic symptoms may be accompanied by petechial or purpuric skin lesions that become gangrenous. In untreated cases mortality is high (e.g. 25–75%). *Waterhouse–Friderichsen syndrome* is a severe, fulminating form of the disease involving cyanosis, coma and haemorrhages in the skin and kidneys; it can be fatal within hours. [Prevention of meningococcal meningitis: RMM (1998) *9* 9–37.]

Routes of transmission include droplet infection and head wounds.

Necrotizing fasciitis   Various organisms, including *Staphylococcus aureus*, enterobacteria and group A streptococci, have been isolated from tissues affected by this condition. Necrotizing fasciitis is characterized by necrosis of the superficial fascia and fascial oedema, with an infiltrate of polymorphonuclear leukocytes; toxin-based systemic effects may be severe. Identification of streptococcal involvement may be assisted by a PCR assay for the *speB* gene (encoding exotoxin B in group A streptococci) [JCM (1998) *36* 1769–1771]. [Fatal necrotizing fasciitis in Japan caused by *Photobacterium*: JCM (2004) *42* 1370–1372.]

Plague   *Yersinia pestis* (containing genes for the Yop virulon [PNAS (2000) *97* 8778–8783]). There are three main clinical forms of plague.

Bubonic plague, the commonest form, is characterized by sudden onset, fever, prostration, haemorrhages, and the development of *buboes*: swollen, inflamed, necrotic lymph nodes. Septicaemia may be followed by meningitis or secondary pneumonia.

Primary septicaemic plague is characterized by sudden onset with high fever, haemorrhages, vomiting and bloody diarrhoea, but no buboes.

Bubonic plague and primary septicaemic plague are transmitted by the bite of a flea: mainly the rat flea (especially *Xenopsylla cheopsis*); person-to-person transmission may occur via the human flea (*Pulex irritans*), but this is a less effective vector.

Primary pneumonic plague follows droplet infection. It involves high fever, prostration, severe pneumonia, and frequently delirium and coma; the sputum contains large numbers of plague bacilli.

Mortality rates in untreated cases: e.g. bubonic plague 25–50%; other types: ~100%. Early chemotherapy (with e.g. tetracyclines, streptomycin or chloramphenicol) lowers mortality rates significantly. Transferable (plasmid-mediated) multidrug resistance has been reported in *Y. pestis* [NEJM (1997) *337* 677–680].

[Molecular and cell biology aspects of plague: PNAS (2000) *97* 8778–8783. Dynamics of plague in the rat population, and transmission of the disease to humans: Nature (2000) *407* 903–906. Bubonic plague (a metapopulation model of a zoonosis): PRS (2000) *267* 2219–2230. Rapid mAb/F1-based test for bubonic/pneumonic plague: Lancet (2003) *361* 211–216.]

Pneumonia   Various, e.g. *Streptococcus pneumoniae*, *Haemophilus influenzae* type b, *Staphylococcus aureus*, *Klebsiella pneumoniae*. Fever, difficulty in breathing, chest pain, cough. Droplet infection etc.

Pseudomembranous colitis   A potentially fatal colitis, with watery diarrhoea, cramps, fever and the formation of pseudomembranous patches of inflammatory exudate, mainly on the colonic mucosa. It may follow chemotherapy with e.g. clindamycin or some broad-spectrum β-lactam antibiotics (3rd-generation cephalosporins have a high propensity to incite the disease). Pseudomembranous colitis is linked to the activity of *Clostridium difficile*, which generally produces toxins A and B; toxin-mediated damage to gut epithelial cells appears to involve disruption of tight-junction proteins. Toxin $A^- B^+$ strains of *C. difficile* can cause pseudomembranous colitis, indicating that assay for the A toxin alone is insufficient for laboratory diagnosis of *C. difficile* infection [JCM (2000) *38* 1696–1697]. An inadequate IgG response may be an important factor that promotes disease or relapse [BPRCG (2003) *17* 475–493].

Q fever   *Coxiella burnetii*. Acute form: e.g. fever, muscular pain, pericarditis, respiratory symptoms; chronic form: endocarditis, osteomyelitis, miscarriage. Inhalation, ingestion of contaminated milk, mother–fetus transmission (rare), transfused blood etc. [PCR and serology in early diagnosis of acute Q fever: JCM (2003) *41* 5094–5098.]

Scalded skin syndrome   Strains of *Staphylococcus aureus*, primarily of phage group II, that produce an exfoliative toxin (ET). Mainly in neonates and young children: fluid-filled blisters (containing *S. aureus*), or rash, followed by loss of surface layers of skin. ETs (serine proteases) apparently cleave desmoglein-1, a desmosomal glycoprotein involved in intercellular adhesion [FEMSIMM (2003) *39* 181–189].

Scarlet fever   *Streptococcus pyogenes*. Commonly in children: sore throat, fever, swelling of cervical lymph nodes, rash. Droplet infection, ingestion of contaminated milk etc.

Syphilis   *Treponema pallidum*. A lesion (*chancre*) at the site of infection (typically genital mucosa) followed by a skin rash and, after months/years in untreated cases, lesions in e.g. heart, central nervous system etc. Infection by direct, particularly sexual, contact.

Tetanus   *Clostridium tetani* (virulent strains). Sustained, involuntary muscular contraction (section 11.3.1.3). Typically, wound contamination. Treatment with intrathecal antitetanus immunoglobulin gave better results (in terms of clinical progression) than did intramuscular administration [BMJ (2004) *328* 615].

Toxic shock syndrome   (TSS) Strains of *Staphylococcus aureus* and *Streptococcus pyogenes* which produce certain *superantigens* (section 11.5.4.1). The toxic shock syndrome toxin 1 (TSST-1) of *Staphylococcus aureus* binds preferentially to T cells that express the Vβ2 chain, while some streptococcal toxins bind preferentially to T cells that express several types of Vβ chain (including Vβ2). The massive release of cytokines (resulting from polyclonal activation of T cells by a superantigen) is believed to account for the observed symptoms of TSS: fever, skin lesions, vomiting, diarrhoea and hypotension; in young children, encephalitic symptoms may be common owing to the ability of cytokines to cross the blood–brain barrier. Bacteraemia is uncommon (staphylococcal TSS) or common (streptococcal TSS). In the acute phase of staphylococcal TSS a higher proportion of Vβ2 T cells as well as raised levels of certain cytokines (e.g. IL-2, IL-4, TNF-α, TNF-β, IFN-γ) have been reported. Mortality rates can be high.

In infants/children, infection may occur via burns, scalds etc. In menstruating women the source of infection may be tampons.

Trachoma   *Chlamydia trachomatis* (Plate 11.5, page 390; see also Appendix). Initially a follicular conjunctivitis. Discharging follicles form in the conjunctival tissue, which becomes scarred. Contraction of scarred tissue leads to inturned eyelids with consequent abrasion of the cornea by the eyelashes, causing ulceration, scarring, impairment/loss of vision. A thin fibrovascular membrane (*pannus*) forms over the cornea. Chemo-therapy typically includes tetracyclines. Infection is contaminative [flies and trachoma: Lancet (2004) *363* 1093–1098].

Tuberculosis (pulmonary)   Usually *Mycobacterium tuberculosis* but sometimes *M. bovis*; *M. africanum* causes tuberculosis primarily in Africa.

Infection can occur by inhalation of pathogen-contaminated aerosols (droplet infection) and may be latent (section 11.2.3) or active. In active disease the pathogen *invades* alveolar macrophages (section 11.2.2.2) and continues to survive and grow. Granulomatous lesions (called *tubercles*) develop within the lungs (Plate 11.3). (The pathogen may spread to other parts of the body; extrapulmonary tuberculosis occurs in a proportion of patients, being more common in those who are HIV-seropositive.)

Tuberculosis is a major cause of death. The rising incidence of the disease (in both developed and less-developed countries) has been attributed to factors such as HIV infection (increasing patients' susceptibility) and inefficient treatment/health-care systems.

Multidrug-resistant (MDR) strains of *M. tuberculosis* and *M. bovis* have been reported. Fatal MDR tuberculosis caused by *M. bovis* has been reported in HIV-infected patients [Lancet (1997) *350* 1738–1742].

Among animals, many warm-blooded and cold-blooded species are susceptible to

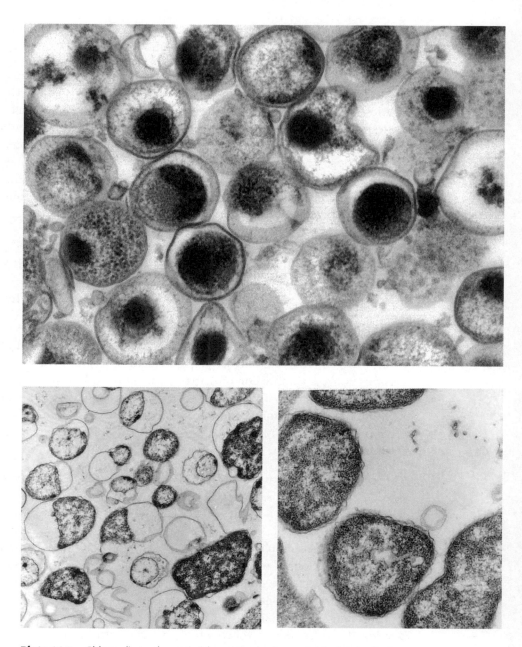

Plate 11.5 *Chlamydia trachomatis* (elementary bodies and reticulate bodies).

Top. Elementary bodies (the non-dividing, infectious form of the organism) obtained as a purified preparation in a sucrose gradient. Note the electron-dense (dark) appearance of the condensed nuclear material.

Lower right. Reticulate bodies (the metabolically active, ribosome-rich, dividing and non-infectious form infection by one or more species of *Mycobacterium*. *M. bovis* is the most important species among domestic animals, causing tuberculosis in cattle and pigs (less commonly in sheep, horses, dogs and cats etc.).

[Mycobacterial disease (reviews): BCID (1997) *4* 1–238. Tuberculosis (seminar): Lancet (2003) *362* 887–899.]

Tularaemia *Francisella tularensis*. Onset may be sudden, with chills, fever, headache, nausea, vomiting, prostration. In *ulceroglandular* tularaemia an ulcer forms at the site of infection and there is regional lymphadenopathy. *Glandular* tularaemia resembles bubonic plague. *Pulmonary* tularaemia may resemble legionellosis. Septicaemia, or severe gastrointestinal disease (following infection by ingestion), may be fatal. The pathogen may disseminate to various organs.

Infection may occur via wounds or mucous membranes (e.g. contact with an infected animal or carcass), or by the bite of any of various vectors (e.g. biting flies, ticks). [Exposure of laboratory workers to *Francisella tularensis* despite a bioterrorism procedure: JCM (2002) *40* 2278–2281.]

In animals, *F. tularensis* can cause fatal (usually tick-borne) septicaemia in sheep, pigs and calves. In horses, tularaemia can involve e.g. fever and oedema of the limbs; dyspnoea and incoordination may also occur in foals.

Typhoid *Salmonella typhi*. Fever, transient rash, intestinal inflammation, sepsis, sometimes with tissue necrosis and intestinal haemorrhage. A carrier state (section 11.7) is recognized, *S. typhi* typically localizing in the gall bladder. Oral–faecal route, e.g. contaminated food or water.

Typhus (classical, epidemic) *Rickettsia prowazekii*. Headaches, sustained fever and a rash which may haemorrhage, muscular pain. Contamination of a wound or 'bite' with louse faeces containing *R. prowazekii*; possibly, inhalation of dried, infected louse faeces. [Monoclonal antibodies for detecting *R. prowazekii* in lice and louse faeces: JCM (2002) *40* 3358–3363.]

Tyzzer's disease *Clostridium piliforme*. In foals: a necrotizing hepatitis and colitis with fever and, in some cases, jaundice and scouring; onset sudden, and death may occur in hours/days. The disease also affects e.g. rodents, cats and rhesus monkeys.

---

of the organism) within a vacuole (= phagosome; pale background); parts of the boundary of the vacuole can be seen in the upper left and lower left corners of the photograph. The nuclear material is more diffusely distributed (less electron-dense) than it is in elementary bodies. The cell in the lower right-hand corner of the photograph is dividing.

Lower left. A vacuole ~24 hours post-infection. This particular strain of *C. trachomatis* gives rise to 'ghost' forms.

Photographs courtesy of Professor Roberto Cevenini, Sezione di Microbiologia, Dipartimento di Medicina Clinica Specialistica e Sperimentale, Università di Bologna, Italy.

Urethritis   Various, e.g. *Neisseria gonorrhoeae* (in gonorrhoea), or, in non-gonococ-
cal (= non-specific) urethritis, *Chlamydia trachomatis, Mycoplasma hominis*. Inflammation of the urethra. Commonly, infection involves sexual contact.

Whipple's disease   *Tropheryma whipplei* (formerly: *T. whippelii*); previously resistant
to culture, this organism was cultured in fibroblasts [NEJM (2000) *342* 620–625] and
its complete genome sequence has been published [Lancet (2003) *361* 637–644].

Whipple's disease is a malabsorption syndrome characterized by diarrhoea, steator-
rhoea, lymphadenopathy, involvement of the central nervous system, and infiltration
of the intestinal mucosa by macrophages containing PAS-positive material (i.e. material positive in the periodic acid–Schiff reaction). Diagnosis involves histology of
intestinal tissue and confirmation by PCR using e.g. duodenal tissue or cerebrospinal
fluid (CSF) [diagnosis and treatment of Whipple's disease: Drugs (1998) *55* 699–704].

[Whipple's disease (review): Lancet (2003) *361* 239–246.]

Whooping cough   *Bordetella pertussis, B. parapertussis*. Paroxysms of coughing, each
followed by an inspiratory 'whoop'. Droplet infection.

12 Applied bacteriology I: food

12.1 BACTERIA IN THE FOOD INDUSTRY

Bacteria are used for: (i) fermentations in the dairy industry; (ii) the processing of raw materials in the manufacture of coffee and cocoa; (iii) the manufacture of food additives; (iv) other processes, such as vinegar production. (Bacterial involvement in the production of food for farm animals is considered in Chapter 13.)

12.1.1 Dairy products

The manufacture of butter, cheese and yoghurt involves a *homolactic fermentation* (section 5.1.1.1) in which the lactose in milk is metabolized to lactic acid.

- *Butter.* 'Cultured creamery butter' is usually made from pasteurized cream to which a *lactic acid starter* culture (section 12.1.1.1) has been added; the starter contains e.g. *Lactococcus lactis* as the main lactic acid producer, together with *L. lactis* subsp *diacetylactis* and/or *Leuconostoc cremoris* as the main contributors of diacetyl. Lactic acid and diacetyl provide flavour, diacetyl giving the characteristic buttery odour and taste. Following fermentation, in which the pH falls to about 4.6, the cream is cooled and churned to destabilize the fat globules. The aqueous phase (*buttermilk*) is strained off, and the butter (often salted) may be stored for long periods at e.g. –25°C. 'Sweet cream butters' are made without a starter.

- *Cheese.* Cheese is made by the fermentation and coagulation of milk, the different types of cheese reflecting differences e.g. in the source of milk (cows', goats' etc.), the type of milk (whole milk, skimmed milk etc.) and the types of microorganisms used – commonly species of *Lactococcus* and *Lactobacillus*. Fermentation lowers the pH, thus helping in the initial coagulation of milk protein; additionally, minor products of fermentation (e.g. acetic and propionic acids) give characteristic flavours. (In Swiss cheeses, such as Emmentaler and Gruyère, the typical flavour of propionic acid is due to the use of *Propionibacterium* spp.) The cheese (coagulated protein, fat) is separated from the aqueous phase (*whey*) and may then undergo various processes such as salting and ripening.

- *Yoghurt.* This product is usually made from pasteurized low-fat milk that is high in milk solids. The milk is inoculated with *Lactobacillus bulgaricus* and *Streptococcus thermophilus* and incubated at 35–45°C for several hours; the pH falls to about 4.3, coagulating the milk proteins. The bacteria act co-operatively: *L. bulgaricus* breaks down proteins to amino acids and peptides – which stimulate the growth of *S. thermophilus*; formic acid produced by *S. thermophilus* stimulates the growth of *L. bulgaricus*, which forms most of the lactic acid.

12.1.1.1   Starter cultures

The purpose of a starter culture is to ensure that fermentation is initiated reliably with appropriate microorganisms. The choice of microorganisms in a given starter is made e.g. on the basis of the characteristics required in the final product.

As well as their basic function, some starters have been designed specifically to deal with problems such as potential contamination of the product with the pathogen *Listeria monocytogenes* (which has been found in certain types of cheese). Another problem is that starter strains may be susceptible to infection by certain bacteriophages (Chapter 9) – leading to slow (or absent) fermentation and consequent economic loss. For example, to improve the safety of cottage cheese, one starter contains organisms that produce the bacteriocin (section 10.1) lacticin 3147, an inhibitor of *L. monocytogenes* [JAM (1999) *86* 251–256]. A genetically engineered starter has also been used to control *L. monocytogenes* in cheddar cheese [AEM (1998) *64* 4842–4845]. Mechanisms for resistance to phage have been engineered in *L. lactis* [AEM (2001) *67* 608–616].

12.1.2   Coffee and cocoa

The manufacture of coffee from ripe coffee fruits requires the initial removal of a sticky mucilaginous mesocarp from around the two beans in each fruit. The outer skin of the fruit is disrupted, and the whole is left to ferment. The mucilage is degraded by the fruit's own enzymes and by microbial extracellular enzymes. As well as e.g. yeasts (single-celled fungi), the important organisms in this process include pectinolytic species of e.g. *Bacillus* and *Erwinia*. After fermentation, the beans are washed, dried, blended and roasted.

Cocoa is made from the seeds (beans) of the cacao plant, the fruit of which is a pod containing up to 50 beans covered in a white mucilage. The mucilage is fermented by yeasts, producing ethanol, and the ethanol is oxidized to acetic acid by certain aerobic bacteria; other organisms are also present. Once free of mucilage, the beans – which darken during the week-long fermentation – are dried and roasted.

12.1.3 Food additives

Monosodium glutamate, the ubiquitous 'flavour enhancer', is manufactured from L-glutamic acid; this latter product is obtained commercially from certain bacteria – e.g. strains of *Corynebacterium glutamicum* grown aerobically on substrates such as molasses or hydrolysed starch. These organisms form glutamic acid from 2-oxo-glutaric acid; because they lack the appropriate enzyme, commercial strains of *C. glutamicum* cannot metabolize 2-oxoglutaric acid in the TCA cycle (Fig. 5.10), so that maximum conversion to glutamic acid can occur. To allow secretion of the glutamic acid, the cell envelope is made more permeable by certain procedures; this is necessary to prevent feedback-inhibition of glutamic acid synthesis.

Xanthan gum is an extracellular polysaccharide slime synthesized by strains of *Xanthomonas campestris*. It has many commercial uses – e.g. as a gelling agent, a gel stabilizer, a thickener and a crystallization inhibitor in various foods.

12.1.4 Vinegar

Vinegar is made by the *acetification* of various ethanol-containing products – e.g. wine, cider, beer. Manufacture involves a carefully controlled process in which the ethanol is oxidized, aerobically, to acetic acid by species of *Acetobacter*. The vinegar may be aged, filtered, bottled and pasteurized.

One type of Japanese rice vinegar (*komesu*) is made from polished rice, while another type (*kurosu*) is prepared from unpolished rice; however, *Acetobacter pasteurianus* is important in both processes [AEM (2001) *67* 986–990].

A synthetic form of vinegar, made chemically (e.g. by the carbonylation of methanol), is called *non-brewed condiment*. It is cheaper than bacterially produced vinegar.

12.2 FOOD PRESERVATION

The various methods of food preservation aim to prevent or delay microbial and other forms of spoilage, and to guard against food poisoning; such methods therefore help the product to retain its nutritive value, extend its shelf-life and keep it safe for consumption. Physical methods of preservation (e.g. refrigeration) are generally preferred to chemical methods.

12.2.1 Physical methods of food preservation

12.2.1.1 *Pasteurization and UHT processing*

Pasteurization is a form of heat treatment used e.g. for milk, vinegar and certain foods; its object is to kill certain pathogens and spoilage organisms. Milk is held at a minimum temperature of 72°C for at least 15 seconds (high-temperature, short-time [HTST] pasteurization). Pasteurization kills the causal agents of many milk-borne diseases (such as salmonellosis and tuberculosis) as well as much of the natural milk microflora; it also inactivates certain bacterial enzymes (e.g. lipases) which would otherwise cause spoilage. The causal agent of e.g. Q fever (*Coxiella burnetii*) is not necessarily killed by HTST pasteurization.

In the UHT process, milk is typically pumped between heated plates and achieves a temperature of approx. 141°C for a holding time of 3–4 seconds. The UHT ('ultra heat-treated') milk is collected and packaged aseptically, and it normally has a shelf life (without refrigeration) of about 6 months. In an alternative method, heat is supplied by injecting steam under pressure; this procedure is particularly useful for viscous products such as custard.

Samples are taken at frequent intervals to test for contamination; rapid results are obtained e.g. by the Lumac system described in section 12.3.3.1.

12.2.1.2   Canning

Typically, suitably prepared food is put into metal containers ('cans' or 'tins') which are then exhausted of air, sealed, and heated – usually to well over 100°C. After heating, the cans are generally cooled by water; the cooling water must be microbiologically clean because, on rare occasions, the normally efficient seal (Fig. 12.1) has allowed contamination of food during the cooling process.

What determines the temperatures and times used in canning? To answer this we first look briefly at the way in which a lethal temperature inactivates a population of cells or endospores (Fig. 12.2); this will also help to make clear the meanings of some important parameters used in the food industry: the D, z and F values.

After processing, canned food must not contain *Clostridium botulinum* (or any other species) capable of growing and producing botulinum toxin (section 11.3.1.2) under the conditions of storage. To achieve this, heating must be carried out at a temperature, and for a time, that will depend on the type and pH of the food and the conditions under which the canned food is to be stored. 'Neutral' foods (e.g. potatoes,

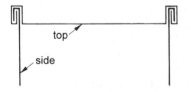

Figure 12.1 Double-seaming, as used in the canning process (section 12.2.1.2). The edge of the can-end is wrapped around the body flange to form a seam of five thicknesses of metal. Not shown (but often present) are corrugations in the body and ends of the can; these help to relieve strain during the heating process.

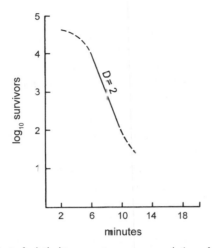

Figure 12.2   The effect of a lethal temperature on a population of cells or endospores. In an idealized response, the number of survivors falls exponentially with time, at a given temperature, i.e. the graph is linear when survivors are plotted on a logarithmic scale. In practice, there are commonly variations from the linear response – for example, the non-linear 'shoulder' and 'tail' effects (*dashed line*).

Using this idea of 'logarithmic death', we can define three important parameters used in the food industry.

D value: the time (in minutes), at a given temperature, for a population of a particular type of bacterium or endospore to fall by 90% (i.e. by one unit on the $\log_{10}$ scale). In the linear part of the graph, the population has fallen by 1 log unit in 2 minutes so that (in this case) D=2. Notice that, in the linear part of the graph, the actual number of organisms killed in a given period of time depends on the initial number present; for example, 9000 are killed in 2 minutes when the population falls from 10000 to 1000, and 900 are killed in 2 minutes when the population falls from 1000 to 100. That is, the *proportion* of cells/spores killed in a given time remains constant in the linear part of the graph.

The temperature at which a D value is obtained (by experiment) may be shown as a subscript: e.g. D_{70}=the value at 70°C. For cells of *Staphylococcus aureus*, D_{70} is typically less than 1 minute (much longer in some strains), and for *S. epidermidis* it s about 3 minutes; a D_{60} of 1 hour has been reported for *Salmonella seftenberg*, though for *Escherichia coli* the D_{60} is typically only a few minutes.

z value: the amount by which the temperature (°C) must be increased in order to decrease a given D value by 90% (i.e. to one-tenth of its value). For example, if – as in the graph – the temperature used to kill particular cells/spores gives a D value of 2, then the z value would be the increase in temperature needed to give a D value of 0.2 for the same type of cells/spores.

F value: the time (in minutes) required for adequate heat treatment at 121.1°C *calculated from* the time required for adequate heat treatment at another temperature (characterized by a particular D value), assuming a z value of e.g. 10°C. F values are used e.g. to compare heat treatments at different temperatures.

D, z and F values form a *basis* for designing a suitable heat treatment. Other factors also have to be considered – e.g. the time taken for heat to penetrate the contents of the can.

carrots, mushrooms) require – as a minimum – a 12D cook (= *botulinum cook*), i.e. heating, at a given temperature, for a time equal to 12 times the D period of *C. botulinum* endospores at that temperature. After such heating, the probability of contamination by *C. botulinum* is regarded as negligible.

Foods whose pH is below 4.6 may be given less heat treatment because it is generally understood that *C. botulinum* does not grow or form toxin below this pH. (However, one early report [Nature (1979) *281* 398–399] described growth and toxin production in a laboratory medium (not food) at values of pH down to 4.0.)

Large (catering-size) cans of *cured* meat (e.g. ham) may be processed at relatively low temperatures (e.g. 70°C in the centre of the can for several minutes) but such cans *must* be stored under refrigeration; this is because safety (i.e. lack of growth/toxin production by *C. botulinum*) depends on the combined effects of curing salts and low temperature.

Sweetened, condensed milk is not heated after can-sealing because the high content of sugar (low water activity: section 3.1.3) inhibits the growth of most bacteria.

The range of heat treatment of canned foods is summarized in Table 12.1.

Spoilage of canned foods. Heat treatment generally inactivates at least some of the enzymes which would otherwise cause spoilage in stored, canned foods. However, spoilage is sometimes caused by heat-stable enzymes and/or by certain bacteria, or endospores, which survive heat treatment. For example, the endospores of *Bacillus stearothermophilus* may survive a botulinum cook, and may cause spoilage in low-acid foods if cans are stored at temperatures above about 30°C; in such spoilage, the food is soured but the cans are not swollen (such cans being called *flat sours*). Low-acid foods

Table 12.1   Canned foods: the range of heat treatment

Type of food	Heat treatment	Comment
Potatoes, carrots, mushrooms and other 'neutral' foods	Botulinum cook as a minimum	Treatment necessary to kill the endospores of *Clostridium botulinum*; inactivation of spoilage organisms may involve further heating
Acidic produce	Minimum may be less than the botulinum cook	It is generally reported that *C. botulinum* does not grow or form toxin below pH 4.6 (but see text)
Catering-size cans of *cured* meat (e.g. ham)	Moderate heating (e.g. 70°C at the centre of the can for several minutes)	*Must* be stored under refrigeration
Sweetened, condensed milk	Nil	Low water activity (due to high concentration of sugar) inhibits the growth of most bacteria; pre-pasteurization of the milk helps to prevent spoilage by fungi

are often heated for more than the (minimum) botulinum cook in order to avoid this type of spoilage.

Canned foods usually remain safe and edible for some years if stored correctly.

12.2.1.3   Refrigeration

Temperatures such as 0–5°C (used for short-term storage) can delay spoilage by inhibiting the metabolism and growth of contaminating organisms and/or the activity of their extracellular enzymes; even so, psychrotrophic organisms (section 3.1.4) may cause spoilage. Note that some bacterial pathogens (e.g. *Listeria monocytogenes*, *Yersinia enterocolitica*) can continue to grow at 4°C.

Freezing (used for long-term storage) may kill some contaminants, and it also reduces the amount of available water (section 3.1.3). At sub-zero temperatures, such as −5 to −10°C, certain *fungi* may become important spoilage agents of e.g. meat.

12.2.1.4   Dehydration

Dehydration reduces the available water (section 3.1.3) to a point at which contaminants will not grow. The process may involve evaporation by heating – as e.g. in the manufacture of dried milk. Alternatively, the amount of available water may be reduced by adding sodium chloride (as in salted fish products) or syrups (as in preserved fruits).

12.2.1.5   Ionizing radiation

Ionizing radiation (e.g. high-energy electrons or *gamma*-radiation) is used in some countries for treating foods such as chicken and fish products, strawberries and spices.

12.2.2   Other methods of food preservation

12.2.2.1   Acidification

The traditional method of *pickling* preserves food by virtue of the lowered pH. This can be achieved either by adding acids (usually lactic acid, sometimes vinegar) or, in some cases, by fermenting the food; *sauerkraut* is cabbage which has been subjected to a natural lactic acid fermentation involving species of *Lactobacillus* and *Leuconostoc*.

12.2.2.2   Preservatives

Food preservatives are chemicals which can inhibit contaminants; some inhibit fungi as well as bacteria. They include benzoic acid (used e.g. in fruit juices, cordials), nitrites and sorbic acid. The lantibiotic *nisin* is active against Gram-positive bacteria, and can inhibit the transition from endospore to vegetative cell; it is used e.g. in some cheeses and in certain canned foods. Nisin and carbon dioxide are reported to act synergistically against the food-borne pathogen *Listeria monocytogenes* [AEM (2000) *66* 769–774]. The use of preservatives is typically subject to governmental regulations.

The traditional method of *curing* (particularly pig meat) involves permeating the meat, at e.g. 4°C, with a solution that includes sodium chloride (which reduces water activity), and sodium nitrite (which, under appropriate conditions, inhibits bacterial growth and the germination/outgrowth of endospores).

In *smoking* (bacon, fish etc.) the food is exposed for hours/days to wood smoke; this reduces water activity in the food, and permeates the food with certain antimicrobial substances (e.g. phenolic compounds) present in the smoke.

12.3   FOOD POISONING AND FOOD HYGIENE

'Food poisoning' usually refers to acute *gastroenteritis* resulting from the ingestion of food contaminated with certain pathogens and/or toxins (Table 12.2); traditionally, the term has also included food-borne cases of botulism (sections 11.3.1.2 and 11.11).

Listeriosis (section 11.11) does not fit the description food poisoning (above); however, the disease can be fatal, and food seems to be the main source of infection. Although *Listeria monocytogenes* can be isolated from various foods (e.g. some cheeses), an accurate assessment of the risk of listeriosis is hindered e.g. by the occurrence of some weakly pathogenic and even non-pathogenic strains of this species; only a proportion of food-borne strains may be pathogenic for humans [JCM (1999) *37* 103–109].

12.3.1   Bacterial food poisoning

Where do the pathogens come from? In some foods (e.g. poultry, eggs, shellfish) pathogens may be present at source but, in general, contamination may occur anywhere between the source and point of consumption; in the kitchen, pathogens are frequently transferred from one food to another (= *cross-contamination*), particularly from raw meat or poultry to cooked foods.

Sometimes the pathogen must *grow* on or in the food to form enough cells or toxin to cause disease. However, growth may not be necessary when, for example, disease can be caused by relatively few cells of the pathogen – e.g. 100 cells of *Shigella*, or less than 1000 cells of *Campylobacter* (Plate 12.1).

Food poisoning may involve an *infection* of the gut (= *food-borne infection*) and/or the action of toxins (= *food-borne intoxication*).

The risk of food poisoning involves the following factors.

Table 12.2 Food poisoning: some of the main types[1]

Causal agent	Foods commonly implicated[2]	Incubation period[3]; main symptoms; comments
Bacillus cereus[4]	Unrefrigerated meat dishes; contaminated spices Cooked rice, stored without refrigeration	8–16 hours; diarrhoea, abdominal pain, nausea 1–7 hours; vomiting, abdominal pain, nausea
Campylobacter jejuni, C. coli	Undercooked meat, poultry; unpasteurized milk	1–7 days; abdominal pain, diarrhoea – watery or bloody; sometimes preceded by influenza-like symptoms; the pathogen cannot grow below 25°C
Clostridium botulinum	Meat, fish; home-canned mushrooms	See 'botulism' in section 11.11
Clostridium perfringens	Cooked, unrefrigerated meat dishes	8–24 hours; abdominal pain, diarrhoea; large numbers of bacteria must be ingested, the toxin being released when sporulation occurs in the gut
Escherichia coli (EIEC, EHEC/VTEC, ETEC, EPEC)	Undercooked meat; milk (and water)	See Table 11.2
Salmonella serotypes[3]	Poultry (including cross-contamination from raw poultry); eggs	12–48 hours; diarrhoea, abdominal pain, vomiting, fever
Staphylococcus aureus	Various (ham, custards, desserts, sandwiches, poultry etc.)	<1–6 hours; nausea, vomiting, abdominal pain, often diarrhoea; the staphylococcal enterotoxins (section 11.3.1.5) can withstand 100°C for some time; the source of contamination is usually a food-handler
Vibrio parahaemolyticus	Shellfish	12–24 hours, but may be less than 2 hours; diarrhoea, abdominal pain[5]
Yersinia enterocolitica	Meat, milk	Fever, diarrhoea, abdominal pain, vomiting; the pathogen can grow at 4°C and below

[1] Other food-poisoning pathogens include species of Shigella, Listeria monocytogenes and some viruses (e.g. rotaviruses).
[2] Only a few examples are given. Pathogens/toxins can occur in unexpected places; for example, botulism has been caused by hazelnut yoghurt in which the hazelnut component had been insufficiently heat-treated. Salad vegetables and bottled water are unexpected risk factors for Campylobacter infections [EID (2003) 9 1219–1225], and iceberg lettuce has been a source of Yersinia pseudotuberculosis [JID (2004) 189 766–774].
[3] The approximate period between ingestion of contaminated food and the start of symptoms.
[4] Produces two forms of disease, each caused by a distinct type of toxin.
[5] [Molecular, serological and virulence characteristics of V. parahaemolyticus: AEM (2003) 69 3999–4005.]

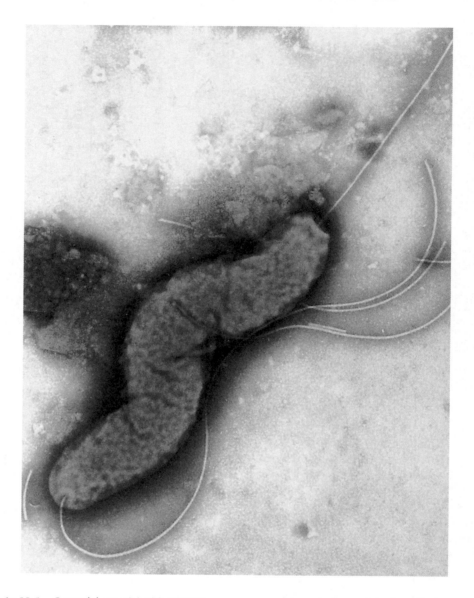

Plate 12.1   *Campylobacter jejuni* (×60000): a common causal agent of acute infectious diarrhoea in all age groups (particularly young adults) in the developed countries, and (primarily) in infants and very young children in the developing countries. The pathogen is associated with animals and animal products, especially poultry; cross-contamination of other foods (section 12.3.1) appears to be an important mode of transmission. Photograph courtesy of Dr Alan Curry, PHLS, Withington Hospital, Manchester, UK.

1. The initial level of contamination on/in the food. In the production/manufacture of food, the greater the initial contamination the less likely it is to be reduced to safe levels by routine processing. In prepared foods, contamination may be above or below the level needed to cause disease. Phages (Chapter 9) can help to reduce contamination on e.g. fresh-cut fruit and on poultry skin [AEM (2003) *69* 4519–4526; (2003) *69* 5032–5036].

2. The type of processing, storage, distribution and preparation involved in the production of food. The risk of food poisoning depends on the ease with which a pathogen can gain access to, and grow in or on, food at each stage of the food chain, and on the success or failure of those processes used to eliminate pathogens. With meat and poultry, major food-poisoning pathogens should ideally be eliminated at farm level, i.e. in livestock, because control of contamination is much more difficult after slaughter and distribution of the products; the reason for this is that distribution can involve dissemination of pathogens to multiple recipients.

3. The dose/response relationship, particularly the smallest dose (of cells or toxin) able to cause disease. In fact, susceptibility to pathogens varies widely with age, acquired immunity and state of health; 'special-risk' groups, in which the risk is greater and the disease may be far more serious, include infants, the very old, pregnant women, and the immunocompromised. Because 'dose/response' can vary, acceptable (i.e. safe) levels of contamination are often based on epidemiological studies, limited experimental data, and a consensus of opinion.

12.3.1.1 *Detecting food-poisoning pathogens and toxins*

Often, an episode of food poisoning will have largely resolved before the pathogen could be cultured from either suspect food or the patient's specimens. Nevertheless, foods and specimens may be examined for pathogens or toxins e.g. for epidemiological or forensic purposes. Clearly, growth media appropriate for suspected pathogen(s) must be used, and it may also be necessary initially to disrupt a sample of food in order to make the organisms accessible (see section 14.8.3).

Some pathogens (e.g. *E. coli* O157) may cause disease even when the infecting dose is very low (e.g. < 100 organisms). Consequently it is important to use the most sensitive method when examining foods for the presence of such pathogens; IMS (section 14.6.1) is useful for improving the recovery of pathogens from various foods (see also Table 14.2).

Highly sensitive methods are likely to be needed for detecting the low concentrations of toxin which can cause food poisoning. For staphylococcal toxins, agglutination methods (sensitivity about 1 ng/ml) are based on the ability of toxin to agglutinate red blood cells (or particles of latex) which have been coated with antitoxin (the molecules of toxin acting as 'bridges' between the RBCs or latex particles); tests based on the ELISA principle (Fig. 11.5) can often detect toxin at or below 0.5 ng/ml.

12.3.2   Food hygiene – domestic

In the kitchen, safe food becomes unsafe in various ways without necessarily showing warning signs (taste, smell, appearance). For example:

1. In general, handling food with unwashed hands may promote infection e.g. by the faecal–oral route. Handling ready-to-eat foods with unwashed hands after handling raw meat or poultry may allow cross-contamination (e.g. with *Salmonella* from poultry) and, particularly if foods are then left for some time at room temperature, growth of the pathogen to dangerous levels.

2. Using knives and/or work surfaces for raw meat and then for cold, pre-cooked foods without a thorough wash between uses.

3. Contamination by pathogens derived from the food handler – often the case with staphylococcal food poisoning.

4. Contamination by flies etc.

5. Undercooking contaminated meat or poultry – e.g. through inadequate thawing of frozen products prior to cooking.

6. Inadequate cooking of eggs stored (correctly) in the refrigerator; as they are initially colder than eggs kept (incorrectly) at room temperature, such eggs require cooking for a longer time.

7. Storage of food at too high a temperature (due e.g. to faulty refrigeration) – allowing bacterial growth and the conversion of low levels of contamination to dangerous levels. To work efficiently, a refrigerator should not be overloaded, i.e. air should circulate around the shelves; the coldest shelves should be 0–5°C. The freezer compartment should not be warmer than −18°C.

   Note that *Yersinia enterocolitica* (and e.g. *Listeria monocytogenes*) can grow at and below 5°C and may therefore increase in number under refrigeration.

8. Consuming food after the 'use by' date – or even before that date if storage conditions have not been appropriate.

12.3.3   Food hygiene – industrial

For food producers and manufacturers the guiding principle in hygiene is the exclusion or elimination of pathogens or the reduction of contamination to safe ('acceptable') levels. This involves e.g. rearing/selecting disease-free animals (for meat, poultry and some fish products), and the use of appropriate and reliable methods of preservation (section 12.2).

Because unsafe food on an industrial scale may affect large numbers of people, the manufacture and sale of food is subject to government regulations. One approach to food safety – used world-wide – is the HACCP system.

12.3.3.1   *The hazard analysis critical control point (HACCP) system*

The HACCP system is essentially the detailed surveillance of a production process, with control of known hazards at specific critical stages. It includes the following phases: (i) objective assessment of hazards, using a flow-chart which shows all aspects of the production process (and which may include the source of raw materials and/or the distribution/sale of final products); (ii) identification of *critical control points* (CCPs), i.e. stages in the process at which preventative or corrective action can be taken effectively against the known hazards; (iii) monitoring at CCPs to ensure that the process continues within pre-determined tolerance limits.

Bacteriological monitoring at CCPs may detect the presence/level of contamination or may detect the presence of a specific organism; either way, rapid assessment is needed to permit rapid corrective action when required. Traditional methods – particularly those based on colony counting – are typically too slow for timely intervention in a modern production process; some rapid methods of assessment are described briefly below.

Lumac® Autobiocounter M 4000 (Perstorp Analytical LUMAC, Landgraaf, The Netherlands). This system is used to detect microbial contamination in samples of UHT milk etc.; the principle depends on the detection of ATP in the sample – the presence of ATP indicating contamination with metabolizing (living) organisms. Samples are pre-incubated (e.g. 30°C/48 hours) before testing to allow growth of any contaminating organisms. After incubation, an aliquot (e.g. 50 μl) of the sample is transferred to an ATP-free cuvette (container) and treated first with an ATPase to hydrolyse the sample's background ATP. The sample is then treated with a reagent which degrades the cytoplasmic membrane of any contaminating cells – allowing leakage of ATP from such cells. ATP is detected by adding reagents (luciferin–luciferase) which produce light (*chemiluminescence*) in the presence of even minute amounts of ATP; emitted light is measured by a photomultiplier (light meter) and the result indicated digitally. The addition of reagents etc. is fully automated, and the test is complete in about 15–20 minutes.

In such an arrangement, each batch of product is held back until the test results become available – being released for distribution when tested samples are found to be satisfactory. Despite the delay (for sample pre-incubation), the manufacturing process itself can be continuous – requiring only that batches be retained pending test results. (The shelf life of UHT milk is about 6 months.)

Impedance monitoring. The food sample is dispersed in a suitable growth medium, and the number of bacteria in the sample is assumed to be proportional to the level of any ongoing metabolism detected in the medium. The level of metabolism is assessed by measuring the decrease in electrical impedance (increase in conductivity) in the medium with time; any fall in impedance is assumed to result from the metabolism of complex molecules and the formation of increased numbers of ions. There are some problems: particular contaminating organism(s) may fail to grow in the medium, and unknown factor(s) may inhibit normal metabolism in other species.

Carbon dioxide monitoring. In some cases, the level of contaminating organisms can be assessed by measuring their output of carbon dioxide; this method is useful e.g. for detecting those organisms which cannot be monitored efficiently by impedance changes.

Direct epifluorescent filter technique (DEFT). This method (see section 14.8.1) is particularly useful for milk.

Limulus *amoebocyte lysate (LAL) test.* This test, which detects or quantifies *lipopolysaccharide* (LPS) (section 2.2.9.2), is used to monitor contamination by Gram-negative bacteria. The test depends on the ability of LPS to coagulate a lysate of the blood cells (amoebocytes) of the horseshoe crab, *Limulus polyphemus*; LPS less than 10^{-9} g/ml may be detected. The LAL test registers *any* LPS-containing bacterium, and even isolated LPS; hence, both living and dead cells can give a positive test.

Monitoring with DNA probes. A labelled DNA *probe* (section 8.5.3) may be able to detect a particular type of bacterium in a food sample provided that the probe is sufficiently specific to the target organism. An extension of this idea is the use of PCR (section 8.5.4) to detect a given organism by using organism-specific primers – as is done in a medical context (section 8.5.4.1).

13 Applied bacteriology II: miscellaneous aspects

13.1 FEEDING ANIMALS, PROTECTING PLANTS

As well as influencing soil fertility (Chapter 10), bacteria make specific contributions to agriculture by helping us to feed farm animals and assisting in the protection of certain crops.

13.1.1 Silage and single-cell protein

13.1.1.1 Silage

Silage-making, the traditional way of preserving grass (and certain other crops), enables animals to be fed during the winter months when vegetation is relatively scarce. Essentially, grass is stored anaerobically and allowed to ferment. Bacteria – e.g. *Lactobacillus* spp – are present on the vegetation and/or in the storage vessel (*silo*); they metabolize plant sugars (e.g. fructose, glucose, sucrose) mainly by a lactic acid fermentation (section 5.1.1.1). The production of lactic acid rapidly lowers the pH (to about 4.0), thereby inhibiting those organisms (particularly *Clostridium* spp) which would otherwise cause putrefaction; this helps to preserve the nutritive value of the crop. The acidity also helps to inhibit growth of *Listeria monocytogenes*, an organism which can grow in incompletely fermented silage (pH > 5.5); infection with *L. monocytogenes* may lead e.g. to abortion/stillbirth.

In *big-bale* silage-making, the vegetation is fermented in large black plastic bags rather than in a silo.

Various silage additives are used e.g. to increase the efficiency of the process and/or to enhance the performance of animals fed on the silage. One common additive includes strains of lactic acid bacteria that carry out a homolactic fermentation (section 5.1.1.1).

Ideas for improved silage additives include (i) use of specific strains of lactic acid bacteria particularly suited to given target crops; (ii) use of lactic acid bacteria that carry out a *heterolactic* fermentation – volatile fatty acids inhibiting the growth of fungi when silage is exposed to air at feeding time; (iii) use of genetically engineered lactic acid bacteria which can utilize polysaccharides in those silage crops which have low levels of soluble carbohydrates Aerobic spoilage of (opened) silage can be inhibited by acetic acid [AEM (2003) *69* 562–567].

13.1.1.2 *Single-cell protein (SCP)*

SCP refers to the cells of certain microorganisms (including bacteria, yeasts and microalgae) grown in large-scale cultures for use as a source of protein in the animal (and human) diet. In the early 1980s, when the price of protein was high, thousands of tons of bacterial protein were being manufactured each year for animal feed; the organism, *Methylophilus methylotrophus* (a methylotroph – section 6.4), was grown on a methanol substrate, and its yield was improved by incorporating an *E. coli* gene (encoding glutamate dehydrogenase) which improved the assimilation of ammonia.

13.1.2 Biological control

'Biological control' generally refers to the use, by man, of one species of organism to control the numbers or activities of another. Such exploitation is used on a commercial scale e.g. in agriculture and forestry. It commonly involves the use of certain microorganisms and/or their toxins to kill or disable the insect pests of certain crop plants; such microorganisms/toxins are called *microbial insecticides* (*bioinsecticides*, *biopesticides*).

Strains of *Bacillus thuringiensis* produce various types of insecticidal crystal protein (ICP) – designated Cry types I–IV (and subtypes); these products are important bioinsecticides which are used world-wide against insect pests on a range of crops. The genes encoding these ICPs are typically plasmid-borne.

Most ICPs are sporulation-specific, i.e. they are formed only in sporulating cells – as a *parasporal crystal* located near the endospore; synthesis is tied to sporulation because transcription of the relevant ICP genes requires sporulation-specific sigma factors such as σ^E – analogous to σ^E in *Bacillus subtilis* (see Fig. 7.14). The genes of CryIII ICPs differ in that they are transcribed during vegetative growth; these ICPs are overexpressed in mutant strains which are blocked in the phosphorelay (Fig. 7.14).

Crystals applied to crops are soon degraded, necessitating repeat application. To avoid this problem, *B. thuringiensis* has been modified so that toxin accumulates within the toxin-producing cells – this (partly protected) intracellular toxin still being highly active.

Certain strains of *Bacillus thuringiensis*, *B. sphaericus* and *Clostridium bifermentans* form toxins which kill mosquitoes; efforts have been made to develop a product suitable for the control of mosquito-borne diseases such as malaria, filariasis and yellow fever [see e.g. AEM (2003) *69* 4111–4115].

[Biological control by Gram-positive bacteria: FEMSML (1999) *171* 1–19.]

Among Gram-negative bacteria, *Pseudomonas fluorescens* can reduce frost injury in some crops by using nutrients otherwise available to ice-nucleation bacteria (see section 10.4).

13.2 BIOMINING (BIOLEACHING)

For many years, chemolithotrophic bacteria (e.g. species of *Thiobacillus*) have been used commercially to extract certain metals from low-grade ores. Today, such *biomining* (= *bioleaching*) is of greater interest because many of the sources of richer ore have been exhausted. In 1990, an estimate of the world's production of bioleached copper, for example, was over 1 million tons per year.

Copper can be recovered e.g. from low-grade ores containing the mineral chalcopyrite ($CuFeS_2$). In one process, a liquor containing sulphuric acid and chemolithotrophic bacteria is repeatedly recycled through a mound of crushed ore; iron (Fe^{2+}), leached from the ore, is oxidized, by bacteria, to Fe^{3+} – resulting in the ongoing solubilization of iron. Solubilization of iron releases sulphur – which is oxidized to sulphate. '*Leptospirillum ferrooxidans*' (an iron oxidizer) and *Thiobacillus thiooxidans* (a sulphur oxidizer) may carry out these processes. Nutrients are provided by the liquor and/or by the ore itself. Copper in the leachate (up to 5 grams per litre) is periodically removed (e.g. by electrolysis). Bacterial metabolism maintains a temperature of about 40–50°C, and the convective upflow of air preserves the necessary aerobic conditions.

An examination of the sulphur chemistry of bacterial leaching has suggested that the leaching of pyrite involves a cyclical degradative process in which the intermediate products include thiosulphate and polythionates; ferric ions, produced by bacterial oxidation of ferrous ions, have an active role in oxidizing both pyrite and thiosulphate – the lithotrophic metabolism of leaching bacteria maintaining a constant supply of these ions.

Leaching of copper ore at very low pH (advantageous because it inhibits precipitation of ferric iron) has been achieved with strains apparently corresponding to *Thiobacillus thiooxidans* and '*Leptospirillum ferrooxidans*' – identification of the leaching organisms involving characterization of the spacer region between 16S and 23S rRNA (see Fig. 16.6) in DNA extracted from the leachate [AEM (1997) *63* 332–334].

In recent years bioleaching has been used (e.g. in South Africa, Brazil, Australia) as a preliminary process in the extraction of gold from recalcitrant arsenopyrite ores; gold cannot be solubilized from these ores in the usual way (by the use of cyanide) because it is embedded in a matrix of pyrite and arsenopyrite (which protects the gold from

Plate 13.1   Bioleaching for gold (section 13.2): cells of *Thiobacillus* among particles of crushed arsenopyrite ore at the Faïrview mine in Transvaal, South Africa. Gold-bearing arsenopyrite concentrate is ground to a fine powder, mixed with water, acid and nutrients, and passed through a series of highly aerated bio-oxidation tanks containing chemolithotrophic bacteria. These bacteria solubilize the arsenopyrite matrix – thus exposing the gold which can then be extracted by cyanide. As a result of bioleaching, the amount of gold recovered may be as much as twice that recoverable without bioleaching. Bioleaching is a low-energy process compared with the traditional methods of treating recalcitrant ores: roasting (high-temperature oxidation) and pressure leaching (aerated acid-digestion under pressure). Photograph reproduced by courtesy of Alex Hartley.

cyanide). The crushed ore is therefore subjected to bacterial action (Plate 13.1); arsenopyrite (FeAsS) is solubilized by oxidation (to $FeAsO_4$ and H_2SO_4), thus exposing the gold – which can then be extracted by cyanidation. Gold is separated from the gold–cyanide complex by adsorption to carbon (e.g. charcoal). The spent, cyanide-containing liquor can be detoxified by bacteria, cyanide being oxidized to CO_2 and urea, and urea being oxidized to nitrate.

[Microbiology of metal leaching (symposium report): FEMSMR (1993) *11* 1–267. Mining with microbes (review): Biotechnology (1995) *13* 773–778. Predominant types of bacteria in commercial bioleaching processes: Microbiology (1999) *145* 5–13.]

13.3 BIOLOGICAL WASHING POWDERS

'Biological' washing powders usually contain enzymes called *subtilisins*, produced by species of *Bacillus*. One of these, 'subtilisin Carlsberg' (obtained from *B. licheniformis*), can hydrolyse most types of peptide bond (in proteins) and even some ester bonds (in lipids). It is stable over a wide range of pH, and its stability does not depend on Ca^{2+}; this latter feature is important because washing powders often include agents which 'soften' water by chelating ions such as Ca^{2+}.

13.4 SEWAGE TREATMENT

Sewage includes domestic wastes (e.g. from drains and water-closets) and often varying amounts of agricultural and/or industrial effluent; it contains substances in suspension, in solution, and in colloidal form.

If discharged to rivers or lakes etc. sewage can be harmful in various ways. It can, for example, be a source of infection – promoting the spread of water-borne diseases such as cholera. Another problem is its content of dissolved organic matter; in metabolizing such nutrients, the large numbers of sewage bacteria can quickly use up the available oxygen in a locally polluted region – particularly in slow-moving or static waters. This can mean death for fish and other oxygen-dependent animals living in these waters. Additionally, such anaerobiosis permits the growth of sulphate-reducing bacteria (section 5.1.1.2; Fig. 10.3) and other organisms whose metabolic products include sulphide and other malodorous substances. Two major aims of sewage treatment are therefore (i) to eliminate (or reduce the numbers of) pathogens, and (ii) to diminish the oxygen-depleting ability of the sewage, i.e. to diminish its *biological oxygen demand* (BOD).

13.4.1 Aerobic sewage treatment

Raw (untreated) sewage first undergoes *primary* treatment which normally includes time in a settlement tank for removal of some of the particulate matter (separated as sludge). The so-called 'settled sewage' (i.e. the liquid) is then subjected to *secondary* treatment.

One form of secondary treatment is the familiar *trickle filter* (= biological filter) (Plate 13.2, *top*) in which sewage is sprayed, via holes in a horizontal rotating arm, onto a thick layer of crushed rock enclosed by a circular wall; percolating through the crushed rock, the sewage makes close contact with surfaces that bear a biofilm containing large numbers of ciliates (protozoa) and other organisms – including bacteria such as *Zoogloea ramigera*. The sprayed sewage carries with it dissolved oxygen, so that some organic matter in the sewage can be oxidized by the sewage

organisms and by those in the biofilm on the rock surfaces; the process is a controlled form of mineralization (section 10.3.4). After treatment, the effluent has a much lower BOD, i.e. when discharged to a river etc., it will take less oxygen from the water. As the sewage percolates, large numbers of bacteria are consumed by the protozoa on the rock surfaces.

Another form of aerobic secondary treatment is the *activated sludge* process. Settled sewage enters a vessel containing activated sludge, i.e. a mass of organisms consisting mainly of bacteria (e.g. *Acinetobacter* spp, *Alcaligenes* spp, *Sphaerotilus natans, Zoogloea ramigera*) and protozoa (ciliates, flagellates and amoebae). Effluent and sludge are vigorously agitated and aerated for e.g. 6–12 hours so that much of the soluble organic matter in the effluent is oxidized or assimilated by the biomass; the BOD is thus greatly reduced. Final effluent of good quality requires efficient *flocculation* (= aggregation) of the organisms, which is enhanced by cell-surface hydrophobicity [Microbiology (1998) *144* 519–528]; this permits high-level clarification by sedimentation. Both composition and function of the sludge microflora can be manipulated by adding *N*-acyl-L-homoserine lactones (AHLs) (section 10.1.2) [Environmental Microbiology (2004) *6* 424–433].

[Structure of microbial communities in sewage treatment plants: FEMSME (1998) *25* 205–215.]

During the activated sludge process the mass of sludge increases as a result of microbial growth, and some sludge is kept for treating the next batch of sewage.

More efficient secondary treatment is given by the *biological aerated filter* (BAF) (shown in Plate 13.2, *bottom*). This consists essentially of a submerged bed of finer granular material (coated with biofilm – see Plate 13.2, *bottom, inset*) through which the sewage passes, downwards, while air is pumped in at or near the base of the bed; the use of small granules allows the system to function as a mechanical filter (for fine particulate matter) as well as a process for mineralizing the dissolved organic matter in the sewage. A BAF contains up to five times more biomass in the biofilm of the filter bed than that in a trickle filter of equivalent size so that, for a given treatment capacity, the BAF can be much smaller.

Under appropriate flow conditions (allowing adequate aeration), nitrifying bacteria (section 5.1.2; Fig. 10.2) can become established and functional in the BAF's biofilm. This is important because nitrification is useful for the elimination of nitro-

Plate 13.2 Sewage treatment (aerobic): old and new technology (see section 13.4.1). *Top*. Trickle filters at Bodmin, Cornwall, UK. *Bottom*. BAF units at St Austell, Cornwall. *Bottom, inset*: bacteria (approx. ×11000) colonizing a particle of 'Biocarbone' filter medium used in some types of BAF unit.

Photographs of treatment plants in Cornwall courtesy of Mr Brian Lessware ABIPP, South-West Water, Exeter, UK.
Electronmicrograph of bacteria on a Biocarbone particle courtesy of Anjou Recherche – OTV, Compagnie Générale des Eaux, Maisons Laffitte, France.

Table 13.1 Sewage treatment: comparison of three aerobic processes [1]

Process	Mechanical sieving action	Enhanced aeration for efficient reduction of BOD	One-stage high performance [2]	In-built nitrification potential [3]
Trickle filter	+/−	−	−	−
Activated sludge	−	+	−	+/−
BAF	+	+	+	+

[1] See section 13.4.1.

[2] In terms of lowering BOD *and* clarification of effluent.

[3] In trickle filters, the available oxygen is used primarily for the oxidation of carbon. Those activated sludge plants which are specially designed to include nitrification tend to be larger than those which are not so designed; nitrification requires longer exposure to the sludge. Nitrification in BAF units depends on the maintenance of good aeration and appropriate loading of the system; under such conditions nitrifying bacteria can colonize the filter bed.

gen from sewage; if nitrification can be achieved, the effluent can then be subjected to denitrification (sections 5.1.1.2 and 10.3.2; Fig. 10.2). Because the BAF can also be operated *anaerobically* (to encourage denitrification) a high proportion of the nitrogen in sewage can be eliminated by operating aerobic and anaerobic reactors in series.

Quite recently, another process (i.e. other than conventional denitrification) has been found to eliminate nitrogen anaerobically in nature and in wastewater treatment plants. This process – called *anammox* [FEMSMR (1999) *22* 421–437] – involves the oxidation of ammonium to gaseous nitrogen with the involvement of nitrite as terminal electron acceptor. The anammox reaction may well be exploited in the design of future wastewater treatment plants.

Periodic cleaning of the BAF is achieved by backwashing.

The three forms of aerobic sewage treatment are compared in Table 13.1.

Removal of ammonia from sewage helps to avoid eutrophication in rivers etc. receiving such effluents; *eutrophication* means over-enrichment of water with nutrients suitable for algal and cyanobacterial growth – conditions which may lead to aquatic pollution. One development is a reactor capable of high-level nitrification in effluents containing $> 0.5\,\mathrm{g\,N\,l^{-1}}$ ammonia – almost all the ammonia being converted to nitrite at the plant's operating temperature (35°C); molecular methods indicate that the dominant nitrifier is *Nitrosomonas* [FEMSME (1998) *27* 239–249].

13.4.2 Anaerobic sewage treatment

Sewage containing a relatively high content of solids – e.g. farm wastes, sludge from some aerobic treatment processes – can be treated by *anaerobic digestion*. In this process, complex organic matter is broken down to simple substances which include a high proportion of gaseous products; it involves a wide range of bacteria which, collectively, carry out a spectrum of metabolic activities. Essentially, sewage is digested in a tank at about 35°C. Polymers, such as polysaccharides, are degraded by extracellular enzymes, and the resulting subunits (sugars etc.) are fermented (e.g. by species of *Bacteroides* and *Clostridium*) to products which include acetate, lactate, propionate, ethanol, CO_2 and hydrogen. Methanogens (section 5.1.2.2) produce methane from acetate and from the CO_2 and hydrogen; much of the bulk of sewage carbon is eliminated via CO_2 and methane.

Anaerobic digestion can yield a final product which is relatively odorless and rich in microbial biomass – a useful agricultural fertilizer. The gaseous product (*biogas, sewer gas*) may contain more than 50% of methane, and it can contribute most or all of the energy needs of the treatment process.

13.4.3 Sewage treatment and the environment

The technology for proper sewage treatment has been available (and used) for many years. Despite this, outfalls (long pipes) are still being built to convey raw (untreated) sewage into coastal waters; for those responsible, this may seem a 'cheap option', but for the *environment* it is certainly a most expensive option.

13.5 WATER SUPPLIES

Supplies of fresh (i.e. non-saline) water are obtained mainly from rivers (surface waters) and *aquifers* (underground layers of water-bearing rock); sub-surface water is called *groundwater*. Surface waters are typically much more polluted. The extent of reliance on surface waters varies geographically; thus, e.g. British Columbia (Canada) and Scotland rely heavily on surface waters, while Austria, Denmark and Portugal obtain almost all their supplies from groundwater.

Water intended for public supply must be treated to (i) eliminate pathogens (those causing water-borne diseases such as cholera), and (ii) eliminate, or decrease to safe levels, any harmful substances which may be present. Treatment also aims at a final product which is acceptable in terms of clarity, taste and odour.

13.5.1 Large-scale (urban) water supplies

Water undergoes various processes before entering the mains distribution system; the lower the quality of the 'raw' water (i.e. the water to be made potable) the more extensive the treatment. Thus, different combinations of processes are used for treating different types of water.

Plate 13.3   Water supply from surface waters: damming a river can help to cope with fluctuations in river flow and demand for water. This small (44-metre high) dam on the West Okement River at Meldon, Dartmoor (Devon, UK), was completed in 1972; it created a reservoir of 23 hectares with a capacity of 3091 million litres. The catchment area is 1660 hectares of high moorland. The reservoir yields 24.5 million litres daily and serves a population of about 200000 in nearby towns and rural areas. Water is pumped to the treatment works using energy from a hydro-electric plant added in 1987.

13.5.1.1 Treatment of groundwater

Groundwater (obtained from bore-holes and springs) is generally of good quality and may need little more than aeration, rapid sand filtration and disinfection (see later); however, appropriate treatment may be required if it contains significant levels of e.g. nitrates (see later).

13.5.1.2 Treatment of surface waters

Initially, raw water may be stored for a number of days; this e.g. allows sedimentation of some particulate matter. The water may then be microstrained through rotating stainless-steel mesh drums (pore size ~30 μm).

Certain types of raw water (e.g. water from polluted rivers – and also groundwater) may contain low levels of dissolved oxygen, and such water must be aerated e.g. by a cascade or fountain process which achieves good air–water contact; aeration avoids certain problems during subsequent treatment.

Much of the remaining particulate matter can be removed by adding a coagulant (such as aluminium sulphate) which causes the particles to aggregate (forming *flocs*). To remove the flocs (*clarification*) the water is passed upwards through a *floc blanket clarifier*: a tank within which the flocs form a 'sludge blanket' below a layer of clarified water; the process is continuous, the clear water overflowing at the top of the tank and passing on to the next stage of treatment.

Water containing only fine particles (and dissolved substances) may be passed downwards through a *rapid sand filter*: a bed of sand (grain size ~1 mm) which acts essentially as a mechanical sieve; when the filtration rate falls, air is blown upwards through the sand (to dislodge the particulates) and this is followed by water (back-washing) to remove the solids.

A *slow sand filter* contains finer sand, the upper layer of which supports a biofilm of microorganisms (including bacteria and algae); thus, as well as acting as a mechanical sieve, the slow sand filter also effects biological purification e.g. by mineralizing some of the dissolved organic matter and by removing some nitrogen and phosphorus as biomass. Slow sand filters can eliminate certain taste- and odour-causing substances, and may also reduce the levels of any cyanobacterial toxins which may be present. Dangerous levels of these toxins are produced by 'blooms' of e.g. *Anabaena*,

---

Storage of water in a reservoir may improve water quality by allowing time for (i) bacterial mineralization of dissolved organic matter, and (ii) sedimentation of some particulate matter; this lowers the cost of treatment. However, under certain conditions, impounded water may support dense populations of algae/cyanobacteria, causing e.g. potential problems with toxins (see text). Moreover, deep, semi-static water can become thermally stratified; the lower, colder layer is then likely to become anaerobic and support the growth of sulphate-reducing (sulphide-producing) bacteria. Algal growth is sometimes controlled with algicides; stratification is often prevented by pumping water from base to surface.

Aphanizomenon, Microcystis, Nodularia and *Oscillatoria* which can develop in bodies of water (including reservoirs) if conditions are right; such toxins may cause gastro-enteritis, liver disease and other conditions in man and animals. [Ecological and molecular investigations of cyanotoxin production (review): FEMSME (2001) *35* 1–9.]

The biofilm (which includes e.g. nitrifying bacteria – section 5.1.2) operates under aerobic conditions; hence, a slow sand filter will function correctly only if the water has been adequately aerated (see above).

Slow sand filters are not back-washed; instead, the uppermost layer is periodically skimmed off and must be replaced to maintain the depth of the filter.

Before disinfection (usually by chlorine) the pH of the water may need to be adjusted chemically. (pH affects the activity of chlorine.)

Disinfection. Water is disinfected, usually with chlorine, before entering the distribution system. Appropriate levels of chlorine are particularly effective against vegetative pathogenic bacteria of faecal origin – such as *E. coli* and species of *Campylobacter, Clostridium, Salmonella, Shigella* and *Vibrio*; hence, efficient chlorination is an excellent measure against cholera, typhoid and a range of other diseases.

Being a strong oxidizing agent, chlorine reacts with (and is 'used up' by) various impurities in the water; thus, some chlorine is initially lost in satisfying this *chlorine demand*. Once the demand has been satisfied, addition of further chlorine to the water will leave *free residuals* of chlorine; because chlorine reacts with water, the free residuals will include Cl_2, HClO and ClO^-. The recommended 'free residual' level of chlorine is usually about 0.5 parts per million (p.p.m.) (= 0.5 mg/litre). Chlorine is more effective in water at pH < 7 because dissociation of hypochlorous acid (HClO) – the more efficient disinfecting agent – tends to be suppressed under acidic conditions.

If the water contains ammonia, chlorine and ammonia will combine to form *chloramines* (= *combined residuals*). The chloramines are disinfectants which decompose slowly, releasing chlorine; they are less effective but more persistent than chlorine. Ongoing addition of chlorine to the water will increase levels of chloramines to a maximum (which depends on the initial level of ammonia). Further addition of chlorine will oxidize the chloramines – so that levels of combined residuals will fall; after the *breakpoint* (Fig. 13.1), any chlorine added to the water will contribute to free residuals – so that to achieve free residuals the dose of chlorine must be beyond the breakpoint (= *breakpoint chlorination*) (Fig. 13.1).

A minority of water treatment plants add ammonia (as well as chlorine); in such cases, disinfection depends on combined residuals rather than on free residuals.

Superchlorination involves the use of relatively high concentrations of chlorine to eliminate undesirable odours and tastes; excess chlorine is then removed by adding sulphur dioxide (*sulphonation*) until the normal residual level of chlorine is reached.

Testing for free residuals can be carried out e.g. with tablets of DPD (*N,N*-diethyl-*p*-phenylenediamine). Chlorine levels are indicated by the degree of intensity of the red

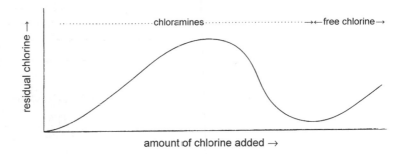

Figure 13.1   Chlorination of water supplies (at about pH 7): chlorine–ammonia interaction and the breakpoint (section 13.5.1.2). The chlorine initially added may oxidize various substances/ions; this loss of chlorinating power (reflecting the *chlorine demand*) is shown by the shallowness of the slope at the start of the graph. Chlorine subsequently added reacts with ammonia to form monochloramine (NH_2Cl) and, as further chlorine is added, a proportion of dichloramine ($NHCl_2$). The rise in concentration of chloramines (=*combined residuals*) is indicated in the graph as a rise in 'residual chlorine'. As more chlorine is added, a point is reached at which chlorine starts to oxidize the chloramines (to N_2 and HCl) – so that their concentration (and that of 'residual chlorine') falls; after the combined residuals (chloramines) have reached their lowest level (the *breakpoint*), continued addition of chlorine brings about a rise in the level of *free residuals* (i.e. chlorine and the products of its reaction with water – section 13.5.1.2).

coloration produced with DPD; the intensity of coloration (and, hence, level of chlorine) is determined by comparison with a colour chart or 'comparator'.

Note that fully treated water may still contain pathogens. For example, certain viruses (e.g. Norwalk virus) and protozoa (e.g. thick-walled oocysts of *Cryptosporidium*) can survive routine levels of chlorination.

Ozone (compared with chlorine) is a stronger disinfectant, but it lacks residual action; its use can be followed by low-level chlorination to provide residual disinfection.

Tests for the efficiency of disinfection. Standards for the bacteriological quality of treated drinking water are provided by the World Health Organization (WHO), the US Environmental Protection Agency (USEPA), the European Commission (EC) and by appropriate bodies in various countries. Certain bacteria – particularly coliforms (see Appendix) and 'faecal streptococci' – are used as indicators of faecal pollution; their presence in treated drinking water points to a failure of the treatment process or to contamination following treatment. Table 13.2 lists some recommended standards.

Coliforms and faecal streptococci are used as indicator organisms because they are common intestinal bacteria; if they are present in water samples then sewage-borne pathogens are *potentially* present in the water. It is not appropriate to test water samples routinely for each of the (many) different types of sewage-borne pathogen

Table 13.2   Bacteriological standards for treated drinking water: coliforms (see Appendix) and 'faecal streptococci'

	Maximum number			
	E. coli or thermotolerant coliforms[1]	*Total coliforms*	*Faecal coliforms*[1]	*'Faecal streptococci'*[2]
EC		0/100 ml	0/100 ml	0/100 ml
UK		0/100 ml	0/100 ml	0/100 ml
WHO	0/100 ml			

[1] 'Thermotolerant' and 'faecal' are synonymous; these coliforms produce acid and gas from lactose at 44°C in 24 hours.
[2] 'Faecal streptococci' refers to *Enterococcus faecalis* and related organisms.

because e.g. a given pathogen may occur only intermittently. Moreover, normally, the indicator bacteria greatly outnumber pathogens, and they occur in large numbers in faeces (e.g. *E. coli* ~10^8 cells/gram); hence, it is possible to detect very low levels of faecal pollution.

Traditional tests for indicator bacteria include *membrane filtration*: a known volume of sample is filtered through a membrane filter (pore size e.g. 0.2 μm) and the membrane is then incubated, face upwards, on a pad saturated with an appropriate liquid selective medium; indicator bacteria (if present) form colonies on the membrane.

In the *multiple-tube test*, aliquots of sample are added to each of a number of tubes containing a lactose-based medium (e.g. MacConkey's broth – Table 14.1) and a Durham tube (Fig. 16.3); on incubation (37°C), any tube which has received at least one viable coliform (in the aliquot) will give a 'positive' test (lactose fermentation): acidification (shown by the pH indicator) and gas production (gas in the Durham tube). The number of positive and negative tubes is then used, in conjunction with statistical tables, to indicate the *most probable number* (MPN) of coliforms in the water sample. This is actually a *presumptive coliform count* because a 'positive' result in any given tube could be due to certain spore-forming bacteria (which can also ferment lactose and form gas); this is particularly important when testing *chlorinated* water supplies because spore-formers are more resistant than coliforms to chlorine. Confirmation that the result is due to *E. coli* – a thermotolerant (= 'faecal') coliform – requires two further tests which are carried out at 44°C/24 hours: the *indole test* (see section 16.1.2.5) and the *Eijkman test*. In the Eijkman test, each 'positive' tube is subcultured to a medium (e.g. lauryl tryptose lactose broth) that includes an agent which inhibits spore-forming bacteria; on incubation at 44°C, acid and gas from lactose, and indole from tryptophan, is taken as confirmation of *E. coli*.

[Microbiological safety of drinking water (emerging pathogens, cyanobacterial toxins, microbial ecology): ARM (2000) *54* 81–127.]

[Developments in microbiological risk assessment for drinking water (with particular reference to *Cryptosporidium parvum*, rotavirus and bovine spongiform encephalopathy): JAM (2001) *91* 191–205.]

The problem of nitrates, pesticides etc. Nitrates occur in surface water and, increasingly, in groundwater; a major source is agricultural fertilizer. The upper limit recommended by the EC/UK (NO_3/litre) is 50 mg/litre; the USEPA value is 44.29 mg/litre (= 10 mg/litre as N). Water containing high levels of nitrate is sometimes blended with low-nitrate waters. Alternatively, the water may be stored for extended periods to permit denitrification (section 10.3.2). Removal of nitrate by ion-exchange processes is also practicable.

Pesticides and some other unwanted substances can be removed by adsorption to activated carbon.

Processes which deal with nitrates, pesticides etc. (and with hardness) are referred to as *tertiary treatment*.

13.5.1.3   *Problems in the distribution system*

Microbial growth/slimes in pipelines etc. can lower water quality and reduce pumping efficiency. Such growth has been linked with the presence of organic carbon in the water, and the increased availability of such carbon due to oxidation by chlorine and ozone. However, workers in Finland have shown that microbial growth can be low – despite high levels of carbon – if the concentration of phosphorus is low; this suggests that a more stringent removal of phosphorus from the water might permit more cost-effective ways of controlling the problem [Nature (1996) *381* 654–655]. [Drinking water biofilms: FEMSME (1997) *22* 265–279.]

Pathogens may enter treated water via defective pipes, pumps, valves, service reservoirs/water towers etc., and may survive/grow if the level of residual disinfectant is inadequate. Such events are likely to involve a complex interaction of diverse parameters [coliform re-growth in drinking water: AEM (1996) *62* 2201–2211].

13.5.2   Small-scale water supplies

Small rural supplies (e.g. from streams or wells) can be disinfected by ultraviolet radiation after passage through an efficient fibre filter. Alternatively, disinfection of the filtered water can be effected by calcium hypochlorite tablets in a 'chlorinator' fitted in series with the incoming supply; the tablets gradually dissolve, releasing chlorine.

Small amounts of water (on expeditions etc.) can be disinfected by boiling or by tablets of *halazone* (*p*-carboxy-*N*,*N*-dichlorobenzenesulphonamide).

Novel methods have been used to improve the microbiological quality of drinking water in rural areas of some developing countries. For example, multi-layered local fabric, when used as a filter, can retain a high proportion of plankton-associated *Vibrio cholerae*, and this procedure is likely to reduce the incidence and severity of cholera in countries such as Bangladesh.

In South America, solar radiation has been used to reduce the bacterial load in contaminated water by 99.99% [AEM (2004) *70* 1145–1151]. [A plastic solar disinfection pouch for drinking water: AEM (2004) *70* 2545–2550.]

13.6   PUTTING PATHOGENS TO WORK

Pathogens such as *Clostridium botulinum* and *C. tetani* produce highly potent neurotoxins which cause severe or fatal disease. Botulinum toxin, for example, inhibits release of the neurotransmitter *acetylcholine* – with consequent reduction in muscle activity, or muscle paralysis. However, this precise effect of the toxin has been put to good use in the treatment of e.g. hyperactive muscle disorders such as, strabismus ('squint'). [Botulinum toxin in cervical dystonia: Drugs (2002) *62* 705–722. Unlicenced use of botulinum toxin (comments): BMJ (2002) *325* 1188.]

13.7   PLASTICS FROM BACTERIA: 'BIOPOL'

The natural bacterial storage polymer poly-β-hydroxybutyrate (PHB: section 2.2.4.1) is the basis of a range of biodegradable thermoplastics ('Biopol' – trade name of Zeneca, Great Britain). The homopolymer (PHB) is formed under appropriate conditions when *Alcaligenes eutrophus* uses glucose as the sole source of carbon. When the growth medium is appropriately supplemented, *A. eutrophus* forms co-polymers of hydroxybutyrate and hydroxyvalerate; the proportion of hydroxyvalerate (controlled by adjusting the composition of the growth medium) determines the properties of these co-polymers. The intracellular granules of polymer (either PHB or co-polymer) are harvested and purified to a fine, white powder.

Biopol can be used for containers, mouldings, fibres, films and coatings, and can be worked by blow-moulding and injection-moulding processes. While stable in normal use, Biopol is fully degradable after suitable disposal.

Genes encoding enzymes of the PHB biosynthetic pathway in *A. eutrophus* have been modified and expressed in plastids of the green plant *Arabidopsis thaliana* – in which PHB was formed; it was suggested that this may allow the eventual use of such *transgenic* plants for the commercial production of bioplastics. Efforts have been made to develop a transgenic form of the plant oilseed rape (*Brassica napus*) for the commercial production of bioplastics.

13.8   BIOREMEDIATION

Bioremediation means biotechnological (microbe-based) clean-up of pollutants in the environment. Microorganisms, collectively, are metabolically highly diverse, so that (theoretically) any of a wide range of organic pollutants can be degraded given a suitable choice of microorganism(s); moreover, *microbial* degradation of pollutants potentially yields simple inorganics, such as CO_2 and water, whereas other forms of technology (e.g. physical decontamination) may simply transfer the problem from one site to another.

However, while physical methods are typically quick, with outcomes often predictable, biological methods frequently have unknown, unpredictable or unquantifiable effects *in the environment*. Thus, e.g. the *bioavailability* of the pollutant, i.e. its accessibility to microbial populations, may be reduced by adsorption to soil constituents, and this may limit or preclude efficient bioremediation.

For wider acceptance, bioremediation must be shown to be effective and reliable in the environment. The efficacy of a bioremediation process may be assessed e.g. by (i) assaying pollutants of all levels of biodegradability at the polluted site; (ii) assaying (long-lived) breakdown products of pollutants; (iii) chemical assessment of the levels/states of *normal* constituents of the environment; (iv) quantification of specific microbial genes involved in the catabolism of pollutants.

[Bioremediation (review): Microbiology (1998) *144* 599–608.]

Particular types of pollutant:

- *Biphenyls, polychlorinated biphenyls* (PCBs). These compounds constitute an important category of environmental pollutants. They can be degraded by various types of Gram-negative and Gram-positive bacteria. The relevant degradative activity is encoded in genes of the *bph* operon. [Transcription of the *bph* operon in *Pseudomonas pseudoalcaligenes* strain KF707: JBC (2000) *275* 31016–31023.]

- *Pentachlorophenol* (PCP). This pesticide can be degraded by strains of the bacterium *Sphingobium chlorophenolicum* [AEM (2004) *70* 2391–2397].

- *Polyurethane.* Enzymes that degrade polyurethane (*polyurethanases*) are produced by certain microorganisms. Polyurethanase activity in bacteria may be detected by a simple plate assay which tests the ability of an organism to degrade colloidal polyurethane [LAM (2001) *32* 211–214].

- *Heavy metals.* [FEMSMR (2002) *26* 327–338].

Anaerobic processes. Certain pollutants are susceptible to anaerobic degradation. In the future, anaerobic microbial processes may be useful for dealing with types of pollutant that include, for example:

- *Perchlorate.* Various bacteria – including *Wolinella succinogenes* and species of

Dechloromonas and *Dechlorosoma* – can degrade perchlorate to chlorite in a form of anaerobic respiration; the chlorite is then split to form chloride and molecular oxygen. [Perchlorate-reducing bacteria: AEM (1999) *65* 5234–5241.]

- *Benzene.* Degradation of benzene (and similar hydrocarbons) would be useful e.g. if these pollutants were found in fuel-contaminated aquifers under anoxic conditions. Benzene typically resists anaerobic degradation. One possible approach would be to add low levels of chlorite so that the (ubiquitous) perchlorate reducers (see above) could convert the chlorite to chloride and (free) oxygen, thus allowing certain aerobes (e.g. *Pseudomonas* spp) to degrade the pollutant [see Nature (1998) *396* 730]. [Aerobic and anaerobic degradation of toluene by *Thauera*: AEM (2004) *70* 1385–1392.]

- *Terephthalic acid.* This compound (1,4-benzenedicarboxylic acid) and its isomers are used in the manufacture of polyester products. Wastewater containing these compounds has been treated anaerobically, the relevant organisms including bacteria and archaeans [Microbiology (2001) *147* 373–382].

13.9   BIOMIMETIC TECHNOLOGY

Biomimetic technology refers to certain procedures in which biologically based systems are used to achieve results similar to those which are normally obtained by mechanical, electrical and/or chemical methodology. For example, under certain conditions, a strain of *Pseudomonas stutzeri* can form intra-periplasmic crystals of silver in a nanometre range of sizes; when heat-treated (400°C), a layer of these bacteria can give rise to a carbonaceous matrix in which the small particles of silver are homogeneously embedded – a structure which may be useful as a coating for the absorption of solar energy. Such silver–carbon films may also find use e.g. as electrodes in lithium-ion batteries.

[Metal-accumulating bacteria and their potential in materials science: TIBTECH (2001) *19* 15–20.]

14 Some practical bacteriology

14.1 SAFETY IN THE LABORATORY

A student new to bacteriology should be constantly aware that he or she is dealing with living organisms – which may include actual or potential pathogens. Good bacteriology is safe bacteriology, and it is wise to get to know the safety rules of the laboratory before carrying out any practical work; the following rules deserve special attention.

1. While working in the laboratory wear a clean laboratory coat to protect your clothing. Do not wear the coat outside the laboratory.
2. Put *nothing* into your mouth. It is potentially dangerous to eat, drink or smoke in the laboratory. For pipetting, use a rubber bulb (teat) or a mechanical device such as a 'pi-pump'; do not use your mouth for suction. If necessary, use self-adhesive labels.
3. Keep the bench – and the rest of the laboratory – clean and tidy.
4. Dispose of all contaminated wastes by placing them (not throwing them) into the proper container.
5. Leave contaminated pipettes, slides etc. in a suitable, active disinfectant for an appropriate time before washing/sterilizing them.
6. Avoid contaminating the environment with *aerosols* containing live bacteria/spores. An aerosol consists of minute (invisible) particles of liquid or solid dispersed in air; aerosols can form e.g. when bubbles burst, when one liquid is added to another, or when a drop of liquid falls onto a solid surface – things which can happen during many bacteriological procedures (see e.g. Fig. 16.2). Particles of less than a few micrometres in size can remain suspended in air for some time and may be inhaled by anyone in the vicinity; clearly, aerosols can be a potential source of infection. Bacteriological work is sometimes carried out in special cabinets (described later) – partly in order to avoid the risk of infection from aerosols.
7. Report all accidents and spillages, promptly, to the instructor or demonstrator.
8. Wash your hands thoroughly before leaving the laboratory.

Special note 1. When working with certain organisms (e.g. *Mycobacterium tuberculosis*) it is *essential* to use a safety cabinet of suitable type (section 14.3.1).

Special note 2. When accepting clinical specimens, the laboratory *must be* informed by medical staff of any suspicion of a disease caused by a 'hazard group 3' or 'hazard group 4' pathogen (European Union designations) so that investigations can be carried out under appropriate conditions. (In the absence of such liaison, laboratory workers were exposed to *Francisella tularensis* despite adherence to an established laboratory protocol [JCM (2002) *40* 2278–2281].)

14.2   BACTERIOLOGICAL MEDIA

A *medium* (plural: *media*) is any solid or liquid preparation made specifically for the growth, storage or transport of bacteria; when used for growth, the medium generally supplies all necessary nutrients. Before use, a medium must be *sterile*, i.e. it must contain no living organisms. (Methods for sterilizing media are given in Chapter 15.)

Before discussing the different media, it will be helpful to give again the meanings of a few words which are used very commonly in bacteriology; this is best done by giving the following outline of a simple laboratory procedure. To grow an organism such as *E. coli*, the bacteriologist takes an appropriate sterile medium and adds to it a small amount of material which consists of, or contains, living cells of that species; the 'small amount of material' is called an *inoculum*, and the process of adding the inoculum to the medium is *inoculation*. (The tools and procedures used for inoculation are described in sections 14.3 to 14.5.) The inoculated medium is then *incubated*, i.e. kept under appropriate conditions of temperature, humidity etc. for a suitable period of time. Incubation is usually carried out in a thermostatically controlled cabinet called an *incubator*. During incubation the bacteria grow and divide – giving rise to a *culture*; thus, a culture is a medium containing organisms which have grown (or are still growing) on or within that medium.

A liquid medium may be used in a test tube (which is stoppered by a plug of sterile cotton wool, or which has a simple metal cap – see e.g. Fig. 16.3) or in a glass, screw-cap bottle; a *universal bottle* (Fig. 14.1) is a cylindrical bottle of about 25 ml capacity, while a *bijou* is smaller (about 5–7 ml).

Most solid media are jelly-like materials which consist of a solution of nutrients etc. 'solidified' by *agar* (a complex polysaccharide gelling agent obtained from certain seaweeds). A solid medium is commonly used in a plastic *Petri dish* (illustrated in Fig. 16.2) – usually the size which has a lid diameter of about 9 cm. The medium, in a molten (liquid) state, is poured into the Petri dish and allowed to set; a Petri dish containing the solid medium is called a *plate*. (A *vented* Petri dish has three very small projections equally spaced around the inside of the lid; this keeps the lid of the (closed) Petri dish slightly raised, thus facilitating equilibrium between the air/gases inside and outside the Petri dish.)

14.2.1 Types of medium

For many chemolithoautotrophic bacteria the medium can be a simple solution of inorganic salts (CO_2 being used for carbon).

Nutritionally undemanding heterotrophs (such as *E. coli*) need only the common organic substances found in *basal media* (Table 14.1). Many bacteria will not grow in basal media, but may do so after the addition of substances such as egg, serum or blood; media which have been supplemented in this way are called *enriched media*.

A *selective medium* is one which supports the growth of certain bacteria in preference to others. An example is MacConkey's broth (Table 14.1) – in which the bile salts inhibit *non*-enteric bacteria but do not inhibit enteric species; this medium can be used e.g. to isolate enteric bacteria from a mixture of enteric and non-enteric bacteria when both types are present in an inoculum. (To some extent, *all* media are selective in that no medium can give equal support to the growth of every type of bacterium.)

An *enrichment medium* allows certain species to outgrow others by encouraging the growth of wanted organism(s) and/or by inhibiting the growth of unwanted species. Hence, if an inoculum contains only a few cells of the required species (among a large population of unwanted organisms), growth in a suitable enrichment medium can increase ('enrich') the proportion of required organisms. For example, selenite broth (Table 14.1) inhibits many types of enteric bacteria (including e.g. *E. coli*) but does not inhibit *Salmonella typhi*, the causal agent of typhoid. Suppose, for example, that we need to detect *S. typhi* in a specimen of faeces from a suspected case of typhoid. The specimen may contain only a few cells of *S. typhi*, so that it may be difficult or impossible to detect them among the vast numbers of non-pathogenic enteric bacteria. However, if an inoculum from the specimen is incubated in selenite broth, the proportion of cells of *S. typhi* increases to the point at which they can be detected more readily.

A *solid* medium is used, for example, to obtain the *colonies* (section 3.3.1) of a particular species. Many solid media are simply liquid media which have been solidified by a gelling agent such as gelatin or agar; agar is the most commonly used gelling agent because (i) it is not attacked by the vast majority of bacteria, and (ii) an agar gel does not melt at 37°C – a temperature used for the incubation of many types of bacteria. (By contrast, gelatin can be liquefied by some bacteria, and it is molten at 37°C.) One widely used agar-based medium is *nutrient agar* (Table 14.1), a general-purpose medium used for culturing (i.e. growing) many types of bacteria; it can also be enriched and/or made selective by the inclusion of appropriate substances.

Blood agar is an agar-based medium enriched with 5–10% blood; it is used e.g. for

Table 14.1 Some common types of bacteriological medium [1]

Medium	Composition of medium: typical formulation (% w/v in water)
Basal medium	
Peptone water	Peptone (soluble products of protein hydrolysis) 1%; sodium chloride 0.5%
Nutrient broth [2]	Peptone 1%; sodium chloride 0.5%; beef extract 0.5–1%
Nutrient agar	Nutrient broth gelled with 1.5–2% agar
Differential medium	
MacConkey's agar	MacConkey's broth (see below) gelled with 1.5–2% agar
CLED medium	Peptone 0.4%; tryptone 0.4%; meat extract 0.3%; lactose 1%; L-cystine 0.013%; bromthymol blue 0.002%; agar 1.5%
XLD medium	Xylose 0.38%, lactose 0.75%, sucrose 0.75%, L-lysine 0.5%, sodium deoxycholate 0.2%, yeast extract 0.3%, ferric salt 0.08%, thiosulphate 0.7%, phenol red (pH indicator) 0.008% (etc.)
Enriched medium	
Blood agar	Nutrient agar (or similar medium) containing 5–10% defibrinated or citrated blood
Chocolate agar	Blood agar heated to 70–80°C until the colour changes to chocolate brown
Serum agar	Nutrient agar (or similar medium) containing 5% (v/v) serum
Enrichment medium	
Selenite broth	Peptone 0.5%; mannitol 0.4%; disodium hydrogen phosphate 1%; sodium hydrogen selenite ($NaHSeO_3$) 0.4%
Selective medium	
Deoxycholate–citrate agar (DCA)	Meat extract and peptone 1%; lactose 1%; sodium citrate 1%; ferric ammonium citrate 0.1%; sodium deoxycholate 0.5%; neutral red (pH 8.0 yellow to pH 6.8 red) 0.002%; agar 1.5%
MacConkey's broth	Peptone 2%; lactose 1%; sodium chloride 0.5%; bile salts (e.g. sodium taurocholate) 0.5%; neutral red 0.003%
Transport medium	
Stuart's transport medium	Salts; agar 0.2–1.0% (semi-solid or 'sloppy' agar); sodium thioglycollate; methylene blue (as redox indicator)

[1] A major source of information on several thousand microbiological media is *Handbook of Microbiological Media*, ISBN 0-8493-2638-9.
[2] In bacteriology, 'broth' may refer to any of various liquid media, but, when used without qualification, it commonly refers to nutrient broth.

the culture (growth) of nutritionally 'fastidious' bacteria such as *Bordetella pertussis* (causal agent of whooping cough), and also to detect *haemolysis* (section 16.1.4.1). *Chocolate agar* is made by heating blood agar to 70–80°C until it becomes chocolate brown in colour; it is more suitable than blood agar for growing certain pathogens (e.g. *Neisseria gonorrhoeae*).

MacConkey's agar is an example of a *differential medium*, i.e. one on which different species of bacteria may be distinguished from one another by differences in the characteristics of their colonies etc. On MacConkey's agar, lactose-utilizing enteric bacteria (such as *E. coli*) form *red* colonies because they produce acidic products (from the lactose) which affect the pH indicator in the medium; enteric species which do not use lactose (e.g. most strains of *Salmonella*) give rise to colourless colonies.

CLED (cystine lactose electrolyte-deficient) medium is another differential medium. On this medium, colonies of *E. coli* are *yellow* (as the pH indicator becomes yellow below pH 6.0). Colonies of *Salmonella* are blue. This medium inhibits swarming by *Proteus* (owing to the deficiency of electrolyte). It is used e.g. for examining samples of urine. CLED medium supports the growth of some staphylococci and streptococci (and also *Enterococcus faecalis*).

Sorbitol MacConkey agar (SMAC) (medium CM813, Oxoid, Basingstoke, UK) resembles MacConkey's agar but contains sorbitol instead of lactose; it is useful for detecting the pathogenic (EHEC/VTEC) O157 strain of *E. coli* – which does not ferment sorbitol and which therefore forms colourless colonies on this medium. (Most strains of *E. coli* ferment sorbitol and form pink colonies.) Adding potassium tellurite and cefixime suppresses *other* non-sorbitol-fermenters (e.g. *Proteus*).

On *XLD medium* (xylose–lysine–deoxycholate medium: Table 14.1) *Shigella* uses no sugars (i.e. produces no acid) and forms red colonies (phenol red: pH 6.8 yellow → pH 8.4 red). *Salmonella* ferments xylose (i.e. forms acid) – but decarboxylates lysine, giving a *net* alkaline reaction (red colonies); however, unlike *Shigella*, *Salmonella* forms *black-centred* red colonies (Plate 14.3) because it produces hydrogen sulphide that reacts with the ferric salt. *Escherichia* typically ferments all three sugars; it forms yellow colonies but is inhibited by deoxycholate.

Dorset's egg is made by heating (coagulating) homogenized hens' eggs in saline. It is a *maintenance medium*: organisms (e.g. *Mycobacterium tuberculosis*) can be grown on the medium and the culture is then stored under refrigeration.

Sometimes we need to use a medium in which all the constituents, including those in trace amounts, are quantitatively known; such a *defined medium* is prepared from known amounts of pure substances – e.g. inorganic salts, glucose, amino acids etc. in distilled or de-ionized water. A defined medium would be used e.g. when determining the nutritional requirements of a given species of bacterium.

A *transport medium* is used for the transportation (or temporary storage) of material (e.g. a swab) which is to be subsequently examined for the presence of particular organism(s); the main function of the medium is to maintain the viability of those organism(s), if present. A transport medium need not support growth; in fact, growth may be disadvantageous because the waste products formed may adversely affect the survival of the organisms. One such medium, *Stuart's transport medium* (Table 14.1), is suitable e.g. for a range of anaerobic bacteria and for 'delicate' organisms such as *Neisseria gonorrhoeae*.

One problem with agar-containing transport media (such as Stuart's) is that the agar can inhibit attempts to detect pathogens by PCR (section 8.5.4) [JCM (1998) *36* 275–276]. In this respect, transport media containing *gellan gum* (in place of agar) gave results superior to those obtained with agar [JMM (2001) *50* 108–109].

14.2.1.1   BACTEC culture systems

BACTEC™ culture systems (Becton Dickinson) are liquid media that can be used e.g. for culturing the slow-growing pathogen *Mycobacterium tuberculosis*; in these media the pathogen can grow more rapidly (e.g. 1–2 weeks) than it can on solid media. Such media can be used for (i) detecting *M. tuberculosis* and (ii) examining an isolate of the pathogen for susceptibility to antibiotics; in antibiotic-susceptibility testing, an isolate is tested for growth (or lack of growth) in a medium containing a known amount of a given antibiotic.

The earlier systems were radiometric (growth being detected by detection of radio-active carbon dioxide produced from a radioactive substrate in the medium). One recent system, the BACTEC MGIT 960 (MGIT = mycobacteria growth indicator tube) monitors growth by means of a fluorescent sensor system that detects the consumption of oxygen. [Evaluation of the BACTEC MGIT 960: JCM (1999) *37* 748–752.]

14.2.2   The preparation of media

Most media can be obtained commercially in a dehydrated, powdered form. Such media are commonly dissolved in the appropriate volume of water, sterilized, and dispensed to suitable sterile containers; as an alternative, some media are dispensed to containers before sterilization.

For most agar-based media (e.g. nutrient agar) the powdered medium is mixed with water and steamed to dissolve the agar; the whole is then sterilized in an *autoclave* (section 15.1.1.3) and subsequently allowed to cool to about 45°C, a temperature at which the agar remains molten. To prepare a *plate*, some 15–20 ml of the molten agar medium is poured into a sterile Petri dish which is left undisturbed until the agar sets. Blood agar plates are made by mixing molten nutrient agar (at about 45–50°C) with 5–10%, by volume, of (e.g. citrated) blood before pouring the plates.

For some uses (e.g. streaking: section 14.5.2), the surface of a newly made plate must be 'dried' – i.e. *excess* surface moisture must be allowed to evaporate; this is often achieved by leaving the plate, with the lid partly off, in a 37°C incubator for about 20 minutes. Spread plates (section 14.5.2) may also be dried.

To prepare a nutrient agar *slope* or *slant* (Fig. 14.1) the molten agar medium is allowed to set in a sterile bottle or test tube which has been placed at an angle to the horizontal.

Some types of medium cannot be sterilized by autoclaving because one or more of their constituents are destroyed at the temperatures reached in an autoclave. Such

Figure 14.1 A slope (also called a *slant*). The medium (stippled) has a large surface area available for inoculation; the thickest part of the medium is known as the *butt*. Slopes are commonly made of agar-based or gelatin-based media, and a slope is usually prepared in a universal bottle (as shown), in a bijou or in a test tube. Slopes are used e.g. for storing a purified strain of bacteria. A sterile slope is inoculated with an inoculum from a pure culture of the bacterial strain, and the slope is then incubated at a suitable temperature to allow growth; the slope can then be stored in a refrigerator at 4–6°C until needed e.g. as a source of inoculum.

media include e.g. DCA (Table 14.1), which is steamed but not autoclaved, and those media which contain glucose or other heat-labile sugars; in preparing the latter type of medium the sugar solution is sterilized separately by filtration (section 15.1.3) before being added to the rest of the (autoclaved) medium.

14.3 ASEPTIC TECHNIQUE

Instruments and media etc. must be *sterile* before use (section 15.1); if we are not *sure* of their sterility we will simply not know what is happening in our practical work. Additionally, *during* bacteriological procedures, instruments and materials must be protected from contamination by organisms that are constantly present in the environment. *Aseptic technique* involves the pre-use sterilization of all instruments, vessels, media etc., and avoidance of their subsequent contact with non-sterile objects – such as fingers, or the bench top etc.

Vessels containing sterile contents are kept closed except for the minimum time needed for access. Before opening a vessel (e.g. a sterile bottle, or one containing a pure culture), the rim of the screw-cap (or equivalent) is passed briefly through the bunsen flame to prevent any live contaminating organisms from falling into the vessel when the cap is removed; this procedure is called *flaming*, and it is used e.g. whenever an inoculum is withdrawn from a culture, or when a sterile medium is being inoculated. Flaming of the bottle's rim is also carried out immediately before the vessel is closed. Flaming is generally *not used* when working with Petri dishes, and is **never** used when the contents of a vessel are likely to catch fire.

The risk of contamination in the laboratory may be further reduced by treating bench tops etc. with a suitable disinfectant, and by filtering the air to remove cells and spores of bacteria and fungi etc.

14.3.1 Safety cabinets (sterile cabinets)

Some bacteriological work is carried out in a safety cabinet (also called a sterile cabinet). There are three classes of cabinet (I, II and III); the use of a given type of cabinet depends primarily on the nature of the organisms or materials being handled. When handling pathogens, the cabinet must provide a level of *containment* appropriate to the pathogenicity of the organisms and to the consequences of their escape into the community. In other cases the main purpose of the cabinet is to protect media and/or specimens from airborne contamination.

- *Class I cabinets.* The front of a class I cabinet is partly closed by a glass viewing panel – but is open between the lower edge of the glass panel and the base of the chamber, allowing access for work. An external fan draws air into the cabinet, via the open front; the airflow is kept above a certain minimum so that particles (aerosols etc.) within the cabinet do not leave via the front opening. A HEPA filter (high-efficiency particulate air filter), located between the cabinet and the fan, prevents transmission of particles from the cabinet to the environment.

  A class I (or class III) cabinet is used e.g. when working with pathogens of European Union 'hazard group 3' which include *Coxiella burnetii*, *Francisella tularensis*, *Mycobacterium tuberculosis* and species of *Rickettsia*. Such work also requires the use of gloves and the provision of *plenum* ventilation in the laboratory, i.e. a constant inflow of air and the filtering of exhausted air through a HEPA filter. Class I cabinets are common in hospital pathology laboratories.

- *Class II cabinets.* In a class II cabinet, sterile (filtered) air constantly flows down onto the work surface, and air passes to the exterior after further filtration (see Fig. 14.2 for pattern of airflow). The filters are HEPA filters. Class II cabinets are common in the microbiology laboratories of colleges; a major function of these cabinets is to protect materials from airborne contamination.

- *Class III cabinets.* A class III cabinet is a totally closed, gas-tight cabinet in which air is filtered (with HEPA filters) before entry and before discharge to the environment; work is conducted by arm-length rubber gloves fitted into the front panel, and access to the interior is via a separate two-door sterilization/disinfection chamber. Air within the cabinet is maintained at negative pressure.

  A class III cabinet is used for handling 'hazard group 4' pathogens, which include Ebola virus, Lassa fever virus and Marburg virus.

14.4   THE TOOLS OF THE BACTERIOLOGIST

In most cases bacteria can be handled with one of the simple instruments shown in Fig. 14.3. A loop or straight wire is sterilized immediately before use by flaming: the wire portion of the instrument is heated to red heat in a bunsen flame and is then allowed to cool.

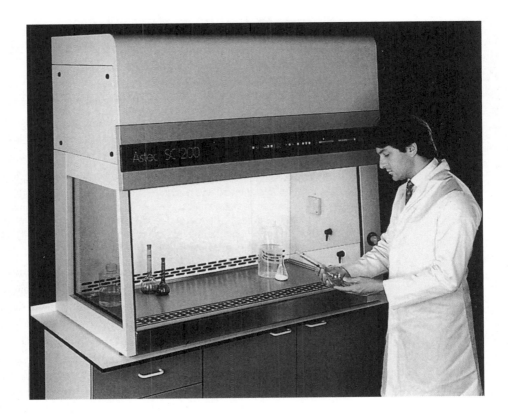

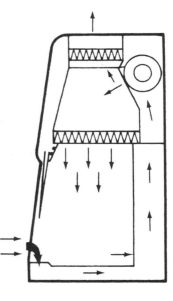

Figure 14.2 *Top*. A class II safety cabinet. *Below*. The pattern of airflow (arrows) during use; filtered air passes downwards onto the working surface, and air passes outwards through a filter at the top of the cabinet. (Courtesy of Astec Environmental Systems, Weston-super-Mare, Avon, UK.)

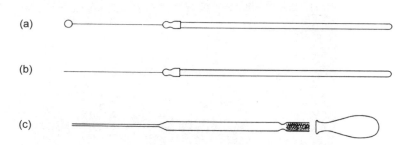

Figure 14.3 Basic tools of the bacteriologist. (a) A loop: a piece of platinum, nickel-steel or nichrome wire, bent into a closed loop at the end and held in a metal handle of about 10–12 cm in length. (b) A straight wire: the metal handle carries a straight piece of wire of about 5–8 cm in length. (c) A Pasteur pipette: an open-ended glass tube, the narrow end of which has an internal diameter of about 1 mm; the wider end is plugged with cotton wool, before sterilization, and a rubber bulb (teat) is fitted immediately before use. For calibration of a Pasteur pipette see Fig. 14.6 (legend).

If a sterile loop is dipped into a suspension of bacteria and withdrawn, the loop of wire retains a small circular film of liquid containing a number of bacterial cells – and this can be used as an inoculum; the size of this inoculum will depend on (i) the concentration of cells in the suspension, and (ii) the size of the wire loop (which often carries 0.01–0.005 ml of liquid) – clearly, two factors which can be controlled. Even smaller amounts of liquid can be manipulated with a straight wire because this picks up only the minute volume of suspension which adheres to the wire's surface.

The loop and straight wire can also be used for picking up small quantities of solid material – e.g. small amounts of growth from a bacterial colony – simply by bringing the wire loop, or the tip of the straight wire, into contact with the material; the *amount* of material which adheres to the wire will be unknown, but usually this is not important. Liquid or solid inocula carried by a loop or straight wire can be used to inoculate either a liquid or a solid medium (section 14.5).

Both the loop and straight wire must always be flamed immediately after use so that they do not contaminate the bench or environment. Spattering, with aerosol forma-tion, may occur when flaming a loop or straight wire containing the residue of an inoculum; for this reason, flaming is often carried out with a special bunsen burner fitted with a tubular hood.

Larger volumes of liquid may be handled by means of Pasteur pipettes or graduated pipettes; suction is obtained either from a rubber bulb (teat) or from a mechanical device – the mouth is never used. Pipettes used in bacteriology are usually plugged with cotton wool (Fig. 14.3), before being sterilized, in order to avoid contamination from the bulb or from the mechanical pipetting device during use. Pipettes are usually sterilized (in batches) inside metal canisters or in thick paper envelopes; when a pipette is removed from the container, only the plugged end should be held so as to avoid contaminating the rest of the pipette. Pasteur pipettes are commonly used once only and then discarded into a jar of suitable disinfectant. Graduated pipettes which have been contaminated with bacteria are immersed in a disinfectant until they are safe to handle, when they can be washed, sterilized and re-used.

14.5   METHODS OF INOCULATION

14.5.1   Inoculating a liquid medium

To inoculate a liquid medium with a *liquid* inoculum, the loop (or straight wire) carrying the inoculum is simply dipped into the liquid medium, moved slightly, and then withdrawn. Inoculation can also be carried out with a Pasteur pipette. With a *solid* inoculum, the loop or straight wire may be rubbed lightly against the inside of the vessel containing the medium – to ensure that at least some of the inoculum is left behind when the instrument is withdrawn.

14.5.2   Inoculating a solid medium

Solid media may be inoculated in a variety of ways, particular methods being used for particular purposes.

Streaking (Fig. 14.4) is used when individual, well-separated colonies are required, and the (liquid or solid) inoculum is known to contain a large number of cells. In this method the inoculum is progressively 'thinned out' in such a way that individual, well-separated cells are left on at least some areas of the plate – usually in the third, fourth or fifth streakings (Fig. 14.4); on incubation, each well-separated cell gives rise to an individual colony.

In *stab inoculation*, a solid medium – e.g. the butt of a slope (Fig. 14.1) – is inoculated with a straight wire by plunging the wire vertically into the medium; the inoculum (at the tip of the wire) is thus distributed along the length of the stab. This procedure is used e.g. for inoculating deep, microaerobic or anaerobic parts of a medium.

A *spread plate* is made by spreading a small volume of liquid inoculum (e.g. 0.05– 0.1 ml) over the surface of a solid medium by means of a sterile L-shaped glass rod (a 'spreader').

A *flood plate* is made by flooding the surface of a solid medium with a liquid inoculum and withdrawing excess inoculum with a sterile Pasteur pipette.

If the inoculum in a flood plate, or a spread plate, contains enough cells, incubation will give rise to a *lawn plate*: a plate in which the surface of the medium is covered with a layer of confluent growth (section 3.3.1).

A plate is sometimes inoculated with a *swab*: typically, a compact piece of cotton wool attached securely to one end of a thin wooden or plastic stick or a piece of wire. A sterile swab is used e.g. for sampling organisms at a given site (such as the throat). After exposure, the swab is drawn lightly across the surface of a plate of suitable medium –

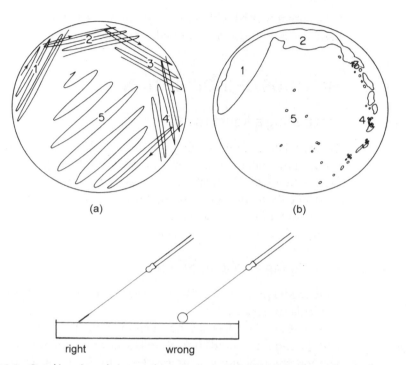

Figure 14.4 Streaking: inoculating a plate to obtain individual colonies. (a) A loop carrying the inoculum is moved from side to side (i.e. streaked) across a peripheral region of the plate, following the path shown at 1. The loop is then flamed and allowed to cool; the *sterile* loop is now streaked across the medium as shown at 2. Streakings 3, 4 and 5 are similarly made, the loop being flamed and cooled between each streaking and after the last. (b) After incubation, those areas of the plate on which large numbers of cells had been deposited show areas of confluent growth, as at 1, 2 and 3; well-separated cells give rise to individual colonies, as at 5.

When streaking with a loop, the plane of the wire loop should *not* be vertical. Starting from the wrong position (see lower diagram), the correct position is achieved by rotating the loop's handle through 90°; in the 'right' position (lower diagram), the loop – *lightly* in contact with the surface of the medium – is moved towards and away from the viewer during streaking.

taking care that all areas of the cotton wool make contact with the medium. The swab may be used to inoculate the entire surface of a plate; alternatively, it may be used to inoculate a small, peripheral area, the inoculum then being further distributed by streaking.

14.6   PREPARING A PURE CULTURE FROM A MIXTURE OF ORGANISMS

Some of the basic techniques of bacteriology can be illustrated by following through a common procedure such as the *isolation* of a particular strain or species from a mixture of organisms. The following describes the isolation of *E. coli* from a sample of sewage (which usually contains a range of enteric and non-enteric organisms).

A loopful of sewage is streaked onto a plate of MacConkey's agar (section 14.2.1), and the plate is then incubated for 18–24 hours at 37°C. (Plates are commonly incubated upside-down, i.e. with lid below; if incubated the right way up, water vapour from the medium may condense on the inside of the lid and drop onto the surface of the medium – with possible disruption of the streaked inoculum.) During incubation, well-separated cells of *any* species capable of growing on the medium will each give rise to an individual colony. After 18–24 hours on MacConkey's agar, *E. coli* forms round, red colonies of about 2–3 mm in diameter – but not all colonies with this appearance will necessarily be those of *E. coli*. The next step is to choose several such colonies for further examination; because *E. coli* is very common in sewage, at least one of the selected colonies is likely to be that of *E. coli*.

Before identification can be attempted it is necessary to ensure that each of the selected colonies contains cells of only one species. There is always the possibility that a given colony – even a well-separated one – may contain the cells of two different species; this may occur if, during streaking, two different cells had been deposited (by chance) very close together on the surface of the medium. To resolve this doubt, each chosen colony is subcultured; *subculturing* is a process in which cells from an existing culture or colony are transferred to a fresh, sterile medium. To subculture a given colony, the surface of the colony is touched lightly with a sterile loop so that a minute quantity of growth adheres to the loop; the growth is then streaked (in this case) onto a plate of sterile MacConkey's agar. (Some bacteria form very small colonies, and in such cases it is often easier to subculture by touching the surface of the colony with the tip of a straight wire; the inoculum is then carried (on the straight wire) to a fresh medium where it is streaked with a sterile loop.) Each plate, inoculated from a single colony, is then incubated. If each red colony had been that of a single species, we should now have several pure cultures – at least one of which is likely to be that of *E. coli*. Each pure culture can now be examined by the identification process outlined in Chapter 16.

14.6.1   Dynabeads® and immunomagnetic separation (IMS)

Magnetic separation is a technique for separating particular types of cell (or molecule) from a mixture of types by allowing the required cells or molecules to adsorb to specifically coated *beads* which are then segregated by means of a magnetic field.

The beads (Dynabeads®, trade name of Dynal, Skøyen, Oslo, Norway) are uniform microscopic spheres, containing a mixture of iron oxides, which are *superparamagnetic*, i.e. they exhibit magnetic properties *only* when placed in a magnetic field. The beads can be coated with any of a variety of ligands, the type of ligand determining the specific use (i.e. the type of cell or molecule which can be separated). Essentially, the beads and sample are mixed (according to specified protocols) so that the beads are dispersed within the sample; the required cells/molecules adsorb to the beads – which are then drawn to one side of the vessel by a magnetic field. After removal of unwanted material the cells/molecules can be washed etc. prior to use. (Note that *permanently* magnetic beads would be unsuitable: they would tend to clump together rather than disperse in the sample.)

Beads coated with specific *antibodies* can be used to isolate particular pathogens (e.g. *E. coli* O157, *Listeria monocytogenes, Salmonella* spp) from food and other samples (see Plate 14.1); the bead–pathogen complex is plated on an appropriate medium. Similarly, antibody-coated beads can be used to isolate particular types of eukaryotic cell from mixtures of cells. This type of procedure is called *immunomagnetic separation* (IMS).

Dynabeads can also be coated with ligands for nucleic acids. For example, Dynabeads Oligo $(dT)_{25}$ beads bind poly-A tails of mRNA molecules and are useful for separating mRNA from total RNA when preparing cDNA libraries (section 8.5.1.2).

Dynabeads® DNA DIRECT™ binds genomic DNA in 'crude' preparations and can provide samples for PCR (section 8.5.4). By effectively concentrating the DNA, this procedure enhances the sensitivity of PCR assays; moreover, the washing of bead–DNA samples is useful for removing inhibitors of the PCR process.

Some applications of Dynabeads are listed in Table 14.2.

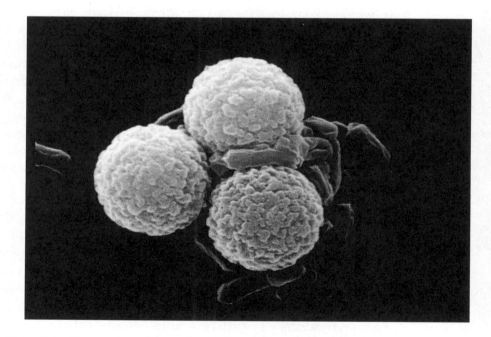

Plate 14.1 Electronmicrograph of cells of *Escherichia coli* O157 bound to Dynabeads® (see section 14.6.1). Photograph courtesy of Dynal A.S., Skøyen, Oslo, Norway.

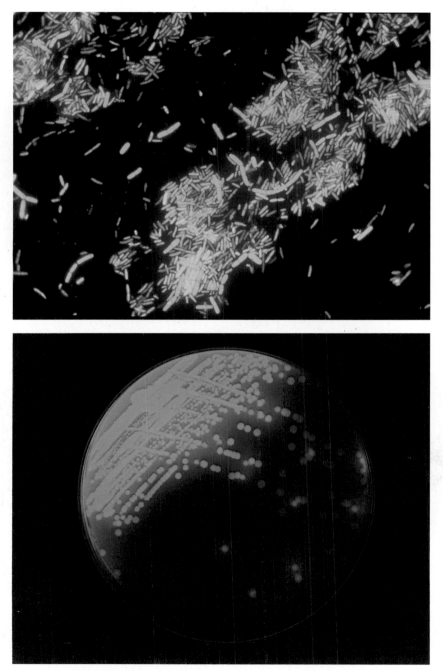

Plate 14.2 *Top.* Living cells of *Bacillus cereus* (Gram positive, orange) and *Pseudomonas aeruginosa* (Gram negative, green) stained with the LIVE *Bac*Light™ Bacterial Gram Stain Kit (see section 14.9.1.1). Reproduced with permission from Molecular Probes Inc., Eugene, Oregon, USA.
Below. Fluorescent colonies of *Escherichia coli* on MUG medium (see section 16.1.2.15). Reproduced with permission from Oxoid Ltd, Basingstoke, UK.

Plate 14.3   *Top*. Colonies of *Listeria monocytogenes* on an Oxoid selective medium containing lithium chloride, antibiotics, aesculin and a ferric salt; products from the bacterial hydrolysis of aesculin have reacted with the ferric salt, forming brown/black compounds that give the colonies and medium a characteristic appearance. Reproduced with permission from Oxoid Ltd, Basingstoke, UK.
Below. Colonies of *Salmonella utrecht* on xylose–lysine–deoxycholate (XLD) medium. On this medium (see Table 14.1 and section 14.2.1) the (red) colonies of salmonellae usually have a black centre because these organisms typically form hydrogen sulphide that reacts with a ferric salt in the medium. *Shigella* forms red colonies on XLD, and most other enterobacteria (including *Escherichia coli*) form yellow colonies. Reproduced with permission from Oxoid Ltd, Basingstoke, UK.

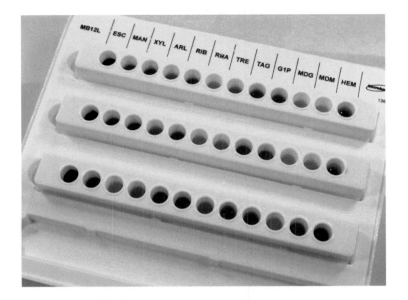

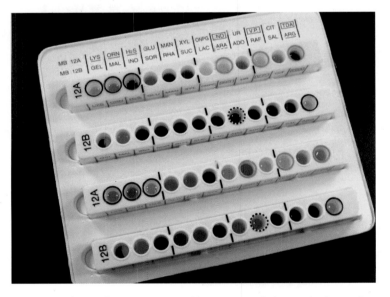

Plate 14.4 *Top*. Oxoid Microbact™ 12L, a multitest system (micromethods: section 16.1.3) for carrying out twelve tests on each of three strains. In this particular system the tests are useful for examining species of *Listeria*; they include a test for hydrolysis of aesculin (American: esculin), and tests for the fermentation of (i.e. acid production from) certain *sugars* (section 16.1.2.4), e.g. mannitol, rhamnose and xylose – these latter tests helping to distinguish between different species of *Listeria*. Reproduced with permission from Oxoid Ltd, Basingstoke, UK.

Below. Oxoid Microbact™ 12A and 12B GNB multitest strips which are used for identifying members of the Enterobacteriaceae and miscellaneous Gram-negative bacteria. The tests include: lysine and ornithine decarboxylases (section 16.1.2.8); hydrogen sulphide (16.1.2.6); sugars (e.g. glucose, sucrose, lactose, mannitol, xylose) (16.1.2.4); ONPG (16.1.2.10) urease (16.1.2.7); citrate, VP test (16.1.2.5). Reproduced with permission from Oxoid Ltd, Basingstoke, UK.

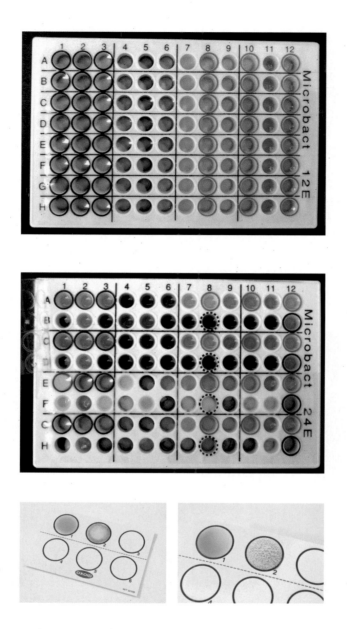

Plate 14.5 *Top* and *centre*. Oxoid Microbact™ 12E and 24E Microplate systems, used e.g. for identifying commonly isolated members of the Enterobacteriaceae. Microbact™ 12E is shown unin-oculated (unused); Microbact™ 24E has been inoculated with four test strains and incubated.

Below. A latex slide-agglutination test (see section 16.1.5.1). In the left-hand panel, a drop of water (used as a control) was mixed with latex suspension in circle 1, and a drop of sample containing the relevant antigen was mixed with latex suspension in circle 2. Clumping, typical of a positive test, is shown in circle 2 in the right-hand panel. Blue-stained latex particles are used to aid visibility. This kind of test is used for the detection of various pathogens, including *Escherichia coli* O157, *Legionella* and *Staphylococcus aureus*; it is also used e.g. for Lancefield grouping of *Streptococcus* (see Appendix). Reproduced with permission from Oxoid Ltd, Basingstoke, UK.

Table 14.2   Dynabeads®: some applications

Application	Reference
Detection of *E. coli* O157 in various foods	LAM (1997) *25* 442–446
Detection of *E. coli* O157 in minced meat	LAM (1998) *26* 199–204
Isolation of *E. coli* O157 from raw meat	LAM (1996) *23* 317–321
Detection of mycobacterial DNA in clinical specimens	JCM (1996) *34* 1209–1215
Isolation of DNA for taxonomic studies on cyanobacteria and prochlorophytes	AEM (1997) *63* 2593–2599
Preparation of human glomerular microvascular endothelial cells for studies on their susceptibility to the *E. coli* verocytotoxin in the presence of TNF-α	Kidney International (1997) *51* 1245–1256
Isolation of *Mycobacterium paratuberculosis* from milk (as an enrichment technique prior to e.g. PCR detection of IS*900* target DNA)	AEM (1998) *64* 3153–3158

14.7   ANAEROBIC INCUBATION

Anaerobic bacteria are incubated under anaerobic conditions. This can be achieved by using an *anaerobic jar* – one form of which is the McIntosh and Fildes' jar: a strong, metal cylindrical chamber with a flat, circular, gas-tight lid. The jar is loaded with a stack of inoculated plates, and the lid is secured with a screw-clamp. The jar is then connected to a suction pump via one of two valves in the lid; after a few minutes the valve is closed. Hydrogen (in a rubber bladder) is then passed into the jar via the other valve; this valve is then closed. Evacuation and re-filling may be repeated several times. On the inside of the lid is a gauze envelope containing a catalyst (e.g. palladium-coated alumina pellets) which promotes chemical combination between hydrogen and the last traces of oxygen. The jar is then placed in an incubator for an appropriate period of time. (Within the jar, plates are stacked the right way up; if stacked upside-down, the agar may be sucked from the base of the Petri dish by the vacuum.)

Another (more modern) form of anaerobic jar is a stout cylindrical vessel of strong, transparent plastic with a flat, gas-tight lid. The jar is loaded with plates; water is then added to a small packet of chemicals which is dropped into the jar immediately before the lid is secured with a screw-clamp. The chemicals liberate hydrogen which, in the presence of a catalyst, combines with all the oxygen in the jar. Because, in this case, there is no vacuum, the plates can be inserted upside-down (i.e. lid-side down) so as to avoid the problem of condensation.

Most anaerobic jars contain a redox indicator which indicates the state of anaerobiosis in the jar. In metal jars the indicator is placed in a small glass side-arm,

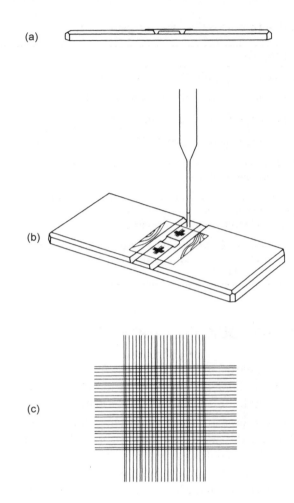

Figure 14.5   A typical counting chamber (haemocytometer). The instrument, seen from one side at (a), consists of a rectangular glass block in which the central plateau lies precisely 0.1 mm below the level of the shoulders on either side. The central plateau is separated from each shoulder by a trough, and is itself divided into two parts by a shallow trough (seen at (b)). On the surface of each part of the central plateau is an etched grid (c) consisting of a square which is divided into 400 small squares, each 1/400 mm². A glass cover-slip is positioned as shown at (b) and is pressed firmly onto the shoulders of the chamber; to achieve proper contact it is necessary, while pressing, to move the cover-slip (slightly) against the surface of the shoulders. Proper (close) contact is indicated by the appearance of a pattern of coloured lines (Newton's rings), shown in black and white at (b).

Using the chamber. A small volume of a bacterial suspension is picked up in a Pasteur pipette by capillary attraction; the thread of liquid in the pipette should not be more than 10 mm. The pipette is then placed as shown in (b), i.e. with the opening of the pipette in contact with the central plateau, and the side of the pipette against the cover-slip. With the pipette in this position, liquid is automatically drawn by capillary attraction into the space bounded by the cover-slip and part of the central plateau; *the liquid should not overflow into the trough*. (It is sometimes necessary to tap the end of the pipette, *lightly*, against the while in plastic jars an indicator-soaked pad is usually visible through the wall of the jar.

Some anaerobes can be grown (without an anaerobic jar) in media such as *Robertson's cooked meat medium* (minced beef heart, beef extract (1%), peptone (1%), sodium chloride (0.5%) and a reducing agent, e.g. L-cysteine or thioglycollate); the medium, sterilized by autoclaving, is stored in screw-cap universal bottles which are sometimes equilibrated under oxygen-free conditions after inoculation and before closure.

Anaerobic cabinets allow samples/cultures to be handled/incubated under oxygen-free conditions with control of e.g. temperature, humidity and CO_2 concentration; items inside the cabinet can be manipulated by means of gas-tight gloves fixed into the front panel.

(See also anaerobic blood culture in section 11.8.1.1.)

14.8   COUNTING BACTERIA

The total number of (living and dead) cells in a sample is called the *total cell count*; the number of living cells is the *viable cell count*. Counts in liquid samples are usually given as the number of cells per millilitre (or per 100 ml).

(See also section 3.5.)

14.8.1   Total cell count

The total cell count in a liquid sample (e.g. a broth culture) can be estimated by direct counting in a counting chamber (Fig. 14.5).

---

central plateau to encourage the liquid to enter the chamber.) A second sample can be examined, if required, in the other half of the counting chamber. The chamber is left for 30 minutes to allow the cells to settle, and counting is then carried out under a high power of the microscope – which is focused on the grid of the chamber. Because the volume between grid and cover-slip is accurately known, the count of cells per unit volume can be calculated.

A worked example. Each small square in the grid is $1/400$ mm^2. As the distance between grid and cover-slip is $1/10$ mm, the volume of liquid over each small square is $1/4000$ mm^3 – i.e. $1/4000000$ ml.

Suppose, for example, that on scanning all 400 small squares, 500 cells were counted; this would give an average of $500 \div 400$ (= 1.25) cells per small square, i.e. 1.25 cells per $1/4000000$ ml. The sample therefore contains 1.25×4000000 cells/ml, i.e. 5×10^6 cells/ml. Several counts may be made and averaged.

If the sample had been diluted before examination (because it was too concentrated), the count obtained must be multiplied by the dilution factor; for example, if diluted 1-in-10, the count should be multiplied by 10.

N.B. The chamber described above is the *Thoma chamber*; in a *Helber chamber* the distance between central plateau and cover-slip is 0.02 mm.

Another method, the *direct epifluorescent filter technique* (DEFT), is used e.g. for counting organisms in milk. Essentially, the milk is passed through a membrane filter, and the cells retained on the filter are stained with a fluorescent dye; ultraviolet radiation is then beamed onto the filter, and the (fluorescent) cells are seen (through a microscope) as bright particles against a dark background. (For DEFT, the milk is pre-treated to disrupt fat globules etc. which would otherwise block the filter.)

The total cell count can also be estimated by comparing the *turbidity* of the sample with that of each of a set of tubes (*Brown's tubes*) containing suspensions of barium sulphate in increasing concentration; the tubes range from transparent (tube 1), through translucent, to turbid and opaque (tube 10). For a given species of bacterium the turbidity of a particular tube corresponds to the turbidity of a suspension of cells of known concentration. The sample is examined in a tube of size and thickness equivalent to those containing the standard suspensions; the turbidity of the sample is matched, visually, with that of a particular tube, and the concentration of the sample is then read from a table supplied with the tubes.

14.8.2   Viable cell count

Most methods of estimating the viable cell count involve the inoculation of a solid medium with the sample (or diluted sample). After incubation, the number of cells in the inoculum can be estimated from the number of colonies which develop on or within the medium. It is always assumed that each colony has arisen from a single cell; the number of cells which actually give rise to colonies depends at least partly on the type of medium used and on the conditions of incubation.

In the *spread plate* (or *surface plate*) method, an inoculum of about 0.05–0.1 ml is spread over the surface of a sterile plate, as described earlier; the plate is 'dried' (section 14.2.2), incubated, and the viable cell count is estimated from (i) the number of colonies, (ii) the volume of inoculum used, and (iii) the degree (if any) to which the sample had been diluted. If a sample is suspected of containing many cells – e.g. 10^6 cells/ml – it can be diluted in 10-fold steps, and an inoculum from each dilution spread onto a separate plate; at least one dilution will give a countable number of colonies.

In the *pour plate* method, the (liquid) inoculum is mixed with a molten agar-based medium (at about 45°C) which is then poured into a Petri dish and allowed to set; on incubation, colonies develop within (as well as on) the medium, and the viable count is calculated as in the spread plate method.

Yet another method for viable count is *Miles and Misra's method* (Fig. 14.6).

A sample likely to contain small numbers of bacteria (e.g. water from a *clean* river) may be passed through a sterile membrane filter of pore size about 0.2 μm – which retains cells on the upper surface; a volume of, say, 100 ml or more may be filtered. The membrane is then placed (cell-side uppermost) onto an absorbent pad saturated with a suitable medium; on incubation, nutrients diffuse through the membrane, and

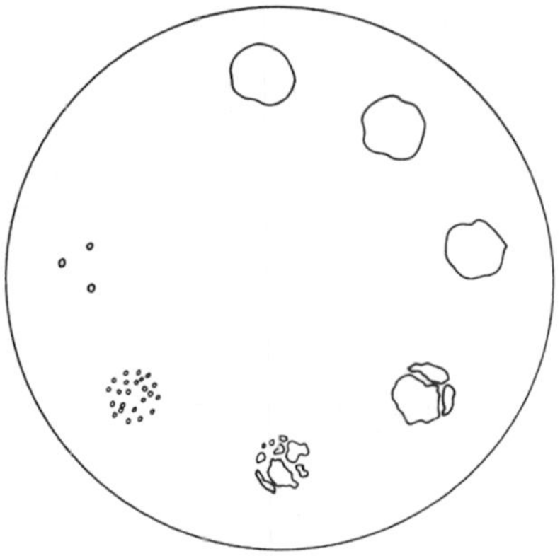

Figure 14.6 Miles and Misra's method (the 'drop method') for estimating viable cell counts. The sample is first diluted, usually in 10-fold steps – i.e. 1/10, 1/100 . . . etc. One drop (of known volume) from each dilution is then placed at a separate recorded position on the surface of a 'dried' plate of suitable medium. The drops are allowed to dry, and the plate is incubated until visible growth develops. Drops which contained large numbers of viable cells give rise to circular areas of confluent growth. Any drop which contained less than about 15 viable cells will give rise to a small, countable number of colonies; by assuming that each colony arose from a single cell, the viable count can be estimated from (i) the number of colonies, (ii) the volume of the drop, and (iii) the dilution factor.

The same Pasteur pipette can be used throughout *if* the drops are taken first from the highest dilution and then from the next highest dilution, and so on.

A Pasteur pipette can be *calibrated*, i.e. the volume of drops delivered by the pipette can be measured. This is done by drawing into the pipette a measured volume (e.g. 1 ml) of water, and counting the number of drops formed when this volume is discharged; if 1 ml is discharged in, say, 30 drops, then each drop is about 1/30 ml.

colonies develop from those cells capable of growth under such conditions. The viable cell count can then be estimated from (i) the number of colonies on the membrane, and (ii) the volume of sample filtered.

14.8.3 Counting cells in (or on) solids

Sometimes we need to count the bacteria in a sample of food or fabric etc. In order to do this, the particles of food etc. must be broken up and/or the bacteria must be separated from the sample; ideally, clumps of bacteria should also be broken up. The sample and a diluent (such as Ringer's solution) can be sealed into a sterile plastic bag which is then subjected to mechanical agitation; in this way, at least some of the bacteria can be brought into suspension in the diluent.

14.9   STAINING

Staining is often used to detect, categorize or identify bacteria, or to observe specific bacterial components; in most cases the cells are killed and 'fixed' before being stained.

Dyes etc. are usually applied to a thin film of cells on a glass microscope slide. Cells from a pure culture may be examined as follows. A loopful of water is placed on a clean slide, and (using the loop) a speck of growth from a colony is mixed ('emulsified') with the water to form a suspension of cells. Using the loop, the suspension is spread over an area of one or two square centimetres and allowed to dry – forming a *smear*. The smear is then fixed by passing it quickly through a bunsen flame twice; it is then ready for staining and subsequent examination under the microscope.

A smear may also be made directly from the sediment of centrifuged urine, or e.g. pus from an abscess.

14.9.1   The Gram stain

The background to this stain is given in section 2.2.9. One of many versions of the procedure is as follows.

A heat-fixed smear (see above) is stained for 1 minute with *crystal violet*; it is then rinsed briefly under running water, treated for 1 minute with *Lugol's iodine* (a solution of iodine and potassium iodide in water), and briefly rinsed again. Decolorization is then attempted by treating the stained smear with a solvent such as ethanol (95%), acetone or iodine–acetone. This is the critical stage: with the slide tilted, the solvent is allowed to run over the smear only for as long as dye runs *freely* from it (about 1–3 seconds); the smear is then *immediately* rinsed in running water. At this stage, any Gram-negative cells will be colourless; Gram-positive cells will be violet. The smear is now counterstained for 30 seconds with dilute *carbolfuchsin* to stain (red) any Gram-negative cells present. After a brief rinse, the smear is blotted dry and examined under the oil-immersion objective of the microscope (final magnification about 1000×).

Some bacteria do not give a clear or constant reaction to the Gram stain – sometimes reacting positively, sometimes negatively. These bacteria are said to be *Gram-variable*. To avoid problems of Gram-variability in taxonomy (classification), certain bacteria are described as *Gram-type-positive* or *Gram-type-negative* according to whether their cell walls are of the Gram-positive type (section 2.2.9.1) or of the Gram-negative type (section 2.2.9.2) respectively.

14.9.1.1   *Gram staining living cells*

The Gram reaction of *living* cells can be determined by a differential fluorescence staining process (LIVE *Bac*Light™ Bacterial Gram Stain Kit: Molecular Probes Inc., Eugene, Oregon, USA). Essentially, bacteria in the logarithmic phase of growth are treated with a mixture of two dyes, each dye being a fluorescent stain for nucleic acids; these dyes are SYTO® 9 and hexidium iodide. In Gram-positive bacteria, nucleic acids are preferentially stained by hexidium iodide – this dye competing successfully for the binding sites. In Gram-negative bacteria, nucleic acids are stained by SYTO® 9 because *this* dye can penetrate the outer membrane (section 2.2.9.2). Under radiation of wavelength c. 480 nm, the bound hexidium iodide produces an orange fluorescence while bound SYTO® 9 fluoresces green (see Plate 14.2, *top*).

14.9.2   The Ziehl–Neelsen stain (acid-fast stain)

'Acid-fast' bacteria (AFB) differ from all other bacteria in that once they are stained with hot, concentrated carbolfuchsin they cannot be decolorized by mineral acids or by mixtures of acid and ethanol; such bacteria include e.g. *Mycobacterium tuberculosis*. A heat-fixed smear is flooded with a concentrated solution of carbolfuchsin, and the slide is heated until the solution steams; it should not boil. The slide is kept hot for about 5 minutes, left to cool, and then rinsed in running water. Decolorization is attempted by passing the slide through several changes of acid–alcohol (e.g. 3% v/v concentrated hydrochloric acid in 90% ethanol). After washing in water, the smear is counterstained with a contrasting stain (such as malachite green), washed again, and dried. Acid-fast cells stain red, others green.

14.9.3   Capsule stain

Bacterial capsules may be demonstrated e.g. by *negative staining* (Plate 2.3: *centre, right*). The cells are emulsified with a loopful of e.g. *nigrosin* solution on a clean slide and are then overlaid with a cover-slip; under the oil-immersion or high dry ($\times 40$) objective lens of the microscope the capsule appears as a clear, bright zone between a cell and its dark background.

14.9.4   Endospore stain

See section 16.1.1.4.

14.9.5   Distinguishing live from dead cells by staining

Live and dead bacteria can be distinguished e.g. by a simple one-step staining technique (LIVE/DEAD® *Bac*Light™ Bacterial Viability Kit: Molecular Probes Inc., Eugene, Oregon, USA). In one protocol, bacteria are treated with a mixture of two dyes, each being a fluorescent stain for nucleic acids; these dyes are SYTO® 9 (green fluorescence) and propidium iodide (red fluorescence). Propidium iodide can enter only those cells with damaged/permeable membranes (a characteristic of dead cells); it is excluded from normal, healthy cells. SYTO® 9 can enter both living and dead cells. Under the fluorescence microscope, live cells therefore fluoresce green; in dead cells, propidium iodide competes with SYTO® 9 for binding sites, and such cells fluoresce red (see cover of book).

14.10   MICROSCOPY

In bacteriology there is often a need to use high magnification (e.g. 1000×), and for this the microscope must have an *oil-immersion* objective lens (magnification about 100×) and a suitable eyepiece (about 10×). When using oil-immersion objectives, a drop of immersion oil fills the space between the lens and the top of the cover-glass (or, as is often the case, between lens and smear). What is the oil for? A powerful objective lens has a very short focal length, and light from the specimen must enter the lens at a *wide* angle; this is possible only if the space between lens and specimen is filled with a material that has a suitable refractive index (Fig. 14.7).

The maximum useful magnification obtainable from a given objective lens is 1000 times its *numerical aperture* (NA) (Fig. 14.7).

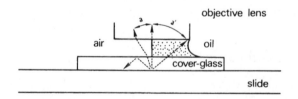

Figure 14.7   Oil-immersion objective lens: the reason for using oil. In the diagram, the (thin) specimen lies between slide and cover-glass, and rays from one point in the specimen are shown travelling towards the objective lens.

When immersion oil – refractive index (RI) approx. 1.5 – fills the space between lens and cover-glass (right-hand side of diagram), rays of light passing upwards from the cover-glass enter a medium whose RI is similar to that of the glass itself (approx 1.5); consequently, each ray will continue on its original path (unrefracted) as though it were still travelling through glass. Such a ray is shown entering the objective lens; note that rays can enter at the widest angle.

When air (RI = 1) separates lens and cover-glass (left-hand side of diagram), a ray whose angle is similar to that shown on the right-hand side of the diagram cannot leave the glass; instead, it is reflected back into the glass via the upper, inner surface of the cover-glass. Rays which make a smaller angle with the vertical (as shown) can enter the objective lens; however, in this case, less light enters the lens. In this mode, some of the (image-forming) rays from the specimen do not enter the lens.

Numerical aperture (NA) is a characteristic of a lens given by:

$$NA = n \times \sin\theta$$

where n is the refractive index of the material between lens and cover-glass, and θ is *half* the maximum angle at which rays enter the lens. In the diagram, θ is the angle a in air, and a' with immersion oil.

The maximum resolving power of a lens is related to its NA.

14.10.1 Köhler illumination

For the best image, the specimen must be illuminated evenly, regardless of any unevenness in the source of light. Köhler illumination, which is optimal, uses a lamp fitted with a condensing lens and an adjustable diaphragm (= *field stop*) that determines the diameter of the illuminating beam; when set up correctly, an image of the lamp's filament is formed in the lower focal plane of the microscope's substage condenser, and rays from *each* point of this image pass through the condenser to emerge as parallel rays which illuminate the specimen and enter the objective lens (Fig. 14.8).

14.10.2 Phase-contrast microscopy

Ordinary (*bright-field*) microscopy can reveal fine detail when different parts of the specimen absorb different amounts of light or (perhaps due to staining) absorb different colours. *Unstained* cells can sometimes be seen – perhaps well enough to be counted – but with little detail.

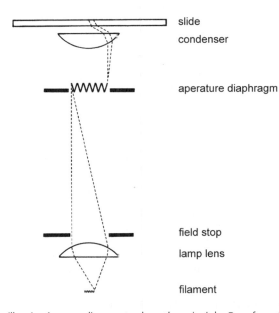

slide
condenser

aperature diaphragm

field stop
lamp lens
filament

Figure 14.8 Köhler illumination: ray diagram to show the principle. Rays from the lamp's filament are focussed to form an image (size exaggerated) in the lower focal plane of the substage condenser; coinciding with this plane is an iris diaphragm (*aperture diaphragm*) which forms part of the substage condenser unit. Because the image is in the condenser's focal plane, rays from any given point in the image emerge from the condenser as parallel rays (which illuminate the specimen on the slide). If a microscope has been set up for Köhler illumination, the edge of the field stop should become visible in sharp focus *in the plane of the specimen* if the field stop is gradually closed.

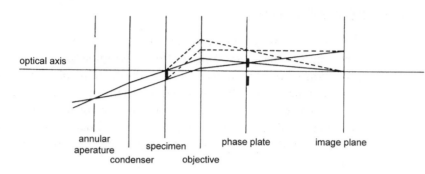

Figure 14.9   Phase-contrast microscopy: simplified diagrammatic scheme to show the principle.

Within a transparent or translucent specimen, the incident light has illuminated a region which causes diffraction; the first-order waves (*dashed line*) are retarded by about one-quarter of a wavelength ($1/4\lambda$). If these first-order waves and the non-diffracted (zero-order) waves interact in a normal (bright-field) microscope, the resultant wave would have an amplitude similar to that of the background waves so that the feature would not be clearly distinguished; the resultant and background waves would differ slightly in phase, but this cannot be detected by the eye.

In a phase-contrast microscope the condenser has an annular (ring-shaped) aperture in its front focal plane so that a *hollow* cone of light can be focused (as a small bright ring) onto a *phase plate* located in the back focal plane of the objective lens (see diagram). The phase plate is a glass disc on which has been deposited a ring of material (e.g. magnesium fluoride) of such thickness that it retards by $1/4\lambda$ the light that passes through it.

In the absence of a specimen all the light passes through the phase ring. When a specimen (e.g. an unstained cell) is examined, the zero-order and background waves (solid lines) pass through the ring but the first-order waves (dashed lines) pass through the phase plate via regions outside the ring; thus, by retarding the zero-order and background waves by $1/4\lambda$, the phase ring brings all the waves into the same phase. In the image plane, interaction between zero-order and first-order waves gives rise to a visible image because the resultant wave has an amplitude greater than that of the background waves (i.e. the image-forming wave is brighter than the background waves). (This is 'negative' or 'bright' phase-contrast microscopy; the opposite effect (image darker than background) is seen in 'positive' or 'dark' phase-contrast microscopy.)

An image of greater clarity is obtained when the condenser contains a green filter.

Compared with bright-field microscopy, the lamp must be more powerful because only a proportion of the lamp's output is used for image-making.

Care must be taken to ensure that the annular diaphragm in the condenser matches the particular phase-contrast objective lens being used.

The small bright ring on the phase plate must coincide *exactly* with the phase ring (see diagram). Adjustment, if necessary, is made when the condenser is in the correct position and the specimen has been brought into focus to the maximum extent possible. Under these conditions, remove the eyepiece and replace it with the specialized 'telescope' used for this purpose; the telescope is focused on the phase plate, and adjustments to the position of the small bright ring can be made e.g. by adjusting centring screws on the condenser (depending on the particular model of microscope).

Diagram from *Dictionary of Microbiology and Molecular Biology*, Singleton & Sainsbury, 3rd edition, 2001, p. 479, by courtesy of the publisher, John Wiley, Chichester, UK.

Phase-contrast microscopy can reveal fine detail e.g. in transparent/translucent, unstained living cells by altering the phase difference between diffracted and non-diffracted light from the specimen (Fig. 14.9).

14.10.3 Epifluorescence microscopy

See section 11.8.2.

14.10.4 Literature on microscopy

Particularly helpful booklets on microscopy (containing both theoretical and practical information) have been produced by the major manufacturers of microscopes. These include *Microscopy from the Very Beginning*, by Friedrich K. Möllring (Carl Zeiss, Germany), and (a more comprehensive treatment) *The Microscope and its Applications*, by Hans Determann and Friedrich Lepusch (Leitz Wetzlar, Germany).

15 Man against bacteria

Bacteria can be a nuisance, or even dangerous, in many everyday situations, and we therefore need methods to eliminate them or to inhibit their activities. Sometimes it is necessary to destroy, completely, all forms of life on a given object – as, for example, when surgical instruments are prepared for use. At other times it may be sufficient merely to eliminate only the potentially harmful organisms. There is also the special problem of inactivating pathogenic bacteria on or within living tissues.

15.1 STERILIZATION

Any procedure guaranteed to kill *all* living organisms – including endospores (section 4.3.1) and viruses – is called a *sterilization* process. (Because sterilization kills *all* organisms, the expression 'partial sterilization' is meaningless.)

Autoclaving (section 15.1.1.3) is regarded as a 'gold standard' of sterilization. Hospitals and microbiology laboratories rely heavily on the autoclave for ensuring sterility; this instrument uses steam under pressure at e.g. 121°C for an appropriate period of time.

Until recently it seemed certain that all the standard procedures for sterilization could, in fact, kill all known organisms, even the most heat-resistant forms of life. Then, in 2003, came the report of an organism that not only survives autoclaving but can actually *grow* at 121°C [Science (2003) *301* 934]. Fortunately, this organism lives in hydrothermal vents on the ocean floor and is not a serious challenge to the autoclave; even so, the existence of such organisms is of importance for the strict definition of a sterilization process. Nevertheless, in most situations it's probably safe to continue to regard autoclaving as a reliable sterilization process.

Ideally, methods for sterilization should be efficient, quick, simple and cheap, and they should be applicable to a wide range of materials. Sterilization is usually carried out by physical methods – most commonly by the use of heat.

15.1.1   Sterilization by heat

The cells of different species of bacteria vary in their susceptibility to heat, and endospores are much more resistant than vegetative cells; vegetative cells generally die rapidly in boiling water, while endospores may survive for long periods of time.

The sterilizing power of heat depends not only on temperature but also on factors such as time, the presence of moisture, and the number and condition of the microorganisms present. Note that the larger the initial population of bacteria, the longer will be the time needed to achieve sterility at a given temperature; this can be seen from Fig. 12.2.

15.1.1.1   *Fire*

Fire is used e.g. for the rapid sterilization of surfaces and loops etc. (sections 14.3, 14.4), while disposable items such as used surgical dressings and one-use syringes may be sterilized – and destroyed – by incineration. However, less destructive methods are used for most other purposes.

15.1.1.2   *The hot-air oven*

This apparatus is used e.g. for the sterilization of heat-resistant items such as clean glassware. In use, a temperature of 160–170°C is maintained for 60–90 minutes; this denatures proteins, desiccates cytoplasm and oxidizes various components in any organisms present. Within the oven, air should be circulated by a fan to ensure that all parts are kept at the required temperature; items should be well spaced in order not to impede the flow of air.

15.1.1.3   *Sterilization by steam: the autoclave*

Steam can sterilize at lower temperatures (for shorter times) than those used in a hot-air oven. At normal atmospheric pressure, steam has a temperature of only 100°C – a temperature at which some endospores can survive for long periods – but, when *under pressure*, steam can reach higher temperatures suitable for sterilization; in fact, there is a definite relationship between the pressure and temperature of *pure* steam, i.e. steam containing no air: the higher the pressure the higher the temperature. When sterilizing with steam, items to be sterilized are placed inside a strong, metal, gas-tight chamber (an *autoclave*). Steam is produced within the chamber (Fig. 15.1) or, in larger autoclaves, is piped in from a boiler; air passes out through a valve until the chamber is filled with pure steam – at which time the valve is closed. The pressure and temperature of the steam rise as heating is continued (Fig. 15.1) or as more steam is piped in. At a pre-determined pressure, an (adjustable) valve opens – thus determining the

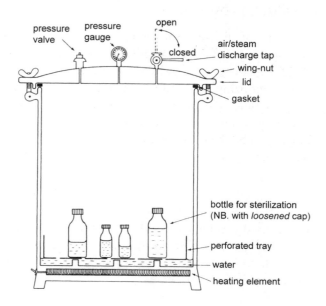

Figure 15.1 A typical, small laboratory autoclave. Water is placed in the bottom of the chamber. Objects for sterilization are placed on the perforated tray which holds them above the water. The lid, with the air/steam discharge tap *open*, is clamped securely in position; the rubber gasket ensures a gas-tight seal. The heating element is switched on and the water boils. Steam fills the chamber, eventually displacing all the air (which leaves via the discharge tap). Pure steam begins to issue vigorously from the discharge tap, which is then closed. As heating continues, water continues to vaporize so that the pressure (and hence temperature) within the chamber increases. Once the desired pressure/temperature has been reached (see text), a pre-set pressure valve opens; steam escapes, thus maintaining the pressure at a steady level. When the appropriate time has elapsed (see text), the heating element is switched off and the autoclave is allowed to cool until the pressure inside the chamber (indicated by the gauge) does not exceed atmospheric pressure.

See text for safety note.

pressure/temperature within the autoclave; steam which escapes (via the valve) is replaced by steam generated in the chamber, or piped in, so that pressure in the autoclave remains constant.

The time allowed for sterilization must be sufficient for all parts of the *load* (i.e. items/materials being sterilized) to reach the sterilizing temperature and to stay at that temperature until any organisms present have been killed.

Typical temperature/time combinations used in autoclaves are (i) 115°C (at a pressure of 0.68 bar [=69 kPa; 10 lb/inch^2] above atmospheric pressure) for 35 minutes; 121°C (1.02 bar [103 kPa; 15 lb/inch^2]) for 15–20 minutes; 134°C (2.04 bar [207 kPa; 30 lb/inch^2]) for 4 minutes. These times may be varied according to the nature of the load and the nature and degree of contamination (see section 15.1.1). Note that the timing does not start until all parts of the load have reached the sterilizing temperature. Note also that an autoclave's pressure gauge may show pressure either in terms of the pressure *above atmospheric* (as given above) or as the *absolute pressure* (abs), i.e. steam pressure + atmospheric pressure; moreover, the gauge may be calibrated in different units in different autoclaves (approx. equivalents: 1 bar = 101 kPa = 14.7 lb in² = 760 mmHg).

For effective sterilization the steam must be saturated, i.e. it must hold as much water in vapour form as is possible for the given temperature and pressure; no air should be present because air upsets the pressure–temperature relationship: an air–steam mixture at a given pressure has a lower temperature than that of pure steam at the same pressure. Hence, *all* air must be purged from the chamber – and from all items within the chamber – before the valve is closed.

Small portable laboratory autoclaves generally resemble the domestic pressure cooker in both principle and mode of use (Fig. 15.1). In larger autoclaves, such as those in hospitals, steam is usually piped to the autoclave chamber from a boiler, and factors such as timing, pressure and steam quality are often controlled automatically. In some models steam is admitted at the top of the chamber so that air is displaced downwards; this is more effective than upward displacement (used in small autoclaves) because steam is lighter than air under these conditions.

In another type of autoclave, air is removed from the chamber by a vacuum pump before steam is admitted; this allows rapid and thorough penetration by the steam of porous materials such as dressings or bed linen – materials which tend to trap air.

Some materials cannot be sterilized by autoclaving; these include water-repellent substances (e.g. petroleum jelly) and substances which are volatile or are heat-labile (i.e. destroyed by heat). Some of these materials (such as petroleum jelly) can be sterilized in a hot-air oven.

Certain materials, which would be damaged by autoclaving, may be sterilized by a combination of steam at reduced pressure (at e.g. 80°C) and formaldehyde; this method kills endospores within about 2 hours, and is used for sterilizing heat-sensitive surgical instruments, plastic tubing, woollen blankets etc.

Safety note. To avoid the risk of explosion of liquid-containing bottles etc.: (i) loosen caps before insertion into the chamber, and (ii) allow time for liquids to cool to ~ 80°C before opening the chamber.

15.1.2 Sterilization by ionizing radiation

Ionizing radiation – e.g *beta*-rays (electrons), *gamma*-rays, X-rays – sterilizes by supplying energy for a variety of lethal chemical reactions in the contaminating organisms. *Gamma*-radiation (typically using a cobalt-60 source) is widely used e.g. for the sterilization of pre-packed biological equipment such as plastic Petri dishes and syringes.

15.1.3   Sterilization by filtration

Filtration through membrane filters of pore size ~0.2 μm is sometimes used e.g. to obtain sterile solutions of heat-labile antibiotics or sugars; the liquid may be drawn through the filter by reduced pressure in the (sterile) receiving vessel, or forced through e.g. by a syringe plunger. The filter itself consists of a thin sheet of cellulose acetate, polycarbonate or similar material. The sterility of the filtrate depends on the absence of viruses – and e.g. nanobacteria (section 2.1.2) – in the pre-filtered sample; the presence of bacteriophages may be important if the filtrate is used in growth media for bacteria.

15.1.4   Sterilization by chemical agents

Chemicals used for sterilization (*sterilants*) are necessarily highly reactive and damaging to living tissues; they therefore require careful handling, and tend to be used only in larger institutions with suitable equipment and personnel.

Ethylene oxide (C_2H_4O) – a water-soluble cyclic ether – is a gas at temperatures above 10.8°C and forms explosive mixtures with air; it is therefore used diluted with another gas such as carbon dioxide or nitrogen. For sterilization, the gas mixture is used in a special chamber, and the temperature, humidity, time, and concentration of ethylene oxide must be carefully controlled. Ethylene oxide is an alkylating agent which reacts with various groups in proteins and nucleic acids; it is used e.g. for sterilizing clean medical equipment, bed linen, and heat-labile materials such as certain plastics.

Other sterilants include glutaraldehyde and β-propiolactone.

'Gas plasma' refers to a process in which hydrogen peroxide is injected into the sterilizing chamber (a dry atmosphere under reduced pressure) and radio-frequency energy converts the hydrogen peroxide to reactive chemical species (which effect sterilization). Conditions are important; failure to sterilize may be due e.g. to (i) low temperature (< 42°C); (ii) the presence of lipids; (iii) the presence of cellulose; or (iv) damp loads. Uses may include e.g. the sterilization of certain medical instruments (such as endoscopes).

15.2   DISINFECTION

Disinfection refers to any procedure which destroys, inactivates or removes *potentially harmful* microbes – without necessarily affecting the other organisms present; it often has little or no effect on bacterial endospores. 'Disinfection' often refers specifically to the use of certain chemicals (*disinfectants*) for the treatment of non-living objects or surfaces, though the term is sometimes also used to refer to antisepsis (section 15.3).

Although chemical disinfection is widely used, physical methods are more suitable for certain purposes.

15.2.1   Disinfection by chemicals

Ideally, disinfectants for general use should be able to kill a wide range of common or potential pathogens. However, any given disinfectant is usually more effective against some organisms than it is against others, and the activity of a disinfectant may be greatly affected by factors such as dilution, temperature, pH, or the presence of organic matter or detergent; to be effective at all, a disinfectant needs appropriate conditions, at a suitable concentration, for an adequate period of time. Some disinfectants (e.g. hypochlorites) tend to be unstable, and some (e.g. pine disinfectants) need solubilization in order to be effective. At low concentrations some disinfectants not only cease to be effective, they can actually be metabolized by certain bacteria – e.g. species of *Pseudomonas* can grow in dilute solutions of carbolic acid (phenol).

Disinfectants which *kill* bacteria are said to be *bactericidal*. Others merely halt the growth of bacteria, and if such a disinfectant is inactivated – e.g. by dilution, or by chemical neutralization – the bacteria may be able to resume growth; these disinfectants are said to be *bacteriostatic*. A bactericidal disinfectant may become bacteriostatic when diluted.

Of the many disinfectants in use, only a few of the common ones are mentioned here.

Phenol and its derivatives (e.g. 'phenolics' such as *cresols* and *xylenols*) can be bactericidal at appropriate concentrations; they appear to act mainly by affecting the permeability of the cytoplasmic membrane. *Lysol* is a mixture of methylphenols solubilized by soap; at 0.5% it kills many non-sporing pathogens in 15 minutes, but endospores may survive in 2% Lysol for days.

Chlorine is widely used for the disinfection of water supplies and for the sanitation of water in swimming pools. It acts (directly, and via hypochlorous acid) as an effective disinfectant, though its activity is decreased by the presence of organic matter and by other substances with which it can react.

Quaternary ammonium compounds (QACs) are cationic detergents used e.g. for the disinfection of equipment in the food and dairy industries. They are bacteriostatic at low concentrations, bactericidal at higher concentrations, and are typically more active against Gram-positive than Gram-negative bacteria. QACs appear to disrupt the cytoplasmic membrane; their activity is inhibited e.g. by soaps, some cations (e.g. Ca^{2+}, Mg^{2+}), low pH and organic matter.

Hypochlorites are highly effective against a wide range of bacteria (including endospores), the undissociated form of HOCl being strongly bactericidal. Sodium hydroxide is commonly used as a stabilizer in commercial hypochlorite disinfectants.

(See also *antiseptics* in section 15.3.)

15.2.1.1   *Testing disinfectants*

The efficacy of a *phenolic* disinfectant can be described by the *phenol coefficient*, i.e. the antimicrobial activity of the given disinfectant – relative to that of phenol – under standardized conditions. The phenol coefficient may be determined by a *suspension test* in which various dilutions of the disinfectant in a liquid medium are allowed to act, at a given temperature, and for specified times, on a specific test organism.

The *Rideal–Walker* suspension test determines that dilution of the test disinfectant which kills a test organism (e.g. *Salmonella typhi* NCTC 786) at a rate equal to that of a standard dilution of phenol, the result being expressed as the *Rideal–Walker coefficient* (dilution factor of test disinfectant/standard dilution factor of phenol).

The *Chick–Martin* suspension test determines the phenol coefficient in the presence of organic matter (e.g. yeast), i.e. under conditions simulating those met with in practice; this is significant in that disinfectants may be inactivated by organic matter.

For disinfectants in general, a *capacity test* is used to determine the ability of a given disinfectant to retain activity in the presence of increasing concentrations of bacteria; bacteria are periodically added to a fixed volume of the disinfectant – which is subcultured, at fixed intervals of time, to detect viable bacteria. The *Kelsey–Sykes* capacity test determines the efficacy of a disinfectant under simulated practical conditions; of various test organisms (e.g. *Pseudomonas aeruginosa* NCTC 6749, *Staphylococcus aureus* NCTC 4163), the one used is that which is known to be most resistant to the disinfectant under test.

A *carrier* test determines the ability of a given disinfectant to disinfect an *object* (the 'carrier') which has been artificially contaminated with microorganisms.

Procedures for testing disinfectants are under review, and emphasis is currently being placed on the efficacy of a disinfectant under 'in use' conditions.

15.2.2   Disinfection by physical agents

Ultraviolet radiation can damage DNA and can be lethal to bacteria under appropriate conditions. It has poor powers of penetration (being readily absorbed by solids), but ultraviolet lamps (wavelength about 254 nm) are used e.g. for the disinfection of air and exposed surfaces in enclosed areas.

The disinfection of milk by *pasteurization* and UHT processing is described in section 12.2.1.1.

A *pulsed electric field* procedure, which may have applications e.g. in the food industry, has been found to reduce significantly the numbers of viable cells of *Mycobacterium paratuberculosis* in milk [AEM (2001) *67* 2833–2836] and to sensitize *Listeria monocytogenes* to heat [AEM (2004) *70* 2289–2295].

15.3 ANTISEPSIS

Antisepsis is the disinfection of *living* tissues; it may be used prophylactically (i.e. to prevent infection) or therapeutically (i.e. to treat infection).

The comments on disinfectants (section 15.2.1) generally apply also to *antiseptics*, i.e. the chemicals used for antisepsis.

Dettol is a general-purpose phenolic antiseptic when used in dilute form, but a domestic disinfectant in more concentrated form; it is based on chloroxylenols.

Hexachlorophene has been used in antiseptic soaps; it is a *bisphenol* (i.e. the molecule contains two phenolic groups) which is much more effective against Gram-positive than Gram-negative bacteria.

Ethanol (70%) is used as a general skin antiseptic.

Soaps generally have little or no antibacterial activity unless they contain antiseptics; however, soap can help to remove bacteria from the skin – along with dirt and grease.

Triclosan, a phenolic antiseptic, 5-chloro-2-(2,4-dichlorophenoxy phenol), is used e.g. in medicated soaps and hand cleansers; it is active against certain bacteria, particularly staphylococci. Triclosan was reported to inhibit fatty acid biosynthesis by inhibiting enoyl-acyl carrier protein reductase (ENR) [Nature (1999) *398* 383–384]. However, *Streptococcus pneumoniae* ENR is apparently insensitive to triclosan (even though *S. pneumoniae* itself is sensitive), and the antiseptic was reported to be partially mineralized by a *Sphingomonas*-like organism [FEMSME (2001) *36* 105–112]. [Possible use as a *systemic* agent: AAC (2003) *47* 3859–3866.]

QACs (section 15.2.1) include cetyltrimethylammonium bromide (*Cetrimide, Cetavlon* etc.) which is used in antiseptic creams.

Iodine (in alcoholic or aqueous solution) is a potent bactericidal and sporicidal antiseptic.

Chlorhexidine is a diguanide used e.g. in alcoholic solution for disinfection of skin; it is not sporicidal but is bacteriostatic/bactericidal (depending on concentration), and may act by damaging the cell membrane and/or inhibiting ATPases. It is inactivated by soaps, anionic detergents and acidity. Chlorhexidine is used e.g. against MRSA (section 15.4.11), although some strains of MRSA exhibit plasmid-encoded resistance to it.

Acridine derivatives are potential antiseptics; aminoacridines, used topically with directed low-power light, provide low-dose bactericidal activity [JAC (2001) *47* 1–13].

15.4 ANTIBIOTICS

Originally, 'antibiotic' meant any microbial product which, even at very low concentrations, inhibits or kills certain microorganisms; the term is now generally used in a wider sense to include, in addition, any semi-synthetic or wholly synthetic substance with these properties.

Like disinfectants, antibiotics may be bactericidal or bacteriostatic, and one which is bactericidal at one concentration may be bacteriostatic at a lower concentration.

No antibiotic is effective against all bacteria. Some are active against a narrow range of species, while others are active against a *broad spectrum* of organisms – including both Gram-positive and Gram-negative bacteria.

In some cases, natural antibiotics (i.e. those produced by microbes) have been chemically modified in the laboratory, forming *semi-synthetic* antibiotics whose spectrum of activity differs from that of the parent drug. In the case of erythromycin, the genes encoding synthesis of the drug's polyketide nucleus have been transferred from an actinomycete (the natural producer) to a strain of *Escherichia coli* – thus facilitating the engineering of new derivatives [Science (2001) *291* 1790–1792; commentary: Science (2001) *291* 1683].

An antibiotic acts at a precise site. The target of a given antibiotic may be e.g. in the cell wall, in the cytoplasmic membrane, in the protein-synthesizing machinery, or in DNA or RNA synthesis. Antibiotics of the same group have similar/identical target sites, and all affect cells in the same way; infrequently, antibiotics from different groups have a common target site – e.g. the target site of oxazolidinones appears to overlap that of chloramphenicol and lincomycin [JB (2000) *182* 5325–5331].

Because a bacterium differs from a eukaryotic cell in many ways (Table 1.1), the toxic effect of an antibiotic on a bacterium is unlikely to be exerted on human or animal cells. This *selective toxicity* permits some antibiotics to be used for treating certain diseases: the pathogen can be attacked without harming the host; clearly, the antibiotic must retain activity in the body for long enough to be effective against the pathogen.

Of the many known antibiotics, relatively few are suitable for treating disease; some of these are described briefly below.

15.4.1   *β*-Lactam antibiotics

These antibiotics (Fig. 15.2) include the penicillins and cephalosporins, the carbapenems, clavams and monobactams; in each case the molecule contains a four-membered nitrogen-containing ring, the *β-lactam ring* (Fig. 15.2). Penicillins and cephalosporins are produced by fungi. Carbapenems, clavams, monobactams and nocardicins are produced by bacteria [production of carbapenems and clavams: TIM (1998) 6 263–208; *in vitro* activity of the broad-spectrum methylcarbapenem *ertapenem*: AAC (2001) *45* 1860–1867].

β-Lactam antibiotics act by disrupting synthesis of the cell envelope in growing cells: they inactivate *penicillin-binding proteins* (PBPs, enzymes involved in peptidoglycan synthesis) and thus inhibit synthesis of the peptidoglycan sacculus (Fig. 2.7). The accompanying cell lysis apparently results from enzymic cleavage of the peptidoglycan; the cell envelope normally contains both hydrolytic and synthetic

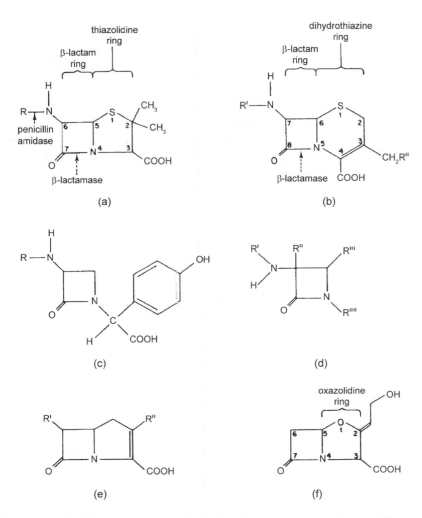

Figure 15.2 β-Lactam antibiotics (section 15.4.). Each formula shows the *nucleus* (= essential framework) of the molecule of one member of the family of β-lactam antibiotics; these antibiotics are grouped together because each contains the β-lactam ring.

The nuclei are those of (a) *penicillins*, (b) *cephalosporins*, (c) *nocardicins*, (d) *monobactams* and (e) *carbapenems*.

The structure at (f) is that of clavulanic acid (a *clavam*). Clavulanic acid is a weak antibiotic, but it inactivates many types of β-lactamase (section 15.4.1.2) – apparently by reacting covalently with a specific sequence of amino acids in the enzyme; it has therefore been used in combination with other β-lactam antibiotics – one antibiotic acting on the bacteria while clavulanic acid protects the first antibiotic from β-lactamases. For example, *Augmentin*® is a mixture of amoxycillin (a penicillin) and clavulanic acid.

The dashed arrow shows the site of action of β-lactamases; these enzymes open the β-lactam ring, thereby inactivating the antibiotic. The site of action of another enzyme, penicillin amidase, is shown at (a).

The formulae are from *Dictionary of Microbiology and Molecular Biology*, 3rd edn, Singleton & Sainsbury, p. 423, by courtesy of the publisher, John Wiley, Chichester, UK.

enzymes which jointly mediate growth of the sacculus (Fig. 3.1), so that selective inhibition of the *synthetic* enzymes (PBPs) may account for lysis. Note that *only* growing cells are killed by these antibiotics.

Other antibiotics (e.g. *vancomycin*) inhibit the synthesis of peptidoglycan at an earlier stage (section 6.3.3.1).

Various β-lactam antibiotics are mentioned in Plate 15.2.

In many β-lactam antibiotics the β-lactam ring is susceptible to cleavage by certain bacterial enzymes (*β-lactamases*: section 15.4.1.2); cleavage destroys the antibiotic, and organisms which produce these enzymes generally show at least some resistance to particular β-lactam antibiotic(s).

15.4.1.1   *Penicillins*

The original penicillins (e.g. *benzylpenicillin*) have low activity against Gram-negative bacteria owing to poor penetration of the outer membrane (section 2.2.9.2); they also have little or no effect against those Gram-positive bacteria which form β-lactamases (section 15.4.1.2). Some semi-synthetic penicillins (such as *cloxacillin, methicillin, nafcillin*) are resistant to a number of different β-lactamases (including those formed by some staphylococci), but they are still poorly effective against Gram-negative bacteria. *Ampicillin* and its derivatives (e.g. *amoxycillin*) combine resistance to some β-lactamases with increased activity against Gram-negative bacteria.

15.4.1.2   *β-Lactamases*

These enzymes inactivate (susceptible) β-lactam antibiotics by hydrolysing the β-lactam ring (Fig. 15.2); they may be encoded by chromosome-, plasmid- or transposon-borne genes. Some β-lactamases are secreted into the medium; others are retained in the cell envelope.

A given β-lactamase inactivates a *particular* range of β-lactam antibiotics, being weakly active, or inactive, against other β-lactams; antibiotics susceptible to one β-lactamase may be unaffected by another. Thus, a bacterium producing a given β-lactamase may exhibit resistance to only *certain* β-lactam antibiotics.

Some β-lactamases are *inducible*, and some β-lactam antibiotics (e.g. cefoxitin, imipenem) are particularly good inducers (see e.g. Plate 15.2). Bacterial resistance due to inducible β-lactamase(s) may be apparent from a disc diffusion test – see e.g. *Staphylococcus aureus* in section 15.4.11.1, and 'enzyme inactivation' in Plate 15.1.

Resistance to β-lactam antibiotics, due to β-lactamases, poses a problem in chemotherapy. In some cases it may be possible to circumvent this problem by using a different type of antibiotic. Another possibility is to use so-called *extended-spectrum* β-lactam antibiotics that are resistant to the β-lactamase(s) produced by the given pathogen. It may also be possible to use a combination of drugs, such as Augmentin®

[review: JAC (2004) *53* (suppl S1)], a combination of amoxycillin and the *β-lactamase inhibitor* clavulanic acid (Fig. 15.2). Unfortunately, the two latter approaches to chemotherapy created a selective pressure under which *β*-lactamases evolved to give rise to so-called *extended-spectrum β-lactamases* (ESBLs) which can now cleave at least some of the newer *β*-lactam antibiotics and resist inactivation by *β*-lactamase inhibitors. (ESBLs appear to have evolved mainly from TEM and SHV *β*-lactamases [TIM (1998) *6* 323–327].)

Resistance to the fourth-generation cephalosporin *cefepime* was linked to hyperproduction of SHV-5 [JCM (1998) *36* 266–268]. Other ESBLs include TEM-50 (which incorporates resistance to inhibitors) [AAC (1997) *41* 1322–1325], and SHV-18 (a plasmid-encoded enzyme from *Klebsiella pneumoniae*) [AAC (2000) *44* 2382–2388].

Plasmids that carry ESBLs often carry other antibiotic-resistance genes, thus limiting the choice of drugs for chemotherapy. [Treatment options for ESBL producers: FEMSML (2000) *190* 181–184.] Infections with strains of *Acinetobacter* and *Pseudomonas aeruginosa* that encode PER-1-type ESBLs are associated with increased levels of fatality [JMM (2001) *50* 642–645].

[ESBLs (characterization, epidemiology etc.): CMR (2001) *14* 933–951.]

The production of *β*-lactamases can be readily detected in the laboratory. One method uses a chromogenic cephalosporin, *nitrocefin* (Oxoid, Basingstoke, UK), which, on cleavage by a *β*-lactamase, changes from yellow to red; a drop of nitrocefin solution added to a colony of a *β*-lactamase producer gives a red coloration either rapidly or after ~30 minutes incubation.

15.4.2 Aminoglycoside antibiotics

These broad-spectrum, typically bactericidal antibiotics include *amikacin*, *gentamicin*, *kanamycin*, *neomycin* and *streptomycin*; they are active against both Gram-positive and Gram-negative bacteria. Aminoglycoside antibiotics act on various bacterial functions, but their main effect results from binding to the 30S ribosomal subunit and interference with protein synthesis. For example, low levels of streptomycin cause misreading of the mRNA (i.e. incorporation of incorrect amino acids), while high levels completely inhibit protein synthesis – apparently by blocking ribosomes specifically at the start of translation (the stage in Fig. 7.9b).

Resistance to aminoglycosides can be due to e.g. (i) mutation(s) in ribosomal proteins of the 30S subunit (affecting binding of the antibiotics); (ii) modification (inactivation) of the antibiotics by bacterial enzymes which carry out *O*-phosphorylation, *N*-acetylation or *O*-adenylation; (iii) reduced uptake.

Side-effects (particularly with neomycin and streptomycin) include dose-dependent damage to the 8th cranial nerve (causing impairment of hearing). Hypersensitivity reactions also occur.

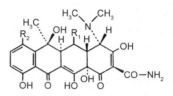

Figure 15.3   Tetracycline (R_1, R_2=H); chlortetracycline (R_1=H, R_2=Cl); oxytetracycline (R_1=OH, R_2=H).

15.4.3   Tetracyclines

These broad-spectrum, bacteriostatic antibiotics (Fig. 15.3) inhibit protein synthesis by binding to ribosomes (in *E. coli*, preferentially to protein S7 in the 30S subunit) and inhibiting the binding of aminoacyl-tRNAs to the A site (Fig. 7.9); they are used for treating human and animal diseases caused e.g. by *Brucella*, *Chlamydia*, *Mycoplasma* and *Rickettsia* – and, interestingly, for the treatment of certain plant diseases such as coconut lethal yellowing (caused by a *Mycoplasma*-like organism).

Bacteria normally accumulate tetracyclines (in an energy-dependent manner). Some bacteria are resistant to these drugs because they can pump them outwards across the cytoplasmic membrane: see TET protein in section 15.4.11.

Tigecycline (GAR-936; the 9-*t*-butyl-glycylamido derivative of minocycline) was designed to evade bacterial defence mechanisms, including efflux [*in vitro* and *in vivo* activities: AAC (1999) *43* 738–744]. [Activity against 1924 clinical isolates: AAC (2003) *47* 400–404.] *Pseudomonas aeruginosa* strain PAO1 is unusual in exhibiting efflux-mediated resistance [AAC (2003) *47* 972–978].

The accumulation and activity of tetracyclines in bacteria is reported to be inhibited by iron [AAC (2004) *48* 1892–1894].

15.4.4   Macrolides, chloramphenicol and streptogramins

All of these antibiotics inhibit protein synthesis by binding to the 50S ribosomal subunit (Fig. 7.9).

Macrolides. The molecule consists of a large ($>$12-membered) lactone ring substituted with one or more sugars or aminosugars; the important macrolide *erythromycin* has a 14-membered ring linked to cladinose and desosamine. Macrolides, which are typically bacteriostatic, cause premature termination of polypeptide synthesis; they bind to peptidyltransferase (Fig. 7.9d) and may inhibit transpeptidation under certain conditions and/or trigger abortive translocation (Fig. 7.9e) leading to dissociation of an incomplete polypeptide. Macrolides are effective against mainly Gram-positive bacteria. In *E. coli*, mutants resistant to erythromycin may have modified protein(s) in the 50S subunit; in *Staphylococcus aureus*, inducible resistance involves an enzyme which methylates a site in the 23S rRNA of the 50S subunit.

Strains of *Streptococcus pyogenes* and *S. pneumoniae* exhibiting the 'M' phenotype contain an efflux-based mechanism (encoded by transposon Tn*1207.1*) that confers resistance specifically to macrolides [AAC (2000) *44* 2585–2587].

The macrolide *azithromycin* inhibits the ability of *Pseudomonas aeruginosa* to produce virulence factors (by inhibiting the organism's quorum sensing mechanism) [AAC (2001) *45* 1930–1933].

The macrolide *clarithromycin* has been used against *Helicobacter pylori*; it is more acid-stable (and absorbed more readily) than e.g. erythromycin.

Chloramphenicol. This small molecule (readily synthesized) inhibits the activity of peptidyltransferase, possibly by preventing normal binding of the aminoacyl-tRNA at the ribosomal A site. Chloramphenicol is a broad-spectrum, bacteriostatic agent, but its use is limited by toxicity (e.g. it affects mitochondrial ribosomes in mammalian cells). It is used e.g. against *Salmonella typhi* and in cases where other drugs are unsuitable. *Pseudomonas aeruginosa* is innately resistant. Bacteria with acquired resistance may encode the enzyme *chloramphenicol acetyltransferase* (CAT), which inactivates the antibiotic; CAT is usually inducible in Gram-positive bacteria (see section 7.8.5) but synthesized constitutively in Gram-negative species.

Streptogramins are composite antibiotics, each one consisting of at least two distinct types of molecule – which include a polyunsaturated macrolactone ring (group A component) and a cyclic hexadepsipeptide (group B component); A and B form a synergistic combination which can be bactericidal for certain (primarily Gram-positive) pathogens. Group A molecules inactivate functional sites on peptidyltranferase, thus inhibiting transpeptidation; group B molecules inhibit normal binding of the growing polypeptide chain at the P site during translation of certain amino acids (e.g. proline), thereby promoting early release of an (incomplete) polypeptide chain.

The streptogramin quinupristin/dalfopristin (RP 59500; Synercid®) has useful activity against certain pathogens (including MRSA) and has a long post-antibiotic effect (section 15.4.12). It is useful e.g. for treating severe infections caused by multi-resistant Gram-positive pathogens.

Resistance to quinupristin/dalfopristin in *Staphylococcus aureus* and *Streptococcus pneumoniae* can result from mutation in the ribosomal protein L22 [AAC (2002) *46* 2200–2207; AAC (2003) *47* 2696–2698].

15.4.5 Polymyxins

Polymyxins are peptides which are active against many Gram-negative bacteria; most Gram-positive bacteria are resistant. In Gram-negative bacteria polymyxins act by increasing the permeability of the cytoplasmic membrane and the outer membrane.

15.4.6   Quinolone antibiotics and novobiocin

Quinolone antibiotics are synthetic agents (each containing a substituted 4-quinolone ring) whose targets include the A subunit of *gyrase* (section 7.2.1) and topoisomerase IV (section 7.2.1). Binding to e.g. gyrase inhibits the normal activity of the enzyme and inhibits DNA replication.

First-generation quinolone antibiotics (cinoxacin, nalidixic acid, oxolinic acid and pipemidic acid) are active mainly against Gram-negative bacteria (though not against *Pseudomonas aeruginosa*). Resistance develops readily.

The later drugs – *fluoroquinolones* such as ciprofloxacin, enoxacin, norfloxacin, ofloxacin and perfloxacin – are active against *P. aeruginosa* and some Gram-positive cocci, including MRSA. Compared with earlier quinolones, they are active at lower *in vivo* concentrations and are more stable; resistance develops less readily. Gemifloxacin has useful activity against *Streptococcus pneumoniae* and *Haemophilus influenzae* [JAC (2004) *53* 144–148].

[Comparison of some newer fluoroquinolones (e.g. gatifloxacin, grepafloxacin, levofloxacin, moxifloxacin, trovafloxacin) against ciprofloxacin-resistant *Streptococcus pneumoniae*: AAC (2001) *45* 1654–1659.]

Resistance to quinolones often results from mutations in the *gyrA* (gyrase) and/or *parC* (topoisomerase IV) genes [e.g. RMM (1998) *9* 87–97]. In *Streptococcus pneumoniae*, mutations in *gyrA* and *parE* (encoding the other subunit of topo IV) increased resistance to certain of the newer fluoroquinolones (grepafloxacin, moxifloxacin, sparfloxacin) but produced little or no increase in resistance to older fluoroquinolones (ciprofloxacin, perfloxacin) – and increased the sensitivity to novobiocin [AAC (2001) *45* 952–955].

Other modes of resistance include reduced permeability of the cell envelope and efflux mechanisms [efflux-mediated resistance to fluoroquinolones in Gram-negative bacteria: AAC (2000) *44* 2233–2241].

The fluoroquinolone *danofloxacin* has been useful against *Mycoplasma bovis* in the treatment of calf pneumonia; little resistance to danofloxacin has developed in *M. bovis* but significant levels of resistance have developed against certain other antibiotics (such as oxytetracycline and the macrolide tilmicosin) [Veterinary Record (2000) *146* 745–747].

[Report of the 7th International Symposium on New Quinolones: JAC (2001) *47* (suppl S1).]

Novobiocin is a coumarin derivative which inhibits DNA synthesis by binding to the B subunit of gyrase and blocking the enzyme's ATPase activity (which is essential for ongoing gyrase activity). Topoisomerase IV is less susceptible than gyrase, apparently due to a one-residue difference at the relevant site [AAC (2004) *48* 1856–1864].

Toxicity and resistance (which develops readily) limit the use of novobiocin.

Figure 15.4   Metronidazole: a nitroimidazole antibiotic used against anaerobic bacteria.

15.4.7   Metronidazole

Metronidazole is a nitroimidazole derivative (Fig. 15.4); it is an effective antimicrobial agent only at low (i.e. highly negative) redox potentials, and is therefore used against anaerobic bacteria (and certain pathogenic protozoa). In sensitive anaerobes, the drug's nitro group is reduced (by electron transport components such as *ferredoxins* – section 5.1.1.2), apparently producing short-lived cytotoxic derivative(s) which can cleave DNA. Studies on the mode of action of metronidazole on *Helicobacter pylori* concluded that the bactericidal effect depends on the reduced (active) form of the antibiotic – rather than on reactive oxygen species formed during its reduction; resistance to metroni-dazole appears to correlate with a failure to reduce the antibiotic *rapidly* [JAC (1998) *41* 67–75].

Metronidazole is used e.g. against *Clostridium difficile* in pseudomembranous colitis, and, prophylactically, in bowel surgery to protect against anaerobes such as *Bacteroides* and clostridia.

15.4.8   Rifamycins

These macrocyclic antibiotics – e.g. rifampicin (= rifampin) – are generally active against Gram-positive bacteria (including mycobacteria and staphylococci) and certain Gram-negative bacteria (e.g. *Brucella*, *Chlamydia*, *Haemophilus*, *Legionella*). Rifamycins bind to the β subunit of DNA-dependent RNA polymerase, inhibiting the initiation of transcription (section 7.5). Resistance to rifamycins can result from mutations in the *rpoB* gene (see also section 15.4.11.2).

[Detecting resistance to rifampicin in *Mycobacterium tuberculosis* by DNA chip technology: JCM (1999) *37* 49–55.]

15.4.9   Sulphonamides, trimethoprim and cotrimoxazole

Sulphonamide antibiotics (Fig. 15.5) are typically bacteriostatic for susceptible Gram-positive and Gram-negative bacteria. Sulphonamides inhibit the synthesis of dihydrofolic acid (DHF), a precursor of the coenzyme tetrahydrofolic acid (THF) – thus inhibiting THF-dependent functions (section 6.3.1); they are believed to act in one or

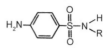

Figure 15.5 The sulphonamide nucleus; R=H in sulphanilamide (*p*-amino-benzenesulphonamide). The therapeutic sulphonamides may be regarded as derivatives of sulphanilamide: e.g. *sulphadiazine* is 4-amino-*N*-2-pyrimidinyl-benzenesulphonamide, and *sulphamethazine* (=*sulphadimidine*) is 4-amino-*N*-(4,6-dimethyl-2-pyrimidinyl) benzenesulphonamide.

both of two ways: (i) as *competitive inhibitors* of the enzyme dihydropteroate synthetase (DHPS, involved in the synthesis of DHF), (ii) as *substrates* of DHPS – giving rise to functionally defective analogues of DHF. These activities appear to reflect the structural similarity between the sulphonamide molecule and that of *p*-aminobenzoic acid (PABA), the latter being a precursor in the synthesis of DHF.

Certain bacteria use an external source of folate, and are therefore not affected by sulphonamides, but most bacteria synthesize their own folate. However, resistance to sulphonamides develops readily and may be due e.g. to (i) the production of higher levels of PABA (thus combatting the competitive inhibition) and/or (ii) the development of a mutant form of DHPS whose affinity for sulphonamides is much lower than that for PABA.

In sensitive cells, growth continues for several generations following exposure to sulphonamides; during this time the existing stock of folate is being used up.

Currently, sulphonamides are used e.g. for urinary-tract infections (much of the drug is excreted in the urine). The most frequently used sulphonamides include sulphadiazine and sulphamethazine.

Trimethoprim, a diaminopyrimidine derivative, inhibits the conversion of DHF to THF by inhibiting the enzyme dihydrofolate reductase (DHFR). DHFR occurs in all cells (microbial and human), but trimethoprim inhibits bacterial DHFRs without significantly affecting human DHFR at the concentrations used.

Cotrimoxazole is a (synergistic) combination of trimethoprim and the sulphonamide sulphamethoxazole – two drugs which block different reactions in the same pathway.

15.4.10 Synergism and antagonism between antibiotics

If two antibiotics, acting simultaneously on an organism, produce an effect which is greater than the sum of their individual effects, the antibiotics are said to be acting *synergistically*. Cotrimoxazole (section 15.4.9) is one example of a synergistic combination of antibiotics. Another example is the combination of the A and B components of streptogramins (section 15.4.4).

Antagonism is the converse of synergism. For example, antibiotics which inhibit growth (e.g. chloramphenicol) antagonize those antibiotics (e.g. β-lactams) which act only when cells are growing. Another form of antagonism is exemplified by those antibiotics which induce bacteria to form antibiotic-inactivating enzymes that affect *other* antibiotics; examples include the antagonism of imipenem and cefoxitin to other β-lactam antibiotics shown in Plate 15.2.

15.4.11   Bacterial resistance to antibiotics

Why are some bacteria not affected by some antibiotics? In some cases a bacterium is resistant because it lacks the target structure of a given antibiotic; for example, species of *Mycoplasma* (which lack cell walls) will not be affected by penicillins – whose ultimate target (peptidoglycan) is a cell-wall component. Some bacteria may not carry out the particular process affected by an antibiotic: sulphonamides, for example, will not affect organisms which normally obtain their folic acid, ready-made, from the environment. Resistance can also be due to the ability of a cell to exclude an antibiotic from the target site; in many Gram-negative bacteria the outer membrane is impermeable to certain antibiotics, and in both Gram-positive and Gram-negative species the cytoplasmic membrane may act as a barrier.

The types of resistance mentioned above can be called *constitutive* or *innate* resistance to a given antibiotic. Additionally, a cell may *acquire* resistance to particular antibiotic(s). Moreover, in some cases, resistance to given antibiotic(s) may be *inducible*, i.e. expressed only in the presence of an inducer; this aspect of resistance is discussed later.

Resistance to antibiotics may involve any of the following mechanisms:

- *Mutation.* The antibiotic-binding site may be modified by mutation so that the antibiotic does not bind and the target (e.g. enzyme) is functional in the presence of otherwise inhibitory concentrations of that antibiotic. (The mutant cell can form a population of cells resistant to the given antibiotic.) Example: in *Mycobacterium tuberculosis*, point mutations in the *rpoB* gene (encoding the β subunit of RNA polymerase) can confer resistance to rifamycins (which target the polymerase). Further example: mutation in ribosomal protein L22 can cause resistance to quinupristin/dalfopristin in *Staphylococcus aureus* [AAC (2002) *46* 2200–2207] and *Streptococcus pneumoniae* [AAC (2003) *47* 2696–2698].

- *Inactivation or degradation of antibiotics by enzymes.* Enzymes that inactivate or degrade one or more antibiotic(s) may be encoded by chromosomal and plasmid-borne genes; some of these enzymes are inducible. Example: β-lactamases (see section 15.4.1.2).

- *Efflux mechanisms* [review: Drugs (2004) *64* 159–204]. Certain transport systems (probably all energy-dependent) can pump out particular antibiotics through the cytoplasmic membrane or through the cell envelope. Some of the systems are multi-protein ABC exporters (section 5.4.1.2), and some are single-protein systems of the major facilitator superfamily. Example: the inducible cytoplasmic membrane protein Tet (= TET) (encoded e.g. by Tn*10*) mediates the efflux of tetracyclines in some Gram-negative bacteria.

• *Diminished permeability.* Any change in the composition of the cell envelope that hinders the uptake of a given antibiotic will result in an increase in the minimum inhibitory concentration (MIC) or a conversion to high-level resistance. Examples: (i) in *Enterobacter aerogenes*, diminished permeability resulting from changes in outer membrane porins (section 2.2.9.2) increases resistance to certain antibiotics [Microbiology (1998) *144* 3003–3009]; (ii) in *Pseudomonas aeruginosa*, membrane permeability regulated by a two-component regulatory system (section 7.8.6) can determine resistance to aminoglycosides and other antibiotics [AAC (2003) *47* 95–101].

• *Increased synthesis of target metabolite.* Increased production of a given metabolite may overcome e.g. competitive inhibition by an antibiotic. Example: production of increased amounts of PABA as a form of resistance to sulphonamides (see section 15.4.9).

• *Acquisition of an exogenous determinant of resistance.* Genes specifying resistance to one or more antibiotics may be acquired through transformation (section 8.4.1), conjugation (8.4.2), conjugative transposition (8.4.2.3) or transduction (9.5). (In some cases, a 'gene cassette' encoding antibiotic resistance can insert by site-specific recombination (section 8.2.2) into a specific sequence of DNA (*integron*) found e.g. in plasmids and transposons; an integron includes the insertion site and also the necessary integrase (recombinase) [integrons and gene cassettes: JAC (1999) *43* 1–4].) Example: *in vivo* transfer (indicated by genetic analyses) of vancomycin resistance from *Enterococcus faecalis* to a multidrug-resistant strain of *Staphylococcus aureus* [Science (2003) *302* 1569–1571].

Exceptionally, *apparent* resistance to an antibiotic may be due to a patient's abnormal metabolism of the drug. For example, treatment of syphilis with a normally adequate dosage of penicillin has failed in a few cases in which the patients were categorized as 'quick penicillin secretors'; in a very few instances, penicillin-sensitive treponemes have been recovered from these patients and have produced typical lesions when inoculated into rabbits.

Inducible resistance to antibiotics. Certain antibiotics – when present above a certain minimum concentration – may induce resistance in a bacterial cell. For example, in some Gram-positive bacteria (including staphylococci) the presence of chloramphenicol induces chloramphenicol acetyltransferase, an enzyme which catalyses acetylation (and hence inactivation) of the antibiotic. (See also section 7.8.5.)

β-Lactam antibiotics can induce several distinct modes of resistance. One mechanism involves β-lactamases (section 15.4.1.2). Another involves expression of the *mecA* gene which mediates high-level resistance to methicillin and other β-lactams (see MRSA, below). The induction of *mecA* may involve a novel mechanism in which, following binding of a β-lactam to a cell-surface receptor, several regulatory proteins undergo cleavage – with subsequent inactivation of the transcriptional repressor of the *mecA* gene (MecI) [Science (2001) *291* 1962–1965].

The ongoing story of bacterial resistance. Although many new antibiotics have been identified in recent decades, bacterial resistance continues to emerge so that, in some cases, therapeutic options are very limited. Some pathogens are resistant to drugs to which they were once invariably sensitive, and multidrug resistance is common.

MRSA refers to methicillin-resistant strains of *Staphylococcus aureus* (which are often 'healthcare-associated' [e.g. JAC (2004) *53* 474–479]). In *S. aureus*, resistance to methicillin may involve different mechanisms, but most attention has been given to the *mecA* gene; this (chromosomal) gene encodes a 78 kDa penicillin-binding protein, designated PBP 2a (= PBP 2'), which has a low affinity for methicillin (and other β-lactams) and which confers high-level resistance to methicillin. Expression of *mecA* is controlled by various regulatory genes; thus, *mecI* encodes a repressor, and *mecR1* an inducer. [Roles of *mecA*, *mecI* and *mecR1*: JAC (1999) *43* 15–22.] Induction of *mecA*-based resistance in *S. aureus* may involve cleavage of MecI [Science (2001) *291* 1962–1965].

Resistance encoded by *mecA* can be lost by mutations in *mecA* or in the *fem* genes (*fem* = factor essential for methicillin resistance). Mutation in the *femAB* operon can cause abnormal cross-links in peptidoglycan (Fig. 2.7), and such cross-links are poor substrates for the (transpeptidase) activity of PBP 2a; because the normal (methicillin-sensitive) PBPs are enzymically more effective than PBP 2a on the abnormal cross-links, the mutant cell is sensitive to methicillin.

MRSA was once commonly sensitive to vancomycin, but reduced susceptibility to the antibiotic has been reported frequently [JAC (1997) *40* 135–136; Lancet (1999) *353* 1587–1588; JCM (2001) *39* 591–595]; more recently, high-level resistance seems to have been acquired from *Enterococcus faecalis* via transposon Tn*1546* [Science (2003) *302* 1569–1571].

The following are some examples of resistance reported between 1996 and 2003: penicillin-resistant pneumococci [Drugs (1996) *51* (suppl 1) 1–5]; multidrug-resistant *Vibrio cholerae* O1 [Lancet (1997) *349* 924]; multidrug-resistant *Mycobacterium bovis* [Lancet (1997) *350* 1738–1742]; low-level resistance to vancomycin in MRSA [JAC (1997) *40* 135–136]; vancomycin-resistant enterococci [EID (2001) *7* 183–187]; resistance to linezolid (an oxazolidinone) in vancomycin-resistant *Enterococcus faecalis* [Lancet (2001) *357* 1179]; high-level resistance to vancomycin in MRSA [Science

(2003) *302* 1569–1571]. Even so, these antibiotics can still be useful [e.g. linezolid (for the treatment of Gram-positive infections): JAC (2004) *53* 335–344, JAC (2004) *53* 345–355].

There is clearly a need to prevent or delay the emergence of bacterial resistance by avoiding the inappropriate use of antibiotics. [Controlling antimicrobial resistance in hospitals (infection control and use of antibiotics): EID (2001*) 7* 188–192.]

Some alternative approaches to antibacterial therapy. Because of the increasing problem of resistance to antibiotics, efforts are constantly being made to find new therapeutic targets. In *E. coli*, one potential target is de-acetylase, an essential enzyme in the biosynthesis of lipid A (see section 2.2.9.2); some candidate antibiotics have useful activity, and the inhibition of lipid A synthesis may have two further benefits: (i) increasing the permeability of the outer membrane to other antibiotics, and (ii) reducing the risk of chemotherapy-associated endotoxic shock [TIM (1998) *6* 154–159].

One novel approach targets the pathogen's virulence factors. In *Pseudomonas aeruginosa*, expression of certain virulence factors is regulated by quorum sensing (section 10.1.2), suggesting the possibility of therapy based on the inhibition of autoinducers. One study found that the production of virulence factors could be inhibited by the antibiotic azithromycin [AAC (2001) *45* 1930–1933]; a second study found that the virulence of *P. aeruginosa* could be attenuated with the quorum sensing inhibitor furanone [EMBO Journal (2003) *22* 3803–3815].

Interest has also been focused on the development of antimicrobial peptides based on those that form part of the innate defence mechanisms of animals and plants. Some of the peptides are being developed commercially. [Antimicrobial peptides: Nature (2002) *415* 389–395.]

Phage therapy is yet another option. Table 9.2 describes some of the earlier work carried out in Poland and the former Soviet Union. [Phage therapy: AAC (2001) *45* 649–659.]

Antisense PNA (peptide nucleic acid: section 7.2.1.1) directed against an essential gene in *Escherichia coli* exhibited potent antibacterial activity; such agents can be useful e.g. for strain selection and maintenance in research, and may have applications in medicine [BioTechniques (2003) *35* 1060–1064].

15.4.11.1   *Antibiotic-sensitivity tests (traditional)*

Tests can be carried out to determine the susceptibility of a pathogen to a range of antibiotics; the pattern of sensitivities of a given strain towards a range of antibiotics is called an *antibiogram*. The results of such tests may enable the clinician to select optimally active antibiotic(s) for chemotherapy (section 11.10) and to avoid antibiotic(s) to which the pathogen is resistant.

One common form of test is the *disc diffusion test*. A plate of suitable agar medium is inoculated from a suspension of a pure culture of the pathogen; the entire surface of the medium is inoculated (often with a swab) so that near-confluent growth (section 3.3.1) will develop on incubation. Before incubation, several small absorbent paper discs, each impregnated with a different antibiotic, are placed at different locations on the inoculated medium. On subsequent incubation, antibiotic diffuses from each disc; if the organism if sensitive to a given antibiotic a zone of growth-inhibition develops around that particular disc. Methods must be standardized: the presence or size (diameter) of a no-growth zone can be interpreted correctly only when the whole procedure has been standardized in terms of type of medium, density of inoculum on the plate etc. Reactions which may be seen in a disc diffusion test are shown in Plate 15.1. The test is suitable for those bacteria which produce visible growth after over-night incubation.

In a disc diffusion test, a no-growth zone (simulating sensitivity) may be formed by organisms which encode inducible antibiotic-inactivating enzymes – e.g. strains of *Staphylococcus aureus* which encode an inducible β-lactamase. Cells in the inoculum close to the disc are killed before adequate amounts of enzyme can be synthesized. Cells further from the disc experience gradually increasing concentrations of the outwardly diffusing antibiotic, and, at a certain distance from the disc, some of the cells will have synthesized enough enzyme to permit survival; these cells give rise to normal-sized or relatively large colonies by using the nutrients forfeited by the inactivated cells in their immediate vicinity. An example of a zone characteristic of enzyme inactivation is shown in Plate 15.1.

Some tests use antibiotic-containing tablets or paper/plastic strips (Plate 15.2).

In a *dilution test*, the organism is tested for its ability to grow in the presence of each of a range of concentrations of a given antibiotic (in either solid (agar-based) or liquid media). The lowest concentration of antibiotic which prevents growth is the *minimum inhibitory concentration* (MIC) of that antibiotic under the given conditions for the given strain. For any particular antibiotic, different strains of an organism may have different MICs.

The *E test* (Plate 15.2) is a diffusion test which can give an MIC. One side of a plastic strip (placed in contact with the solid medium) carries the antibiotic in a concentration gradient, while the other (uppermost) side is graduated with the MIC scale; after incubation, the test is read by noting the lowest concentration of antibiotic which prevents growth.

A *breakpoint test* uses the dilution test technique to determine whether the MICs of given strains are above or below certain chosen test concentration(s) of the antibiotic – thus determining which strains are to be regarded as 'resistant' and which 'sensitive'. The actual choice of test concentrations is influenced by clinical, pharmacological and microbiological factors. The concentrations used in breakpoint tests are not the same in all countries.

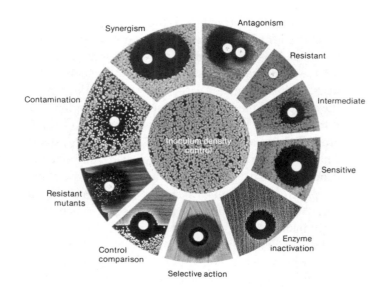

Plate 15.1   Antibiotic-sensitivity testing (disc diffusion test): types of zone which may develop around the discs. (The rationale for the test is given in section 15.4.11.1.)

Resistant. No zone: growth occurs right up to the disc.

Intermediate. A narrow growth-free zone surrounds the disc.

Sensitive. A wide growth-free zone surrounds the disc.

Enzyme inactivation. A narrow growth-free zone surrounds the disc. Unlike the 'intermediate' zone, the edge of the zone is sharply defined and it contains somewhat heavier growth with some normal-sized or relatively large colonies (for explanation see section 15.4.11.1).

Selective action. This result can be obtained e.g. when the inoculum consists of two different strains which differ in their degree of susceptibility to the given antibiotic. Close to the disc, the concentration of antibiotic is high enough to inhibit both strains (narrow growth-free zone). Further from the disc (lower concentration of antibiotic) one strain is still inhibited while the other can grow.

Control comparison. The 'half-zone' obtained with a control strain (one side of the disc) is opposite the 'half-zone' obtained with the test strain (other side of the disc). To make this comparison, the test and control strains are inoculated onto separate halves of the plate and the disc is placed between them.

Resistant mutants. The inoculum contained a small proportion of mutant cells which were able to form colonies under conditions which inhibited non-mutants. The mutants are usually antibiotic-resistant cells. There is, however, another type of mutant which grows *only in the presence* of a given antibiotic, and such mutants would also form colonies in an otherwise growth-free zone; for example, 'streptomycin-dependent' mutants contain non-functional ribosomes which, in the presence of streptomycin, appear to be distorted in such a way that they become functional.

Contamination. The inoculum contained a mixture of organisms, at least one of which is resistant to the antibiotic.

Synergism. The two discs contain different antibiotics. The zone shows an inhibitory effect which is greater than the sum of the effects of each antibiotic acting alone; that is, the antibiotics are acting synergistically.

The BACTEC MGIT™ 960 (Becton Dickinson) is a growth-based system in which antibiotic-susceptibility testing of *Mycobacterium tuberculosis* can be carried out in a liquid medium in 3–14 days [evaluation: JCM (1999) *37* 45–48].

15.4.11.2 DNA-based ('genotypic') detection of antibiotic resistance

For many pathogens, sensitivity to antibiotics can be determined easily and rapidly by traditional methods (section 15.4.11.1). However, such methods may not indicate the *mechanism* of resistance to a given antibiotic – information which may be needed for optimal chemotherapy (see later). Here, we look at the use of some DNA-based methods which have been used for the detection of antibiotic resistance in *Mycobacterium tuberculosis* and *Staphylococcus aureus*. These methods illustrate the diversity of the approaches used. Not considered here is the use of microarrays (section 8.5.15); this technology has been used to detect resistance to rifampicin in *M. tuberculosis* [JCM (1999) *37* 49–55].

Mycobacterium tuberculosis

Resistance to the commonly used drugs is caused by certain mutations; these can be detected by DNA-based methods. Resistance to *isoniazid* may follow mutation in any of several genes (e.g. *ahpC, katG*); however, resistance to *rifampicin* usually involves only one gene, *rpoB*, so that this is easier to detect.

Gene *rpoB* encodes the β-subunit of an enzyme, RNA polymerase (section 7.5). In wild-type (rifampicin-*sensitive*) strains, rifampicin binds to the β-subunit and inhibits the enzyme – thus blocking RNA synthesis and killing the cell. Certain mutation(s) in *rpoB* result in a modified β-subunit which apparently fails to bind rifampicin; the (mutant) enzyme remains functional – so that the cell is resistant to rifampicin.

Mutations may be detected by (i) SSCP analysis, (ii) dideoxy fingerprinting, or (iii) line probe assay. Each of these methods involves an initial PCR amplification (section 8.5.4) of a sequence in the *rpoB* gene; in each method, the products of PCR from each test strain are compared with those from a wild-type (rifampicin-sensitive) strain.

(i) *SSCP analysis.* In this method (see Fig. 8.19c), PCR is used to amplify the sequence {codon 435 → codon 458} in gene *rpoB* because this sequence includes the sites of some common resistance-conferring mutations – and is thus a useful target for SSCP analysis. The dsDNA amplicons are denatured prior to electrophoresis. Using the sequence referred to, 20 of 20 rifampicin-sensitive strains were correctly identified; in another (rifampicin-resistant) strain, analysis suggested the presence of a specific

Antagonism. The two discs contain different antibiotics. Here, the presence of one antibiotic inhibits the activity of the other. Examples of antagonism are seen in Plate 15.2.

Photograph courtesy of Oxoid, Basingstoke, UK.

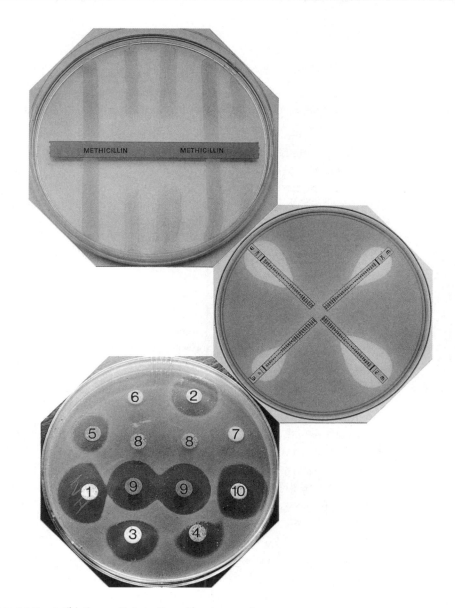

Plate 15.2   Antibiotic-sensitivity testing without paper discs.

Top. Testing five strains of *Staphylococcus aureus* for resistance to methicillin (a penicillin which is resistant to many β-lactamases). The plate is first inoculated with each of the strains, inoculation being carried out in five parallel lines across the plate. These lines are then overlaid, at right-angles, with a paper strip impregnated with methicillin. After incubation (at e.g. 30°C for 18 hours), the result is as shown: the first and fourth strains (from the left) are resistant, the others are sensitive.

Centre. The E test (see section 15.4.11.1 for rationale). Here, a strain of *Pseudomonas aeruginosa* has been tested for its reaction to four different antibiotics; for each antibiotic, the MIC is indicated where the edge of the serine→leucine mutation, and this was subsequently confirmed by sequencing the amplicon [JCM (1995) *33* 556–561].

Silent mutations (Fig. 8.1 f) can interfere with SSCP analysis. Thus, following a silent mutation, a strand from a rifampicin-*sensitive* strain exhibited the *conformation* of a strand from a resistant strain [JCM (1997) *35* 492–494]. False-positive results can also be caused e.g. by deletion mutations in *rpoB*.

(ii) *Dideoxy fingerprinting* (ddF) (Fig. 15.6) includes features of dideoxy sequencing (section 8.5.6), PCR and SSCP.

For each sample, the procedure is: (i) PCR amplification of the target sequence in *rpoB*; (ii) use of the amplicons as *templates* in a modified form of sequencing involving temperature-cycling (as in PCR) – this allowing the use of (dsDNA) amplicons (which form ssDNA templates at higher temperatures). Unlike normal sequencing (Fig. 8.23), only *one* type of dideoxynucleotide is used; the same ddNTP is used for each strain so that the results can be compared. In Fig. 15.6, ddG is used for each strain.

After sequencing, electrophoresis in a *non*-denaturing gel (as in SSCP) is used to detect mutation(s). These may be detected, for example, by changes in electrophoretic mobility of full-length and/or chain-terminated products; such changes in mobility are the result of changes in intra-strand base-pairing due to the mutation(s). Mutations can also be detected as changes in the *number* of types (sizes) of chain-terminated product in a given reaction; thus, more or fewer types of product are formed when

no-growth zone intersects the scale. When compared with other forms of antibiotic-sensitivity testing (including dilution methods), the E test was found to be as reliable [JCM (1991) *29* 533–538]. Since then the E test has been found to be a reliable test for various other bacteria, including *Helicobacter pylori* [JCM (1997) *35* 1842–1846]. For *Mycobacterium tuberculosis,* results correlated well with those obtained by standard methods for isoniazid and rifampicin, but a high proportion of false-sensitive and false-resistant results were reported for streptomycin [JCM (2002) *40* 2282–2284].

Bottom. Testing with tablets. Neo-Sensitabs are antibiotic-containing tablets, rather than impregnated paper discs; they are colour-coded for identification, and – at room temperature – most are stable for a minimum of 4 years. Neo-Sensitabs are used in the standard procedure for bacterial sensitivity testing in Denmark, Holland, Belgium, Norway and Finland, and are also used in other countries.

A strain of *Pseudomonas aeruginosa* has been tested (on a plate of Mueller–Hinton agar) against a range of β-lactam antibiotics: the penicillins piperacillin (1) and ticarcillin (2); the cephalosporins cefsulodin (3), cefotaxime (4), ceftriaxone (5), cephalothin (6) cefamandole (7); imipenem (9 – a carbapenem); and aztreonam (10 – a monobactam). Tablet 8 is *cefoxitin*, a member of the *cephamycins* (=7-α-methoxy-cephalosporins); the 7-α-methoxy group tends to confer decreased antibacterial activity but increased resistance to β-lactamases. [Cefoxitin: RMM (1995) *6* 146–153.]

Note the antagonism of imipenem towards e.g. piperacillin and aztreonam; imipenem has induced the synthesis of β-lactamases in cells of the inoculum – enzymes to which imipenem itself is resistant but to which piperacillin and aztreonam are sensitive. Note also the resistance of this strain of *P. aeruginosa* to several of the antibiotics, including cefoxitin.

Photograph of the methicillin test courtesy of Mast Diagnostics Ltd, Bootle, Merseyside L20 1EA, UK.
Photograph of the E test courtesy of Dr Carolyn Baker, Centers for Disease Control, Atlanta, Georgia, USA.
Photograph of Neo-Sensitabs courtesy of A/S ROSCO, Taastrup, Denmark.

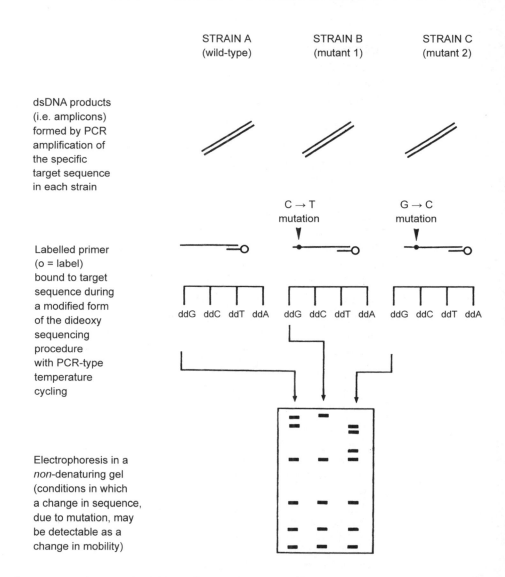

Figure 15.6 The principle of dideoxy fingerprinting (ddF) (section 15.4.11.2) (diagrammatic). Initially, PCR is used to amplify the specific target sequence in each sample of DNA under test. Here, DNA from two mutant strains (B and C) is being compared with that from a wild-type strain (A).

For each strain, the amplicons are used as templates in a modified form of dideoxy sequencing involving only *one* reaction mixture (instead of four) and in which the ddNTP may be any one of the four standard ddNTPs; ddG is used in the diagram. The reaction mixture is subjected to PCR-type temperature cycling (e.g. for 30 cycles) – this permitting the use of amplicons (dsDNA) as templates and avoiding the need to provide samples of ssDNA. Only one type of primer is used, and this binds to, and extends along, one specific strand of the amplicon.

Products from the (ddG) sequencing reaction are examined by electrophoresis in a *non*-denaturing gel. Before electrophoresis, products and templates are separated by brief heating and chilling in a mutation(s) increase/decrease the number of chain-termination sites in the (mutant) template strand – a chain-termination site being a base *complementary* to the ddNTP in a given reaction.

When ddF was compared with SSCP for screening rifampicin resistance in *M. tuberculosis*, it was concluded that ddF could differentiate more easily between silent and resistance-conferring mutations owing to its higher content of *sequence-based* information [JCM (1995) *33* 1617–1623]. The interpretation of ddF results is facilitated by the fact that a number of common resistance-conferring mutations have been *mapped*, i.e. their precise details – base change and location – are known.

(iii) *Line probe assay*. Amplicons of the 'mutation-prone' region of *rpoB* (see above) – here called the *resistance region* – are denatured to single-stranded products, and these are allowed to interact with a set of probes immobilized on a membrane. Some of the probes correspond to sequences within the *wild-type* resistance region; the sequences of these probes overlap so that, collectively, they cover the entire wild-type resistance region. PCR products from wild-type test strains will bind to all of these probes; products from test strains that have mutations in the resistance region will not bind to at least one probe because conditions for hybridization are so *stringent* (section 8.5.4).

Other probes correspond to *mutant* sequences in the resistance region; PCR prod-

---

'loading buffer' containing a denaturing agent (e.g. formamide). Products from all three strains are compared, in different lanes, within the same gel. In the diagram, fragments move down the page, i.e. the smallest at the bottom.

In strain A, each product has a wild-type sequence, and its length and conformation will be reflected in the location of the corresponding band in the gel.

In strain B, the C→T mutation (near one end of the target sequence) has caused the *loss* of a chain-termination site; as a result, strain B lacks the longest chain-terminated product seen in strain A, and this is indicated by the *absence* of a band in the gel. The band for the *full*-length product of strain B is shown here in a (hypothetical) location which differs slightly from that of the full-length product of strain A; such a difference in location may occur if the C→T mutation in strain B has caused the full-length product from that strain to have an altered conformation/mobility. The other chain-terminated products from strain B would be unaffected by *this* mutation (which is beyond their chain-termination points) and would therefore form bands at locations indistinguishable from those of strain A. Had the (C→T) mutation been closer to the primer-binding site, then one or other of the shorter chain-terminated products would have been lost.

In strain C, the G→C mutation has introduced a *new* C-site in the target sequence, i.e. a new chain-termination site in the ddG reaction. This results in an *extra* chain-terminated product – shown by an extra band in the gel. In strain C, the bands formed by the full-length product and the *longest* chain-terminated product are shown here at (hypothetical) locations which differ from those of the corresponding bands of strain A; this type of change may occur because the position of the G→C mutation in the template is such that it *may* cause an alteration in the conformation/mobility of these two products.

Clearly, mutations other than those which remove or add a chain-termination base will not affect the number of bands in the gel. However, such mutations may be detected if they affect the conformation/mobility of *any* product strand in which they occur.

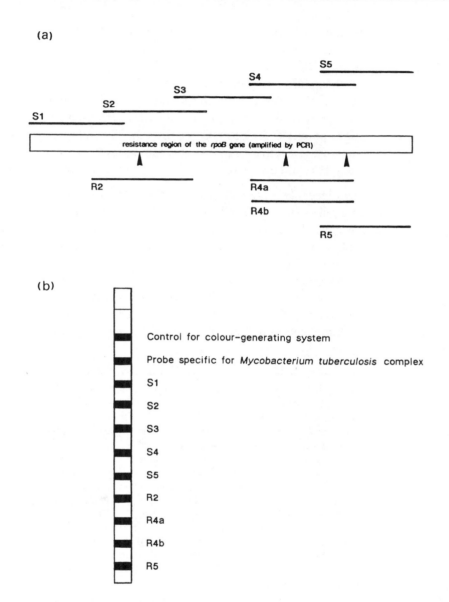

Figure 15.7 Line probe assay (section 15.4.11.2) (diagrammatic): the 'LiPA' assay for rifampicin resistance in *Mycobacterium tuberculosis* (information courtesy of the manufacturer: Innogenetics N.V., Zwijndrecht, Belgium). (a) The 'resistance region' of the *rpoB* gene in *M. tuberculosis* showing those sections covered by each of five wild-type probes (S1–S5) and four mutant probes (R2, R4a, R4b, R5); probes R4a and R4b contain different mutations at the same locus. Mutations covered by the 'R' probes are indicated by arrowheads; they are: (R2) aspartic acid → valine at codon 516, (R4a) histidine → tyrosine at codon 526, (R4b) histidine → aspartic acid at codon 526, and (R5) serine → leucine at codon 531 (codons according to *E. coli* nomenclature). Each type of probe is immobilized in a separate band on a nitrocellulose test strip, as shown at (b).

ucts containing any of the several *common* mutations will bind to one or more of these probes.

Thus, mutations are detected by noting the pattern of binding between single-stranded products and the sets of wild-type and mutant probes; hybridization to any given probe is detected by adding colour-generating reagents.

The line probe assay is described in Fig. 15.7.

In one study [TLD (1995) *76* 425–430], results from line probe assay correctly matched those from conventional tests in 65 out of 67 culture-positive specimens; the two false-negative results may have been due to mutation(s) outside the resistance region or to other mechanism(s) of resistance.

Staphylococcus aureus

In *S. aureus*, high-level ('intrinsic') resistance to methicillin is commonly due to the *mecA* gene (MRSA: section 15.4.11); hence, the detection of such resistance depends on detecting *mecA* rather than on detecting mutation(s).

High-level (*mecA*-based) resistance is important clinically: infections involving these strains may require treatment with vancomycin, and as this is one of the few antibiotics generally effective against *mecA*-based MRSA, its use should be minimized in order to discourage the emergence of resistant strains. Non-*mecA*-based MRSA (section 15.4.11) has been treated successfully with β-lactam chemotherapy [e.g. AAC (1989) *33* 424–428] – hence the importance of distinguishing *mecA*-based from other forms of methicillin resistance.

The *mecA* gene can be detected by PCR (section 8.5.4) with specific primers. Using *multiplex* PCR (section 8.5.4.2) it has been demonstrated that *S. aureus* can be examined – simultaneously – for *mecA* and eight other antibiotic-resistance genes (including genes for resistance to e.g. tetracyclines and aminoglycoside antibiotics) [JCM (2003) *41* 4089–4094].

As mentioned in section 15.4.11, *mecA*-positive strains with *fem* mutations can be phenotypically sensitive to methicillin.

In a given test strain, the target sequence (~ 250 bp in length) is amplified by PCR, the 5′ end of *one* of the primers being labelled with biotin. The (dsDNA) amplicons are denatured and, after addition of a hybridization solution, a test strip is added and the whole incubated at precisely 62°C, with shaking; this permits specific hybridization between probes and (biotinylated) ssDNA from the amplicons. (This stage is sometimes called 'reverse hybridization' because the *probes* are immobilized.)

Unbound DNA is washed away. A streptavidin–alkaline phosphatase conjugate is added and allowed to bind to biotin on the probe-bound DNA; after rinsing, a chromogen is added and is split by alkaline phosphatase to form a purple precipitate – thus marking any probe which has bound amplicons.

Results for rifampicin-sensitive strains will be: positive (purple) staining of the S1–S5 bands, no staining of the 'R' bands. Lack of staining of one (or more) S probes indicates the presence of mutation(s). *If* a given mutation is one of the four common mutations covered by the 'R' probes, then the appropriate 'R' probe will stain purple; for example, the common mutation serine → leucine at codon 531 will cause staining of R5 – and an *absence* of staining of S5.

15.4.12 The post-antibiotic effect

Post-antibiotic effect (PAE) refers to the suppression of bacterial growth (typically for a number of hours) following a short exposure to certain antimicrobial agents – including aminoglycosides, β-lactams, 4-quinolones, macrolides and streptogramins; PAE is a factor when considering the dose/dosage frequency of these antibiotics. Clinically, PAE can be advantageous when it coincides with low concentrations of an intermittently administered antibiotic. A further advantage is that, during the period of suppressed growth, bacteria may be more readily eliminated by the immune system.

Studies on various bacteria/antibiotics have suggested that PAE involves a range of mechanisms.

15.4.13 Activity of antibiotics within eukaryotic cells

Because some pathogens can survive, or even grow, within tissue cells or phagocytes, it is useful to know whether the antibiotics administered to patients can penetrate these cells and act against the intracellular pathogens. Some reported findings are summarized in Table 15.1. The need to consider this issue is illustrated by certain strains of group A streptococci which can invade, and survive within, epithelial cells of the human respiratory tract; although accessible to erythromycin, these particular organisms are resistant to the antibiotic [Lancet (2001) *358* 30–33].

Table 15.1 Activity of antibiotics in eukaryotic cells [1]

Antibiotic	Activity in eukaryotic cells
Aminoglycosides	Generally poor uptake by cells, and often no (or low-level) activity
Fluoroquinolones	Good penetration (e.g. ciprofloxacin accumulates within phagocytes); generally high-level activity
β-Lactams	Uptake generally poor: most do not accumulate in cells; activity often reported absent, although in 2002 amoxycillin was recommended by the CDC as an alternative treatment for chlamydial cervicitis in pregnancy [2]
Macrolides	Good penetration: erythromycin and azithromycin accumulate in tissue cells and phagocytes and are among the drugs used against *Chlamydia trachomatis*
Rifampicin	Accumulates in cells; associated with high-level activity against various types of bacteria

[1] Intracellular accumulation of an antibiotic does not necessarily correlate with high-level activity against a given organism.
[2] *Morbidity and Mortality Weekly Reports* (2002) *51(RR06)* 1–80.

16 The identification and classification of bacteria

16.1 IDENTIFICATION

How do we identify bacteria? Traditionally, an organism is first obtained in pure culture (section 14.6) and then – after certain observations and tests – checked against known, named species until a match is found. This can be quite time-consuming, and rapid methods of identification have been developed for medical and veterinary work (where prompt diagnosis can be crucial) and also e.g. for use in the food industry; rapid, nucleic-acid-based identification is considered in section 16.1.6.

Traditional methods are included here because (i) they are still widely used, and (ii) they are useful for illustrating the types of characteristic which distinguish one bacterium from another.

For traditional tests we need to start with a pure culture; with an impure culture (mixture of organisms) the reaction of one organism may differ from that of another so that, generally, a meaningful result would not be obtained. However, even with a pure culture, it sometimes happens that the characteristics of the unknown organism do not exactly match those of any species in a manual of identification. This can occur, for example, if a mutation (section 8.1) has altered one or more of the organism's typical characteristics. Similarly, a plasmid (section 7.1) may confer characteristics which are not typical of the species to which the unknown organism belongs. Nevertheless, the organism is not always to blame: inexplicable results can sometimes be due to slight variations in the methods and/or materials used in the tests themselves.

Essentially, in the traditional approach, the characteristics of the unknown organism are compared with those of each of a number of known, named species until a match is found. The principle is simple enough, but, in practice, which criteria are used in a given case – and must the unknown organism be compared with each of the thousands of known species of bacteria? Fortunately, the source of an organism often gives clues which, together with a few simple observations and tests, may indicate the possible identity of the organism, or, at least, may serve to narrow the search to one of the major groups of bacteria. For example, if the bacterium comes from sewage, and is found to be a Gram-negative, motile, facultatively fermentative bacillus, the bacteriologist would immediately think of the family Enterobacteriaceae – because this family contains many bacteria of that type, some of which are common in sewage. Identification to species level may then be possible by comparing the characteristics of the unknown organism with those of genera and species of the Enterobacteriaceae.

Clearly, the practice of identification is made easier if the bacteriologist (i) has a knowledge of the types of organism likely to be present in a given environment, and (ii) is familiar with the main distinguishing features of the families, genera and species of common bacteria – information of the type given in the Appendix.

The following characteristics are often the first to be determined because they have the greatest *differential* value, each helping to exclude one or more of the major groups of bacteria: (i) reaction to certain stains, particularly the Gram stain (section 14.9.1); (ii) morphology (coccus, bacillus etc.); (iii) motility; (iv) the ability to form endospores; (v) the ability to grow under aerobic and/or anaerobic conditions; (vi) the ability to produce the enzyme *catalase*. Fortunately, these characteristics are among the easiest to determine.

16.1.1   Preliminary observations and tests

16.1.1.1   *Reaction to stains*

A smear (section 14.9) from a pure culture is generally Gram-stained (section 14.9.1). A capsule stain (section 14.9.3) is used when capsulation is an important differential feature.

16.1.1.2   *Morphology*

Morphology is generally determined by examining a stained smear under the microscope. The smear may be stained either by Gram's method or by a simpler procedure – for example, by treating a heat-fixed smear for one minute with methlyene blue or carbolfuchsin. Usually the stained smear is examined under the oil-immersion objective of the microscope (total magnification about 1000×), although the cells of some species (e.g. *Bacillus megaterium*) are clearly visible under the high-dry lens (total magnification about 400×).

16.1.1.3   *Motility*

Motility (section 2.2.15.1) can often be determined by examining a 'hanging drop' preparation (Fig. 16.1) under the microscope. Even with unstained cells and ordinary

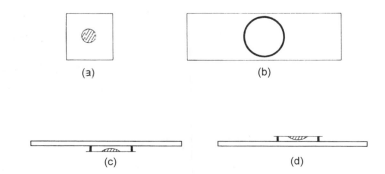

Figure 16.1 The hanging-drop method for determining motility. (a) A drop of culture containing live, unstained bacteria is placed on a clean cover-slip. (b) A ring of plasticine is pressed onto a microscope slide. (c) The slide is inverted and pressed onto the cover-slip. (d) The whole is inverted for examination under the microscope.

(bright-field) microscopy, it is often possible to see whether or not the cells are motile; however, cells can be seen more clearly with dark-field or phase-contrast microscopy. Motility should be distinguished from *Brownian motion*: small, random movements exhibited by any small particle of about 1 μm or less when freely suspended in a liquid medium; these movements (seen e.g. in particles of colloidal clay) are due to bombardment of the particles by molecules of the liquid.

Motility can sometimes be inferred from the way an organism grows on solid media: motile species may tend to spread outwards from the inoculated area as organisms swim in the thin layer of surface moisture.

16.1.1.4 *Endospore formation*

Relatively few bacteria can form endospores (section 4.3.1). If an endospore-former is grown for several days or a week on a solid medium, endospores can usually be detected by treating a heat-fixed smear of the growth with a 'spore stain'. This resembles the Ziehl–Neelsen stain (section 14.9.2) but uses e.g. ethanol for decolorization; vegetative cells are decolorized but endospores retain the (red) dye. Vegetative cells may be counterstained.

Endospores may be detected indirectly by heating a culture to 80°C for 10 minutes – a procedure which endospores usually survive but which kills most types of vegetative cell. Any growth following subculture to a fresh medium suggests the presence of endospores.

The shape of an endospore and its position within the cell are features sometimes used for identification. These observations are best made with the phase-contrast microscope (section 14.10.2, Fig. 14.9) on a thin wet film of unstained cells overlaid with a cover-slip; in such a preparation, the spore-containing mother cells have not undergone shrinkage – as have cells in a heat-fixed smear – so that the spore's relation to the mother cell can be more easily determined. A spore is seen as a bright body (oval or round, depending on species) with a dark margin.

Spores can be distinguished from other intracellular inclusions by differential staining. For example, granules of PHB (section 2.2.4.1), but not endospores, can be stained by dyes such as Sudan black B.

16.1.1.5   Aerobic/anaerobic growth

Whether an organism is an aerobe or anaerobe (section 3.1.6) is easily determined by attempting culture anaerobically (section 14.7) and aerobically.

16.1.1.6   Catalase production

Catalase is an iron-containing enzyme which catalyses the decomposition of hydrogen peroxide (H_2O_2) to water and oxygen; it is formed by most aerobic bacteria, and it de-toxifies hydrogen peroxide produced during aerobic metabolism. The catalase test is used to detect the presence of catalase in a given strain of bacterium. Essentially, bacteria are exposed to hydrogen peroxide, the presence of catalase being indicated by bubbles of gas (oxygen). In the traditional form of the test, a speck of bacterial growth is transferred, with a loop, to a drop of hydrogen peroxide on a slide; however, in a positive test the bursting bubbles will give rise to an aerosol (section 14.1). The author's method (Fig. 16.2) avoids this problem.

Some bacteria (e.g. certain strains of *Lactobacillus* and *Enterococcus faecalis*) produce *pseudocatalase*, a non-iron-containing enzyme which behaves like catalase.

If growth used for the catalase test is obtained from a colony on *blood agar* (section 14.2.1), care should be taken to exclude erythrocytes (red blood cells) from the sample because they contain catalase – and may therefore give a false-positive result.

16.1.2   Secondary observations and metabolic ('biochemical') tests

Once the search for identity has been narrowed to one or a few families, the bacteriologist uses some simple 'biochemical' tests; these tests distinguish between bacteria of different genera and species by detecting differences in their metabolism. For example, a test may distinguish between species which can and cannot ferment a particular carbohydrate, or which produce different products when metabolizing a particular substrate. The following are a few (of many) tests which are frequently carried out in bacteriological laboratories – sometimes as *micromethods* (section 16.1.3).

16.1.2.1   The oxidase test

This test detects a particular type of respiratory chain (section 5.1.1.2): one containing

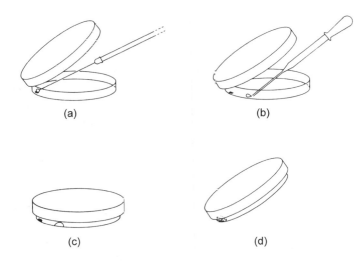

Figure 16.2   The catalase test: a method which avoids contamination of the environment by aerosols. (a) A small quantity of bacterial growth is placed in a clean, empty, non-vented Petri dish. (b) Two drops of hydrogen peroxide are placed in the Petri dish a short distance from the growth. (c) The Petri dish is closed. (d) The closed Petri dish is tilted so that the hydrogen peroxide runs onto the bacterial growth. A positive reaction is indicated by the appearance of bubbles. The lid can be taped to the base before disposal in the proper container.

a terminal cytochrome *c* and its associated *oxidase*. Bacteria which contain such a chain can oxidize chemicals such as *Kovács' oxidase reagent* (1% tetramethyl-*p*-phenylenediamine dihydrochloride); electrons are transferred from this reagent to cytochrome *c* and thence, via the oxidase, to oxygen. When oxidized in this way, the reagent develops an intense violet colour. In the test, a small area of filter paper is moistened with a few drops of Kovács' oxidase reagent, and a small amount of bacterial growth is smeared onto the moist filter paper with a glass spatula or a platinum loop (but *not* a nichrome loop); oxidase-positive species give a violet coloration immediately or within 10 seconds. Oxidase-positive bacteria include e.g. species of *Neisseria*, *Pseudomonas* and *Vibrio*; a negative reaction is given e.g. by members of the family Enterobacteriaceae. (The used test paper should be disposed of safely.)

16.1.2.2   *The coagulase test (for staphylococci)*

Some ('coagulase-positive') strains of *Staphylococcus* produce one or (usually) both of two different proteins called *coagulases*: this test is used to detect coagulase production.

Free coagulase (= true coagulase, or staphylocoagulase) is released into the medium and is detected (in a *tube test*) by its ability to coagulate (i.e. clot) plasma containing an anticoagulant such as citrate, oxalate or heparin; the anticoagulant is necessary because, without it, the plasma would clot spontaneously. In one form of tube test, 0.5 ml of an 18–24 hour broth culture of the strain under test is added to 1 ml of plasma in a test-tube; the tube is kept at 37°C and examined for the presence of a clot after 1, 2, 3 and 4 hours, and at 24 hours. Free coagulase triggers conversion of the plasma protein *fibrinogen* to *fibrin* – which forms the clot. Those coagulase-positive strains which also produce a *fibrinolysin* (an enzyme which lyses fibrin) may not form a clot, or may lyse a clot soon after its formation – hence the need for frequent examination of the tube.

Bound coagulase (= clumping factor) is a protein component of the cell surface; it binds to fibrinogen, resulting in the clumping of cells (*paracoagulation*). (Contrast this with the clotting of plasma.) Bound coagulase is detected by a *slide test*: a loopful of citrated or oxalated plasma is stirred into a drop of thick bacterial suspension on a microscope slide; in a positive test, cells clump within 5 seconds.

Known coagulase-positive and coagulase-negative strains should be used as controls in each form of the test.

Some other bacteria (e.g. strains of *Yersinia pestis*) also form coagulases.

16.1.2.3   The oxidation–fermentation (O–F) test

This is an early form of test (= Hugh & Leifson's test) for determining whether an organism uses oxidative (respiratory) or fermentative metabolism for the utilization of a given carbohydrate (usually glucose). Two test-tubes are filled to a depth of about 8 cm with a peptone–agar medium containing the given carbohydrate and a pH indicator, bromthymol blue, which makes the medium green (pH 7.1). One of the tubes is steamed (to remove dissolved oxygen) and is quickly cooled just before use. Each tube is then stab-inoculated (section 14.5.2) with the test organism to a depth of about 5 cm; in the 'steamed' tube the medium is immediately covered with a layer of sterile liquid paraffin about 1 cm deep (to exclude oxygen). Both tubes are then incubated and later (1–14 days) examined for evidence of carbohydrate utilization, namely, acid production – indicated by yellowing of the pH indicator.

Respiratory organisms (such as *Pseudomonas* species) cause yellowing only in the uncovered ('aerobic') medium. *E. coli* causes yellowing in both media; glucose is fermented in the covered medium, and is attacked first by respiration and then by fermentation in the uncovered medium.

16.1.2.4   Acid/gas from carbohydrates ('sugars')

In some genera the species can be distinguished from one another by differences in the types of carbohydrate ('sugar') which they can metabolize. The range of sugars utilized by any particular organism can be determined simply by inoculating the organism into

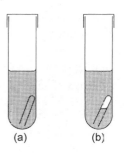

Figure 16.3 The detection of gas produced during growth in a liquid medium. (a) An uninoculated tube of liquid medium containing an inverted Durham tube. (b) A gas-producing organism has been grown in the medium; some of the gas has collected in the Durham tube.

a series of media – each containing a different sugar (as a source of carbon) and an indicator of sugar utilization – and incubating. The medium may be based on peptone-water or nutrient broth; it contains, in addition to the sugar, a pH indicator to detect acidification due to metabolism of the sugar. Such 'peptone-water sugars' or 'broth sugars' are used e.g. in tests on bacteria of the family Enterobacteriaceae. If a particular sugar is metabolized, acid products will be formed, and the acidity is detected by the pH indicator.

Certain bacteria cannot be tested in media containing peptone-water or broth – for example, species of *Bacillus* may form excess alkaline products (from the peptone or broth) so that any acid formed from sugar metabolism would not be detected. For some species the tests may be carried out in media containing inorganic salts and a given sugar. For other types of bacteria the medium must be enriched with e.g. serum, otherwise growth will not occur.

Test media are generally used in test-tubes, or in bijoux (section 14.2), and the tube or bijou may contain an inverted Durham tube (Fig. 16.3) to collect gas that may be formed during the metabolism of the sugar. Gas may be formed e.g. in the mixed acid and butanediol fermentations (Figs 5.5 and 5.6), formic acid being split into carbon dioxide and hydrogen by the *formate hydrogen lyase* enzyme system.

16.1.2.5   IMViC tests

The IMViC tests are a group of tests used particularly for identifying bacteria of the family Enterobacteriaceae. IMViC derives from: *i*ndole test, *m*ethyl red test, *V*oges–Proskauer test and *c*itrate test. (See e.g. *Escherichia* in the Appendix.)

The indole test. This test detects the ability of an organism to produce indole from the amino acid tryptophan. The organism is grown in peptone-water or tryptone-water for 48 hours; to the culture is then added Kovács' indole reagent (0.5 ml per 5.0 ml culture) and the (closed) container is gently shaken. In a positive test, indole (present in the culture) dissolves in the reagent – which then becomes pink, or red, and forms a layer at the surface of the medium.

The methyl red test (MR test). The MR test detects the ability of an organism, growing in a phosphate-buffered glucose–peptone medium, to produce sufficient acid (from the metabolism of glucose) to reduce the pH of the medium from 7.5 to about 4.4 or below. The medium is inoculated and is then incubated for at least 48 hours at 37°C, following which the pH of the culture is tested by adding a few drops of 0.04% methyl red (yellow at pH 6.2, red at pH 4.4); with an MR-positive organism the culture becomes red.

The Voges–Proskauer test (VP test). This test detects the ability of an organism to form acetoin (acetylmethylcarbinol) – see butanediol fermentation, Fig. 5.6. A phosphate-buffered glucose–peptone medium is inoculated with the test strain and incubated at 37°C for 2 days, or at 30°C for at least 5 days. In one form of the test (*Barritt's method*), 0.6 ml of an ethanolic solution of 5% α-naphthol, and 0.2 ml of 40% potassium hydroxide solution, are added sequentially to 1 ml of culture; the (stoppered) tube or bottle is then shaken vigorously, placed in a sloping position (for maximum exposure of the culture to air), and examined after 30 and 60 minutes. Acetoin (if present) is apparently oxidized to diacetyl ($CH_3.CO.CO.CH_3$) which, under test conditions, gives a red coloration (a positive VP test).

The citrate test. This test detects the ability of an organism to use citrate as the sole source of carbon. Media used for the test – e.g. Koser's citrate medium (a liquid), and Simmons' citrate agar – include citric acid or citrate, ammonium dihydrogen phosphate (as a source of nitrogen and phosphorus), sodium chloride and magnesium sulphate. A saline suspension of the test organism is made from growth on a solid medium; using a *straight wire* (section 14.4), Koser's medium is inoculated from the suspension and is then incubated and examined for signs of growth (turbidity) after one or two days. Organisms which grow in the medium are designated 'citrate-positive'. A straight wire is used so that little or no nutrient is carried over in the inoculum; any nutrient carried over from the original medium may permit a small amount of growth in the citrate medium, thus giving a false-positive result. In an alternative method of inoculation, a straight wire is used to transfer a small quantity of growth from the *top* of a colony direct to the test medium.

16.1.2.6 *Hydrogen sulphide production*

Many species of bacteria produce hydrogen sulphide, e.g. by the reduction of sulphate (section 5.1.1.2) or from the metabolism of sulphur-containing amino acids. A sensitive test for sulphide is likely to be positive even for those species which form very small amounts of sulphide. A test of *low* sensitivity can distinguish between those species which form negligible or small amounts of sulphide and those which form large amounts. In one form of test, the organism is stab-inoculated into a tube of solid, gelatin-based medium containing peptone and a low concentration of ferrous chloride; organisms which form a lot of sulphide form visible amounts of black ferrous sulphide. In a more sensitive test, a strip of lead acetate paper is placed above the medium on/in which the test organism is growing; hydrogen sulphide production is indicated by the formation of lead sulphide, which causes blackening of the strip.

16.1.2.7 The urease test

Ureases are enzymes that hydrolyse urea, $(NH_2)_2.CO$, to carbon dioxide and ammonia. [Bacterial ureases (structure, regulation of expression, and role in pathogenesis): MM (1993) *9* 907–913.] Production of urease in enterobacteria can be detected by culture on e.g. *Christensen's urea agar*: a phosphate-buffered medium containing glucose, peptone, urea, and the pH indicator phenol red (yellow at pH 6.8, red at pH 8.4); when grown on this medium, 'urease-positive' strains liberate ammonia which raises the pH and causes the pH indicator to turn red. Urease-positive bacteria include e.g. *Klebsiella pneumoniae* and *Helicobacter pylori*.

16.1.2.8 Decarboxylase tests

These tests detect the ability of an organism to form specific *decarboxylases* – enzymes which decarboxylate the amino acids arginine, lysine and ornithine to agmatine, cadaverine and putrescine, respectively. Three tubes of *Møller's decarboxylase broth*, each including glucose, peptone, one of the amino acids, and the pH indicators bromcresol purple and cresol red, are inoculated with the test organism; each broth is covered with a layer of sterile paraffin (to exclude air), incubated at 37°C, and examined daily for 4 days. Initially the medium becomes acidic (yellow) due to glucose metabolism; if a decarboxylase is *not* formed the medium remains yellow. Decarboxylation of the amino acid produces an alkaline product which subsequently raises the pH, causing the medium to become purple. A *control* medium resembles the test medium but lacks an amino acid; it should become, and remain, yellow.

16.1.2.9 The phenylalanine deaminase test (PPA test)

This test detects the ability of an organism to deaminate phenylalanine to phenylpyruvic acid (PPA). The organism is grown overnight on phenylalanine agar (containing yeast extract, Na_2HPO_4, sodium chloride and DL-phenylalanine); 0.2 ml of a 10% solution of ferric chloride is then added to the growth. PPA, if present, gives a green coloration with the ferric chloride. PPA-positive bacteria include e.g. *Proteus vulgaris*.

16.1.2.10 The ONPG test

The utilization of lactose often involves two enzymes: (i) a galactoside 'permease' (which facilitates the *uptake* of lactose), and (ii) β-D-galactosidase (which splits lactose into glucose and galactose); species such as *E. coli* can usually synthesize both of these enzymes. Certain bacteria which do not utilize lactose, or which metabolize it very slowly, may nevertheless form the enzyme *β*-D-galactosidase; the inability of such organisms to carry out normal lactose metabolism may be due, for example, to an inability to synthesize galactoside permease. To detect the presence of *β*-D-galactosidase in such organisms, use is made of a substance, *o*-nitrophenyl-*β*-galactopyranoside (ONPG), which can enter the cell without a specific permease; once inside the cell, ONPG is cleaved by the galactosidase to galactose and the yellow-coloured *o*-nitrophenol. In the ONPG test, the organism is grown for 18–24 hours in broth containing ONPG; a positive test (*β*-D-galactosidase present) is indicated by the appearance of the (yellow) *o*-nitrophenol in the medium.

16.1.2.11 The phosphatase test

Phosphatases, enzymes which hydrolyse organic phosphates, are produced by a number of bacteria and can be detected by the phosphatase test. The organism is grown for 18–24 hours on a solid medium which includes sodium phenolphthalein diphosphate; this substance is hydrolysed by phosphatases with the liberation of phenolphthalein – a pH indicator which is colourless at pH 8.3 and red at pH 10.0. To detect phenolphthalein (a positive test), the culture is exposed to gaseous ammonia, which causes phosphatase-containing colonies to turn red.

16.1.2.12 The nitrate reduction test

This test detects the ability of an organism to reduce nitrate (see e.g. anaerobic respiration, section 5.1.1.2). The following test can be used e.g. for enterobacteria and pseudomonads. The organism is grown for one or more days in nitrate broth (e.g. peptone water containing 0.1–0.2% w/v potassium nitrate), and the medium is then examined for evidence of nitrate reduction. To test for *nitrite*, 0.5 ml of 'nitrite reagent A' and 0.5 ml of 'nitrite reagent B' are added to the culture; these reagents combine with any nitrite present to form a soluble red azo dye. The *absence* of red coloration could mean that either (i) nitrate had not been reduced, or (ii) nitrite was formed but had been subsequently reduced e.g. to nitrogen or ammonia. To distinguish between these two possibilities, the medium is tested for the presence of nitrate by adding a trace of zinc dust – which reduces nitrate to nitrite; if nitrate is present (i.e. it has not been reduced by the test organism), the addition of zinc will bring about a red coloration because the newly-formed nitrite will react with the reagents present in the medium.

16.1.2.13 Reactions in litmus milk

Many species of bacteria give characteristic reactions when they grow in *litmus milk* (skim-milk containing the pH indicator litmus). A given strain of bacterium may produce one or more of the following effects: (i) no visible change; (ii) acid production from the milk sugar (lactose) indicated by the litmus; (iii) alkali production, usually due to hydrolysis of the milk protein (casein); (iv) reduction (decolorization) of the litmus; (v) the production of an acid clot, which is soluble in alkali; (vi) the formation of a clot at or near pH 7 due to the action of rennin-like enzymes produced by the bacteria; (vii) the production of acid *and* gas which may give rise to a *stormy clot*: a clot which has been disturbed and is permeated by bubbles of gas.

16.1.2.14   *Aesculin hydrolysis*

Some bacteria can hydrolyse aesculin (the 6-β-D-glucosyl derivative of 6,7-dihydroxycoumarin); hydrolysis releases 6,7-dihydroxycoumarin – detected by the formation of a brown/black coloration with soluble ferric salts. In one form of the test, the organism is grown on an agar-based medium which includes aesculin (0.1%) and ferric chloride (0.05%); a brown/black coloration indicates a positive test (Plate 14.3, *top*). Organisms which hydrolyse aesculin include e.g. strains of *Bacteroides*, *Enterococcus* and *Streptococcus*, and *Listeria monocytogenes*.

16.1.2.15   *The MUG test for* Escherichia coli

The MUG test helps to detect *E. coli*. A reagent, 4-methyl-umbelliferyl-β-D-glucuronide (MUG), is added to the (solid) culture medium prior to inoculation. Most strains of *E. coli* contain the enzyme β-glucuronidase which cleaves MUG to a fluorescent compound, 4-methylumbelliferone; this compound (and, hence, the presumptive presence of *E. coli*) is detected as green-blue fluorescence when the culture is exposed to ultraviolet radiation of wavelength 366 nm (Plate 14.2, *bottom*). (Note that external sources of glucuronidase – found e.g. in shellfish samples – can lead to false-positive indications.) An alternative method is to smear a colony onto a filter paper which has been impregnated with MUG and examine it under ultraviolet radiation.

16.1.3   Micromethods (= multitest systems)

In clinical bacteriology, *micromethods* are miniaturized test procedures which are used to carry out, simultaneously, a range of routine biochemical identification tests on a given organism. These procedures involve the use of commercial 'kits' which save time, space and materials; some are mentioned below. (See Plates 14.4, 14.5.)

The *API* system consists of a plastic strip holding a number of microtubes, each containing a different dehydrated medium. Each microtube is inoculated from a suspension of the test organism; where necessary, mineral oil is added to particular microtubes (to exclude air) and the strip is then incubated. Later, reagents are added, where appropriate, to detect particular metabolic products. Separate API test systems are used for enterobacteria, for streptococci, and for anaerobic bacteria (as different tests are appropriate to each group of organisms).

Staph-Ident resembles the *API* system but contains tests that are particularly appropriate for staphylococci.

Enterotube II consists of a tube divided into a sequence of 12 compartments, each containing a different agar-based gel medium. The media are inoculated by passing an inoculum-bearing straight wire axially through the tube.

The *PathoTec* system consists of various test strips, each impregnated with a dehydrated medium appropriate to a given test. Each strip is incubated in a suspension of the test organism (or inoculated directly from a colony) and the test is read after a specified time.

The *Rapidec Staph* test can detect coagulase-positive staphylococci by detecting an enzyme, aurease, which is specific to these bacteria. Apparently, aurease and a test reagent, prothrombin, form a complex that can lyse a substrate used in the test; such lysis produces a compound which is detected by its fluorescence under ultraviolet radiation. [Evaluation of test: JCM (2003) *41* 767–771.]

16.1.4   Haemolysis

When certain bacteria grow on *blood agar* (section 14.2.1), each colony is surrounded by a 'halo' of differentiated medium in which the erythrocytes have been lysed or in which the blood has been discoloured; the lysis of erythrocytes (*haemolysis*) is due to the activity of proteins (*haemolysins*) released by the bacteria. Some species produce glass-clear, colourless haemolysis which contrasts sharply with the opaque red medium; this is formed e.g. by *Streptococcus pyogenes* and by some strains of *Staphylococcus aureus*. (When formed by *Streptococcus*, glass-clear haemolysis is often called β-haemolysis, but when formed by *S. aureus* it may be referred to as α-haemolysis; it is probably best to refer to this type of haemolysis simply as 'clear haemolysis'.)

The so-called *viridans streptococci*, and some other bacteria, form zones of greenish-brown discoloration (*greening*) around their colonies.

For any given haemolytic bacterium, haemolysis – or a particular form of it – may develop only if the organism has been grown on media containing the blood of specific type(s) of animal (e.g. horse, rabbit, man etc.).

16.1.5   Typing

'Typing' means (i) matching an unknown strain with a specific known strain of a given species (a form of identification), *and* (ii) distinguishing between different strains of a given species (a form of classification); *these two procedures use similar criteria (and methods)*. Clearly, (ii) must precede (i).

In sense (i), it should be appreciated that no system of typing can prove that a given unknown strain is *totally* identical to a particular known strain – although it can indicate *non*-identity; this is because any given typing system is designed to reveal similarity, or otherwise, in respect of only one (or a few) characteristics – so that even if an unknown strain is identical to a known strain in *these* characteristics it may well differ in others. Hence, for a given unknown strain, different typing systems may give different results – they frequently do.

Uses of typing. In a medical/veterinary context, typing has various uses. For example:

- Typing is particularly useful in epidemiology. The essential premise here is that various isolates of a given pathogen from within the same *chain of infection*, or *outbreak of disease*, will be progeny derived from the same ancestral cell; such a clonal relationship will be detectable by the high degree of similarity of the genotypes and/or phenotypes of the isolates – as compared with other, randomly acquired isolates of the same species. Thus, if different strains of a pathogen have been stably distinguished (typed), we can often match a fresh isolate of the pathogen with one of the known strains; this may enable a particular case or outbreak of the relevant disease to be linked with a particular source of infection (e.g. by noting the prevalence of particular strains in particular geographical areas). [Evaluation and use of epidemiological typing systems: CMI (1996) *2* 2–11.]

  In a veterinary context, typing has been used e.g. to investigate outbreaks of mastitis, caused by *Pseudomonas aeruginosa*, in dairy herds [AEM (1999) *65* 2723–2729].

- Typing has been used to assess the relative importance of new infection (as opposed to re-activation of existing infection) in cases of tuberculosis. It appears that, in at least some geographical areas, recent transmission, rather than re-activation, is more common than had been supposed [BCID (1997) *4* 173–183].

- Typing is useful for detecting cross-contamination in the laboratory. In one episode, typing revealed that 60 specimens in a reference laboratory had been contaminated with *Mycobacterium tuberculosis* from a (true-positive) specimen [JCM (1999) *37* 916–919].

- Typing is useful in some forensic/medico-legal investigations.

Typing methods. The two main categories of typing involve (i) conventional (phenotypic) methods, and (ii) DNA-based (genotypic) methods. Conventional methods include those based on serology (section 16.1.5.1); susceptibility to phages (section 16.1.5.2); and electrophoresis of enzymes (section 16.1.5.3). Additionally, there are *ad hoc* typing systems applicable only to particular species; for example, strains of *Pseudomonas aeruginosa* have been typed on the basis of variability in the siderophore *pyoverdin* [Microbiology (1997) *143* 35–43].

Genotypic typing methods are considered in section 16.1.6.2.

16.1.5.1 Serological tests

These tests can distinguish between closely related bacteria by detecting differences in their cell-surface *antigens* (section 11.4.2.1). For example, using serological tests, thousands of different strains of *Salmonella* can be distinguished primarily by slight chemical differences in their O antigens (lipopolysaccharide–protein antigens) and H antigens (flagellar antigens); differences in the O antigens, for example, occur in the O-specific chains (section 2.2.9.2). Strains which are distinguished mainly on the basis of their antigens are called *serotypes* (see e.g. *Salmonella* in the Appendix).

In practice, we can detect (and identify) a given serotype by using specific *antibodies* which are known to combine only with the antigens of known serotype(s). Antibodies can be obtained by injecting an experimental animal with antigens from a known serotype; after an interval of time, the animal's serum will contain antibodies to those antigens. Such a serum is called a specific *antiserum*. On mixing this antiserum with cells of the given serotype, combination will occur between cell-surface antigens and their corresponding antibodies in the antiserum; when this happens, the bacterium–antibody complexes commonly form a visible whitish suspended mass, or a sediment, in the test-tube (an *agglutination reaction*). If this same antiserum can agglutinate an unidentified strain, it may be concluded that the unknown strain has antigen(s) in common with the original serotype. Hence, an unknown serotype can be identified by testing it with each of a range of antisera, each antiserum containing antibodies of particular, known serotype(s).

Rapid results can be obtained with *latex slide-agglutination tests* in which the reagent consists of minute (microscopic) particles of latex coated with specific antibodies (whose antigen-binding sites face outwards). Bacteria with the matching cell-surface antigens will bind these particles to each other, causing them to agglutinate into visible clumps; hence, clumping indicates a positive test.

One example of a latex test is that for *Escherichia coli* O157 (product DR620, Oxoid, Basingstoke, UK) (Plate 14.5, *bottom*). A potential problem with (any) serologically based test for *E. coli* O157 is that strains may lose their O157 antigenicity under starvation conditions (even though they remain toxigenic); thus, *environmental* strains of *E. coli* O157 (e.g. those isolated from rivers) may not be detected by serological methods if they have been starved for some time [AEM (2000) *66* 5540–5543].

16.1.5.2 Bacteriophage typing ('phage typing')

This procedure distinguishes between different strains of closely related bacteria by

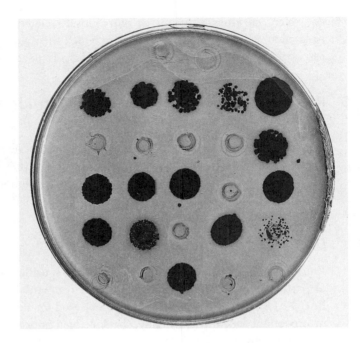

Plate 16.1   Phage typing of a strain of *Staphylococcus* (see section 16.1.5.2). In some cases, the drop of phage suspension has produced a circular area of total lysis in the bacterial 'lawn'. In other cases, small individual plaques are visible in the area covered by the drop. The strain on this plate was inoculated with 27 different phages of the international set, and is sensitive to (i.e. lysed by) 15 of them: phages designated 29, 52, 52A, 79, 80, 95, 6, 42E, 47, 54, 75, 77, 84, 85 and 81. Photograph courtesy of Dr Judith F. Richardson, Laboratory of Hospital Infection, Public Health Laboratory Service, London.

exploiting differences in their susceptibility to a range of bacteriophages (Chapter 9). A *flood plate* (section 14.5.2) is prepared from a culture of a given strain; the plate is 'dried' (section 14.2.2), and a grid is drawn on the base of the Petri dish. Next, the agar over each square of the grid is inoculated with one drop of a suspension of phage – each square being inoculated with a different phage; the drops are allowed to dry and the plate is incubated. Usually, one, two or more of the phages will be found to be lytic for a given strain; lysis (susceptibility to a given phage) is indicated on the lawn plate by the formation of a *plaque* (section 9.1.3) at the point of inoculation of each phage particle. In this way, strains can be defined (and identified) by the range of phages to which they are susceptible. Phage typing is used e.g. for *Staphylococcus aureus* and *Yersinia enterocolitica*. (See Plate 16.1.)

16.1.5.3   *Multilocus enzyme electrophoresis (MEE)*

In this method, strains are distinguished by comparing the electrophoretic mobilities of a range of enzymes from one organism with the electrophoretic mobilities of equivalent enzymes from one or more closely related organisms. Enzymes must be isolated and tested under conditions in which their activity is retained. The enzymes can be identified in gels by the use of specific colour-generating substrates.

16.1.6   Rapid (DNA-based) methods for detection/identification and typing of pathogens

The basic procedures used in these methods are described in Chapter 8: isolation of nucleic acids (8.5.1.4); restriction endonucleases (8.5.1.3); probes (8.5.3); DNA sequencing (8.5.6 and 8.5.6.1); polymerase chain reaction (PCR: 8.5.4); ligase chain reaction (LCR: 8.5.9.1); NASBA/TMA (8.5.9.2).

16.1.6.1   DNA-based detection/identification

Probe-based methods. Species-specific or strain-specific probes permit rapid detection of particular organism(s) in clinical or environmental samples. One example is the PACE 2 test (Gen-Probe, San Diego, USA) used for detecting *Chlamydia trachomatis* in urogenital specimens. Initially, the specimen is processed to expose nucleic acids from any cells of *C. trachomatis* that may be present. Copies of the probe are added, and if *C. trachomatis* is present, probes bind to a specific sequence in the pathogen's rRNA. An added reagent then hydrolyses the label (acridinium ester) on any *un*bound probes; on *bound* probes the acridium ester is protected within the probe–rRNA complex. Addition of further reagents produces a chemiluminescent signal from the acridinium ester label on each bound probe. Emitted light is monitored by a luminometer.

The PACE 2 test, with high sensitivity and specificity, has replaced culture in many laboratories. The PACE 2C test is a *combination probe test* that can detect *Chlamydia trachomatis* and *Neisseria gonorrhoeae* in a single assay; the probe includes sequences specific to each organism. Specimens positive in the PACE 2C test are then tested with separate probes to identify the organism(s) present.

Some probes give a positive reaction with any species of a particular group; one example is the probe for members of the '*Mycobacterium tuberculosis* complex' (which includes e.g. *M. bovis* and BCG as well as *M. tuberculosis*). By contrast, other probes are so specific that they can distinguish between strains of a given species; examples include the probe for *Staphylococcus aureus* strain NBSA [FEMSML (1999) *172* 47–52] and a probe for (multidrug-resistant) *Salmonella typhimurium* DT104 [JCM (1999) *37* 1348–1351].

PCR-based methods. Given very specific primers (see note in item 1, section 8.5.4.1), PCR may permit rapid detection and identification of a given organism; this may be useful e.g. for the diagnosis of certain diseases. After processing the specimen to expose nucleic acids, a small aliquot is subjected to PCR; in some cases the target is a sequence in rRNA (rather than DNA) – this being advantageous in that each cell contains thousands of rRNA molecules but typically only one, or a few, copies of a given chromosomal sequence. The products of PCR (i.e. the amplicons) are then detected/characterized by any of the methods described in section 8.5.4.

Theoretically, detection by PCR is possible if a specimen contains only one accessible copy of the target sequence. However, PCR uses very small aliquots of the specimen (measured in microlitres) so that, in general, a specimen should contain at least 10^3–10^4 cells/ml. If the concentration of target cells is low, it may be possible to improve detection (and, hence, improve sensitivity of the assay) by using a technique such as IMS (immunomagnetic separation: section 14.6.1); IMS can concentrate sequence-specific fragments of DNA [JCM (1996) *34* 1209–1215] or specific whole cells [e.g. *Mycobacterium paratuberculosis* in milk: AEM (1998) *64* 3153–3158].

Detection of specific cells may be improved by *short-term* culture of the specimen prior to use of PCR (or other nucleic-acid-based method) [e.g. JCM (1997) *35* 714–718].

Examples of the use of PCR for detection/identification of bacterial pathogens include:

1. *Mycobacterium malmoense* (an organism that grows particularly slowly and is difficult to identify by conventional methods) [JCM (1999) *37* 1454–1458].
2. *Streptococcus* (group A strains associated with necrotizing fasciitis) [JCM (1998) *36* 1769–1771].
3. *Escherichia coli* O157. This study [JCM (1998) *36* 1801–1804] rapidly and efficiently detected *E. coli* O157. One disadvantage of the assay is that it does not distinguish between toxigenic and non-toxigenic strains of this serotype; however, a modified (multiplex) PCR assay could include primers able to detect DNA encoding shiga-like toxin.
4. *Bordetella pertussis, B. parapertussis*. All culture-positive specimens were also PCR-positive; moreover, PCR detected the pathogens in a significant number of specimens that were culture-negative [JCM (1999) *37* 606–610].
5. MRSA (methicillin-resistant *Staphylococcus aureus*). Rapid detection of MRSA in specimens containing a mixture of staphylococci has been reported with a real-time PCR-based assay [JCM (2004) *42* 1875–1884].

TMA-based methods. Transcription-mediated amplification (TMA) resembles NASBA (section 8.5.9.2) and is used e.g. for detecting *Mycobacterium tuberculosis*. The Amplified *Mycobacterium tuberculosis* Direct Test (AMTDT; Gen-Probe) was approved by the Food & Drug Administration (FDA) in the USA for testing *smear-positive* specimens, i.e. specimens from which can be prepared a smear containing acid-fast bacilli (AFBs: section 14.9.2). [Comparison of original and enhanced versions of AMTDT for respiratory and non-respiratory specimens: JCM (1998) *36* 684–689. False-positive results with AMTDT in patients with *Mycobacterium avium* and *M. kansasii* infections: JCM (1999) *37* 175–178.]

LCR-based methods. The ligase chain reaction is used e.g. for detecting *Chlamydia trachomatis* and *Neisseria gonorrhoeae* (LCx® assays; Abbott Diagnostics). [Improved reproducibility with modified procedure: JCM (2000) *38* 2416–2418.] (Note that the test for *C. trachomatis* has relied on target DNA in a *plasmid* (pCT); rare isolates of *C. trachomatis* do not contain pCT [INFIM (1998) *66* 6010–6013].)

The LCR has been useful for early identification of *Mycobacterium tuberculosis* in BACTEC cultures (section 14.2.1.1), allowing a faster turnaround time for isolates of the *M. tuberculosis* complex than is otherwise possible [JMM (2002) *51* 710–712].

Like other nucleic-acid-based procedures, the LCR is susceptible to malfunctioning as a result of changes in the target sequence. For example, a chromosomal deletion in the region containing the target sequence resulted in the failure of a commercial LCR assay to detect *Mycobacterium tuberculosis* in sputum samples from a patient with smear-positive pulmonary tuberculosis [JCM (2002) *40* 2305–2307].

16.1.6.2   *DNA-based typing*

As discussed (section 16.1.5), typing includes elements of both identification and classification – which use similar types of criteria and methodology. DNA-based typing involves various procedures discussed under Classification (section 16.2); these, in turn, involve basic techniques (such as PCR and sequencing) covered earlier in the book.

DNA-based typing differs from conventional typing in several ways. For example, it may be carried out *directly* from the clinical specimen (without prior isolation of the organism in pure culture) – as has been done e.g. for *Neisseria meningitidis* [JCM (1997) *35* 1809–1812] and *Mycobacterium tuberculosis* [JCM (1997) *35* 907–914]. Another difference is that *silent* mutations (Fig. 8.1), as well as other changes in the target sequence, may interfere with typing.

DNA-based typing methods [review: JCM (1999) *37* 1661–1669] are more numerous than conventional methods, and new forms continue to appear. [Molecular typing of microorganisms (editorial): JMM (2002) *51* 7–10.]

Examples of nucleic-acid-based typing methods include:

☞ **DNA fingerprinting**. See 16.2.2.3.

☞ **PFGE** (pulsed-field gel electrophoresis). See 16.2.2.3 and 8.5.1.4. Typing by PFGE is often regarded as the gold standard among molecular typing methods; compared with other methods it often has superior discriminatory power – e.g. it was more discriminatory than PCR-based methods for typing isolates of methicillin-resistant *Staphylococcus aureus* (MRSA) [JMM (1998) *47* 341–351]. [Typing with PFGE guidelines): JCM (1995) *33* 2233–2239; *Shigella dysenteriae* type 1: JMM (1999) *48* 781–784; *Vibrio parahaemolyticus*: JCM (1999) *37* 2473–2478.]

☞ ***Ribotyping****.* See 16.2.2.3. [Ribotyping *Vibrio parahaemolyticus*: JCM (1999) *37* 2473–2478; *Pseudomonas syringae*: AEM (2000) *66* 850–854.]

☞ ***PCR-ribotyping****.* See 16.2.2.7. Use: e.g. typing *Clostridium difficile* [RMM (1997) *8* (suppl 1) S55-S56].

☞ ***RFLP*** (restriction fragment length polymorphism). Section 16.2.2.6.

☞ ***PCR-RFLP****.* Section 16.2.2.6. Use: e.g. typing isolates of *Staphylococcus aureus* from animals with mastitis [JCM (1999) *37* 570–574].

☞ ***RAPD*** (random amplified polymorphic DNA). Section 16.2.2.4. Use: e.g. typing *Campylobacter* spp [LAM (1996) *23* 167–170]; *Vibrio vulnificus* [AEM (1999) *65* 1141–1144]; *Serratia marcescens* [LAM (2000) *30* 419–421].

☞ ***AP-PCR*** (arbitrarily primed PCR). Section 16.2.2.4.

☞ ***rep-PCR*** (repetitive sequence-based PCR). Section 16.2.2.5. Use: e.g. typing *Vibrio parahaemolyticus* [JCM (1999) *37* 2473–2478]; *Listeria monocytogenes* [JCM (1999) *37* 103–109].

☞ ***AFLP fingerprinting****.* Section 16.2 2.8.

☞ ***Spoligotyping****.* A method for simultaneously detecting and typing strains of *Mycobacterium tuberculosis* [original description: JCM (1997) *35* 907–914]. Amplicons from PCR-amplified spacer sequences in the direct repeat (DR) locus are hybridized to a standard set of oligonucleotides representing spacers from a reference strain; for a given test strain, the pattern of matching and mis-matching is the *spoligotype*. Use: e.g. checking the validity of IS*6110*-based typing [JCM (1999) *37* 788–791]; investigating laboratory cross-contamination [JCM (1999) *37* 916–919]. [Spoligotype database (biogeographic distribution): EID (2001) *7* 390–396. Improved discrimination: JCM (2002) *40* 4628–4639.]

☞ ***MLST*** (multilocus sequence typing). A method permitting long-term, global-scale tracking of virulent/antibiotic-resistant pathogens, with information available on the Internet. Characterization and classification of strains is based, on nucleotide sequences in *specific* alleles. [MLST (review): TIM (1999) *12* 482–487. MLST of methicillin-resistant clones of *Staphylococcus aureus* (MRSA) and methicillin-sensitive clones: JCM (2000) *38* 1008–1015.] MLST data may also have taxonomic value [JB (2004) *186* 1518–1530].

☞ ***SRF*** (subtracted restriction fingerprinting). Digestion of chromosomal DNA with two types of restriction endonulease yields fragments whose *end*-sequences are of three types: AA, BB and AB. The fragments are differentially end-labelled with either biotin or digoxigenin, and the two types of (biotin-labelled) fragment are removed (i.e. 'subtracted'); the third type of fragment is subjected to gel electrophoresis and the bands of fragments detected via the digoxigenin label. [SRF (method): BioTechniques (2003) *34* 304–313.]

☞ *SSCP analysis* (description: Fig. 8.19(c), legend). Use: e.g. discriminating between strains of *Salmonella* [JCM (2002) *40* 2346–2351].

16.2   THE CLASSIFICATION (TAXONOMY) OF PROKARYOTES

Ideally, biological (taxonomic) classification should be *phyletic*, i.e. based on the natural (evolutionary) relationships between organisms. This is traditional in higher organisms – in which evolutionary relationships can be deduced from structural and other features. In prokaryotes, however, the 'simple' structure offers too few clues for phyletic classification, and these organisms have been classified traditionally on the basis of observable (phenotypic) characteristics of the type described in section 16.1.

The modern, phyletic classification of prokaryotes is based on *molecular* criteria: organisms are classified primarily according to differences (and similarities) in the sequences of nucleotides in their nucleic acids. It's appropriate to mention here a consensus form of taxonomy called *polyphasic taxonomy*. This bases classification on the maximum amount of information available, including both phenotypic and genotypic data.

Data consisting of sequences of nucleotides are interpreted, for example, on the principle of *parsimony*, Essentially, the nucleotide sequence of a highly conserved gene is determined for each of a number of organisms; when these sequences are compared, any differences in nucleotides are assumed to have arisen *in the smallest number of steps* from a common ancestor. The organisms can then be arranged in a dendrogram showing their proposed evolutionary relationships. The reader is referred to other texts for details of the theory of phyletic classification.

Section 16.2 deals with the *methods* used for classifying bacteria. Note that methods of this kind can be used for two distinct purposes:

• to look for phyletic relationships among a given group of organisms, i.e. molecular taxonomy

• to distinguish between organisms (especially different strains of a given species) in order subsequently to help *identify* unknown strains by matching them to known, characterized strains

First we look briefly at an earlier molecular criterion: the GC ratio (GC%).

16.2.1   The GC ratio (GC%) of DNA

In DNA, the GC ratio (= GC%) is the amount of guanine and cytosine as a percentage of total nitrogen bases:

(guanine + cytosine)/(guanine + cytosine + adenine + thymine) × 100%

The GC% of a sample of DNA can be estimated by various physical methods, e.g. by measuring its density in a *density gradient column* (section 8.5.1.4). GC% has been used as a crude yardstick for comparing DNA from different organisms; however, similar values do not necessarily mean a close taxonomic relationship, though widely differing values suggest the absence of such a relationship. Among bacteria, GC% values range from about 24 to 76. Some GC% values are given in the Appendix.

16.2.2 Classification based on sequences of nucleotides

In heredity, the features of an organism are maintained from generation to generation because the chromosome is faithfully copied at each cell division. However, during long (evolutionary) periods of time, new sequences of nucleotides – and hence, new organisms – emerge from pre-existing ones. Thus, a chromosome contains more than just the blueprint of an organism: it contains details of history and development which may be revealed when sequences from different organisms are compared. Organisms have therefore been classified according to the unique sequences of nucleotides in their chromosomes.

If we examine a particular nucleotide sequence – of either DNA or RNA – then differences in the corresponding sequence in different organisms may indicate major evolutionary divergence between the organisms (e.g. at kingdom level) or perhaps minor variation (e.g. at species level); whether a major or minor taxonomic divergence is indicated depends e.g. on the particular sequence examined and on the apparent relative stability of that sequence during evolution – differences in a stable sequence seeming to be more significant than those in a less stable sequence.

How do we compare the nucleotide sequences of different cells? Commonly, the cells are initially *lysed* (broken open) and their nucleic acids isolated by suitable techniques (section 8.5.1.4). In some cases we isolate DNA, in others RNA. Sequences are then compared by methods such as those discussed in the following sections and outlined in Figs 16.4 and 16.6.

Because classification necessarily involves characterization and comparison of organisms, certain of these methods are also useful in identification; methods which distinguish between closely related strains are particularly useful for typing (section 16.1.6.2).

16.2.2.1 *DNA–DNA hybridization*

In this method, DNA from two different organisms is compared by measuring the extent to which the two samples can hybridize (*hybridization* referring to base-pairing between strands from different organisms); the greater the degree of hybridization the closer the relationship between the two organisms (Fig. 16.4a). The method is useful e.g. for investigating relationships at the species level; organisms whose

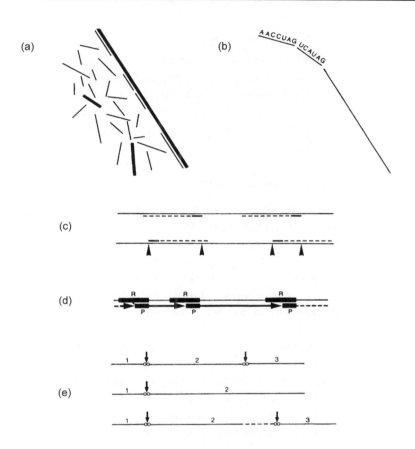

Figure 16.4 Some methods for comparing sequences of nucleotides from different organisms (diagrammatic, to show principles).

(a) DNA–DNA hybridization (section 16.2.2.1). In one form of this method, single-stranded (heat-denatured) chromosomal DNA from strain A is bound to a sheet of nitrocellulose (or other material); in the diagram this DNA is shown as a long thick line. Chromosomal DNA from strain B is broken into fragments, by suitable methods, and the fragments are denatured by heating; similar, single-stranded fragments are also made from *labelled* DNA of strain A. Under suitable conditions, the *un*labelled DNA of strain A (bound to nitrocellulose) is then exposed to the fragments of strain B (short thin lines) and (simultaneously) to the (labelled) fragments of strain A (short thick lines); this allows fragments to *hybridize* (by base-pairing) with complementary sequences in the bound DNA.

The fragments from strains A and B compete for sequences on the bound DNA. However, the concentration of B fragments is *very* much higher than that of A fragments; consequently, if strains A and B are very similar, few (if any) A fragments will hybridize. Conversely, many A fragments will hybridize if strains A and B are very different. Thus, the similarity of strains A and B is indicated by the number of A fragments which hybridize – this being determined by measuring the label after all unbound fragments have been washed away. Clearly, 'controls' are necessary; in one control, only labelled A fragments are used (no B fragments) – this giving a measure of the maximum amount of binding by A fragments.

Results are meaningful only when the fragments hybridize *stably*; stability is subsequently assessed by monitoring the dissociation of fragments from the bound DNA as the temperature is gradually increased.

DNA shows >70% stable hybridization would be classified in the same species.

Results from DNA–DNA hybridization (and from other sequence-based methods) can differ markedly from those of earlier, traditional classifications. For example, in a traditional classification of *Listeria*, the two species *L. grayi* and *L. murrayi* were grouped close to the type species, *L. monocytogenes*; however, according to DNA–DNA hybridization, these species are rather distant from *L. monocytogenes* – hybridization between *L. grayi* and *L. monocytogenes*, for example, being <25% [IJSB (1993) *43*

(b) 16S rRNA oligonucleotide cataloguing (section 16.2.2.2). In this method, 16S rRNA is cleaved by particular enzymes – e.g. RNase T1, which cleaves RNA specifically at Gp↓N (where Gp is guanosine 3'-monophosphate, and N is the next nucleotide); the resulting fragments form the 'catalogue'. Strains are compared by comparing their catalogues.

(c) Arbitrarily primed PCR (AP-PCR) (section 16.2.2.4). Copies of an arbitrarily chosen primer bind, at various locations, to each strand of (heat-denatured) chromosomal DNA under *low*-stringency conditions (section 8.5.4); the primers bind at 'best-fit' sequences, albeit with mismatches. In some cases, two primers will bind relatively efficiently, on opposite strands, at locations a few hundred bases apart; if strand elongation can occur efficiently from these primers, and if elongation is time-limited, the resulting PCR products will be two short single-stranded fragments of DNA.

In the diagram, the two strands of chromosomal DNA are shown as long parallel lines. On the right, two primers (short lines) have bound, close together, on opposite strands; on the left, another two primers have bound, further apart, on opposite strands. Strand elongation from each primer (dashed line) has produced the four fragments shown. (Note that a fragment synthesized on one strand contains a copy of the best-fit sequence of the *other* strand.) Another cycle of low-stringency PCR is used to produce more copies of the fragments.

Subsequently, many cycles of PCR are carried out under higher stringency, using copies of the same primer. Under these conditions, primers bind to best-fit sequences (rather than elsewhere) on the fragments formed by low-stringency PCR, though (due to the higher stringency) primers may not bind to best-fit sequences on *all* of the fragments – so that only a proportion of the fragments formed under low stringency may be amplified under higher stringency. The distance between each pair of arrowheads shows the length of the fragments amplified under higher stringency. On electrophoresis, the fragments from a given strain form a characteristic pattern of bands (the 'fingerprint') (see e.g. Plate 16.2).

(d) Repetitive sequence-based PCR (REP-PCR) (section 16.2.2.5). In this method, PCR (section 8.5.4) involves primers that bind to REP sequences. In the diagram, three REP sequences (R) are shown in one strand of chromosomal DNA. A primer (P) has bound to each REP sequence, and elongation has proceeded from left to right; an arrowhead marks the end of a newly formed fragment. Note that elongation from a given primer cannot continue beyond the next primer. (The fragments do not join together because the reaction mixture does not contain the type of enzyme (a ligase) which could make such a join.) After a number of cycles of PCR the different-sized molecules are separated by electrophoresis, yielding a fingerprint characteristic of the given chromosome.

(e) Restriction fragment length polymorphism (RFLP) (section 16.2.2.6). Horizontal lines represent related DNA duplexes. The top duplex has two sites for a given restriction endonuclease; enzymic cleavage (arrow) at each of these two sites produces three restriction fragments. In the centre duplex, the second cleavage site has been lost through mutation; enzymic cleavage produces only two fragments, fragment 1 being the same as before. In the lower duplex, a new short sequence of nucleotides (dashed line) has been inserted; enzymic cleavage of this duplex produces three fragments, but fragment 2 is longer than that in the top duplex. Electrophoresis of the fragments from each duplex will give different fingerprints.

26–31]. Why do such results differ? In traditional classification organisms are compared mainly on the basis of a limited number of *phenotypic* (observable, measurable) features such as enzymes, subcellular structure, motility and metabolic products. By contrast, DNA–DNA hybridization (which examines the *whole* chromosome) compares (i) sequences that encode the phenotypic features, *and* (ii) various sequences in the chromosome which are *not* expressed phenotypically by the cell (e.g. REP sequences – section 16.2.2.5). Thus, compared with traditional methods, hybridization compares organisms on a broader basis.

16.2.2.2 *16S rRNA: a record of species and kingdoms*

16S rRNA is useful for phyletic classification because (i) it occurs in all bacteria, and (ii) it contains highly conserved (stable) sequences of nucleotides as well as more variable sequences. Feature (ii) permits classification at both ends of the taxonomic spectrum; thus, divergence at the level of higher taxonomic ranks (e.g. kingdom, domain) may be evident when comparing the highly conserved sequences of different organisms, while divergence at the strain or species level may be seen when comparisons are made of the more variable regions.

16S rRNA was used early in the change from traditional to genotypic forms of taxonomy. In *16S rRNA cataloguing*, 16S rRNA from a given organism was cleaved, enzymatically, into small pieces (oligonucleotides) in each of which the nucleotide sequence was determined (Fig. 16.4b); the resulting 'catalogue' characterized the organism, and different organisms could be compared, and classified, on the basis of their catalogues. By this means, bacteria were originally divided into two kingdoms: Eubacteria and Archaebacteria (see Appendix). Subsequently it was possible to sequence the entire molecule of 16S rRNA, and this improved method revealed two main groups of archaebacteria: (i) sulphur-dependent archaebacteria (e.g. *Sulfolobus*), and (ii) methanogens, extreme halophiles, and the wall-less species *Thermoplasma acidophilum* [SAM (1985) 6 251–256]. Later, Eubacteria and Archaebacteria were each elevated to the taxonomic rank *domain* and re-named Bacteria and Archaea, respectively. Some groupings and species within these two domains are shown in Fig. 16.5. [Bacterial phylogeny based on 16S and 23S rRNA sequence analysis (review): FEMSMR (1994) *15* 155–173.]

As mentioned earlier, 16S rRNA analysis is also used at lower taxonomic levels. For example, in a given genus, 16S rRNAs of different species have been found generally to differ by at least 1.5%. An exception to this has been found in certain species of *Bacillus*; although clearly distinguishable by DNA–DNA hybridization, these species were shown to have virtually identical 16S rRNAs. A possible explanation is that these species have diverged recently (on an evolutionary time-scale) and their 16S rRNAs have had insufficient time to change [IJSB (1992) *42* 166–170].

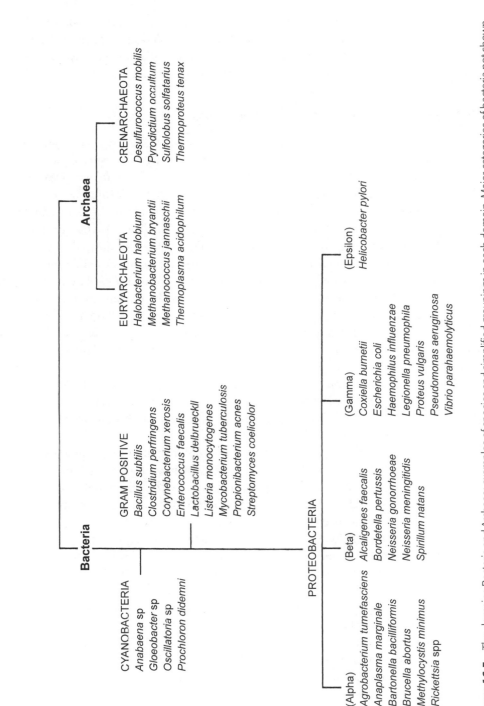

Figure 16.5 The domains Bacteria and Archaea: examples of species and simplified groupings in each domain. Major categories of bacteria not shown include the spirochaetes and the large *Cytophaga/Flexibacter/Bacteroides* group; moreover, the Gram-positive bacteria are divided into 'low GC%' and 'high GC%' groups.

Taxonomic studies often use the 16S rRNA *gene* (rather than 16S rRNA itself). Many copies of the gene are needed for analysis, and in some cases the gene has been amplified by PCR (section 8.5.4). PCR-amplified genes have been particularly useful when bacteria cannot be cultured by standard techniques.

16.2.2.3   *DNA fingerprinting; PFGE; ribotyping*

DNA fingerprinting (also called *restriction enzyme analysis* (REA) and *chromosomal fingerprinting*) is used e.g. for identifying/classifying at the species/strain level. In the original method, chromosomal DNA from the test strain was cleaved by certain restriction endonuclease(s) (Table 8.1) and the fragments (of different lengths) separated by gel electrophoresis; the fragments were denatured (i.e. made single-stranded) within the gel and then blotted (Fig. 8.13) and stained. The pattern of stained bands of fragments (the *fingerprint*) reflects e.g. the number and location of cutting sites, for the given enzyme(s), and is characteristic for a given organism; different organisms can therefore be classified/identified by their fingerprints. (*Note.* The term *fingerprinting* refers also to some newer methods that generate fingerprints in other ways.)

One problem with the original method is that it can yield too many fragments – thus giving a complex fingerprint that is difficult to interpret. One solution is to use a 'rare-cutting' enzyme (such as *Not*I: Table 8.1); the fewer, larger fragments can be separated by PFGE (section 8.5.1.4; see also 16.1.6.2). [Comparison of PFGE with five other molecular methods for typing *Staphylococcus aureus* (MRSA): JMM (1998) *47* 341–351.] One disadvantage of PFGE is that it is susceptible to endogenous nucleases (which can degrade target DNA) [see e.g. JCM (1999) *37* 2473–2478]. A further problem is that it takes 2–3 days.

Another solution to the problem of complex fingerprints is to use a labelled *probe* (section 8.5.3) that binds only to those (few) blotted fragments of the chromosome which contain the probe's target sequence; only *these* fragments are made visible (by the probe's label) so that the fingerprint consists of only a small number of bands. One such probe is labelled rRNA – which binds only to those fragments of the chromosome that contain the genes for rRNA; this method (using a labelled rRNA probe, or a labelled cDNA probe) is called *ribotyping* (Fig. 16.6a). Many bacteria (though not e.g. *Mycobacterium tuberculosis*) contain multiple copies of the rRNA genes, so that probe-binding sequences will occur on a small number of chromosomal fragments – thus giving a fingerprint consisting of a small number of bands. In one study, ribotyping revealed differences among strains of *Legionella pneumophila* serogroup 1, i.e. strains identical by routine serotyping.

[Manual and automated ribotyping (example): JCM (2003) *41* 27–33.]

Strains defined by ribotyping are called *ribotypes*.

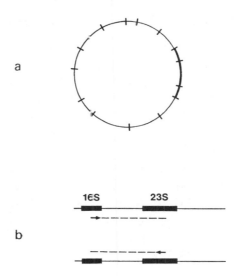

Figure 16.6 Ribotyping and PCR ribotyping (diagrammatic).
(a) Ribotyping (section 16.2.2.3). Chromosomes from bacteria of a given strain are cut into fragments by a restriction endonuclease; one such chromosome is shown with bars representing the locations of the cuts. The genes which encode rRNAs are shown as a thickened line on the right-hand side (relative length exaggerated for clarity); these genes have been cut into three fragments. All the fragments from the chromosomes are subjected to electrophoresis and blotting and are then exposed to labelled rRNA probes; three bands of DNA – corresponding to the three fragments referred to above – will bind the probe and will be detected by the probe's label. A radioactive label can be detected by *autoradiography* (for the basic idea see Fig. 8.10). Ribotyping thus reflects the distribution of cutting sites (of the given enzyme) among the rRNA genes of the strain examined. Bacteria of a different strain, with different cutting sites in their rRNA genes, would yield fragments of correspondingly different lengths – detected as bands in different relative positions on the autoradiograph.

(b) PCR ribotyping (section 16.2.2.7). In most or all bacteria the three types of rRNA (section 2.2.3) are encoded by genes which form an operon (section 7.8.1); in e.g. *E. coli*, the order of transcription is: 16S rRNA, 23S rRNA and 5S rRNA. In many bacteria (though not e.g. in *Mycobacterium tuberculosis*) the chromosome contains multiple copies of the rRNA operon. The diagram shows that part of an rRNA operon which encodes the 16S and 23S molecules; the two strands of the duplex are shown, separated, as thin parallel lines, and the *coding* regions of the two genes are shown as thick lines (▬▬▬▬). The nucleotide sequence between the coding regions is called the *spacer* region. Primers (small arrows) are designed such that they bind to certain highly conserved sequences within the 16S and 23S coding regions; the PCR products formed by elongation (dashed lines) will therefore include the spacer region. The length of the spacer can vary in different copies of the rRNA operon within the same chromosome; hence, after electrophoresis, more than one band may be obtained – though, for a given strain, this will be reproducible. In general, variation in the length of spacer DNA among different isolates can be exploited for typing purposes.

16.2.2.4   Methods using PCR with an arbitrary primer

Several very similar methods – commonly used for typing - involve PCR (section 8.5.4) with primers of arbitrary (random) sequence; they amplify random (but discrete) sequences of chromosomal DNA. Unlike basic PCR, which uses two types of specific primer, these methods use only one type of primer (as explained in Fig. 16.4c).

These methods have several advantages for typing: (i) the entire chromosome is potentially available for examination, and (ii) there is no requirement for prior knowledge of the genome – so that, potentially, any isolate can be typed. One disadvantage is that, although results are reproducible with a standardized technique in a given laboratory, comparable results will not necessarily be obtained by other laboratories unless *identical* procedures are used; comparability of fingerprints depends not only on the primer but also e.g. on the use of a specific polymerase [NAR (1993) *21* 4647–4648] and on the procedure used for preparing the sample DNA [NAR (1994) *22* 1921–1922].

AP-PCR (arbitrarily primed PCR). As a representative of these methods, AP-PCR is described in Fig. 16.4c and is compared with the other methods in Table 16.1. Results from an AP-PCR assay are shown in Plate 16.2.

DAF (direct amplification fingerprinting). This method differs from the other two e.g. in that it uses very short primers (Table 16.1).

RAPD (random amplified polymorphic DNA). This method closely resembles AP-PCR and is considered by some authors to be the same method; it differs e.g. in that somewhat shorter primers are used (Table 16.1). When a group of strains is compared by RAPD analysis it is usual for the comparison to be repeated with each of a range of different primers; different primers give different PCR products (because they bind to

Table 16.1   Typing by PCR with arbitrary primers: comparison of methods

Method	Primers (typical length, in nucleotides)	Temperature[1] (°C)	Electrophoresis (type of gel used)	Stain/bands[2]
AP-PCR[3]	20–50	40	Agarose	Ethidium bromide/few
RAPD[4]	10–20	36	Agarose	Ethidium bromide/few
DAF[5]	5–8	30	Polyacrylamide	Silver/many

[1] These are examples, only, of the types of temperature used for primer binding, though higher temperatures generally correlate with longer primers.
[2] Ethidium bromide fluoresces under ultraviolet radiation, and this effect is used to detect the bands of PCR products (which are stained with the dye). The bands formed in DAF tend to be very close together; fluorescence from such bands would tend to merge (obscuring the presence of some individual bands) so that silver staining is used instead.
[3] Arbitrarily primed PCR.
[4] Random amplified polymorphic DNA analysis.
[5] Direct amplification fingerprinting.

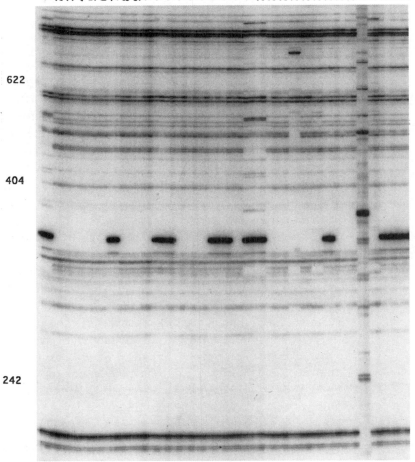

Plate 16.2 Arbitrarily primed PCR (AP-PCR) (section 16.2.2.4) used for comparing and classifying 33 strains of methicillin-resistant *Staphylococcus aureus* from patients in San Diego hospitals. For each strain, the short pieces of DNA (copied from the chromosome) have been separated by gel electrophoresis (moving from top to bottom in the photograph); the scale on the left gives the size of the pieces in terms of the number of bases they contain. (Note that, during electrophoresis, the smaller pieces of DNA have moved further.) One of the strains (29) is obviously different from the rest. Other strains can be grouped together according to shared sequences. Photograph courtesy of Dr Michael McClelland, California Institute of Biological Research, La Jolla, California, USA.

different 'best-fit' sequences in target DNA) so that this approach increases the chance of detecting differences between the strains.

16.2.2.5 Repetitive sequence-based PCR (rep-PCR)

In rep-PCR, PCR is used to generate fingerprints by copying *particular* sequences in the chromosome (rather than random sequences, as in section 16.2.2.4). The method can be used only with those bacteria in which the chromosome contains multiple copies of a specific sequence of nucleotides – such as the *REP sequence* in *Escherichia coli* and other enterobacteria. REP sequences occur in non-coding parts of the chromosome.

Other repetitive sequences include ERIC (= enterobacterial repetitive intergenic consensus sequence, also called intergenic repeat unit) and SERE (= *Salmonella enteritidis* repetitive element).

The abbreviation rep-PCR is used generically to include e.g. REP-PCR, ERIC-PCR and SERE-PCR.

In e.g. REP-PCR, primers are designed to bind to different variant forms of the REP sequence that occur in different strains. REP-PCR yields DNA products in a range of sizes (see Fig. 16.4d), the differing lengths apparently reflecting differences in the distance between consecutive REP sequences in a given chromosome; when separated by gel electrophoresis, and stained, these molecules give a characteristic fingerprint – so that different organisms can be compared and classified by their fingerprints.

rep-PCR has been used e.g. for typing strains of *Vibrio parahaemolyticus* [JCM (1999) *37* 2473–2478], viridans streptococci [JCM (1999) *37* 2772–2776], and *Listeria monocytogenes* [JCM (1999) *37* 103–109].

16.2.2.6 Restriction fragment length polymorphism (RFLP) and PCR-RFLP

Related sequences of nucleotides (e.g. variant forms of a given part of a chromosome) can be compared by exposing them to the same restriction endonuclease(s). Clearly, with given enzyme(s), identical sequences of nucleotides will yield identical fragments. However, different fragments would be formed if, for example, mutation in one sequence had destroyed (or created) a restriction (cutting) site, or if nucleotide(s) had been inserted or deleted (Fig. 16.4e). Electrophoresis, and staining, of fragments from a given sequence will yield a characteristic fingerprint, so that different sequences can be compared by comparing their fingerprints. (The fragments formed in this method are themselves often called RFLPs.)

RFLP analysis, as described above, is used e.g. for identification/classification at the species/strain level.

In conventional RFLP analysis one of the problems is to prepare a sufficient quantity of the particular sequence. Another potential problem is the effect of methylation (section 7.4) in the DNA isolated from cells: this may affect the activity of the restriction enzymes chosen for the analysis. PCR-RFLP analysis can overcome these problems. A specific target sequence (typically 1–2 kb long) is initially chosen and amplified by PCR. The amplified product (non-methylated, because synthesized *in vitro*) is then subjected to RFLP analysis as described above. PCR-RFLP has been used e.g. for analysing the 23S rRNA genes of *Campylobacter jejuni* [LAM (1996) *23* 163–166] and for typing isolates of *Staphylococcus aureus* from cows and sheep with mastitis [JCM (1999) *37* 570–574].

16.2.2.7　PCR ribotyping

Conventional ribotyping (section 16.2.2.3) is rather time-consuming. As an alternative, PCR can be used to look for differences between strains by detecting differences in a specific part of the rRNA-encoding region of the chromosome. This method, called PCR ribotyping, is described in Fig. 16.6b.

If the test strains are identical by this method it may (or may not) be possible to demonstrate differences between the strains by *subtyping* their PCR products. In one study, 40 isolates of *Clostridium difficile* gave identical PCR products (PCR ribotype 1); subtyping by RFLP analysis (section 16.2.2.6) revealed no difference between the strains, indicating a high degree of homogeneity [RMM (1997) *8* (suppl 1) S55–S56].

16.2.2.8　AFLP DNA fingerprinting

This method of typing involves (i) digesting the chromosome with two types of restriction endonuclease (RE), (ii) adding a short adaptor sequence to both ends of each fragment, and (iii) amplifying *certain* fragments by PCR (see Fig. 16.7).

The fragments are mixed with two types of adaptor molecule (A and B), each adaptor having *one* sticky end corresponding to the cutting site of one or other of the REs; ligation therefore produces the following kinds of sequence: A-fragment-A, A-fragment-B, B-fragment-A and B-fragment-B. Note that, in each *adaptor* molecule, the nucleotide immediately adjacent to the overhang is chosen such that, following fragment–adaptor ligation, the cutting site of the RE is *not* generated; this avoids the cycle ligation–restriction–ligation–restriction . . . etc.

PCR primers are designed to be complementary to the adaptor molecules (including the restriction site); importantly, the 3' end of each primer extends one (or a few) nucleotide(s) *beyond* the restriction site – so that a given fragment will be amplified only if these terminal 'selective' nucleotide(s) are aligned with *complementary* bases in the fragment. To ensure specificity, high-stringency conditions (section 8.5.4) are used in the earlier cycles of PCR.

[Review of AFLP fingerprinting: JCM (1999) *37* 3083–3091. Simplified protocol: BioTechniques (2000) *28* 622–623.]

(a)   ----NNNG-3'   5'-AATTGNNNN-3'
              ----NNNCTTAA-5'   CNNNN-5'

(b)   ----NNNGAATTGNNNN-3'
                   TCTTAACNNNN-5'

Figure 16.7   AFLP fingerprinting (section 16.2.2.8) (diagrammatic).
(a) On the left is shown one end of a fragment which has been cut by *Eco*RI (Table 8.1). ('N' is a nucleotide, i.e. A, T, C or G.) On the right is one of the two types of adaptor molecule; this molecule has a sticky end (5'-AATT) corresponding to the *Eco*RI restriction site.

(b) The fragment-end and the adaptor molecule, shown at (a), have been ligated, and strand separation has subsequently occurred during PCR. The upper strand has bound a primer (5'-NNNNCAATTC**T**). The 3'-end of the primer (**T**) can be extended by PCR only if it pairs with the complementary base – in this case adenine (A) – in the fragment strand. The same argument applies to each primer-binding site so that only certain primers will be extended. PCR products are examined by gel electrophoresis, the bands of products (forming the fingerprint) being made visible via the primer's label.

Appendix Minidescriptions of some genera, families, orders and other categories of bacteria

The following minidescriptions give the essential features of many of the bacteria mentioned in the text; they are intended simply for the rapid orientation of the reader. Further details of these and many other bacteria (and other microorganisms) can be found in *Dictionary of Microbiology and Molecular Biology* [Singleton & Sainsbury (3rd edition, 2001); John Wiley: Chichester, UK; ISBN 0 471 49064 4]; entries in the dictionary also include terms, tests, techniques, biochemical pathways, and topics in genetics, molecular biology, medicine and immunology.

The difference between 'Gram-positive' and 'Gram-type-positive' etc. is explained in section 14.9.1.

The GC% (section 16.2.1) gives the range for the genus or other category.

The *type species* is the species which is regarded as the 'permanent representative' of a given genus.

Acetobacter Genus. Gram-type-negative. Ovoid cells/rods, 0.6–0.8 × 1–4 μm. Non-motile or flagellate. Strictly aerobic. Opt. 25–30°C. Chemoorganoheterotrophic. Respiratory. Many strains can oxidize ethanol to acetic acid/CO_2; used in vinegar production. Sugars probably metabolized mainly via the HMP and TCA pathways. GC% 51–65. Type species: *A. aceti.*

Acinetobacter Genus. Gram-type-negative. Rods, 0.9–1.6 × 1.5–2.5 μm. Non-motile. Strictly aerobic. Opt. usually 33–35°C. Chemoorganoheterotrophic. Respiratory. Most strains can grow on minimal salts together with acetate, ethanol or lactate; few strains use glucose. Oxidase −ve. Found in soil, water; opportunist pathogens in man. [Taxonomy and epidemiology: RMM (1995) 6 186–195.] GC% ca. 33–47. Type species: *A. calcoaceticus.*

Actinobacillus Genus. Gram negative. Rods/cocci, ca. 0.3–0.5 × 0.6–1.4μm; non-motile. Facultatively anaerobic. Opt. ca. 37°C. Chemoorganoheterotrophic. Respiratory and fermentative.

Complex nutrients needed. No gas from the fermentation of glucose, lactose etc. Found in man, animals; can be pathogenic. GC% 40–43. Type species: *A. lignieresii*.

Actinomyces Genus. Gram-type-positive. Rods or filaments, often branched; non-motile. No spores. Typically anaerobic/microaerophilic. Opt. ca. 37°C. Chemoorganoheterotrophic. Primarily fermentative. Carbohydrates fermented anaerogenically. Found in warm-blooded animals, e.g. in the mouth; can be pathogenic. GC% ca. 57–73. Type species: *A. bovis*.

Actinomycetales Order. Gram-type-positive. Most genera aerobic. Cocci, rods, mycelium (depending on genus); typically non-motile. Spores in many genera. Found in soil, composts, water etc.; some species symbiotic in plants, some pathogenic in man, other animals or plants. Many genera, including *Actinomyces*, *Arthrobacter*, *Corynebacterium*, *Mycobacterium* and *Streptomyces*.

Aeromonas Genus. Gram negative. Rods or coccobacilli, 0.3–1.0 × 1–3.5 μm; singly, pairs, chains, filaments. Some species motile (usually monotrichous); psychrotrophic species non-motile. Facultatively anaerobic. Chemoorganoheterotrophic. Respiratory and fermentative. Sugars, organic acids used as carbon sources. Oxidase +ve. Found in marine and fresh waters; *A. salmonicida* (opt. 22–25°C) is parasitic/pathogenic in fish. There is evidence for a pathogenic role (e.g. diarrhoeal disease, wound infections) in man [RMM (1997) *8* 61–72]. GC% 57–63. Type species: *A. hydrophila*.

Agrobacterium Genus. Gram negative. Rods, 0.6–1 × 1.5–3 μm, capsulated; motile. Aerobic. Opt. 25–28°C. Chemoorganoheterotrophic. Respiratory. Glucose metabolized mainly via Entner–Doudoroff and HMP pathways (section 6.2). Found in soil; most strains can induce tumours in plants, pathogenicity being plasmid-encoded. GC% 57–63. Type species: *A. tumefaciens*.

Alcaligenes Genus (taxonomically unsettled). Gram negative. Rods, coccobacilli, cocci; motile. Aerobic. Chemoorganoheterotrophic, some strains chemolithotrophic. Respiratory. Acetate, lactate, amino acids etc. used as carbon sources. Oxidase +ve. Found in soil, water, vertebrates etc. GC% ca. 56–70. Type species: *A. faecalis*.

Alteromonas Genus. Gram negative. Rods, 0.7–1.5 × 1.8–3 μm, some pigmented (yellow, orange, violet etc.); monotrichously flagellate. Aerobic. Chemoorganoheterotrophic. Respiratory. Carbon sources include acetate, alcohols, amino acids, sugars. Found in marine waters. GC% ca. 38–50. Type species: *A. macleodii*.

Anabaena Genus. Gram-type-negative. Filamentous cyanobacteria (q.v.); trichome: spherical, ovoid or cylindrical cells. Gas vacuoles. Heterocysts. *A. flos-aquae* can form 'blooms' (section 10.1.1) and produce toxins (*anatoxins*).

Anaplasma Genus. Gram negative. Cocci, about 0.3 μm in diameter. The organisms grow within erythrocytes (red blood cells) in ruminants; they occur world-wide.

Aphanizomenon Genus. Gram-type-negative. Filamentous cyanobacteria (q.v.); trichomes: individual cells cylindrical, end cells are often tapering colourless 'hair cells'. Gas vacuoles. Heterocysts. May form 'blooms' (section 10.1.1) in fresh and brackish waters; some strains produce toxins.

Aquaspirillum Genus. Gram negative. Cells: typically helical, rigid, 0.2–1.4 × 2–30 μm (longer in some species). Motile. Aerobic/facultatively anaerobic. Chemoorganoheterotrophic/ chemolithoautotrophic. Respiratory. Carbon sources: e.g. amino acids, not usually carbohydrates. Some species (e.g. *A. peregrinum*) can fix nitrogen anaerobically. Typically oxidase +ve. Found in various freshwater habitats. GC% 49–66. Type species: *A. serpens*.

Archaea Domain. See section 1.1.1 and Fig. 16.5.

Archaebacteria Kingdom. Differ from Eubacteria (q.v.) (and from eukaryotic organisms) e.g. in nucleotide sequences in 16S rRNA (section 16.2.2.2), and in the chemistry of the cytoplasmic membrane and cell wall (which lacks peptidoglycan). Archaebacteria are often found in 'harsh' environments – many being e.g. extreme thermophiles or halophiles (sections 3.1.4 and 3.1.7). Genera include e.g. *Desulfurococcus, Halobacterium, Thermoproteus*. Now reclassified as the domain Archaea.

Azotobacter Genus. Gram negative. Rods/coccobacilli/filaments; motile or non-motile. Cysts (section 4.3.3). Aerobic. Chemoorganoheterotrophic. Carbon sources: e.g. sugars, ethanol. Can fix nitrogen (section 10.3.2.1). Most strains are oxidase +ve. Found e.g. in fertile soils of near-neutral pH. GC% 63–68. Type species: *A. chroococcum*.

Bacillus Genus. Gram-type-positive. Rods, often 0.5–1.5 × 2–6 μm, typically motile. Endospores (section 4.3.1). Aerobic or facultatively anaerobic, depending on species. Respiratory or facultatively fermentative. Most species chemoorganoheterotrophic; many can grow on nutrient agar. Some species (e.g. *B. polymyxa*) can fix nitrogen (section 10.3.2.1). *B. schlegelii* can grow chemolithoautotrophically. Found e.g. as saprotrophs in soil and water; some species cause disease in man and other animals (including some insects). GC% 30–70. Type species: *B. subtilis*.

Bacteroides Genus. Gram negative. Rods or filaments (some pigmented), non-motile or motile. Anaerobic. Characteristically fermentative; some strains can carry out anaerobic respiration. Most species use sugars; others use peptones. Found e.g. in the alimentary tract in warm-blooded animals; some species are opportunist pathogens. GC% 28–61. Type species: *B. fragilis*. (As indicated e.g. by the GC% range, the genus is heterogeneous; some believe that it should contain only *B. fragilis* and closely-related organisms [IJSB (1989) *39* 85–87].)

Bartonella Genus. Gram-negative, oxidase-negative. Rods. *B. bacilliformis* is the causal agent of Oroya fever (section 11.3.3.3); *B. clarridgeiae* has been linked to cat-scratch disease [JCM (1997) *35* 1813–1818]. The genus now includes bacteria formerly classified in the genus *Rochalimaea*. The organisms can be grown e.g. on agar media enriched with sheep blood in the presence of 5% CO_2. [DNA-based classification: JCM (2002) *40* 3641–3647.]

Bdellovibrio Genus. Gram negative. Cells: vibrioid, 0.2–0.5 × 0.5–1.4 μm, each with one sheathed flagellum. Aerobic. Respiratory. Chemoorganoheterotrophic. Predatory: grow within the periplasmic space, and digest other bacteria (e.g. *Aquaspirillum serpens, Escherichia coli, Pseudomonas* spp). Found e.g. in soil and sewage. GC% 33–51. Type species: *B. bacteriovorus*.

Beggiatoa Genus. Gram negative. Trichomes. Aerobic/microaerophilic/anaerobic. Respiratory.

Typically chemoorganoheterotrophic; carbon sources: e.g. acetate, but hexoses (e.g. glucose) are not used. Found in various aquatic habitats.

Beijerinckia Genus. Gram negative. Rods, typically $0.5–1.5 \times 1.7–4.5$ μm, motile or non-motile. Aerobic. Respiratory. Chemoorganoheterotrophic; sugars (e.g. glucose) are used as carbon sources. Can fix nitrogen (section 10.3.2.1). Opt. 20–30°C; no growth at 37°C. Found e.g. in soil and on leaf surfaces. GC% 55–61. Type species: *B. indica*.

Bordetella Genus. Gram negative. Coccobacilli, approx. $0.2–0.5 \times 0.5–2$ μm, non-motile or motile. Aerobic. Respiratory. Carbon sources: e.g. amino acids; sugars not used. Enriched media are needed for culture. Found e.g. as parasites/pathogens of the mammalian respiratory tract. GC% 66–70. Type species: *B. pertussis*.

Borrelia Genus (order Spirochaetales – q.v. for basic details). Cells: about $0.2–0.5 \times 3–20$ μm. Anaerobic/microaerophilic. Some species can be grown in complex media. Found as parasites/pathogens in man and other animals. Type species: *B. anserina*.

Brucella Genus. Gram negative. Rods, coccobacilli or coccoid cells, about $0.5–0.7 \times 0.6–1.5$ μm; non-motile. Aerobic. Respiratory. Chemoorganoheterotrophic. Complex, enriched media are needed for culture. Most strains are oxidase +ve; typically urease +ve. Found, typically, as intracellular parasites/pathogens of animals, including man. GC% 55–58. Type species: *B. melitensis*.

Burkholderia Genus. Gram negative. Rods. Aerobic. *B. cepacia* (formerly *Pseudomonas cepacia*) can cause disease in plants and also in man. Type species: *B. cepacia*. [*B. cepacia* (medical, taxonomic and ecological issues): JMM (1996) *45* 395–407.]

Campylobacter Genus. Gram negative. Cells: spiral, typically $0.2–0.5 \times 0.5–5$ μm (see Plate 12.1); motile, with a single unsheathed flagellum at one or both poles. Microaerophilic, needing 3–5% CO_2 for growth. Respiratory. Chemoorganoheterotrophic; carbon sources: amino acids or TCA cycle intermediates (Fig. 5.10) but not carbohydrates. Oxidase +ve. Found e.g. in the reproductive and intestinal tracts in man and other animals, [*Campylobacter* spp (genotyping): AEM (2000) 66 1–9.] GC% 30–38. Type species: *C. fetus*. (*C. pylori* was transferred to the genus *Helicobacter* [IJSB (1989) *39* 397–405].)

Caulobacter Genus. Gram negative. Complex cell cycle (section 4.1). Some strains are pigmented. Aerobic. Respiratory. Chemoorganoheterotrophic. Found in certain soils and waters. GC% 64–67. Type species: *C. vibrioides*.

Cellulomonas Genus. Gram-type-positive. Rods/filaments/coccoid cells, non-motile or motile. Aerobic/facultatively anaerobic. Respiratory and fermentative. Chemoorganoheterotrophic; starch and cellulose are attacked. Found e.g. in soil. GC% 71–77. Type species: *C. flavigena*.

Chlamydia Genus. Gram negative. Cells: vary within development cycle, but are non-motile, coccoid, pleomorphic, 0.2–1.5 μm in diameter, with an outer membrane; little/no peptidoglycan [see AAC (1999) *43* 2339–2344]. Obligate intracellular parasites/pathogens in man (e.g. trachoma, genitourinary disease) and other animals. Cultured in e.g. chick embryos and cell cultures. Infective form: the eosinophilic, non-dividing *elementary body* (EB)which may contain a cytotoxin [PNAS (2001) *98* 13984–13989]. After endocytosis, the phagosome does not fuse with lysosomes; EBs convert to basophilic *reticulate bodies* (RBs) which divide within a membrane-bounded *inclusion*. After 2–3 days the RBs have formed EBs, completing the cycle. (EBs and RBs are shown in Plate 11.5.) GC% 41–44. Type species: *C. trachomatis.*

Chlorobium Genus. Gram negative. Rods or vibrios, about 1–2 μm long, non-motile. Obligately anaerobic. Primarily photolithoautotrophic (electron donors: e.g. sulphide). Chlorophyll occurs in *chlorosomes* (section 2.2.7). Found e.g. in sulphide-rich mud.

Clostridium Genus. Gram-type-positive. Cells: typically rods, about 0.3–1.9 × 2–10 μm, motile or non-motile. Endospores (section 4.3.1). Obligately anaerobic (or, in a few cases, aerotolerant). Chemoorganoheterotrophic. Typically fermentative, though some strains (of e.g. *C. perfringens*) can carry out nitrate respiration (section 5.1.1.2). Growth often poor in/on basal media. Found e.g. in soil and in the intestines of man and other animals; some species pathogenic. GC% 22–55. Type species: *C. butyricum.* (According to a study of 16S rRNA gene sequences, the genus *Clostridium* is very heterogeneous, and it was proposed that some of the species be placed in five newly created genera: *Caloramator, Filifactor, Moorella, Oxobacter* and *Oxalophagus* [IJSB (1994) *44* 812–826].)

coliform In general: any Gram-negative, non-sporing, facultatively anaerobic bacillus which can ferment lactose within 48 hours with the formation of acid and gas at 37°C. For water bacteriologists (in the UK): any member of the Enterobacteriaceae which grows at 37°C and which normally possesses the enzyme β-galactosidase [see Report 71 (1994) HMSO, London; ISBN 0 11 753010 7]. *Escherichia coli* is a typical coliform. (See also Table 13.2.)

Corynebacterium Genus. Gram positive. Rods, often curved/pleomorphic, non-motile. Facultatively anaerobic. Chemoorganoheterotrophic. Respiratory and fermentative. Found e.g. in soil and vegetable matter; some species parasitic/pathogenic in man and other animals. GC% 51–59. Type species: *C. diphtheriae.*

Coxiella Genus. Gram negative. Rods (highly pleomorphic), 0.2–0.4 × 0.4–1 μm, non-motile. Endospores (section 4.3.1). The sole species, *C. burnetii*, is an obligate intracellular parasite/pathogen in vertebrates and arthropods; it undergoes a developmental cycle. GC% about 43.

cyanobacteria ('blue-green algae') Non-taxonomic category. Photosynthetic bacteria which differ from most other phototrophic prokaryotes (i) in having chlorophyll *a*, and (ii) in carrying out *oxygenic* photosynthesis (section 5.2.1.1) (cf. *Prochloron*). Cells: Gram-type-negative, single or e.g. in trichomes (according to species). No flagella; some exhibit gliding motility (section 2.2.15.1). Some have gas vacuoles (section 2.2.5). Depending on e.g. pigments, cells may appear blue-green, yellowish, red, purple or almost black etc. Some species form akinetes, heterocysts and/or hormogonia (section 4.4).

Typically photolithoautotrophic, fixing CO_2 via the Calvin cycle (section 6.1.1), and respiratory, using oxygen as terminal electron acceptor. Some can grow as chemoorganoheterotrophs, carrying out e.g. anaerobic respiration or even fermentation. Some can carry out facultative *anoxygenic* photosynthesis, using photosystem I (Fig. 5.11) with e.g. sulphide as electron donor.

Found in a wide range of aquatic and terrestrial habitats; some form 'blooms' (section 10.1.1). (See also section 13.5.1.2.)

Desulfomonas Genus. Gram negative. Rods, non-motile. Anaerobic. Respiratory: sulphate respiration (section 5.1.1.2) using e.g. pyruvate as electron donor. Found e.g. in the human intestine. GC% 66–67. Type species: *D. pigra*.

Desulfurococcus Genus. Archaea. Cocci, about 1 μm diameter, motile or non-motile. Anaerobic, thermophilic, chemolithoautotrophic and/or chemolithoheterotrophic. Respiratory (sulphur respiration: section 5.1.1.2). Found e.g. in Icelandic solfataras. GC% 51.

Dichelobacter Genus. Gram negative. Rods. Anaerobic. *D. nodosus* (formerly *Bacteroides nodosus*) is a causal agent of foot rot in sheep.

Enterobacteriaceae Family. Gram-negative, non-sporing, facultatively anaerobic bacilli, typically 0.3–1 × 1–6 μm; motile (most peritrichously flagellate) or non-motile. Cells occur singly or in pairs. Chemoorganoheterotrophic; typically grow well in/on basal media (section 14.2.1). Carbon sources include sugars. Respiratory *and* fermentative. Oxidase –ve. All except a few strains are catalase +ve. Found e.g. as parasites, pathogens or commensals in man and other animals, and as saprotrophs in soil and water.

Genera (and species) are differentiated e.g. by biochemical tests such as IMViC tests (section 16.1.2.5), the urease test (section 16.1.2.7), and the decarboxylase tests (section 16.1.2.8). Typically, lactose is fermented by e.g. *E. coli*, *Klebsiella pneumoniae* and some strains of *Citrobacter*, but not by *Salmonella*, *Shigella*, *Proteus* or *Yersinia*; a strain which normally does not ferment lactose may do so if it acquires a Lac plasmid (which encodes the uptake and metabolism of lactose). Genera include e.g. *Citrobacter*, *Enterobacter*, *Erwinia*, *Escherichia*, *Klebsiella*, *Proteus*, *Salmonella*, *Serratia*, *Shigella* and *Yersinia*.

Enterococcus Genus. See notes under *Streptococcus*.

Erwinia Genus (family Enterobacteriaceae – q.v. for basic details). Saprotrophic, or pathogenic in plants and animals. Typically motile. Acid (little/no gas) from sugars. Opt. 27–30°C. GC% 50–58. Type species: *E. amylovora*.

Escherichia Genus (family Enterobacteriaceae – q.v. for basic details). The following refers to *E. coli*. Cells: single or in pairs, typically motile (peritrichously flagellate) and fimbriate (section 2.2.14.2); see also items on Plates 2.1, 2.3 and 8.1. Opt. 37°C. Respiratory under aerobic conditions; fermentation (section 5.1.1.1) or e.g. nitrate respiration (section 5.1.1.2) carried out anaerobically. Glucose is fermented (usually with gas) via the mixed acid fermentation (Fig. 5.5). *Typical* reactions as follows. IMViC tests (section 16.1.2.5): +, +, –, –; citrate +ve in strains containing the Cit plasmid; urease –ve; H$_2$S –ve; lactose +ve (acid and gas). Found e.g. as part of the normal microflora of the intestine in man and other animals; some strains can be pathogenic (Table 11.2). GC% 48–52. Type species: *E. coli*. Proposed new diarrhoeagenic species: *E. albertii* [IJSEM (2003) *53* 807–810].

Eubacteria Kingdom. Includes prokaryotes not classified in the Archaebacteria (q.v.) – e.g. all the cyanobacteria and the anoxygenic photosynthetic bacteria, all enterobacteria and pseudomonads, Gram-positive bacteria and the mycoplasmas. Eubacteria differ from members of the Archaebacteria e.g. in their 16S rRNA (section 16.2.2.2) and in the chemistry of their cytoplasmic membrane and cell wall. The medically important prokaryotes – and those species most likely to be encountered in an introductory course in bacteriology – are eubacteria. Now re-classified as the domain Bacteria (see section 1.1.1).

Francisella Genus. Gram negative. Cocci. coccobacilli or rods (depending on species and conditions), non-motile. Aerobic. Chemoorganoheterotrophic; carbohydrates metabolized slowly, without gas. Opt. 37°C. Oxidase −ve. Catalase weakly +ve. Found as parasites/pathogens of man and other animals. GC% 33–36. Type species: *F. tularensis* (formerly *Pasteurella tularensis*). (See also section 11.11.)

Gardnerella Genus. Gram-type-negative (?). Rods (pleomorphic), about 0.5 × 1.5–2.5 μm. Obligately anaerobic and facultatively anaerobic strains. Chemoorganoheterotrophic; growth occurs only on enriched media. Oxidase −ve. Catalase −ve. Opt. 35–37°C. Found in the human genital/urinary tract; (see Bacterial vaginosis: section 11.11). GC% about 42–44. Type species: *G. vaginalis* (formerly *Haemophilus vaginalis*).

Haemophilus Genus. Gram negative. Rods/coccobacilli (pleomorphic), often about 0.4 × 1–2 μm, or filaments; non-motile. Facultatively anaerobic. Respiratory and fermentative. Chemoorganoheterotrophic; growth occurs on enriched media, e.g. chocolate agar (section 14.2.1). Typically, glucose, but not lactose, is fermented. Aerobic growth in *H. influenzae* needs X factor (haemin) and V factor (NAD) – both found in lysed RBCs. Opt. 35–37°C. Found as parasites/pathogens in man and other animals. GC% 37–44. Type species: *H. influenzae*.

Hafnia Genus (family Enterobacteriaceae – q.v. for basic details). Often motile at 25°C, frequently not at 35°C. Typically MR −ve, VP +ve at 22–25°C. Usually lactose −ve, citrate −ve. Lysine and ornithine decarboxylase +ve. Grows on DCA and in KCN media. Found in man, animals, soil and water. Associated with diarrhoea [JCM (1994) *32* 2335–2337] (may cause EPEC-like attaching-effacing lesions: see Table 11.2) and also associated with e.g. liver abscesses, septicaemia, peritonitis and pneumonia [*H. alvei* from human extra-intestinal specimens: JMM (2001) *50* 208–214]. Type species: *H. alvei*.

Halobacterium Genus. Domain Archaea. Rods or filaments, motile or non-motile. Gas vacuoles (section 2.2.5) common. Extremely halophilic. Facultatively anaerobic. Aerobic metabolism: chemoorganoheterotrophic and respiratory, with e.g. amino acids or carbohydrates as carbon sources. Oxidase +ve. Some strains obtain energy from a purple membrane (section 5.2.2). Found e.g. in evapourated brines, salted fish etc. GC% 66–68. Type species: *H. salinarium* (formerly *H. halobium*).

Helicobacter Genus. Gram-negative. Cells helical, motile with several sheathed flagella (section 2.2.14.1; Plate 2.1, *top*, *left*). Several species. *H. pylori* [IJSB (1989) *39* 397–405] is microaerophilic, chemoorganoheterotrophic, urease +ve; pathogenic: associated with e.g. gastric cancer [Nature (2000) *404* 398–402] and peptic ulcer disease. *H. pylori* has also been detected in liver tissue in cases of e.g. primary sclerosing cholangitis [JCM (2000) *38* 1072–1076]. [Genome and main features of *H. pylori*: Nature (1997) *388* 539–547. Diagnosis and treatment of *H. pylori* infection: Lancet (1997) *349* 265–269.]

Klebsiella Genus (family Enterobacteriaceae – q.v. for basic details). Cells: single, pairs, short chains; capsulated. Non-motile. Often MR −ve, VP +ve. Found e.g. in soil, water, and as parasites/pathogens in man and other animals. GC% 53–58. Type species: *K. pneumoniae*.

Kurthia Genus. Gram positive. Rods or filaments; rods are peritrichously flagellate. Aerobic.

Respiratory. Chemoorganoheterotrophic; amino acids, alcohols, fatty acids used as carbon sources. Found e.g. on meat and meat products. GC% about 36–38. Type species: *K. zopfii.*

Lactobacillus Genus. Gram positive. Rods or coccobacilli, singly or in chains; typically non-motile. Anaerobic, microaerophilic or facultatively aerobic; usually catalase −ve. Chemo-organoheterotrophic, using e.g. sugars as carbon sources. Characteristically fermentative, lactic acid being formed from glucose by homolactic fermentation (section 5.1.1.1) or by a heterolactic fermentation in which mixed products, including lactic acid, are formed. Found e.g. on vegetation, as part of the natural microflora in man, and in various fermented food products. GC% about 32–53. Type species: *L. delbrueckii.*

Lactococcus Genus. See notes under *Streptococcus.*

Legionella Genus. Gram negative. Rods/filaments, $0.3–0.9 \times 2– >20$ μm; motile. Aerobic. Chemoorganoheterotrophic, using amino acids (non-fermentatively) for carbon and energy; growth occurs e.g. on blood agar supplemented with L-cysteine and iron. Catalase +ve. Oxidase −ve, or weakly +ve. Urease −ve. Gelatinase +ve. Opt. 35–37°C. Found e.g. in various aquatic habitats (thermally polluted streams etc.); most/all species pathogenic for man (e.g. Legionnaires' disease: section 11.11). Sequencing of the *mip* gene is reported to distinguish between 39 species of *Legionella* [MM (1997) *25* 1149–1158; JCM (1998) *36* 1560–1567]. GC% 39–43. Type species: *L. pneumophila.*

Leuconostoc Genus. Gram positive. Cells: coccoid, about 1 μm in diameter, in pairs or chains; non-motile. Facultatively anaerobic. Fermentative and respiratory; anaerobically, glucose is fermented mainly to lactic acid, ethanol and CO_2. Found e.g. in various dairy products and fermented drinks. GC% about 38–44. Type species: *L mesenteroides.*

Listeria Genus. Gram positive. Rods or coccobacilli, about $0.5 \times 0.5–2$ μm; usually motile at 25°C, apparently always non-motile at 37°C. Aerobic, facultatively anaerobic. Chemo-organoheterotrophic. Catalase +ve. Oxidase −ve. Urease −ve. Sugars are fermented (acid, no gas). Aesculin is hydrolysed. Growth occurs in up to 10–12% sodium chloride. Found in soil, decaying vegetation, certain foods, and as pathogens in man and other animals. [Pathogenicity of *L. monocytogenes* (a public health perspective): RMM (1997) *8* 1–14. Typing of *L. monocytogenes* by rep-PCR: JCM (1999) *37* 103–109.] GC% 38. Type species: *L. monocytogenes.*

methanogens Non-taxonomic category; includes all those organisms which can produce methane. All methanogens are obligately anaerobic members of the domain Archaea; they occur e.g. in river mud and in the rumen of cows and other ruminants. The names of some genera are mentioned in section 5.1.2.2 and in Fig. 16.5.

Methylococcus Genus. Gram negative. Cocci, about 1 μm in diameter; non-motile. Aerobic/microaerophilic. Obligately methylotrophic (section 6.4); methane can be used as sole source of carbon and energy. Found e.g. in mud, soil. GC% about 63. Type species: *M. capsulatus.*

Mobiluncus Genus. Gram-variable. Rods, motile. Anaerobic. Associated with bacterial vaginosis (section 11.11). GC% 49–52%. Type species: *M. mulieris.*

Mycobacterium Genus. Gram positive. Rods, $0.2–0.8 \times 1–10$ μm, coccoid forms, branched rods or fragile filaments; some strains pigmented. Non-motile. Acid-fast (section 14.9.2) during at least some stage of growth. Aerobic or microaerophilic. Respiratory. Typically chemoorganoheterotrophic, though some strains may be chemolithotrophic; typically not nutritionally fastidious, though growth in at least some can be stimulated e.g. by serum or egg-yolk. Found e.g. as free-living saprotrophs in soil and water, or on plants, and as parasites/pathogens of man and other animals. *M. tuberculosis* [genome sequence: Nature (1998) *393* 537–544] is a causal agent of tuberculosis (section 11.11). [Pathogenesis of *M. tuberculosis* (review): Cell (2001) *104* 477–485.] GC% ~62–70. Type species: *M. tuberculosis*.

Mycoplasma Genus. Cells: pleomorphic, ranging from coccoid (about 0.3–0.8 μm in diameter) to branched filamentous forms; some capable of gliding motility (section 2.2.15.1). No cell wall. Facultatively or obligately anaerobic. Chemoorganoheterotrophic; growth occurs on complex media, and all species need cholesterol or related sterols. Catalase −ve. Found as parasites/pathogens e.g. in the respiratory and urogenital tracts in man and other animals. [Attachment organelle and cytoadherence proteins of *M. pneumoniae*: JB (2001) *183* 1621–1630.] GC% ~23–40. Type species: *M. mycoides*.

nanobacteria Minute bacteria, classified in the α-2 subgroup of Proteobacteria; they are generally about 0.2–0.5 μm, but smaller cells have been seen under the electron microscope (accounting for the ability of these organisms to pass through membrane filters of pore size 0.1 μm). Nanobacteria are found e.g. in humans and bovines. The organisms form a cell-surface carbonate apatite deposit at pH ~7, and it has been suggested that they may contribute to the development of kidney stones [PNAS (1998) *95* 8274–8279].

Neisseria Genus. Gram-type-negative. Typically cocci, 0.6–1 μm in diameter; non-motile. Aerobic. Chemoorganoheterotrophic; some species need enriched media (e.g. chocolate agar). Oxidase +ve. Found e.g. as parasites/pathogens of man and other animals. GC% about 46–54. Type species: *N. gonorrhoeae*.

Nitrobacter Genus. Gram negative. Rods, about 0.6–0.8 × 1–2 μm; usually non-motile. Reproduce by budding (section 3.2.2). Obligately aerobic. Some strains obligately chemolithoautotrophic (nitrifying bacteria: section 5.1.2. Fig. 10.2), others facultatively chemoorganoheterotrophic. Opt. 25–30°C. Found e.g. in soil. GC% about 61. Type species: *N. winogradskyi*.

Nitrosococcus Genus. Gram negative. Cocci, about 1.5 μm in diameter; motile or non-motile. Obligately aerobic. Obligate chemolithoautotrophs, oxidizing ammonia to nitrite (section 5.1.2; Fig. 10.2). Found e.g. in soil.

Nostoc Genus. Gram-type-negative. Filamentous cyanobacteria (q.v.). Heterocysts. Gas vacuoles in at least some species. Free-living, and in symbiotic associations with various eukaryotes – e.g. cycads (see section 10.2.4.1), lichens and liverworts.

Oscillatoria Genus. Gram negative. Filamentous cyanobacteria (q.v.); trichomes: motile, composed of flattened, disc-shaped cells. Gas vacuoles. Hormogonia. Found in various aquatic and terrestrial habitats. GC% 40–50.

Pasteurella Genus. Gram negative. Cells coccoid, rod-shaped/pleomorphic, about 0.3–1 × 1–2 μm, occurring singly, in pairs or short chains. Non-motile. Facultatively anaerobic. Chemoorganoheterotrophic. Opt. growth temperature: 37°C. Catalase +ve. Generally oxidase +ve. Found as parasites/pathogens in man and other animals. GC% 40–45. Type species: *P. multocida.*

Pelodictyon Genus. Gram negative. Rods/coccoid forms which may form chains/three-dimensional networks; non-motile. Gas vacuoles. Anaerobic. Phototrophic. Found e.g. in sulphide-rich mud.

Prochloron Genus. Gram negative. Cells contain chlorophylls *a* and *b* and carry out oxygenic photosynthesis. Taxonomy uncertain: prochlorophytes, which include e.g. *P. didemni* (found on warm-water sea-squirts), resemble cyanobacteria in having chlorophyll *a*, but differ in also having chlorophyll *b* and in lacking certain typical cyanobacterial pigments. Possibly related to ancestral chloroplasts. [Photosynthetic machinery in prochlorophytes: FEMSMR (1994) *13* 393–414.]

Propionibacterium Genus. Gram-positive. Pleomorphic branched/unbranched rods or coccoid forms; non-motile. No spores. Anaerobic. Chemoorganoheterotrophic. Fermentative: hexoses (e.g. glucose) or lactate fermented mainly to propionic acid. Growth occurs e.g. on yeast extract–lactate–peptone media. Found e.g. in dairy products and in the body's microflora (Table 11.1). Can be pathogenic [*P. acnes* (pathogenic potential in man): RMM (1994) *5* 163–173; *P. acnes* (in brain abscess, case report): JINF (1997) *34* 269–271]. GC% about 57–67. Type species: *P. freudenreichii.*

Proteus Genus (family Enterobacteriaceae – q.v. for basic details). Motile, often swarming (section 4.2). Typically H_2S +ve, urease +ve. Growth requires nicotinic acid. Found e.g. in soil, polluted waters, and the mammalian intestine; some species (e.g. *P. mirabilis*) can be pathogenic. GC% 38–41. Type species: *P. vulgaris.*

Pseudomonas Genus. Gram negative. Rods, 0.5–1 × 1.5–5 μm; most species have one/several unsheathed, typically polar flagella per cell, though *P. mallei* is non-motile (i.e. it lacks flagella), and some species have sheathed flagella (section 2.2.14.1). Aerobic or facultatively anaerobic. Respiratory; many species can carry out nitrate respiration (section 5.1.1.2). Typically chemoorganoheterotrophic and nutritionally highly versatile; many strains will grow on inorganic salts with an organic carbon source, while some can grow chemolithoautotrophically. Catalase +ve. Commonly oxidase +ve. Found e.g. in soil and water, and as pathogens in man, other animals, and plants. GC% 58–70. Type species: *P. aeruginosa.*

Pyrodictium Genus. Domain Archaea. The organisms grow as a network of filaments associated with 'discs', each disc being 0.3–2.5 μm in diameter. Anaerobic. Energy obtained by sulphur-dependent metabolism. Chemolithoautotrophic. Thermophilic (section 3.1.4). Halotolerant. Found in an underwater volcanic region.

Rhizobium Genus. Gram negative. Rods, 0.5–0.9 × 1.2–3 μm; motile. Aerobic. Chemoorganoheterotrophic; carbon sources include sugars. Found e.g. in soil and in root nodules (section 10.2.4.1). GC% 59–64. Type species: *R. leguminosarum.*

Rickettsia Genus. Gram negative. Rods, 0.3–0.6 × 0.8–2 μm; non-motile. Obligate intracellular parasites/pathogens in vertebrates (including man) and arthropods (ticks, mites etc.). Apparently respiratory, with glutamate as the main energy subtrate; glucose is not used. Opt. 32–35°C. GC% about 29–33. Type species: *R. prowazekii*.

Ruminococcus Genus. Gram-type-positive. Cocci, about 1 μm in diameter, in pairs or chains. Anaerobic. Chemoorganoheterotrophic; typically heterofermentative, forming e.g. acetic and formic acids from carbohydrates. Many strains can use cellulose. Found in the rumen; organisms resembling *Ruminococcus obeum* have been found in (human) faeces [AEM (2002) *68* 4225–4232]. GC% 40–45. Type species *R. flavefaciens*.

Salmonella Genus (family Enterobacteriaceae – q.v. for basic details). Typically motile. *Typical* reactions as follows. IMViC tests (section 16.1.2.5): −, +, −, +; glucose (acid and gas at 37°C) +ve; lactose usually −ve (but the ability to ferment lactose can be plasmid-encoded); H_2S +ve; urease −ve; lysine and ornithine decarboxylases (section 16.1.2.8) +ve. Salmonellae can grow on basal media and e.g. on MacConkey's agar and DCA (Table 14.1); enrichment media include e.g. selenite broth (Table 14.1). Found e.g. as pathogens in man and other animals. GC% 50–52. Type species: *S. choleraesuis*.

Special note. The genus *Salmonella* was named after the American bacteriologist D. E. Salmon. The correct pronunciation of the genus is accordingly 'Salmon-ella'.

Unlike most bacteria, the salmonellae are often identified and named as *serotypes* (section 16.1.5.1) rather than as species. In the *Kauffmann–White classification scheme* there are about 2000 named serotypes; each serotype is defined by its O and H antigens (section 16.1.5.1) and, in some serotypes, by the *Vi antigen*: a polysaccharide antigen in a *microcapsule* (section 2.2.11) associated with virulence for particular host(s). Each serotype is given an *antigenic formula* which lists, in order, the organism's O, Vi (if present) and H antigens; in many serotypes the H antigens can switch, owing to *phase variation* (see Fig. 8.3c), so that the formula of such a serotype includes two alternative H antigens (or two alternative *sets* of H antigens). For example, the antigenic formula of *S. typhimurium* is 1,4,[5],12:i:1,2. This means: O antigens 1, 4, 5 ([ ] indicates variable presence) and 12, phase 1 H antigen 'i', and phase 2 H antigens 1 and 2; O antigen 1 is underlined to show that that antigen is present as a result of *phage conversion* (section 9.4).

It was suggested that all serotypes of *Salmonella* be considered as members of a single species, *Salmonella enterica*. (The name of the type species, *S. choleraesuis*, was not chosen because this was considered likely to cause confusion.) Subspecies of the proposed species *S. enterica* are listed in the table below [see JMM (1992) *37* 361–363].

Salmonella enterica: subspecies

Subspecies name	Subspecies number
enterica	I
salamae	II
arizonae	III(a)
diarizonae	III(b)
houtenae	IV
bongori	V
indica	VI

Subspecies I contains most of the salmonellae that are pathogenic in man and warm-blooded animals.

In this proposal, the serotypes in subspecies I are referred to by *name* – e.g. *S. enterica* subsp *enterica* serotype Typhimurium. A less cumbersome form would be e.g. *Salmonella* serotype Typhimurium, *Salmonella* Typhimurium, or simply Typhimurium [see JMM (1992) *37* 361–363].

For all other serotypes (subspecies II–VI), it has been suggested that individual serotypes be designated by a combination of (i) subspecies *number* and (ii) antigenic formula.

Some have accepted this scheme; others have not.

Serratia Genus (family Enterobacteriaceae – q.v. for basic details). Usually motile. Some strains form a red pigment, *prodigiosin*. *Typical* reactions: MR –ve; VP +ve (at 30°C, but may be –ve at 37°C); citrate +ve; lactose +ve or –ve, according to species. Glucose is fermented by the Entner–Doudoroff pathway (Fig. 6.2). Found e.g. in soil and water, on plants, and in man and other animals. GC% 52–60. Type species: *S. marcescens*.

Shigella Genus (family Enterobacteriaceae – q.v. for basic details). Non-motile. Sugars fermented usually without formation of gas. MR +ve; VP –ve; citrate –ve; hydrogen sulphide –ve; lysine decarboxylase –ve. Found e.g. as intestinal pathogens of man and other primates (see sections 11.2.2, 11.3.1.6 and 11.3.3.2). GC% 49–53. Type species: *S. dysenteriae*.

Simonsiella Genus. Gram negative. Flat, multicellular filaments, the outer face of each terminal cell being rounded (Plate 2.1: *top*, *right*). Gliding motility. Chemoorganoheterotrophic. Found e.g. in the mouth (human and animal).

Spirochaetales Order. Gram-negative, typically helical, motile cells which have a characteristic structure (section 2.2.14.1; Plate 2.4, *bottom*); 0.1–3 × 5–250 μm, according to species. The spirochaetes include both free-living and pathogenic species, anaerobic and aerobic species. Chemoorganoheterotrophic. Respiratory and/or fermentative. Genera include *Borrelia*, *Leptospira*, *Spirochaeta* and *Treponema*. [Colonic spirochaetes (*Brachyspira*) of medical and veterinary significance: JMM (2004) *53* (4) 263–350.]

Staphylococcus Genus. Gram positive. Cocci, about 1 μm in diameter, often in clusters, some containing orange or yellow carotenoid pigments; non-motile. Facultatively anaerobic. Chemoorganoheterotrophic. Carbon sources include various sugars. Commonly halotolerant (section 3.1.7). Catalase +ve. The staphylococci are divided into coagulase +ve and coagulase –ve strains (section 16.1.2.2), the former including *S. aureus* and *S. intermedius*, the latter including *S. epidermidis* (formerly *S. albus*). Found e.g. as commensals and pathogens of man and other animals. GC% about 30–39. Type species: *S. aureus*.

Stenotrophomonas Genus. Gram-negative. Rods, motile. Most strains require cystine or methionine. Strongly lipophilic. One species: *S. maltophila*, formerly *Xanthomonas maltophila*. Found e.g. in soil, water. Long-term infection may affect the prognosis in cystic fibrosis (section 11.3.2.1). [PCR-based identification of *S. maltophila*: JCM (2000) *38* 4305–4309.]

Streptococcus Genus. Gram-positive. Cocci, typically about 1 μm in diameter, often in pairs or chains. Non-sporing. Capsulation common. Facultatively anaerobic (the strictly anaerobic Gram-positive cocci are found in genera such as *Peptococcus*, *Peptostreptococcus* and *Sarcina*).

Catalase −ve. Chemoorganoheterotrophic. Typically fermentative, sugars being metabolized usually without gas. Found e.g. as commensals and pathogens of man and other animals. GC% 34–46. Type species: *S. pyogenes.*

Streptococci can be classified and identified e.g. by *Lancefield's grouping test*. This involves extraction and identification of certain cell-envelope-associated carbohydrates ('C substances'); typically, only one type of C substance occurs in a given strain. To extract the C substance, a cell suspension in 0.5 ml saline can be autoclaved for 15 minutes at 121°C and the supernatant used for the test. In one form of the test, the extract is layered onto an antiserum (containing antibodies to a given C substance) in a small tube; the test is repeated – with antisera to different C substances – until a positive result is obtained, i.e. a whitish precipitate at the extract–antiserum interface. All strains whose C substance reacts with a given antiserum are placed in the same Lancefield group. Lancefield groups are designated A, B, C . . . etc.; strains of *S. pyogenes* belong to group A.

Group A streptococci are causal agents in various diseases (e.g. cellulitis, impetigo, necrotizing fasciitis, scarlet fever, toxic shock syndrome: section 11.11). Some group A streptococci can survive within human respiratory epithelium [Lancet (2001) *358* 30–33]. *S. agalactiae* (group B) causes bovine mastitis and e.g. meningitis in human infants [antibiotic susceptibility in clinical isolates: AAC (2001) *45* 2400–2402]. *S. pneumoniae* (not typable by the Lancefield system) can cause e.g. human bronchitis, meningitis, otitis media, sinusitis and pneumonia. *S. suis* is associated with various diseases in pigs (e.g. bronchopneumonia, reproductive diseases) [serotypes 3–28 associated with pig diseases: Veterinary Record (2001) *148* 207–208]; serotype 2 (the 'group R' streptococcus) can cause disease in people exposed to infected pigs/pork.

Some former species of *Streptococcus* were transferred to other genera. For example, *S. faecalis* and *S. faecium* were transferred to *Enterococcus* (*E. faecalis* and *E. faecium*, respectively) [IJSB (1984) *34* 31–34]. Both species usually grow at 10°C and 45°C, survive 60°C/30 minutes, and can grow in 6% salt (NaCl); they differ e.g. in that *E. faecalis*, but not *E. faecium*, can obtain energy from pyruvate, citrate and malate.

S. lactis was transferred to *Lactococcus* as *L. lactis* [validation of the genus *Lactococcus*: IJSB (1986) *36* 354–356]. Lactococci are cocci/coccoid forms which occur singly, in pairs or chains. Growth occurs at 10°C but not at 45°C. Fermentative, L(+) lactic acid being the main product from glucose metabolism. Lactococci are found e.g. in dairy products (section 12.1.1). Type species: *L. lactis.*

Streptomyces Genus (order Actinomycetales – q.v.). Gram positive. Mycelium (section 2.1.1), part of which fragments to form chains of spores (section 4.3.2; Fig. 4.3). Aerobic. Respiratory. Chemoorganoheterotrophic; carbon sources include glucose, lactate and starch. Antibiotics formed by *Streptomyces* species include chloramphenicol (section 15.4.4) and streptomycin (section 15.4.2). Found e.g. in soil and as pathogens of plants. GC% 69–78. Type species: *S. albus.*

Sulfolobus Genus. Domain Archaea. Cocci, coccoid or irregularly-shaped cells in which the cell wall consists of only an S layer (section 2.2.12). Thermophilic (growth occurs between 50 and 90°C). Acidophilic (section 3.1.5). Aerobic and facultatively anaerobic. Energy is obtained by the (respiratory, aerobic) oxidation of sulphur (or Fe^{2+}) and/or by sulphur respiration (section 5.1.1.2) in which elemental sulphur is used as terminal electron acceptor. Obligately heterotrophic or facultatively autotrophic. Found e.g. in certain hot springs.

Thermoproteus Genus. Domain Archaea. Rods, filaments, about 0.5 × 1–80 μm. Anaerobic. Energy obtained by sulphur respiration (see also *Sulfolobus*). Thermophilic. Autotrophic and/or heterotrophic (carbon sources include glucose, ethanol, formate). Found e.g. in Icelandic solfataras.

Thiobacillus Genus. Gram negative. Rods, about 0.5 × 1–3 μm; typically motile. Obligately aerobic or (some) facultatively anaerobic. Respiratory; energy commonly obtained by the oxidation of sulphur and/or reduced sulphur compounds. Obligately or facultatively chemolithoautotrophic. Found e.g. in soil, mud, hot springs. GC% about 50–68. Type species: *T. thioparus*.

Treponema Genus (order Spirochaetales – q.v. for basic details). Cells: 0.1–0.4 × 5–20 μm. Anaerobic or microaerophilic. Some species can be grown in complex media; others (including *T. pallidum*) cannot, and are grown e.g. intratesticularly in rabbits. Found e.g. as parasites/pathogens in man and other animals. GC% 25–54. Type species: *T. pallidum*.

Tropheryma Genus. Gram-positive. *T. whipplei* (formerly *T. whippelii*) is associated with Whipple's disease (section 11.11); previously uncultured, the organism was cultured in fibroblasts [NEJM (2000) *342* 620–625]. [Genome: Lancet (2003) *361* 637–644.]

Vibrio Genus. Gram negative. Rods, curved (vibrios) or straight, 0.5–0.8 × 1.4–2.6 μm; motile, flagella typically sheathed (section 2.2.14.1). Facultatively anaerobic. Typically oxidase +ve. Chemoorganoheterotrophic; glucose is fermented by the mixed acid fermentation (Fig. 5.5), usually without gas. All species can grow at 20°C, most at 30–35°C, and some at 40°C. Some species tolerate high pH (e.g. *V. cholerae* can grow at pH 10). Found e.g. in various aquatic habitats (freshwater, estuarine and marine) and as pathogens in man, fish and shellfish. GC% 38–51. Type species: *V. cholerae*.

Wolbachia Genus. Gram negative. Cells: coccoid/rods, similar to those of e.g. *Rickettsia*. Intracellular/extracellular parasites of invertebrates (keds, mosquitoes, ticks etc.), but not obviously pathogenic.

   Strains of *Wolbachia* are obligate endosymbionts in the filarial nematode worm *Onchocerca volvulus* – the organism which causes human onchocerciasis (one form of which is the disease known as *river blindness*). Treatment of onchocerciasis with the tetracycline *doxycycline* depletes *Wolbachia* [Lancet (2001) *357* 1415–1416], inhibiting the worm's development and fertility, and interrupting transmission of the disease. [Onchocerciasis: BMJ (2003) *326* 207–210.]

Xanthomonas Genus. Gram negative. Rods, 0.4–0.7 × 0.7–1.8 μm, typically containing yellow pigment(s); some strains form extracellular slime. Motile. Aerobic. Chemoorganoheterotrophic. In strains of *X. campestris*, glucose is metabolized e.g. via the Entner–Doudoroff pathway (Fig. 6.2). Oxidase −ve (or weakly +ve). Catalase +ve. Found e.g. as pathogens in plants. GC% 63–71. Type species: *X. campestris*.

Yersinia Genus (family Enterobacteriaceae – q.v. for basic details). Cells: 0.5–0.8 × 1–3 μm.

Most species are motile 30°C, non-motile at 37°C; *Y. pestis* is non-motile. Growth occurs on basal media. VP −ve at 37°C (+ve in some species at 25°C); MR +ve; acid (little/no gas) from glucose; lactose typically −ve. Urease −ve (e.g. *Y. pestis*) or +ve (e.g. *Y. enterocolitica*). Opt. growth temperature: 28–29°C; *Y. enterocolitica* is psychrotrophic and can grow at 4°C. Found e.g. as parasites/pathogens in man and other animals. GC% 46–50. Type species: *Y. pestis*.

Zoogloea Genus. Gram negative. Rods, monotrichously flagellated, 1–1.3×2.1–3.6 μm, typically in masses in a polysaccharide matrix. Aerobic. Respiratory. Chemoorganoheterotrophic. Oxidase +ve. Catalase +ve. Opt. growth temperature: 28–37°C. Found e.g. in organically polluted freshwater habitats and in aerobic sewage-treatment plants. GC% about 65. Type species: *Z. ramigera*.

Index

A site (cf a ribosome)   154
*Aat*II   223
abasic site   161
ABC excinuclease   162
ABC protein   104
ABC transporters   103, 336
   exporters   104
   importers   104
abortive transduction   292
Ace toxin (of *Vibrio cholerae*)
   331
acellular vaccines (ACVs)   353
acetification   395
Acetobacter   63, 395, 513
acetoin   88, 488
acetyl-CoA   83, 93
N-acetyl-L-cysteine   374
N-acetylglucosamine   24
acetylmethylcarbinol (acetoin)
   88, 488
N-acetylmuramic acid   24, 107
acetyl phosphate   83
acid-fast bacteria   445, 497
acid shock   174
acid-shock proteins   174
acid tolerance response   174
acidophiles   46
Acinetobacter   513
acridine disinfectants   457
ActA protein   325
actin   323
actin-based motility   42, 337
Actinobacillus   513
Actinomyces   514
Actinomycetales   514
actinomycetes   6
Actinoplanes   77
activated sludge   413

acute-phase proteins   340
ACVs (acellular vaccines)   353
N-acyl-L-homoserine lactones
   298, 413
Ada protein   161
adaptation (to changed
   conditions)   47
adaptation (in chemotaxis)   182,
   183
adaptive response (DNA repair)
   161
adaptive response (acquired
   immunity)   348
adaptor (DNA)   233, 234
ADCC   357
adenine   133
   formula   133
adenosine   133
adenosine 5'-triphosphate   82
S-adenosylmethionine   145, 160
adenylate cyclase   168, 331, 361
adhesins   30, 35, 38, 320, 358
adhesion (in infection)   320
adhesion site   29
ADP-ribosylation   281, 335
Aer protein (in *E. coli*)   101
aerobactin   366
aerobes   46
aerogenic fermentation   86
Aeromonas   514
aerosol   425
aerotaxis   42, 101
aesculin hydrolysis   491
AFBs   445, 497
affinity chromatography   224
affinity maturation   351
affinity tail (of fusion protein)
   235

AFLP DNA fingerprinting   511
agar   62, 426
   reasons for use   427
agarose gel   225
agglutination (in serology)   494
aggressin   335
agr locus   181
agriculture
   and the nitrogen cycle   307
   wheat (nitrogen fixation in)
      308
Agrobacterium   114, 189, 514
   crown gall   114
   type V secretion system   114
AHLs   298, 413
AhpC   177, 473
AIDA   111
akinetes   77, 79
alanine biosynthesis   124
alanine (spore germinant)   75
alarmone   174
Alcaligenes   94, 514
   Biopol   422
alginate   77
   in *Azotobacter* cysts   77
   in cystic fibrosis   336
AlgU (in cystic fibrosis)   336
AlkA protein   161
alkalophiles   46
allele   350
allergies (to antibiotics)   349
allolactose   165
Alteromonas   279, 514
*Alu*I   223
amber codon   154
Ames test   196
amikacin   461
aminoacyl-tRNA   151

aminoglycoside antibiotics 461, 480
ammonia
  anammox reaction 95
  assimilation 304, 305
  in nitrification 95
  in sewage effluents 414
ammonification 305
amoxycillin 460
amphitrichous flagella 31
ampicillin 460
amplicon 239, 240, 248, 263, 264
AmpliTaq Gold™ DNA polymerases 246
AmpliWax™ 246
AMPPD 237
AMTDT 497
Anabaena 16, 80, 189, 306, 514
  azollae 302
  blooms 16
anabolism 81
anaerobes 46
anaerobic cabinet 441
anaerobic digestion (sewage) 414
anaerobic incubation 439
anaerobic jar 439
anaerobic respiration 94, 95
anaerogenic fermentation 86
anammox 95, 305, 306, 414
anaphylactic shock 349
anaphylatoxins 342, 345, 346
Anaplasma 131, 514
anaplerotic sequences 122
androphages 291
anergy 366
annealing (of primers) 240
anoxygenic photosynthesis 98
antagonism 295
  between antibiotics 466
anthrax 382
antibiogram 379
antibiotic-sensitivity tests 470
antibiotics 457 *et seq.*
  activity in eukaryotic cells 480
  alternative approaches 470
  amikacin 461
  aminoglycosides 461, 480
  amoxycillin 460
  ampicillin 460
  antagonism 466, 472

antibiotics (*cont.*)
  Augmentin® 460, 461
  azithromycin 463, 470
  benzylpenicillin 460
  breakpoint test 471
  broad-spectrum 458
  carbapenems 458, 459
  cefixime 429
  cefoxitin 475
  cephalosporins 458, 459
  chloramphenicol 462
  chlortetracycline 462
  ciprofloxacin 464
  clarithromycin 463
  clavams 458, 459
  clavulanic acid 460, 461
  clindamycin 388
  cloxacillin 460
  cotrimoxazole 465
  cycloserine 126
  danofloxacin 464
  disc diffusion test 471
  doxycycline (in *Wolbachia*) 526
  E test 471, 474
  enzyme inactivation 472
  erythromycin 458, 462
  fluoroquinolones 464, 480
    ciprofloxacin 464
    gatifloxacin 464
    gemifloxacin 464
    grepafloxacin 464
    levofloxacin 464
    moxifloxacin 464
    trovafloxacin 464
  GAR-936 462
  gatifloxacin 464
  gentamicin 461
  grepafloxacin 464
  imipenem 475
  inducible resistance 468, 471
  isoniazid 177
  kanamycin 461
  β-lactams 458
    β-lactamase inhibitors 461
  levofloxacin 464
  linezolid 469, 470
  macrolides 462, 480
  methicillin 460, 469
  metronidazole 465
  MIC 471
  monobactams 458, 459
  moxifloxacin 464

antibiotics (*cont.*)
  nafcillin 460
  nalidixic acid 464
  neomycin 461
  nocardicins 458, 459
  norfloxacin 464
  novobiocin 464
  ofloxacin 464
  oleandomycin 105
  oxazolidinones 469
  oxytetracycline 462
  penicillins 458, 459, 460
  polymyxins 463
  post-antibiotic effect 480
  quinolones 464
  quinupristin/dalfopristin 463
  resistance to 467
    DNA-based detection 473 *et seq.*
    inducible resistance 468, 469, 471
    mechanisms 467
    transposon-encoded 468
  rifampicin 465, 480
  rifamycins 465
  sensitivity tests 470
  streptogramins 462, 463
  streptomycin 461
  sulphonamides 465
  Synercid® 463
  synergism 466, 472
  target sites 458
  teicoplanin 126
  tetracyclines 462
    doxycycline (in *Wolbachia*) 526
  tigecycline 462
  trimethoprim 465
  trovafloxacin 464
  vancomycin 126, 469
antibodies 349
  allotype 350
  cleavage by pepsin 350
  cleavage by papain 350
  idiotype 350
  isotype 350
  monoclonal 355
antibody-dependent cell-mediated cytotoxicity (see ADCC)
antibody formation 349
anticodon 151
antigen 349

antigen-presenting cell 351
antigenic formula (*Salmonella*) 523
antigenic variation 362
antioxidant enzymes 177
antiparallel 134, 135
antiport 102
antirestriction 146, 294
antisense PNA 470
antisepsis 457
antiseptics 457
antiserum 494
anti-sigma factors 175
antitoxins 382
AP endonuclease 163
AP-PCR 508
AP site 161
APC 351
aperture diaphragm 447
Aphanizomenon 418, 514
API system 491
apoptosis 356, 362
APT paper (for blotting) 229
apurinic site 161
apyrimidinic site 161
AqpZ protein (*E. coli*) 20
aquaporins 20
Aquaspirillum 30, 515
    peregrinum 12, 13
aquifer 415
ara operon 166
arbitrarily primed PCR 508
Archaea 1, 2, 3, 505
    cell cycle 61
    cell wall 28
    cytoplasmic membrane 21
    enzymes 81
    flagellum 33
    FtsZ ring 54
    gene expression 164
    Halobacterium 16, 30, 47, 99, 131
    lipids 21
    Methanobacterium 96
    Methanosarcina 16
    Methanothermus 96
    MIP channels 20
    protein transport 115
    Pyrococcus 158
    Pyrodictium 45
    RadA 197
    Sulfolobus 30, 46, 118
    Thermococcus litoralis 158

Archaea (*cont.*)
    Thermoplasma acidophilum 46
    wall-less 29
Archaebacteria 2, 515
arginine decarboxylase 439
arizonae (*Salmonella*) 523
ARMS 244, 245
Arrhenius plots 67
arsenate respiration 94
artificial chromosomes 227
aseptic technique 431
AsiA protein 187
ASP (acid-shock protein) 174
aspartate biosynthesis 124
assimilatory nitrate reduction 305
assimilatory sulphate reduction 309
asymmetric division (of cells) 69
asymmetric PCR 245
asymmetric septum 74, 75
ATP 85, 90, 99
    formula 82
ATP-binding cassette 103
ATPase 90, 91
ATR 174
attaching and effacing lesions 322, 333, 369
attenuator control (of operons) 166
Augmentin® 460, 461
aurease 492
autocatalytic
    inteins 158
    introns 189
autoclave 451
autodisplay 272
autoinducer 298
autoradiography 222
    in DNA sequencing 260, 261
autotransporter 110, 272
autotrophs, autotrophy 117
auxotroph 195, 196, 215
axial filament (of endospore) 74
azithromycin 299, 463, 470
Azolla 308
Azorhizobium caulinodans 308
Azospirillum 101
Azotobacter 14, 515
    beijerinckii (use of PHB) 15
    vinelandii (cysts) 77
Azotobacteriaceae 306
aztreonam 474, 475

B cell (B lymphocyte) 349, 366
BAC (see bacterial artificial chromosome)
bacillus 6, 7
Bacillus 515
    anthracis 382
    brevis 156
    cereus (food poisoning) 401
    endospore formation 74, 75, 78, 79, 180, 184, 185
    licheniformis 411
    phosphorelay 180, 181
    quorum sensing 300
    schlegelii 118
    shape determinants 8, 9
    spo genes 53
    stearothermophilus 30, 398
    subtilis 8, 9, 49, 74, 75, 78, 79, 129, 180, 184, 185, 187, 205, 206, 300
    thuringiensis 14, 408
    transformation 205, 206, 300
    tRNA-directed transcription antitermination 187
back mutation 196
*Bac*Light™ 444, 445
BACTEC™ 430, 473
bacteria
    acid-fast 445
    antibiotic resistance 467, 473 *et seq.*
    capsules 29
    cell cycle 49
    cell envelope 10
    cell wall 10, 21–28
    chemotaxis 42, 182, 183
    chromosome 10, 11, 13, 14, 128
    circadian rhythms 11
    classification 4, 5
    coenocytic 43
    counting 440–443
    counting chamber 440
    cysts 77
    cytoplasm 10, 14
    cytoplasmic membrane 10, 17–21, 50
    cytoskeleton 8, 11
    desiccation 45
    dimensions during growth 49
    energy requirements 81
    fimbriae 26, 27, 35–38
    flagella 31–34

bacteria (*cont.*)
  gas vacuoles   16, 17
  genomic library   215
  Gram stain   444
  growth   44
    diauxic   68
    measurement of   68
    synchronous   67
  identification   481 *et seq.*
  linear chromosomes   129
  meaning of term   1, 2
  mesophilic   46
  morphology (determination)
      482
  morphology (evolutionary
      aspects)   8
  motility   39, 482
  multicellular   12, 43
  naming   4, 5
  nucleoid   10, 11, 13, 14
  pigmentation   62
  plasmid-less   131
  prokaryotic nature   1
  psychrophilic   46
  psychrotrophic   46
  recombinant, in environment
      314
  reproduction   44
  resistance to antibiotics   467,
      469, 473 *et seq.*
  resting cells (cysts, spores)   73
  ribosomes   10, 14
  shape   6
    determinants   8, 50
  size   9
  staining   444
  storage granules   10, 15
  taxonomy   5
  thermoduric   46
  thermophilic   45
  wall-less   29
Bacteria   1
  Proteobacteria   505, 521
bacterial artificial chromosome
    227
bacterial food poisoning   400
bacterial vaginosis   383
bactericidal   455
bactericidal/permeability-
    increasing protein (see
    BPI protein)
bacteriochlorophylls   96
bacteriocins   295

bacteriological media   426
bacteriological warfare   4
bacteriophage (see phage)
bacteriophage conversion (see
    phage conversion)
bacteriophage typing   494
bacteriorhodopsin   99
bacteriostatic   455
bacteriuria   374
Bacteroides   129, 515
    conjugative transposition   212,
      213
  fragilis   26, 27
  in human microflora   318
bactoprenol   125, 126
*Bae*I   146, 223, 224
BAF (see biological aerated filter)
balanced growth   64
*Bam*HI   223, 230
    star activity   221
Bartonella   131, 338, 515
basal body (of a flagellum)   31
basal medium   427
base analogues   192
base excision repair   162
base-pair   134, 135
base-pairing   134, 135
base plate (phage)   280
batch culture   63–65
BCG   353, 354
*Bcl*I   223
Bdellovibrio   300, 515
Beggiatoa   40, 515
Beijerinckia   516
benzene degradation   424
benzylpenicillin   460
β-clamp   140
β-galactosidase (see under 'g')
bfa mutants   71
BFP   37, 321, 369
*Bgl*II   223
bgl operon   170
biased random walk   42
big-bale silage   407
bijou   426
binary fission   61
binding protein-dependent
    transporters   104
binomial   5
bio genes   200
bioavailability (of pollutants)
    423
biochemical tests   484

biofilms
  medical   300, 323
  in sewage treatment   412, 413
  in water supply systems   418
biogas   415
bioinsecticides   408
bioleaching   409
biological aerated filter (BAF)
    413
biological clock   188
biological containment   314
biological control   408
biological filter   411
biological oxygen demand   411,
    413
biological warfare   4
biological washing powders
    411
biologicals   382
bioluminescence   298
biomass   66
biomimetic technology   424
biomining   409
biopanning   273
biopesticides   408
bioplastics   422
Biopol   422
bioremediation   423
bioterrorism   4
biotin (as a probe label)   237
biphenyls   423
bisphenol   457
bisulphites (mutagens)   192
bleach (as disinfectant)   455
blood agar   427, 428, 484, 492
blood culture   375
  anaerobic   375
bloom   16, 297, 417
blotting
  electro-   229
  immuno-   229
  Northern   229
  Southern   229
  Southwestern   271
  Western   229
blue-green algae   97
blunt-ended DNA   221, 233
BOD (see biological oxygen
    demand)
Bordetella   105, 130, 362, 392,
    516
  anti-pertussis vaccines   353
  detection by PCR   242

Borrelia 516
   burgdorferi (genome) 11, 129
   flagella 33
   Lyme disease 370, 386
   OspC protein 370
   PCR 250
   structure 34, 35
   swimming 39
botulinum cook 398
botulinum toxins 334, 383
   therapeutic uses 422
botulism 331, 383, 400
bound coagulase 486
BPI protein 343
Brachyspira (Spirochaetales) 524
branch migration 198
Braun lipoprotein 23, 27
breakpoint chlorination 418, 419
breakpoint test (antibiotic) 471
bright-field micropscopy 447
broad-range primers 242, 314
broad-spectrum antibiotics 458
bromcresol purple 489
5-bromouracil 192
bromthymol blue 428, 486
broth 428
   Luria–Bertani 206
   MacConkey's 428
   nutrient 428
   selenite 428
broth sugars 487
Brown's tubes 442
Brownian motion 483
Brucella 131, 516
brucellosis 383
bubonic plague 387
Buchnera
   genome 128, 129
budding 61
bullous impetigo 385
bundle-forming pili 37, 321, 369
Burkholderia cepacia 516
   cystic fibrosis 336
   melioidosis 386
   quorum sensing 299
burst size (of phage) 285
butanediol fermentation 88, 89
butt (of a slope) 431
butter 393
buttermilk 393
butyrous growth 62
bystander lysis 345

C-reactive protein 340
C ring (of a flagellum) 31, 32
C substances 525
C_1 compounds 126
cadherins 321, 324
caesium chloride gradient 224
cag pathogenicity island 339, 370
CagA protein 339
Cairns-type replication 140
calcium
   in motility 42
   in endospores 74, 75, 180
   regulatory role 47
   in transformation 206
calcium dipicolinate 74, 75, 180
Caloramator (see *Clostridium*, page 517)
calprotectin 347
Calvin cycle 118, 119
cAMP 168, 331
cAMP phosphodiesterase 168
cAMP-receptor protein 168
Campbell-type integration 254
Campylobacter 400, 516
   coli 401
   jejuni 401, 402
canning 396
cap (flagellar) 33
CAP 168
capacity test (disinfectants) 456
capsid 278
capsules 29
   antiphagocytic 29
   stain 445
carbapenems 458, 459
carbolfuchsin 444
carbon assimilation 118–121
   in autotrophs 118
   in heterotrophs 119
carbon cycle 302, 303
carbon dioxide fixation 117
carbon dioxide monitoring 406
carbon metabolism 117
carbon monoxide 118
carboxydobacteria 118
carboxysomes 17
carcinogens 196
cardiolipin 19
caries 321
carrier (disinfectant tests) 456
carrier (of pathogens) 373
carrier gene 235

caspase 1 362
cat gene 178
CAT (see chloramphenicol acetyltransferase)
cat scratch disease (*Bartonella*) 515
catabolism 81
catabolite activator protein 168
catabolite control protein A 170
catabolite repression 168
catalase 484, 485
   katG-encoded 177, 473
catenation (of chromosomes) 53
catenins 324
catheterization (and infection) 319, 320
Caulobacter 69–72, 516
   cell cycle (control) 59–61
cccDNA 135
CcdA, CcdB 144
CckA 60
CcpA 170
CCPs 405
CD1 357
$CD4^+$ T cell 347, 351, 352
$CD5^+$ B cell 351
$CD8^+$ T cell 356
CD14 342, 357
CD16 357
CD28 351
CD29 357
CD40, CD40L 351
CD44 357
CD55 345
cDNA 217
   libraries 217
   Okayama–Berg method 220
cDNA expression library 219
   screening 219
cefixime 429
cefoxitin 475
cell cycle 49
   Caulobacter 59
   checkpoints 57, 71
   control 57
   in archaeans 61
   in bacteria 49
cell division
   asymmetric 61
   binary fission 61
   budding 61
   multiple fission 61

cell division (*cont.*)
  ternary fission 61
cell envelope 10, 26, 27
  synthesis of 50
cell-mediated immunity 355
cell shape 6, 50
cell wall 10, 21–28
  archaeal 28
  Gram negative 23
  Gram positive 22
cellulitis 383
cellulolytic bacteria 120
Cellulomonas 120, 516
cellulose utilization 120
centrifugation (of nucleic acids)
      224
cephalosporins 458, 459
cephamycins 475
cerebrospinal fluid (CSF) 381
Cetavlon 457
Cetrimide 457
CFI, CFII 332
CFT 377
chain termination 260, 261
chancre 388
chaperones
  in fimbrial assembly 36
  in heat shock 171
  in protein synthesis 107, 156
checkpoints 57, 71, 175
cheese 393, 394
chemiluminescence 237, 405
chemoattractant 42, 182
chemoeffector 42, 182
chemokine 342
chemolithoautotophs 117
chemolithotrophs 83
chemoorganoheterotrophs 117
chemoorganotrophs 83
chemorepellent 42
chemostat 66
chemotaxis 42, 182, 183
chemotherapy 380
chemotroph 83
CheY~P 182, 183
chi (χ) site 170, 198
chi structure 198
Chick–Martin test 456
chips (DNA) 276
Chlamydia 516
  cell wall 23
  elementary bodies 390
  reticulate bodies 390

Chlamydia (*cont.*)
  trachomatis 389
    genome 129
    LCR-based test 265, 498
    PACE 2C test 496
    type III protein secretion 109
chloramines 418
chloramphenicol 462, 463
chloramphenicol acetyltransferase
      178, 463
chlorhexidine 457
chlorinator 421
chlorine (disinfectant) 455
chlorine (in water supplies) 418
chlorine demand 418
Chlorobiaceae 118
Chlorobiineae 98
Chlorobium 517
chlorobium vesicles 17
chlorophylls 81, 96
chlorosomes 17, 98
chlortetracycline 462
chocolate agar 428
cholera 331, 383
  O139 Bengal 279, 288, 383
  toxin 331
  vaccines 353
Christensen's agar 489
chromatography 224
  affinity 224
  spun column 224, 228
chromosomal fingerprinting 506
chromosome
  bacterial 10, 11, 13, 14, 128,
      129
  complete genome sequences
      129
  size 13, 128
  number per cell 13, 129
  bacterial artificial 227
  cccDNA 135
  decatenation and partition 53
  linear 129, 141
  replication 52, 56, 58, 59
  yeast artificial 227
cI protein 289
ciprofloxacin 464
circadian rhythms 11, 188
CIRCE 172
circularly permuted DNA 281
Cit plasmid 131
citrate test 488
citric acid cycle 92, 93

β-clamp 140
clarithromycin 463
class I MHC molecule 346, 347,
      356
class II MHC molecule 347,
      351, 352, 356
class I safety cabinet 432
class II safety cabinet 432
class III safety cabinet 432
class switching 351
class I transposon 204
class II transposon 204
classification 4, 5, 500
clavams 458, 459
clavulanic acid 459, 460, 461
cleared lysate 227
CLED medium 428, 429
clindamycin 388
clonal expansion 357
cloning (molecular cloning)
      214, 227
cloning PCR products 248
closed circular DNA 135
Clostridium 10, 331, 334, 517
  argentinense 383
  barati 383
  bifermentans 408
  botulinum 383, 396, 398
  difficile 388
  fervidus (use of smf) 102
  nitrogen-fixing species 302
  perfringens 179, 385, 401
  piliforme 391
  tetani 334, 388
  thermocellum 120
cloxacillin 460
clue cells (vaginosis) 383
clumping factor 486
CMI 355
CMP 133
CO dehydrogenase 118
coagulase test 485
cobalt-60 (in sterilization) 453
coccobacillus 6
coccus 6, 7
cocoa 394
coding strand (of a gene) 149,
      268
codon 150
  amber 150
  initiator 151
  nonsense 154
  ochre 154

codon (*cont.*)
  opal 150
  stop 154
codon bias 271
coenocytic bacteria 43
coenzymes 122, 123, 308
coffee 394
cognate help 351
cointegrate 202, 203
colanic acid 29
cold-shock response 172
ColE1 plasmid 142
colicins 296
coliform 517
  faecal 420
colon microflora 318
colonization factors 332
colony 62
colony hybridization 219
combination probe test 496
commensal 301
communities (microbial) 295
comparative genomics 130
compatible solute 48
  as thermoprotectant 49
competence (in transformation) 205
complement 343
  action of trypsin 346
  anti-bacterial roles 343
complement activation (fixation) 342
  alternative pathway 342, 343–345
  classical pathway 344, 345
  lectin pathway 344, 345
  salvage (C2a) pathway 344, 346
complement-fixation test 377
complementary (bases, strands) 135
complementary DNA (cDNA) 217
complementation 263
complete medium 195
complete transduction 292
comS gene 206
ComX 205, 300
concatemer 281, 290
confluent growth 62
conjugate (dye-linked antibody) 377
conjugate vaccine 354, 359

conjugation 207
  in Gram-positive bacteria 207
  in Gram-negative bacteria 208
  in situ? 311
  nic 209
  oriT 209
  relaxosome 209
  repliconation 209
  surface-obligatory 211
  T-strand 208
  universal 211
conjugational junction 210, 211
conjugative transposition 212
conjugative transposons 212
conjunctivitis 384
consensus sequence 147, 148
consortia 10
continuous culture 66
convertase 344
cooked meat medium 441
copper leaching 409
co-protease 170, 171
copy DNA 217
copy number (of plasmids) 144
core (of endospore) 74
core (phage) 280
core oligosaccharide 23, 27
cortex (of endospore) 74
Corynebacterium 517
  cell shape determinants 8
  diphtheriae 291, 335, 384
  Elek plate 375, 376
  glutamicum (morphogenesis) 8
  glutamicum (industrial use) 395
  glutamicum (porins) 23
  in human microflora 318
  porins 23
cos site 290
cosmid 227, 231
co-transduction 293
co-translation 150
cotrimoxazole 465
cottage cheese 394
counterselection 294
countertranscript RNA 143
counting bacteria 440–443
counting chamber 440
coupling sequence 212
Coxiella burnetii 73, 328, 396, 517
  after phagocytosis 328

CpxA–CpxR system 155
CR3 (macrophage receptor) 328
crack entry (in nodulation) 308
Crenarchaeota 505
cresol red 489
cresols 455
critical control points 405
critical dilution rate 67
cro protein 289
Crohn's disease 330, 372
cross-contamination (of food) 400, 404
crown gall 114
CRP 168
Cryptosporidium (in water) 419
crystal violet 444
CSF (cerebrospinal fluid) 381
CSF (pheromone) 300
CspA 173
CSPD 237
CTB–WC vaccine 353
CtrA 60, 186
ctRNA 143
CTXΦ phage 279, 288, 293, 353
culture 63, 426
  batch 63–65
  blood 375
  continuous (open) 66, 67
  definition 63
  of *Mycobacterium tuberculosis* 374
  semi-quantitative 374
curing (foods) 400
cut-and-paste transposition 202, 203
cutting DNA 221
cyanobacteria 517
  blooms 16
  carbon dioxide fixation 118
  photosynthesis 96, 97
  toxins 297
cyanogen bromide 235
Cyanothece 307
cyanotoxins 297, 417, 418
cycads (*Nostoc*) 521
cycles of matter 302
cyclic AMP 168, 331
cyclolysin 105, 361
cycloserine 126
cyst 77
cystic fibrosis 336
  Stenotrophomonas 524

cystitis   384
cytidine   133
cytoadherence proteins
        (*Mycoplasma*) 521
cytochalasins   323
cytochrome *c*   485
cytochromes   81, 91
cytokines   340, 346, 363, 364,
        365
cytoplasm   10, 14
cytoplasmic membrane   10,
        17–21, 50
    archaeal   21
    bacterial   10
    permeability to H$^+$ and Na$^+$
        102
    plasmalogens   19
cytosine   133, 134
    formula   133
cytoskeleton   8, 11
cytotoxic T cells   356

D value (in food processing)
        397
D$_{60}$ value   397
DacA protein (of *E. coli*)   56
DAF (decay-accelerating factor)
        345
DAF (direct amplification
        fingerprinting)   508
dairy products   393
dam   416
Dam-directed mismatch repair
        161
dam gene   186
Dam methylation   186, 204, 252
dAMP   133
danofloxacin   464
dark reaction (of
        photosynthesis)   97
DCA   428, 431
DCDS   209
dCMP   133
ddF   475, 476
DDMR   161
de-acetylase   25, 470
death (logarithmic)   397
death phase   65
decarboxylase tests   489
decatenation   53
decay-accelerating factor   342,
        345
Dechloromonas   424

Dechlorosoma   424
decoding region (rRNA)   152
defective phage   293
defensins   319, 340
defined medium   429
DEFT   406, 442
degradosome   11, 157
DegS   156
dehydration (food
        processing)   399
Deinococcus radiodurans   13, 14,
        129
deletion mutation   194
denaturing gradient gel
        electrophoresis
        (see DGGE)
denitrification   305, 306, 307
    in sewage effluents   414
density gradient column   224
dental caries   321
dental plaque   321
deoxyadenosine   133
deoxycholate–citrate agar   428,
        431
deoxycytidine   133
deoxyguanosine   133
deoxynucleotidyl transferase   234
deoxyribonucleic acid (see DNA)
deoxyribonucleotide   132
deoxythymidine   133
de-repressed plasmid   201
desiccation   45
Desulfomonas   517
Desulfotomaculum   73
Desulfovibrio   94, 95, 129
Desulfurococcus   518
Desulfuromonas   94, 309
Dettol   457
DGGE   245
dGMP   133
DHF   123, 465
DHFR   466
DHPS   466
diacetyl   393, 488
diapedesis   342
diauxic growth   68
diazotrophs   306
Dicer   190
Dichelobacter   518
dichlone   297
dideoxy fingerprinting   475, 476
dideoxy method (of DNA
        sequencing)   260, 261

dideoxyribonucleotides   132,
        260, 261
    in DNA sequencing   260, 261
differential display   276
differential medium   428
differentiation   69
digoxigenin (as a probe label)
        237
dihydrofolate reductase   466
dihydrofolic acid   123, 465
dilution rate (open culture)   66
dilution test (antibiotic)   471
dinitrogen   304
1,2-dioxetanes   237
diphosphatidylglycerol   19
diphtheria   335, 384
    toxin   291, 335, 375, 376
dipicolinate   74, 75
diplococci   7
direct amplification
        fingerprinting   508
disc diffusion test   470
disease   316
disinfectant   454
disinfection   454
    of water supplies   418, 419
    testing disinfectants   456
display (of heterologous
        proteins)   272
display (phage)   272, 273
dissimilatory reduction of nitrate
        to ammonia   94, 306
dissimilatory sulphate reduction
        309
divergent transcription   149, 164
DivK, DivL   60
DNA
    amplification (*in vitro*)   239,
        263 *et seq.*
        LCR   263, 265, 266
        NASBA   263–266
        PCR   239 *et seq.*
        SDA   266
    antiparallel   134, 135
    base-pairing   134, 135
    blunt-ended   221
    chips   276
    circularly permuted   281
    degraded (PCR)   250
    double helix   135
    double-stranded (ds)   135
    duplex   134, 135
    excision repair systems   161

DNA (*cont.*)
  base excision repair   162
  mismatch repair   161
  nucleotide excision repair
    162
  fingerprinting   506
  helix   135
  iSDR   141
  melting   138
  methylation   145, 185
  modification   145
  monitoring   160
  overwound   135
  packaging (phage)   288
  phosphodiester bond   134
  plasmid (isolation)   226, 228
  polarity of strands   134, 135
  Pyrosequencing™   259
  relaxed   135
  repair   160
  replication   52, 138
  restriction   145
  semi-conservative replication
    141
  separating supercoiled from
    linear   228
  sequencing   258–262
  sticky ends   221
  strands   135
  structure   132
  supercoiled   136
  template strand   139
  terminally redundant   281
  underwound   135
  vaccines   355
DNA-binding proteins   149,
  271
DNA chip technology   276
DNA–DNA hybridization   501
DNA fingerprinting   506
DNA glycosylase I   161
DNA glycosylase II   161
DNA helicase   139, 140, 162
DNA ligase (see ligase)
DNA methylation   145, 185
DNA packaging (phage)   288
DNA polymerase
  (thermostable)   239
DNA polymerase I   237
DNA polymerase II   171
DNA polymerase III   52, 138,
  191
DNA polymerase IV   171

DNA polymerase V   171
DNA repair   160
  excision repair systems   161
DNA replication   52, 138
  Cairns-type   140
  in phage M13   291
  in plasmids   142
  initiation of   52, 138
  lagging strand   139, 140
  leading strand   139, 140
  linear chromosomes   141
  Okazaki fragments   139, 140
  primase   139
  primer   139
  primosome   139
  protein priming   142
  replication fork   139, 140
  rolling circle   143, 212
  template   139
  ter site   141
DNA sequencing   258–262
DNA vaccines   355
DnaA box   138
dnaA gene   138
DnaA protein   52, 138
DnaB protein   139, 140
DnaG protein   139
dnaJ gene   172
DnaJ protein   172
dnaK gene   156
DnaK protein   157, 171, 172
dnaN gene   138
DNase I footprinting   272, 274,
  275
docking chain   51
domain (supercoiled DNA)   136
domain (taxonomic)   1, 504
donor conjugal DNA synthesis
  209
donor suicide   203
Dorset's egg   429
dot genes   328
double helix   135
doubling time   61, 64
downstream   147
downstream box   267
DPD   418
*Dpn*I   146, 252
DRNA   94, 306
droplet infection   373
drying (of plates)   430
Dsb proteins   155, 270
dsbA gene   155

dsbB gene   155
dTMP   133
DtxR   335
Dukoral®   353
duplex DNA   134
Durham tube   487
dUTP poisoning   241
Dynabeads®   437
dysentery (bacterial)   337, 384
  non-bacterial   384

E-cadherin   321, 324
E. coli (see *Escherichia coli*)
E protein   143
E test   471, 474
EAEC   333
EAF plasmid   369
EaggEC   333
ear microflora   318
*Eco*RI (methylase)   145
*Eco*RI (RE)   145, 216, 221, 223
  star activity   221
*Eco*RII   146
edge (of colony)   62
EF-Tu   154
efflux mechanisms (antibiotic
  resistance)   467, 468
EHEC   332, 335
  phage H-19B   278
  quorum sensing   299
  SMAC medium   429
EIEC   332, 384
Eijkman test   420
El Tor biotype   353
electroblotting   229
electron transport chain   90
electrophoresis (of nucleic
  acids)   224
electrophoresis mobility-shift
  assay   271
electroporation   206
  E. coli   206
  Helicobacter pylori   207
Elek plate   375, 376
elementary body (*Chlamydia*)
  390, 516
elevation (of colony)   62
ELISA   378
ELISPOT assay   329
elongation
  in protein synthesis   154
  in RNA synthesis   148
elongation factors   154

Embden–Meyerhof–Parnas
    pathway (see EMP
    pathway)
Emmentaler cheese   393
EMP pathway   84, 85
EMSA   271
emulsification (of bacterial
    growth) 444
Enbrel®   382
end-product efflux   100
endergonic reactions   81
endolysins (phage)   285
endonuclease   157
    AP   163
    in RNA degradation   157
    rare cutting   223
    restriction   221
    S1   233
    UvrABC   74, 75
endospores   73, 74, 75, 180, 184,
        185, 483
    location, shape   483, 484
endotoxic shock   338
endotoxins   338
energy   81
energy-converting metabolism
        84
energy taxis   101
enriched medium   427
enrichment medium   427
enteroaggregative E. coli   333
Enterobacter   89, 468
Enterobacteriaceae   518
enterobactin   366
enterochelin   366, 367
Enterococcus   207, 315, 468, 469,
        484
    formerly Streptococcus   525
enterohaemorrhagic E. coli (see
        EHEC)
enteroinvasive E. coli (see EIEC)
enteropathogenic E. coli (see
        EPEC)
enterotoxigenic E. coli (see ETEC)
enterotoxin   331, 334
enterotoxin F (staphylococcal)
        366
Entner–Doudoroff pathway   120,
        121
EnvA   56
environment
    bioremediation   423
    cycles of matter   302

environment (cont.)
    greenhouse effect   312
    recombinant bacteria   314
EnvZ   179
enzyme-coupled probes   237
enzyme inactivation (of an
        antibiotic)   472
enzyme-linked immunosorbent
        assay (see ELISA)
EPEC   321, 322, 333, 369
    binding to epithelium   321
    quorum sensing   299
Epicurian Coli® XL1-Red   251
epifluorescence microscopy   377
epilithon   311
Epulopiscium fishelsoni   9
era gene   55
ERIC   510
error-prone repair   170
ertapenem   458
Erwinia   310, 394, 518
erysipelas   384
erythromycin   458, 462
Eσ^{70} 148
ESAT-6   329
ESBLs   461
Escherichia   518
Escherichia albertii   518
Escherichia coli
    acid stress   175
    adhesion site   29
    aerotaxis   101
    AIDA-1 autotransporter   111
    ammonia assimilation   304
    aquaporin   20
    batch culture   63, 65
    branching in   9
    cell cycle   49
    cell envelope   26, 27, 50
    chemotaxis   42, 182, 183
    chi (χ) site   170, 198
    chromosome length   13
    chromosome replication   138
    chromosome sequence   129
    classification   518
    competence (transformation)
        206
    conjugation   208
    D_{60} value   397
    Dam methylation   186
    degradosomes   11
    DNA repair   160
    doubling time   61

Escherichia coli (cont.)
    electroporation   206
    enteroaggregative (see EaggEC)
    enterohaemorrhagic (see
        EHEC)
    enteroinvasive (see EIEC)
    enteropathogenic (see EPEC)
    enterotoxigenic (see ETEC)
    extracytoplasmic oxidation
        101
    fermentation   86
    fimbriae   35, 36, 37
    flagellar genes   176
    flagellar origin   31
    FtsZ ring   54, 55
    fumarate respiration   94
    genome sequence   129
    gluconeogenesis   123, 125
    glutamate synthesis   304
    growth curve   65
    α-haemolysin   105
    heat shock   171
    IMViC tests   518
    in human microflora   318
    integration of phage λ   200
    introns   189
    iron chelation   366
    isolation from sewage   436,
        437
    Kdp system   179
    lac operon   165
    lysis by phage T4   280
    lysogeny by phage λ   289
    mechanosensitive channels
        20
    melibiose uptake   102
    microflora   318
    morphology   12, 13
    MscL   20
    MUG test   491
    mutator genes   191, 192
    O157 serotype (see O157
        E. coli)
    oppBCDF genes   115
    osmoregulation   20
    overproduction of recombinant
        proteins   267
    oxidative stress   177
    Pap pili   37
    pathogenesis   332, 333, 384
    peptidoglycan   25
    3-for-1 growth model   50,
        51

Escherichia coli (*cont.*)
  phages
    H-19B 278
    λ 289
  potassium pump 103
  promoters 147
  proton–lactose symport 103
  PTS transport 105, 106, 107
  REP sequence 510
  restriction endonucleases 145
  ribosomal proteins 15
  rrn operon 159
  sodium motive force 102
  SOS system 170
  sulphide assimilation 309
  T4 infection 280
  topoisomerases 136
  transformation 206
  uropathogenic (see UPEC)
  Z ring (FtsZ ring) 54, 55
esculin (see aesculin)
EspA 321
EspB 110, 321, 333
EspE 110
etanercept 382
ETEC 332, 353
ethanol (from metabolism) 87, 88
ethanol (disinfectant) 457
ether bonds (bacterial cleavage) 304
ethidium bromide 228, 508
ethylene oxide 454
etridiazole 308
Eubacteria 518
eukaryote–prokaryote split 1
eukaryotes
  characteristics 2
  exons 219
  introns 219
  mature mRNA 219
Euryarchaeota 505
eutrophication 414
evolution 192
excision repair systems 161
  base excision repair 162
  mismatch repair 161
  nucleotide excision repair 162
  UvrABC endonuclease 162
exfoliative toxin 388
exochelin 367
exon 219

exonuclease III 163
exonucleases (in RNA degradation) 157
exospores 75
exosporium (of endospore) 74
exotoxin 330
exponential growth 64
export (meaning of term) 103
export apparatus (of a flagellum) 32
expression library 219
expression vector 219, 227, 232, 267
extein 158
extended-spectrum β-lactamases 461
extracytoplasmic oxidation 100
extremophiles 2
extrusion-capture model 54
eye (microflora) 318

F⁻ cell (in conjugation) 208
F⁺ cell (in conjugation) 208
F′ donor (in conjugation) 211
F pilus 26, 27, 28, 208
F plasmid 144, 208
  finO locus 201
F value (in food processing) 397
Fab fragment (of an antibody) 350
F(ab′)₂ portion 350
factor I 344
facultative aerobe 47
facultative anaerobe 47
FAD 83
faecal coliforms 420
faecal–oral route (of infection) 317
faecal pollution tests 419, 420
faecal streptococci 419, 420
family 4
Fc portion (of an antibody) 350
fed batch culture 66
femAB operon 469
Fenton reaction 177
FepA protein 366
fermentation 84
  aerogenic 86
  anaerogenic 86
  butanediol 88, 89
  heterolactic 86
  inorganic 95
  lactic acid 86

fermentation (*cont.*)
  mixed acid 86, 87
ferredoxins 91
FHA 321, 368
fibrillum 38
fibrin 336, 486
fibrinogen 486
fibrinolysin 336
field stop 447
filament (flagellar) 31
filamentous haemagglutinin (see FHA)
Filifactor (see *Clostridium*, page 517)
filtration
  of water supplies 417
  in water testing 420
  for sterilization 454
fim genes 37, 38, 362
FimA protein 38
fimbriae 26, 27, 35–38, 368
  type I (1) 37
  type II (2) 37
  type III (3) 37
  type IV (4) 37
  K88, K99 37
  P 37, 187
  in vaccines 353
  longus 37
fimbrillin 35
FimC protein 36
FimD protein 36
FimF protein 38
FimG protein 38
FimH protein 37, 38
fingerprint (DNA) 506
  AP-PCR 508
  in AFLP 511
  in DAF 508
  in RAPD 508
  in REA 506
  in RFLP 510
  with arbitrary primers 508
finO locus (in F plasmid) 201
fis gene 52, 138
Fis protein 52, 138
FISH 378
FkpA 155, 156
flagella 10, 30–34
  assembly 31, 71, 175
  basal body 31
  export apparatus 32
  filament 31

flagella (*cont.*)
  genes   176
  hook   31
  in the Archaea   33
  in spirochaetes   33
  periplasmic   33
  phase variation   200, 201
flagellar motor   32
flagellin   31, 175
flagellum (see flagella)
flaming   431
flat sours   398
flavin adenine dinucleotide (see
    FAD)
Flexibacter filiformis   41
FlgE   176
FlgH   176
FlgI   176
FlgM   175, 176, 187
FlhC   176
FlhD   176
FliA   175, 176
FliC   176
FliD   33, 176
FliF   176
FliG   176
FliM   176
FliN   176
floc blanket clarifier   417
flood plate   435
flow cytometry   68
fluid mosaic model   18
fluorescence energy resonance
    transfer (see FRET)
fluoroquinolones   464, 480
folate (folic acid)   466
folding (proteins)   155
food additives   395
food bacteriology   393
food hygiene   400, 404–406
food poisoning   334, 385, 400
  toxins (detection of)   403
food preservation   395
  acidification   399
  canning   396
  curing   400
  dehydration   399
  freezing   399
  ionizing radiation   399
  pasteurization   395
  pickling   399
  preservatives   399
  refrigeration   399

food preservation (*cont.*)
  smoking   400
  UHT processing   396
food-borne infection   400
food-borne intoxication   400
footprinting   272, 274, 275
forespore   74, 75
formaldehyde
  assimilation   126
  in sterilization   453
formate hydrogen lyase system
    86, 487
N-formylmethionine   151
fossil microorganisms   298
frame-shift mutation   194
Francisella   328, 391, 426, 519
Frankia   302
free coagulase   485
free-living bacteria   295
free residuals (chlorine)   418
freezing (food)   399
FRET   277
fructose 6-phosphate   85
FtsH protease   172
FtsY   108
ftsZ gene   54, 60, 185
FtsZ   54, 55, 57, 74, 75, 235
fumarate respiration   94
fumaric acid   93
fur gene   174
Fur protein   174
furanone   299
fusiform bacilli   6, 7
fusion proteins   234, 273
Fusobacterium necrophorum   243

gadCB genes   175
gal operon   168
β-galactosidase   165, 166, 225,
    235, 489
  in coliforms   517
  in ONPG test   489
gamma-radiation (for
    sterilization)   453
gap (in DNA)   197
gapped shuttle vector   228, 233
GAR-936 tetracycline   462
Gardnerella   383
gas gangrene   385
  toxin gene regulation   179
gas plasma   454
gas production (detection)   487
gas vacuoles   16

gastritis (*Helicobacter*)   338, 339
gastroenteritis   400
gatifloxacin   464
Gb$_3$ (on vascular endothelium)
    335
GC%   500
GC ratio   500
gel electrophoresis   224
gelatin (in media)   427, 431
gellan gum   249, 430
gemifloxacin   464
Gen-Probe test for
    Mycobacterium
    tuberculosis   497
gene   147
  autoregulation   164
  cassette   468
  cloning   214
  coding (sense) strand   149
  expression   147
  expression (in Archaea)   164
  expression (regulation of)
    163
  flagellar (transcription)   175
  fusion   55, 234
  induction   163
  nomenclature   192
  priority repair   163
  product (gp)   214
  pseudogenes   159
  repression   163
  silencing   190
  transcription   163
  transfer   205
gene cassette   468
gene conversion   158
gene expression (regulation)
  by circadian rhythms   188
  by DNA methylation   185
  by recombination   177
  by rate of decay of mRNA
    178
  by sigma factors   186
  by signal transduction   179
  by temperature   187
  by transcription
    antitermination   187
  by translational attenuation
    178
  by translational
    frame-shifting   181
gene fusion   234
gene transfer   205

general secretory pathway 107
generalized transduction 292
generation time 64
genetic code 150
genetic engineering 213
genetics 128
genome 191
genomic library 215, 218
genomics 130
genotype 193
gentamicin 461
genus 4, 5
geosmin 297
germ warfare 4
germinal centres 351
germination (of endospores) 75
GFP (see green fluorescent
    protein)
gliding motility 40
global warming (see greenhouse
    effect)
Gloeobacter 97
GlpF protein (of *E. coli*) 20
gluconeogenesis 123
glucose (PTS transport) 105,
    106, 107
glucose effect 168
glucose permease 106
glucose 6-phosphate 85, 121
glutamate (in *Rickettsia*) 523
glutamate synthase 304
glutamine synthetase 304
glutaraldehyde 454
glutathione *S*-transferase 235
glyceraldehyde 3-phosphate 120,
    121
glycerol facilitators 20
glycine betaine 48, 49
glycine↔serine 123
glycolysis 84
N-glycosylases 160
glycosylation 262, 288, 294
GMP 133
GOGAT 304, 307
gold leaching 409, 410
gonococcus 385
gonorrhoea 385
gp 214
Gram-negative bacteria 21, 22
Gram-positive bacteria 21, 22
Gram stain 444
    for living cells 444
Gram-type-negative 444

Gram-type-positive 444
Gram-variable 444
gramicidins 156
granzymes 357
gratuitous inducer 165
green fluorescent protein 55,
    235
green manure 308
green photosynthetic bacteria
    98
greenhouse effect 312
    methanotrophs 127
greening 492
grepafloxacin 464
Griffith's serogroups 30
groeL gene 156
GroeL 157, 171
groeS gene 156
GroeS 171
groundwater 415, 417
growth 44
    Arrhenius plot 67
    at 121°C 45
    balanced 64
    colony formation 62
    conditions for 44
    confluent 62
    curve 63
    death phase 65
    diauxic 68
    energy for 45
    exponential 54, 65
    lag phase 63, 65
    log phase 64, 65
    logarithmic 64, 65
    measuring 63
    optimum growth temperature
        45
    oxygen requirement 46
    rate (versus cell size) 49
    Ratkowsky plot 67
    specific growth rate 66
    stationary phase 65
    synchronous 67
    temperature 45
    unbalanced 63
GrpE 172
Gruyère cheese 393
GSP 107
GST 235
GTP 93, 122
GTPase (mammalian) 324
guanine 133

guanosine 133
gut epithelium (infection of)
    371
gyrase 137
    CcdB activity 144
    gyrase-targeted antibiotics
        464

H antigens 494, 523
HACCP system 404
haemagglutination 36
haemocytometer 440
α-haemolysin (*E. coli*) 105
haemolysins 492
haemolysis 492
haemolytic uraemic syndrome
    332, 335
Haemophilus 519
    adhesion 320
    antibody response to 354
    capsule 359
    in microflora 318
    influenzae type b vaccine 354
    lysis 345
    paracytosis 328
    proteome 160
    transformation 205
    transposon mutagenesis 255
    V factor 519
    vaccine 354
    X factor 519
haemorrhagic colitis 332
Hafnia 519
hair cells (*Aphanizomenon*) 514
halazone 421
Halobacterium 16, 30, 47, 99,
    131, 519
halophiles 47
halotolerant 47
hanging-drop method 482
harpins 110
head (phage) 279, 280, 282, 283
headful mechanism 283, 292
heat-shock proteins 171
heat-shock response 171
Helber chamber 441
helicase 139
helicase I 209
helicase II 162
Helicobacter 12, 13, 338, 519
    antigenic variation 363
    electroporation 207
    iron acquisition 368

Helicobacter (*cont.*)
  Lewis antigens   338
  pylori   338, 363
    E test   475
    flagella   31
    genome   129
helix (DNA)   135
Helmstetter–Cooper model   56
helper phage   233
helper T cell   351
hemolysis (see haemolysis)
HEPA filter   432
heterocyst   77, 79, 307
heteroduplex   198
heterolactic fermentation   86
  in silage   408
heterologous expression (of a
      gene)   158
heteropolymer   125
heterotrophs   117
hexachlorophene   457
hexidium iodide   445
hexose monophosphate pathway
      120, 121
Hfr donor   209
Hib vaccine   354
high-dry objective   445
high stringency (in PCR)   241
*Hin*dIII   220, 223, 224
hinge region   350
his operon   167
histamine   342, 345, 349
histidine kinase   179
histidine uptake   104
HlyA   361
HMP pathway   120, 121
H-NS protein   13, 187, 203
holdfast   70
holin   285
Holliday junction   198
holoenzyme   148
homing (intein)   158
homing (intron)   189
homolactic fermentation   86,
      393
homologous recombination   197
homology (in nucleic acids)   197
homopolymer   125
homopolymer tail   234
homoserine lactones   298, 413
hook (of a flagellum)   31
horizontal gene transfer   369
hormogonia   77, 80

hot-air oven   451
hotspots (mutation)   161
hot-start PCR   246
*Hpa*I   223, 224
HPr kinase   170
htpR gene   172
HTST pasteurization   396
HU protein   13
Hugh & Leifson's test   486
human genome   130
humoral immunity   355
HUS   332, 335
hyaluronate lyase   335
hyaluronic acid   29, 359
hybridization   501
hybridoma   355
hydrogen (from metabolism)
      87, 88
hydrogen hypothesis   1
hydrogen sulphide production
      94, 310, 488
hydroxyl radical   176
hydroxylamine   235
hygiene (food)   404
hypersensitivity   349
hypha   6, 7
Hyphomicrobium   61, 72, 126
hypochlorites   455

iatrogenic infection   319
ICAM-1, ICAM-2   342
ICE   362
ice-nucleation bacteria   310
iceberg lettuce (pathogens in)
      401
icm genes   328
icosahedron   279
ICP   408
IcrF   188
IcsA protein   111, 325, 332, 337,
      353
identification of bacteria   481 *et
      seq.*
IdeR   177
idiolites   66
idiotype (of antibody)   350
IF-1, IF-2, IF-3   151
IFN-α 346
IFN-β 346
IFN-γ 329, 346, 347, 352, 357,
      365
Ig   350
IgA   350

IgD   350
IgE   350
IgG   350
IgM   350
IHF   209
IL-1 etc. (see interleukin   1 etc.)
imipenem   475
immune cytolysis   343
immunization (passive)   355
immunoblotting   229
immunofluorescence
      microscopy   377
immunoglobulins   350
immunomagnetic separation (see
      IMS)
impedance monitoring   405
impetigo   385
IMS   437
IMViC tests   487
in situ hybridization   378
in vivo expression technology (see
      IVET)
Inaba serotype   353
inclusion bodies   269
incubation   63, 426
  anaerobic   439
incubator   426
indicator organisms   419
indole test   487
inducer exclusion   169
inducible antibiotic resistance
      468, 471
inducible nitric oxide synthase
      (see iNOS)
inducible stable DNA
      replication   141
induction (of genes)   163
induction (of phage)   289
infection
  adhesion in   320
  gut epithelium   371
  routes of   317
  SIRS   330
infectious jaundice   386
inflammation   341
infliximab   382
inhibitors (PCR)   249
initiation factors
  in protein synthesis   151
initiator codon   151, 232, 269
inlA gene   324
inoculation   63, 426
  methods for   435

inoculum 426
inorganic fermentation 95
iNOS 347, 348
insecticidal crystal protein 408
insertion–duplication
   recombination 254
insertion sequence 199, 201
inside-to-outside mechanism
   22
int gene 289
integration host factor 209
integrins 320, 342
integrons 468
intein 158
intein homing 158
interferons 329, 346, 347, 348,
   357, 365
interleukin 1 340, 342, 352, 364
interleukin 2 352, 364
interleukin 4 347, 352, 364
interleukin 5 352
interleukin 6 340, 342, 352
interleukin 8 324, 340, 343, 364
interleukin 10 352, 364
interleukin 12 352, 364
interleukin 18 365
internalin A 321, 324, 368
intimin 321, 368
intron 217, 219
   bacterial 189
   homing 189
inv-spa genes 324, 369
invasin 368
invasion (of mammalian cells)
   323, 341
   paracytosis 328
   transcytosis 328
   trigger mechanism 323, 324
   zipper mechanism 325, 326,
      327
invasomes 323
inverse PCR 244
inverted repeat 199
iodine (as antiseptic) 457
ionizing radiation 399, 453
ions (sequestration) 347
IpaA protein (*Shigella*) 324
IpaB protein 110, 362
IPTG 165, 215, 232, 267
iron (and pathogenicity) 335,
   366, 367, 368
iron chelators 367, 368
iron–sulphur proteins 91

IS (insertion sequence) 199,
   201
IS*1* 203
IS*10* 203
IS*900* 203, 439
IS*903* 201
IS*6110* 203, 499
iSDR 141
ISH (*in situ* hybridization) 378
isolating nucleic acids 222
isolation 436
isoniazid 177
   resistance to 177
isopropyl-β-D-thiogalactoside
   165
isopsoralen (in PCR) 247
isopycnic centrifugation 224
isoschizomer 221
isotype (of antibody) 350
iteron 143
IVET 360, 361

J chain (in IgA) 350
Jarisch–Herxheimer reaction
   339
jumping gene 199
junctional pore complex 40

kanamycin 461
katG gene 177
KatG 177
Kauffmann–White classification
   523
Kawasaki syndrome 385
kb 141
kdp system
   in osmoregulation 48, 179
kdpABC operon 179
Kelsey–Sykes test 456
kidney stones (nanobacteria)
   521
killer cells 357
killing temperature 397
kilobase (kb) 141
KinA, KinB 181
Klebsiella 306, 519
   butanediol fermentation 89
knock-out mice 258
Köhler illumination 447
Koser's citrate medium 488
Kovács' indole reagent 487
Kovács' oxidase reagent 485
*Kpn*I 223

Krebs cycle (TCA cycle) 92, 93
Kurthia 519

L-form cells 6, 47
L ring (of a flagellum) 31, 32
label (probe) 237
lac genes 165
lac operon 165, 166, 215
β-lactam antibiotics 458
β-lactam ring 458, 459
β-lactamases 459, 460
   extended spectrum 461
   inducible 460
   inhibitors 460
   nitrocefin test for 461
   site of action 459
lactic acid 86, 87, 88
lactic acid bacteria 86
lactic acid fermentation 86
lactic acid starter 393
Lacticin 3147 296, 394
Lactobacillus 129, 394, 484, 520
   in human microflora 318
   in silage 407
Lactococcus 20, 100, 105, 189,
   393
   formerly *Streptococcus* 525
lactoferrin 367
lactose 393, 489, 491
lacZ fusion 235, 236
LAD (leukocyte adhesion
   deficiency) 342
lag phase (growth) 63
lagging strand (DNA) 139
LAL test 406
LAMPs 341
Lancefield's grouping test 525
lantibiotics 295, 296
las (in *P. aeruginosa*) 299
latency (in tuberculosis) 329
latex slide-agglutination tests
   494
Latin binomial 5
lawn plate 435
LB broth 206
LBP (see lipopolysaccharide-
   binding protein)
LCR (see ligase chain reaction)
LCx® assays 265, 498
leader peptide 166
leader sequence 166
leading strand (DNA) 139
lecithin 18

lecithinase 325
lectin pathway 345
LEE pathogenicity island 369
Legionella 385, 520
    antiphagocytic activity 328
    mip gene 520
    tissue-destructive protease 385
    typing 520
legionnaires' disease 385
Leloir pathway 120
lens
    high-dry 445, 482
    numerical aperture 446
    oil-immersion 446
leprosy 386
Leptonema illini 33, 39
Leptospira interrogans 130
Leptospirillum ferrooxidans 409
leptospirosis 386
lethal temperature 397
lettuce (source of pathogens) 401
leucine-responsive regulator protein 149
leucine-rich repeat 371
leucocidins 361
Leuconostoc 393, 520
leukocyte adhesion deficiency (LAD) 342
levofloxacin 464
Lewis x, y (*Helicobacter pylori*) 338
LexA protein 170
library
    cDNA 217
    genomic 215, 218
    screening 215
lichens (*Nostoc*) 521
lif gene 296
ligase (DNA) 216, 218
ligase (RNA) 294
ligase chain reaction 263, 265, 266, 498
ligation 233
LightCycler™ hybridization probes 238
light-harvesting complex 97
light reaction 96
light-up probes 238
Limulus amoebocyte lysate test 406
lincomycin 548

line probe assay 477, 478
linear chromosomes 129
linear plasmids 131
linezolid 469, 470
linked recognition 351
linkers 232, 233
lipid A (in LPS) 23, 25
lipo-oligosaccharides 27
lipopolysaccharide-binding protein 342
lipopolysaccharides (see LPS)
Listeria 520
    aesculin (esculin) hydrolysis 491
    control in cheese 394
    in silage 407
    invasion of mammalian cells 324
    iron acquisition 368
    low-temperature growth 399
    nisin 400
    and NK cells 357
    non-pathogenic strains 400
    sortase 19
listeriolysin O 325
listeriosis 386, 400
litmus milk 490
liverworts (*Nostoc*) 521
LktA toxin 362
LmrA transporter 105
lockjaw 334
locus (genetics) 192
log phase (growth) 64, 65
$\log_2$ scale 64, 65
logarithmic death 397
Lon protease 172
longus 37
loop 432, 434
lophotrichous flagella 31
LOS 27
low stringency (in PCR) 241
low-temperature steam–formaldehyde sterilization 453
LPS 23, 25, 342, 371
    Lewis antigens (*Helicobacter*) 338
lpxA gene 25
lpxC gene 25
Lrp 149
LT-I, LT-II toxins 332
luciferin–luciferase 405
lugol's iodine 444

Lumac® Autobiocounter M 4000 405
Luria–Bertani broth 206
Lyme disease 370, 386
lysine decarboxylase test 489
lysis protein (phage-encoded) 285
lysogeny 288
    and SOS system 289
Lysol 455
lysosome 341
lysostaphin 25, 223, 296
lysozyme 25, 28, 223, 340
lysU gene 172
lytic cycle (of phage T4) 280

M cells 324, 340
M phenotype (streptococcal) 463
M proteins 30
mAbs 355
MAC 343, 345
MacConkey's agar/broth 428
macrocapsule 29
macrolide antibiotics 462, 480
macrophages 341
    CR3 receptor 328
    effect of toxins, LPS etc. 335, 364
    effect of YopJ 361
    invasion of 325, 341
    killed by *Shigella flexneri* 361
macropinocytosis 324
magnetic separation 437
magnetosome 14
maintenance medium 429
malic acid 93
mannitol (PTS transport) 106
mannose-binding lectin 345, 346
mannose-sensitive haemagglutination 36
MAP 151
marginal zone B cells 351
MASPs 346
mast cells 349
Maxam–Gilbert ladder 274
MBL (mannose-binding lectin) 346
Mbl protein (in cytoskeleton) 8
McIntosh & Fildes jar 439
MCP 182, 183
MCS 232

mecA gene   469, 479
  induction   469
mechanosensitive channels   19,
      20, 49
mecI, mecR1   469
medium   426, 428
  basal   427
  blood agar   427, 428, 484, 492
  chocolate agar   428
  Christensen's urea agar   489
  CLED   428, 429, 489
  complete   195
  defined   429
  differential   428
  enriched   427
  enrichment   427
  MacConkey's   427, 428
  maintenance   429
  minimal   195, 217
  Møller's decarboxylase broth
      489
  PRAS   375
  preparation of   430
  pre-reduced   375
  selective   427
  Simmons' citrate agar   488
  SMAC   429
  solid   427
  transport   428, 429
  XLD   428, 429
  xylose–lysine–deoxycholate
      428, 429
MEE   495
melibiose symport (in *E. coli*)
      102
melioidosis   386
melting (DNA)   138
membrane attack complex   343,
      345
membrane filtration
  in water testing   420
  for sterilization   454
membrane fusion protein
      105
memory cells   352, 357
meningitis   386
  conjugate vaccine   354
mesophilic   46
messenger RNA (see mRNA)
metabolism   81
metachromatic granules   16
metals (bioleaching)   409
methane monooxygenase   127

methane production   95
  in anaerobic sewage
      treatment   415
  in marshy soils   96
Methanobacterium   96
Methanococcus   28, 54, 96, 108
methanofuran   96
methanogens   95, 520
Methanolobus   96
Methanothermus   96
methicillin   460, 469, 474
  MRSA   464, 469
methionine aminopeptidase
      151
methionine deformylase   151
methyl-accepting chemotaxis
      proteins   182, 183
methyl red test   488
methylases   145
methylation (of DNA)   145
  as a control mechanism   185
methylene blue   428
N^5,N^{10}-methylene-THF   123
Methylococcus   126, 127, 520
Methylomonas   126, 127
Methylophilus   126, 127, 408
methylotrophy   126
methyltransferases   145
metronidazole   465
mfd gene   163
MFP   105
MHC class I molecule   346, 347,
      356
MHC class II molecule   347,
      351, 352, 356
MHC restriction   356
MIC   471
microaerophilic   47
microarray   276
microbial insecticides   408
microbial mat   298
microcapsule   26, 27, 29
  Vi antigen   523
microcins   296
Microcystis   418
microflora   295
  human   318
micromethods   491
micrometre   9
micron   9
microorganisms   1
  fossil   298
microplasmodesmata   43, 80

microscopy   446
  epifluorescence   377
  in the pathology laboratory
      373
  Köhler illumination   447
  literature   449
  oil-immersion lens   446
  phase-contrast   447, 448
microstrainer (water supplies)
      417
Miles & Misra's method   443
milk   393, 395, 398
minC   56
MinCD   56
minD   56
mineralization   310
minimal medium   195, 217
minimum infective dose   317
minimum inhibitory
      concentration   471
MIP channel   20
mip gene (*Legionella*)   520
mismatch repair   161
mis-sense mutation   194, 195
mitomycin C   289
mitosis-like partition (plasmid
      R1)   144
mixed acid fermentation   86, 87
MLST   499
MMO   127
mobile genetic elements   158,
      212
mobility-shift assay   271
Mobiluncus   383
modification (of DNA)   145
modulins   338, 363
molecular beacon probes   238
molecular biology   129, 191
molecular chaperones   156
molecular cloning   214
Møller's decarboxylase broth
      489
monobactams   458, 459
monoclonal antibodies   355
  against *E. coli* O157:H7   355
  against *Vibrio cholerae* O139
      384
Monod equation   66
monosodium glutamate   395
monotrichous flagella   31
Moorella (see *Clostridium*, page
      517)
morphogenes   56

morphology (determination of) 482
morphology (in evolution) 8
mosquitoes (biological control) 408
most probable number (MPN) 420
MotA 176
MotB 176
motility 39, 482
    actin-based 42, 337
    gliding 40
    calcium-dependent 42
    twitching 40
mouth microflora 318
moxifloxacin 464
mpl gene 115
MPN 420
MppA 115
MR test 488
MreB protein 8, 54, 144
MRHA 36, 37
mRNA 147, 150, 157
    decay, and gene expression 178
    degradation 157
    differential display 276
    polyadenylation 157
    polycistronic 159, 164, 178
MRSA 469
    DNA-based detection 479, 497
    mecA gene 469, 479
    mechanism of resistance 469
    MLST data 499
    resistance to vancomycin 469
    Synercid® 463
    typing 498, 499
MS ring (of a flagellum) 31, 32, 33, 70
MscL channel 20, 49
MSHA 36, 37
muc genes 336
MucA 336
mucolytic agent 374
MUG test (for *E. coli*) 491
mukB gene 53
MukB protein 53, 54
multicopy plasmids 144
multilocus enzyme electrophoresis 495

multilocus sequence typing (MLST) 499
multiple cloning site (MCS) 232
multiple fission 61
multiple-tube test 420
multiplex PCR 246
murein 22
mut gene products 161, 162
mutagen 192
mutagenesis 251
    mutagen-free 251
    oligonucleotide-directed 252
    random 251
    signature-tagged 256, 257
    site-directed 252
    site-specific 252, 253
    SOS 170
    transposon 255
mutant 192
    isolation of 193
    streptomycin-dependent 472
mutation 191
    back 196
    deletion 194
    frame-shift 194
    hotspots 161
    methylcytosine hotspots 161
    mis-sense 194, 195
    nonsense 194
    null 185
    phase-shift 194
    pleiotropic 155
    point 192
    polar 164
    rate 193
    silent 195, 475
    spontaneous
    transition 161
    transversion 161
mutation rate 193
mutator genes 191
mutator strains 251
Mutazyme® 251
MutH 162
MutU 162
mutualism 301
MutY glycosylase 161
mxi-spa genes 324
mycelium 6
mycetocytes 301
mycetome 301
Mycobacterium 520
    avium 327, 328

Mycobacterium (*cont.*)
    bovis 329, 469
    leprae 129, 177
        pseudogenes 160
    malmoense 497
    paratuberculosis 203, 439, 456, 497
    peptidoglycan 25
    tuberculosis (see below)
Mycobacterium tuberculosis
    AMTDT 497
    antibiotic resistance 177, 389, 473
    BACTEC™ culture 430
    complex 329, 496
    culture from sputum 374
    ELISPOT assay 329
    ESAT-6 329
    genome 128, 129
    intein 158
    invasion of macrophages 328
    iron acquisition 367
    IS*6110* 203
    isolation of DNA 223
    isoniazid resistance 177
    latent infection 329
    LCR-based test 265, 498
    mechanosensitive channel 20
    mycolic acids 22
    nitric oxide 348
    oxidative stress 177
    rpoB gene 465, 473
    rRNA genes 506
    Ziehl–Neelsen stain 445
mycobactin 367
mycolic acids 22
Mycoplasma 29, 521
myeloperoxidase 343
Myxobacterales 300
Myxococcus 40, 41
MZ B cells 351

N-terminal amino acid 152
NA 446
Na⁺/melibiose transport 102
NAD 82
    V factor for *Haemophilus* 519
NAD dehydogenase 91
NADP(H) 82
nafcillin 460
nalidixic acid 464
naming bacteria 4
nanobacteria 9, 521

NASBA 263–266
nasopharynx (microflora) 318
natural killer cells 357
Ndd protein 281
necrotizing fasciitis 387, 497
needle complex (type III
          secretion) 110
    in *Salmonella typhimurium*
          110
    in *Yersinia enterocolitica* 110
negative control (operon) 166
negative staining 26, 27, 445
Neisseria 521
    adhesins 369
    autotransporters 111
    capsule 359
    chocolate agar 428
    fimbriae 36, 37
    gonorrhoeae 265
    iron acquisition 355, 367
    lipo-oligosaccharides 27
    meningitidis 129, 359
    Opa+ strains 328
    PACE 2C test 496
    transcytosis 328
    transformation 205
neomycin 381, 461
Neo-Sensitabs 474, 475
nephelometry 68
nested deletions 259
nested PCR 245, 246
neurotoxin 331, 334
neutral red 428
neutrophils 341
NFκB 324, 371, 372
nic 209
nick (in DNA) 198, 209, 228
nick translation 236, 237
nicotinamide adenine
          dinucleotide (see NAD)
nif genes 306
nigrosin 445
Nile blue A 15
nisin 296, 400
nitrate assimilation 305
nitrate reduction test 490
nitrate respiration 94
nitrates (in water supplies) 421
nitric oxide 347
nitrification 305, 306, 307
    in sewage effluents 413, 414
nitrification inhibitor 308
nitrifying bacteria 95

nitrite utilization
    in anammox 95
Nitrobacter 61, 95, 521
nitrocefin 461
Nitrococcus 95
nitrogen cycle 304
    and agriculture 307
nitrogen fixation 306
    in agriculture 307
    in wheat 308
    in wild rice 302
nitrogenase 79, 306, 307
Nitrosococcus 95, 521
Nitrosomonas 95
nitrous acid (mutagen) 192
NK cells 357
NO 347
nocardicins 458, 459
Nod proteins 372
Nodularia 418
nodules (nitrogen-fixing) 302,
          308
    in non-leguminous plants
          308
non-brewed condiment 395
non-O1 serogroups 383
nonsense codon 154
nonsense mutation 194
norfloxacin 464
Northern blotting 229
nosocomial disease 317
Nostoc 80, 306, 521
*Not*I 223, 506
novobiocin 464
ntr genes 306
nucleases 157
nucleation site (of Z ring) 55
nucleic acids 132 *et seq.*
nucleic acid sequence-based
          amplification (see
          NASBA)
nucleoid 10, 11, 13
nucleoprotein filament 197
nucleoside 132
nucleotide 132
nucleotide-binding
          oligomerization domain
          proteins (see Nod
          proteins)
nucleotide excision repair 162
null mutation 185
numerical aperture (of lens)
          446

NusA 173
nutrient agar, broth 427, 428
nutrients (for growth) 44

O antigens 494, 523
    modification by phage 279
O1 serogroup (*Vibrio cholerae*)
          383
O126:H27 (*E. coli*) 333
O139 Bengal (*Vibrio cholerae*)
          279, 288, 353
O157 (*E. coli*) 400, 438, 439,
          497
    latex test 494
O157:H7 (*E. coli*) 129, 299, 332,
          335, 355
    monoclonal antibodies 355
    phage-mediated detection
          235
    SMAC medium for isolation
          429
O-specific chains 23, 27
objective lens (oil immersion)
          446
obligate aerobe 46
ochre codon 154
O-F test 486
ofloxacin 464
Ogawa serotype 353
oil-immersion lens 446
Okayama–Berg method 220
Okazaki fragments 139, 140
oleandomycin 105
oligo(dT)-cellulose 224
oligonucleotide cataloguing
          504
oligonucleotide-directed
          mutagenesis 252
oligopeptide permease
          (*opp*-encoded) transport
          system 115
Omp 28
OmpC 28, 179
OmpF 28, 179
OmpR 179
OmpT 111, 272
ONPG test 489
opa genes 363
Opa adhesins (*Neisseria*) 363
Opa+ strains 328
opal codon 154
open complex 148
open culture 66

operons   164
   ara   166
   attenuator control   166
   bgl   170
   definition   164
   his   167
   infC-rpmI-rplT   167
   kdpABC   179
   lac   165, 166
   negative control   166
   polar mutation   164
   positive control   166
   promoter control   166
   puf   178
   rrn   159
   translational control   167
   trp   167
oppBCDF   115
opportunist pathogens   317
opsonization   341
optimum growth temperature
      45
orf18   213
oriC   138
origin (of DNA replication)   138
oriMs   141
oriT   209
ornithine decarboxylase test
      489
Oroya fever   338
Oscillatoria   521
   anaerobic photosynthesis   97
   blooms in reservoirs   418
osmoregulation   48
osmotic lysis   21
outer membrane   23, 25–28
outfall   415
outgrowth (of spore)   75
overproduction (of proteins)
      267
overwound helix   135, 136
oxaloacetic acid   93
Oxalophagus (see *Clostridium*,
      page 517)
oxazolidinones   469
oxidase test   484
oxidation-fermentation test   486
oxidative burst   341, 343
oxidative phosphorylation   91
oxidative stress   176
Oxobacter (see *Clostridium*, page
      517)
oxoG   161

oxoglutaric acid   93
oxolinic acid   464
oxygen (for growth)   46
oxygenic photosynthesis   97
OxyR   177
oxytetracycline   462
ozone disinfection (water)   419

P fimbriae   37, 187, 369
P ring (flagellar)   31, 32
P site (on ribosome)   151, 232
PACE 2C test   496
packaging (phage DNA)   288
packet (of cells)   9
PAE   480
PAI-1   369
*Pal*I   223
palindromic sequence   233, 234
palisade form   9
PAMPs   371
panning   273
pannus   389
Panton–Valentine leucocidin
      361
Pap pili   37
PAPs   157
paracoagulation   486
paracytosis   328
parasites   301
parasporal crystal   408
parC gene   53, 144, 464
parE gene   464
Park nucleotide   125
parsimony, principle of   500
partitioning (of plasmids)   144
partitioning (of chromosomes)
      53
partitioning complex (plasmid
      R1)   144
partner gene   235
passive immunization   355
Pasteur pipette   434
Pasteurella   521
   haemolytica   362
   LktA toxin   362
   multocida   129
pasteurization   395
pathogen   316
   detection/characterization   373
pathogen-associated molecular
      patterns   371
pathogen-host interactions   370
pathogenesis   330

pathogenicity
   conferred by phages   278
   conferred by plasmids   131,
      316
pathogenicity islands   369
   cag   339, 370
   LEE   369
   mobile   370
   PAI-1   369
   SPI-1   369
   TCP-encoding   331
pattern-recognition receptors
      340, 371
pbpA, pbpH   9
PBPs (see penicillin-binding
      proteins)
pBR322 (plasmid)   230
PCBs   423
pcnB gene   143
PCP   423
PCR   239 *et seq.*
   amplicon   239, 240
   arbitrarily primed   508
   ARMS   244, 245
   asymmetric   245
   broad-range primers   242
   cloning PCR products   248
   controls   241
   decontamination
      isopsoralen method   247
      UNG method   247
   degraded DNA   250
   detection of products   242
   in diagnosis   243, 496
   Dynabeads®   438
   extraneous DNA inactivation
      247
   facilitators   249
   hot start   246
   immunomagnetic separation
      438
   in identification/typing   243,
      498
   inhibitors   249
   inverse   244
   isopsoralen   247
   multiplex   246
   nested   245, 246
   PCR-RFLP   511
   PCR ribotyping   511
   probes used in   238
   QPCR   249
   quantitative   249

PCR (*cont.*)
  real-time 249
  reconstructive 250
  REP-PCR 246
  reverse transcriptase 247
  ssDNA from PCR products 248
  Stoffel fragment 239
  stringency of conditions 241
  Taq polymerase 239
  threshold cycle 250
  touchdown 247
  in typing 498
  UNG 247
  uses 242
PCR-RFLP 511
PCR ribotyping 511
pCT plasmid 498
pellicle 63
Pelodictyon 9, 61, 522
pelvic inflammatory disease 385
penicillin amidase 459
penicillin-binding proteins 9, 19, 25, 55, 458, 469
penicillins 458, 459, 460
  allergy 349
pentachlorophenol 423
pentose phosphate pathway 120, 121
PEP
  formula 85
  in PTS 105–107
peptic-ulcer disease 338
peptide deformylase 151
peptide nucleic acid (PNA) 137
peptidoglycan 9, 22, 24, 371
  biosynthesis 50, 125, 126
  biosynthesis (3-for-1 model) 50, 51
  recycling 115
  structure 22, 24
  transport through 115
peptidyl-prolyl isomerase 156
peptidyltransferase 152, 462
Peptococcus (see *Streptococcus*, page 524)
peptone water 428
peptone water sugars 487
Peptostreptococcus (see *Streptococcus*, page 524)
perchlorate degradation 423, 424
perforins 356

periplasmic binding protein 104
periplasmic flagella 33
  in *Treponema pallidum* 34, 35
periplasmic gel 23
periplasmic permeases 106
periplasmic region 10
permease 106
peroxynitrite 348
pertussis toxin 353
pesticides (in water supplies) 421
Petri dish 426
  illustration 485
  non-vented 485
  vented 426
Peyer's patches 337
PFGE 225, 498
Pfu polymerase 239
pH indicators (in media)
  bromcresol purple 489
  bromthymol blue 428, 486
  cresol red 489
  litmus 490
  methyl red 488
  neutral red 428
  phenol red 428, 489
  phenophthalein 490
phage 278 *et seq.*, 403
  burst size 285
  conversion 279, 291, 523
  display 272, 273
  helper 233
  in recombinant DNA technology
    cosmid 231
    library 273
    M13 253
    phagemid 229, 233
    T4 ligase 234
  lysogeny 288
  temperate 278
  therapy 286, 287, 382, 470
  typing 494
  virulence determinant 278
  virulent 278
phage conversion 279, 291, 523
phage CTXΦ 279, 288, 293, 353
phage display 272, 273
phage f1 279, 291
phage f2 291
phage fd 282, 283, 291
phage H-19B 278

phage L54a 291
phage λ 200, 250, 279, 282, 283, 285, 289, 290
phage library 273
phage M12 279, 283, 291
phage M13 253, 291
phage MS2 291
phage Mu 279, 292
phage MV-L3 279
phage P1 288
phage P22 288, 293
phage φ29 279, 282, 283
  DNA packaging 288
  DNA replication 142
phage φX174 279
phage PM2 279
phage Qβ 282, 283, 291
phage T1 282, 283
phage T2 282, 283, 294
phage T3 282, 283
phage T4 279–284, 294
phage T6 282, 283, 294
phage T7 282, 283
phage therapy 286, 287, 382, 470
phage titre determination 250
phage typing 494
phagemid 229, 233
phagocyte 341
  activity of antibiotics within 480
phagocytosis 341
  avoiding phagocytosis 341, 359
phagolysosome 341
phagosome 324, 341
pharmacokinetics 381
phase 1, phase 2 antigens (*Salmonella*) 523
phase-contrast microscopy 447, 448
phase-shift mutation 194
phase variation 201, 362
phasins 15
PHB (see poly-β-hydroxybutyrate)
phenol (as disinfectant) 455
phenol coefficient 456
phenol red 428, 489
phenolphthalein 490
phenotype 193
phenylalanine deaminase test 489

pheromones 207, 299
phoA 255
PhoE porin 28
phosphatase test 490
phosphatidylcholine 18
phosphatidylethanolamine 18
phosphatidylglycerol 18
phosphodiester bond 134
phosphoenolpyruvate (see PEP)
phosphoenolpyruvate-dependent
    phosphotransferase system
    (see PTS)
phosphogluconate 121, 125
phosphorelay 180, 181
Photobacterium 298
photolithotrophs 99
photolyase 163
photoorganotrophs 99
photophosphorylation 97
photosynthesis 96
    anoxygenic 98
    oxygenic 97
phototrophs 83, 96
phyletic classification 500
phytoplankton 302
pickling 399
PilC 37
pilE, pilS 363
pili 38
Pilimelia 77
p*Inv* 332
pilin 35
piperacillin 475
plague 387
plant tumour (*Agrobacterium*)
    114
plaque (dental) 321
plaque (phage) 285, 495
plaque assay 285
plasma cells 351
plasma membrane 10
plasmalogens 19
plasmid 131
    absence 131
    Cit 131
    ColE1 142
    copy number 144
    de-repressed 201
    EAF 369
    expression vector 232
    F 201
    isolation 226, 228
    linear 131

plasmid (*cont.*)
    mitosis-like partition 144
    multicopy 144
    partition 144
    pBR322 230
    pCT 498
    R 131
    R1 partition 144
    replication 142
        relaxed 145
        stringent 145
    TOL 131
    vector (in genomic library)
        216, 232
plasmid-independent
    conjugation 212
plasmin 336
plasminogen 336
plastics from bacteria (Biopol)
    422
plate 426
plcB gene 325
pleiotropic mutation 155
plenum ventilation 432
pleomorphism 6
plugging pipettes 434
plus strand (of a gene) 149
pmf 90, 91, 92
    in aerotaxis 101
PNA 137
    antisense PNA 470
pneumonia 388
pneumonic plague 387
point mutation 192
polar mutation 164
polarity (of DNA strand) 134,
    135
pollutants (environmental) 423
poly(A)polymerase 157
polyacrylamide gel 225
polyadenylation
    of bacterial mRNA 143, 157
    of eukaryotic mRNA 157, 217
polychlorinated biphenyls 423
polycistronic mRNA 159, 164,
    178
polyclonal activation 351, 366
poly-*β*-hydroxybutyrate 15
    biodegradable plastic 422
    staining 15
    synthesis 126
polylinker 232
polymer 125

polymerase (see DNA polymerase
    and RNA polymerase)
polymerase chain reaction (see
    PCR)
polymyxins 463
polynucleotide phosphorylase
    157
polypeptides 149
polyphasic taxonomy 500
polyphosphate 16
poly(A)polymerase 157
polyribosome 154
polysaccharide capsules 29
polysome 154
polyurethanases 423
porins 27, 28
    in Gram-positive bacteria 23
positive control (operons) 166
post-antibiotic effect 480
potassium pump 103
pour plate 442
PPA test 489
ppGpp 53, 173, 174
PpiA 155
PQQ 101
PRAS medium 375
PRD 107, 169
predators 300
pre-mRNA (eukaryotic) 219
pre-reduced media 375
pre-septal peptidoglycan 55, 56
preservatives 399
prespore 74
presumptive coliform count 420
primary production 302
primase 139
primer (in DNA synthesis) 139
primer-dimers 240
primers (in PCR) 239 *et seq.*
    arbitrary 508
    broad-range 242
primosome 139
pRNA 288
probenecid 381
probes 236
    acridinium ester label 496
    combination 496
    enzyme-coupled 237
    in diagnosis/identification
        496
    labelling 237
    LightCycler™ hybridization
        238

probes (*cont.*)
  light-up 238
  molecular beacon 238
  nick translation 236, 237
  PACE 2C 496
  TaqMan® 238
probiotics 380
processivity 141
Prochlorococcus 98, 129
Prochloron 98, 522
Prochlorophyta 97
Prochlorothrix 98
prodigiosin 524
prohead 283
prokaryote–eukaryote split 1
prokaryotes
  classification 500
  two domains 1
promoter 147
  strength 164, 267
proof-reading (DNA) 160, 239
prophage 288
prophylaxis 355
propidium iodide 445
β-propiolactone 454
Propionibacterium 522
  dairy products 393
  in microflora 318
prostheca 69
proteasomes 11, 348, 356
protein A 359, 363
  use of 222
protein display 272
protein folding 155
protein priming 142
protein secretion 103–115
  Tat system 113, 114, 115
  types I, II, III etc. 103–114
protein synthesis 149
  chaperones 156
Proteobacteria 505, 521
proteome 160
proteomics 160
Proteus 86, 522
  swarming 72
proton ATPase 90, 91
proton/lactose symport 103
proton motive force (see pmf)
protoplasts 21
prototroph 195, 196
ProU transport system 48
PRRs 340, 371
PSI, PSII 97, 98

pseudocatalase 484
pseudogenes 159
pseudoknot (RNA) 167
pseudomembranous colitis 388
Pseudomonas 522
  carboxydovorans 118
  cystic fibrosis 336
  denitrification 94
  extracytoplasmic oxidation 101
  flagella 31
  ice nucleation 310
  introns 189
  iron uptake 367
  membrane permeability 180
  motility 39
  mucoid strains 336
  pyoverdin 367
  quorum sensing 299
  stutzeri 424
  swimming speed 39
pseudomurein 28
*Pst*I 223, 224
psychrophilic bacteria 46
psychrotrophic bacteria 46
PTS 105–107, 168
PTS-regulated domain 107, 169
puf operon 178
pullulanase 109
pulsed electric field disinfection 456
pulsed-field gel electrophoresis 225
purine bases 132
purple membrane 99
purple photosynthetic bacteria 98
pvdS gene 367
pyoverdin 367
pyrC gene 178
pyrimidine bases 132
Pyrococcus furiosus 241
Pyrodictium 45, 522
pyrophosphate 139, 259
pyrophosphotransferase 174
Pyrosequencing™ 259
pyrroloquinoline quinone 101
pyruvic acid 84
  formula 85
pyuria 374

Q fever 388, 396
QACs 455, 457

Qiagen kit (plasmid isolation) 226
QPCR 249
QSY dyes 277
quantitative PCR 249
quarantine 380
quaternary ammonium compounds 455, 457
quick penicillin secretors 468
QuikChange™ 252
quinolobactin 367
quinolone antibiotics 464
quinones (in ETCs) 91
quinupristin/dalfopristin 463
quorum sensing 298, 470
  N-acyl-L-homoserine lactones 298
  agr locus (*Staphylococcus*) 181
  AHLs 298
  azithromycin as inhibitor 299
  biofilms 300, 323
  furanone 299, 470
  type III secretion 299
  virulence factors (regulation) 299, 359

R bodies 14
R plasmids 131
RadA 197
random amplified polymorphic DNA 508
random walk 39
RapA, RapB 181
RAPD 508
rapid sand filter 417
Rapidec Staph test 492
rare-cutting enzyme 221, 223
Ratkowsky plots 67
RBS 232
RE 145, 221
REA 506
reaction centres (PSI, PSII) 97, 98
reactive lysis 345
reading frame 232
readthrough 232
real-time PCR 249
RecA 170
RecA* 170
RecBCD 197, 199
RecF pathway 197, 199
RecJ 162, 198

recognition sequence (of RE)
    221
recombinant DNA technology
    213, 266
recombinase 199, 200
recombination 196
    homologous 197
    hotspots 198
    insertion–duplication 254
    RecBCD pathway 197
    RecF pathway 197
    site-specific 199
recombinational regulation 177,
    199
reconstructive PCR 250
RecQ 198
reductive pentose phosphate
    cycle 118
reductive TCA cycle 118
refrigeration (food) 399
regulons 167
relA gene 174
relapsing fever 362
relaxed DNA 135
relaxed (plasmid replication)
    145
relaxed response (to starvation)
    174
relaxosome (relaxation
    complex) 209
release factor (protein synthesis)
    153, 154
Remicade® 382
Rep protein 143
REP sequence 510
REP-PCR 510
repair (DNA) 160
repetitive sequence-based PCR
    510
replacement vector 227
replica plating 196
replicase 283
replication (of DNA) 52, 138,
    142
replication factory 52, 139
replication fork 139, 140
replicative transposition 202,
    203
repliconation 209
replisome 52, 139
reporter dye 238
reporter gene 236
repressor titration 215

reproduction (of bacteria) 44
reservoir (of infection) 372
reservoir (water) 416
resistance to antibiotics 467,
    473 et seq.
resolvase 203
respiration 84, 89–94
    anaerobic 94
respiratory chain 91, 92
response regulator 179
resting cells 73
restriction (DNA) 145
    antirestriction 146, 294
    methyl-dependent (DpnI) 146
    non-classical 146
restriction endonucleases 145,
    216, 221, 223
    isoschizomers 221
    rare cutting 221, 223
    star activity 221
restriction enzyme analysis 506
restriction fragment length
    polymorphism (see RFLP)
reticulate body (Chlamydia)
    390, 516
retrieval vector 228
reverse electron transport 100
reverse transcriptase 219, 220
reverse transcriptase PCR 247
RFLP 510
    PCR-RFLP 511
Rhizobium 302, 522
rhizosphere 313
rhl (in P. aeruginosa) 299
rho factor 148
rho-dependent termination 148
rho-independent termination
    148
Rhodobacter capsulatus 178, 263
Rhodopseudomonas 309
Rhodospirillales 98, 306
ribonucleic acid (see RNA)
ribonucleosides 132
ribonucleotides 132
ribose 5-phosphate 121
ribosomal RNA (see rRNA)
ribosome binding site 232
ribosomes 10, 14
    composition 14
    and growth rate 158
ribotyping 506
    PCR ribotyping 511
ribozyme 189

ribulose 1,5-bisphosphate 119
ribulose monophosphate
    pathway 127
ribulose 5-phosphate 119, 121
rice vinegar 395
rice water stools 331
Rickettsia 131, 257, 391, 522
Rideal–Walker test 456
rifampin 465, 480
rifamycins 465
river blindness 526
RNA 138, 188
    helicase 157
    interference (RNAi) 190
    ligase 294
    messenger (mRNA) 147, 150,
        157
        fate 157
        polyadenylation 157
    polymerase 147
    pRNA 288
    pseudoknot 167
    ribosomal (rRNA) 14
    synthesis 147
    thermometer 172
    total 217, 224
    transfer (tRNA) 151
RNA I, RNA II 142
RNA III 181
RNA interference (RNAi) 190
RNA polymerase 147, 465, 473
    rifamycins 465, 473
RNA synthesis 147
RNA thermometer 172
RNAi 190
RNase A 226
RNase E 11, 157, 189
RNase H 219, 220, 247, 264,
    265
RNase L 190
RNase P 189
RNase T1 503
Robertson's cooked meat
    medium 441
Rochalimaea (see Bartonella, page
    515)
rods 6
rolling (of white blood cells)
    342
rolling circle mechanism 143,
    212
root nodules 302
RP59500 463

rpoB gene   465, 473
rpoD gene   172
rpoH gene   172, 186, 188, 270
rpoN (= *ntrA*) gene   306
rpoS gene   174, 186
rrn operons   159
rRNA   14, 158, 159
rRNA oligonucleotide
      cataloguing   504
Rsd protein   187
rtPCR   247
RTX toxins   361
RuBisCO   17, 118
ruffling   323
rumen   301
Ruminococcus   120, 301, 523
      obeum   318
RuMP pathway   127
RuvB, RuvC   198

S gene (phage λ)   285
S layer   30
S1 endonuclease   233
sacculus   22, 25, 51
safety (in the laboratory)   425
safety cabinet   431, 433
      class I, II, III   432
*Sal*I   221, 223
salami   86
salicin   170
Salmonella   86, 523
      adaptation to acidity   48, 174
      antigenic formula   523
      D_{60} value   397
      dublin   337
      enterica   523
      food poisoning   401
      histidine uptake   104
      invasion of cells   323
      iron   367
      needle complex   110
      pathogenicity islands   369
      phase variation   200, 201
      siderophore (enterochelin)
            367
      typhi   323, 337, 391
      typhimurium
            Ames test   196
            strain DT104   496
            lysogeny (phage P22)   288
            needle complex   110
            SsrA–SsrB system   179
      typhoid   337, 391

Salmonella (*cont.*)
      vaccines   353
      XLD medium   429
salmonelloses   337
      vaccines   353
salvage pathway (complement
            fixation)   344, 346
salvage transport   115
sand filters (water supplies)   417
Sanger's chain termination   260,
            261
saprotrophs   300
Sarcina   7, 524
*Sau*3AI   223
sauerkraut   399
scaffolding proteins   283
scalded skin syndrome   388
scarlet fever   388
SCP   408
screening libraries   215, 219, 222
SDA   266
SDS   226
sec-dependent pathway   107
SecA   107
SecB   107, 155
secondary metabolites   66
secondary response (immune)
            352
secretion (of proteins)   103
secretion sequence (of a
            protein)   105
secreton   109
secretory component (of IgA)
            350
secretory IgA   350
SecYEG   107
selectin   342
selection synchrony   67
selective medium   427
selective toxicity   458
selenate respiration   94
selenite broth   427, 428
self-splicing
      inteins   158
      introns   189
semi-conservative replication
            141
semi-quantitative culture   374
semi-solid agar   428
sen gene   332
sense strand (of DNA)   149
sepsis   330
septic shock   338

septicaemia   330
septum   54
      at mid-cell site   54
sequencing (of DNA)   258–262
SERE   510
serine   124
serine ↔ glycine interconversion
            123
serine pathway   127
serological tests   494
serotype   494
      serotype-converting phages
            279
Serratia   499, 524
      butanediol fermentation   89
      swarming   72
serum agar   428
serum amyloid A   340
sewage treatment   411
      activated sludge process   413
      aerobic   411
      anaerobic   414
      BAF   413
      biological aerated filter   413
      biological filter   411
      nitrification   413, 414
      primary   411
      secondary   411
      settled sewage   411
      trickle filter   411, 412
sewer gas   415
shape determination   8
shellfish   401, 491
ShET toxin   332
shiga toxin   334
shiga-like toxins   332, 334
Shigella   524
      anaerogenic fermentation   86
      antimacrophage activity   361,
            362
      dysentery   337, 384
      in food   400
      invasion of cells   324
      serotype-converting phages
            279
      shiga toxin   334
      vaccines   353
      XLD medium   429
shigellosis   338
Shine–Dalgarno sequence   151,
            232, 267
short patch repair   162
shuttle vector   227

siderophilins 355, 367
siderophore 177, 366
sIgA 350
sigF gene 329
σ^E 156, 184, 185
σ^F 75, 184, 185
σ^H 184, 185, 186
σ^S 174, 175, 186
σ^{28} 176
σ^{32} 148, 156, 172, 186, 188
σ^{54} 306
σ^{70} 148, 187
sigma factor 148, 175, 186
signal recognition particle 108
signal sequence 107, 154
signal transduction 179
signature-tagged mutagenesis 256, 257
significant bacteriuria 374
silage 407
silent mutation 195, 475
Simmons' citrate agar 488
Simonsiella 12, 13, 524
single-cell protein 408
single-strand binding protein 138
single-strand conformation polymorphism (see SSCP)
Sinorhizobium meliloti 49
siRNA 190
SIRS 330
site-specific mutagenesis 252, 253
site-specific recombination 199, 200
skin (microflora) 318
slant 431
slide agglutination tests (latex) 494
slide test (coagulase) 486
slime layers 29
slope 431
sloppy agar 428
slow sand filter 417
SLTI, SLTII 332
*Sma*I 223
SMAC medium 429
small interfering RNA 190
smc gene 54
Smc (SMC) protein 54
smear 444
smf 102
smoking (bacon, fish etc.) 400

soaps 457
sodA gene 177
sodium dodecyl sulphate 226
sodium motive force (see smf)
sodium polyanetholesulphonate 249
sodium pump 102
Soj protein 54
solar disinfection (of water) 422
somatic hypermutation 352
SopB 337
sorbitol MacConkey agar 429
sortase 19
SOS box 170
SOS mutagenesis 170
SOS system (in *E. coli*) 170
Southern blotting 229
Southern hybridization 229
Southwestern blotting 271
soxR/soxS genes 177
specialized transducing particle 293
specialized transduction 293
specific epithet 5
specific growth rate 66
Sphaerotilus natans 413
Sphingobium chlorophenolicum 423
spirillum 6, 7
Spirillum 7, 505
Spirochaeta 9
Spirochaetales 524
spirochaetes 6, 7
periplasmic flagella 33
splenic fever 383
Spo0A 181, 184, 185
Spo0A~P 181, 184, 185
Spo0F 181
Spo0J 53
SpoIIAB 75
SpoIIIE 75
spoilage (of canned foods) 398
spoligotyping 499
sporangium 77
spore stain 483
spores 73
spread plate 435, 442
spreader 435
spreading factor 335
SPS (in PCR) 249
spun column chromatography 224, 228

SRF (subtracted restriction fingerprinting) 499
*Srf*I 223, 248
SRP 108
srrAB genes (*Staphylococcus*) 181
SSB proteins 138
SSCP 245, 473, 500
silent mutation (effect of) 475
SSR 199
SsrA–SsrB system (*Salmonella typhimurium*) 179
ST-I, ST-II toxins 332
STa, STb toxins 332
stab inoculation 435
staining 444
acid-fast 445
capsule 445
distinguishing live from dead cells 445
endospore 483
flagella 31
Gram 444
Gram staining living cells 444
negative 445
Ziehl–Neelsen 445
staphylocoagulase 485
Staphylococcus 524
albus 524
aurease 492
aureus (adhesins) 368
aureus (*agr* locus) 181
aureus (biofilms) 300
aureus (exfoliative toxin) 388
aureus (inducible β-lactamase) 460
aureus (methicillin-resistant) 469, 479
aureus (MRSA) 469, 479
aureus (sortase) 19
coagulase test 485
D_{70} value 397
enterotoxins 334
epidermidis 524
food poisoning 334, 401
halotolerance 47
in microflora 318
intermedius 524
lysostaphin 25
MRSA 469, 479, 509
osmoregulation 48
peptidoglycan 25

Staphylococcus (*cont.*)
  phage typing 494, 495
  protein A 359, 363
    use of 222
  Rapidec Staph test 492
  scalded skin syndrome 388
  sortase 19
  srrAB genes 181
  toxic shock syndrome 366
  toxins 334, 366
  xylosus
    catabolite repression 170
star activity (of REs) 221
STARFISH 335
start site 147
starter culture 394
stationary phase (of growth) 65
steam (in sterilization) 451, 453
STEC 332
stem peptide (in peptidoglycan)
  25
Stenotrophomonas maltophila
  336, 524
sterilants 454
sterile cabinet 431, 433
  class I 432
  class II 432
  class III 432
sterilization 450
  autoclave 451–453
  chemical methods 454
  hot-air oven 451
  physical methods 451–454
sticky ends 200, 221, 233
STM 257
Stoffel fragment 239
stomach (microflora) 318
stop codon 154
storage granules 10, 15
stormy clot (in litmus milk)
  491
STP 293
straight wire 434
strain 4
strand (of DNA) 135
strand displacement amplification
  (see SDA)
streaking 435, 436
streptavidin 237
Streptococcus 524
  conjugation 208
  conjugative transposition 212
  erysipelas 384

Streptococcus (*cont.*)
  Griffith's serogroups 30
  group R 525
  hyaluronate lyase 335
  hyaluronic acid 359
  in microflora 318
  introns 189
  M protein 30
  mutans 321
  pneumoniae 129, 177, 205,
    336
  pyogenes 129, 359, 388
  thermophilus 394
  transformation 205
streptogramins 462, 463
streptokinase 266, 268, 269, 336
Streptomyces 243, 525
  porins in cell wall 23
  spores 75
streptomycin 461
streptomycin-dependent mutant
  472
streptozotocin 195
strict aerobe 46
stringency (in PCR) 241
stringent (plasmid replication)
  145
stringent response (to
    starvation) 174
stromatolites 297, 298
Stuart's transport medium 428,
  429
stuffer 227
Stx1, Stx2 332
subculturing 437
substrate-level phosphorylation
  85
subtilisins 411
subtracted restriction
    fingerprinting (see SRF)
succinic acid 87, 93
Sudan black B 484
sugars 486
suicide microbes 314
sulA gene 57, 170
Sulfolobus 30, 525
  acidophilic 46
  carbon dioxide fixation 118
  classification 504, 505
  use of sulphur 310
sulphate-reducing bacteria 94,
  309
sulphate respiration 94, 309

sulphide production 94, 309
sulphonamides 465
sulphonation 418
sulphur cycle 308, 309
sulphur-reducing bacteria 94
superantigens 363
superchlorination 418
supercoiled DNA 136
superinfection immunity
    (phage) 288
superoxide 176
superoxide dismutase 177
surface-obligatory conjugation
  211
surface plate 442
surface tension 40, 211
suspension test (disinfectants)
  456
Svedberg unit 14
swab 435
swarm cells 69, 70
swarming 72
Swiss cheese 393
symbionts 301
symbiosis 301
symport 102
synapsis 198
synchronous growth 67
Synechococcus 129
  biological clock 188
  motility 42
Synercid® 463
synergism (between antibiotics)
  466
Synsorb P^k 335
syphilis 388
systemic disease 317
systemic inflammatory response
  syndrome (see SIRS)
SYTO® 9 445, 446

t gene (phage T4) 285
T cell (T lymphocyte)
  in antibody formation 351
  CD4 347, 351, 352
  CD8 356
  cytotoxic 356
  $\gamma\delta$ 340, 356
  helper 351
  naïve 352
  polyclonal activation 366
  receptor 351
  Th1, Th2 352, 364

T cell receptor   351
T-DNA   114
T-strand   208
T4 ligase   234
Tag protein   161
tail fibres (phage)   280
tailing   233, 234
  homopolymer tail   234
  oligo(dC) tail   234
Taq polymerase   239
TaqMan® probes   238, 250
Tat (protein secretion)   113, 114, 115
TATAAT   147
taxonomy   500
  parsimony, principle of   500
  polyphasic   500
TbpA, TbpB   355, 367
TCA cycle   92, 93
TCP (toxin co-regulated pili)   37, 331
TCR (T cell receptor)   351
TD antigens   351
T-DNA   114
TE   199
teichoic acids   22
teicoplanin   126
tellurite   429
temperate phages   278
temperature (growth)   45
temperature (lethal)   397
template (in DNA replication)   139
temporal temperature gradient gel electrophoresis (see TTGE)
ter site   141
terminal deoxynucleotidyl transferase 234
terminal electron acceptor   90
terminally redundant DNA   281
terminator   148
ternary fission   61
terrorism, biological   4
TET protein   468
tetanospasmin   334
tetanus   334, 388
tethering (of white blood cells)   342
tetracyclines   462
tetrahydrofolate   123, 465
tetra-tetra   25, 51
Th1 cells   352, 364

Th2 cells   352, 364
Thauera   94
Thermoactinomyces   73
thermocycler   240
thermoduric bacteria   46
Thermomicrobium   45, 46
thermophilic bacteria   45
Thermoplasma   29, 46, 504
thermoprotectant   49
Thermoproteus   28, 526
thermoregulation (of genes)   187
thermostable DNA polymerase   239
Thermus aquaticus   239
THF   123, 465
thiamine pyrophosphate   123
Thiobacillus   526
  acidophilic   46
  bioleaching   409
  ferrooxidans   95, 101
    reverse electron transport   100
  in sulphur cycle   95, 309, 310
thiol-activated cytolysins   325
Thiovulum majus   39
Thoma chamber   440, 441
three-for-one model   51
threshold cycle (in PCR)   250
thylakoids   17
thymine   133
thymine dimer   163
thymus-(in)dependent antigens   351
TI antigens   351
Ti plasmid   114
tigecycline (GAR-936)   462
TipA   187
Tir   321
tissue-destructive protease   385
TLRs (Toll-like receptors)   371
TMA   266, 497
Tn3   203, 204
Tn5   204, 258
Tn7   205
Tn10   186, 203, 204, 468
Tn501   204
Tn916   212
Tn951   204
Tn1207.1   463
Tn1546   499
Tn1681   204
Tn1721   204
Tn5253   212

TNF   342, 352, 364, 382
Tn*phoA*   255
TOL plasmid   131
Toll-like receptors   371
toluene degradation   424
tooth decay   321
topoisomerase IV   53, 137, 464
topoisomerases   53, 136
total cell count   441
total RNA   217, 224
touchdown PCR   247
toxic shock syndrome   366, 389
toxic-shock-syndrome toxin-1   181, 366, 389
toxicoinfection   383
toxin co-regulated pili (TCP)   37, 331
toxins (food poisoning)   334, 385, 400, 403
toxin-mediated pathogenesis   330
toxoid   355
ToxR, ToxS, ToxT   331
TPP   123
trachoma   389
transconjugant   207
transcript   147
transcription   147
  divergent   149
transcription antitermination   187
transcription factors   149
transcription-mediated amplification (of RNA) (see TMA)
transcription-repair coupling factor   163
transcytosis   328
transductant   292
transduction   292
transfer RNA (see tRNA)
transferrin   355, 366, 367
transformation   205
  competence   205, 300
  laboratory-induced competence   206
transgenic plants (for bioplastics)   422
transglycosylation   125
transition mutation   161
transition state (sporulation)   180

translation (in protein synthesis) 151
translational attenuation 178
translational control (operons) 167
translational frame-shifting 181
translesion synthesis (of DNA) 171
translocation (of gut bacteria) 319
translocation (protein synthesis) 153, 154
translocon 107
transmission of disease 372
  river blindness (see *Wolbachia*) 526
transpeptidation
  in peptidoglycan synthesis 126
  in protein synthesis 154
transport medium 428, 429
transport systems 102
  ABC exporters 104
  ABC importers 104
  autotransporters 110
  general secretory pathway 107
  PTS system 105
  salvage transport 115
  sec-dependent system 107
  Tat system 114
  type I secretion 105, 112
  type II secretion 107, 112
  type III secretion 109, 112, 321, 324, 325, 337, 369
  type IV secretion 110, 112, 272
  type V secretion 111, 113
transposable element 199
transposase 199, 203, 204
transposition 199, 202
  conjugative 212
  cut-and-paste 202, 203
  preferred site of insertion 204, 205
  symbol for inserted TE 205
transposon 199
  Tn*3* 203, 204
  Tn*5* 204, 258
  Tn*7* 205
  Tn*10* 186, 203, 204, 468
  Tn*501* 204
  Tn*915* 212
  Tn*951* 204

transposon (*cont.*)
  Tn*1207.1* 453
  Tn*1546* 499
  Tn*1681* 204
  Tn*1721* 204
  Tn*5253* 212
  Tn*phoA* 255
transposon mutagenesis 255
transversion mutation 161
travellers' diarrhoea 332
TRCF 163
Treponema pallidum 33, 34, 35, 388, 526
tricarboxylic acid cycle (see TCA cycle)
trichome 7, 43
trickle filter 411
triclosan 457
trigger mechanism 323, 324
trimethoprim 465
Trk system 48
tRNA 151
tRNA-directed transcription antitermination 187
Tropheryma whipplei 129, 392, 526
trovafloxacin 464
trp operon 167
true coagulase 485
trypsin
  action on complement 346
trypticase soy broth 375
Tsr protein 101
TSST-1 366, 389
TTGE 245
tube test (coagulase) 486
tubercle 357, 358
tuberculosis 389
  ELISPOT assay 329
  ESAT-6 329
  infliximab 332
  latent 329
  tubercles 357, 358
tularaemia 391
tumbling (in motility) 39, 182, 183
tumour (plant) (*Agrobacterium*) 114
tumour necrosis factor (see TNF)
twitching motility 40
two-component regulatory systems 179
  agr locus 181

two-component regulatory systems (*cont.*)
  Agrobacterium 114
  cell-cycle regulation 59, 61
  ComX-activated 205
  CpxA–CpxR 155
  crown gall 114
  essential 59
  exotoxigenesis 180
  in cell cycle 59, 61
  in gene expression 179
  in quorum sensing 298
  membrane permeability 180
  protein folding 180
  srrAB genes (*Staphylococcus*) 181
  SsrA-SsrB (*Salmonella*) 179
  transformation 180
Ty21a vaccine 353
type species 513
types I–IV (1–4) fimbriae 37
types I–IV protein secretion systems (see transport systems)
typhoid 391
typhoid Mary 373
typhus 391
typing 492
  AFLP 511
  AP-PCR 508
  Griffith's 30
  Lancefield 525
  meaning of term 492
  MEE 495
  MLST 499
  nucleic-acid-based 498
  PFGE 498
  phage 494
  pyoverdin variability 494
  RAPD 508
  rep-PCR 510
  RFLP 510
  serological 494
  spoligotyping 499
  SRF 499
tyrocidins 156
Tyzzer's disease 391

UHT (milk) 396, 405
ultraviolet radiation (disinfection) 456
ultraviolet radiation (mutagen) 192

umuC gene   170
umuD gene   170
unbalanced growth   63
uncultivable bacteria   314
underwound helix   135, 136
UNG   161, 247
universal bottle   426
universal conjugation   211
UP element   148
UPEC   130, 186, 319, 321, 339, 369
upstream   147
uptake-signal sequence   205
uracil   133
uracil-*N*-glycosylase   161, 247
urea   489
urease test   489
ureases   489
urethra (microflora)   318
urethritis   392
uropathogenic *E. coli* (see UPEC)
uroplakins   321, 340
usher protein   36
UTIs   319, 339
UvrABC excinuclease   162
UvrD   162

V factor (*Haemophilus*)   519
VacA (*Helicobacter pylori*)   339
vaccination   352
vaccine   352
  ACVs   353
  BCG   353, 354
  conjugate   354, 359
  CTB-WC anti-cholera   353
  DNA   355
  ETEC   353
  Hib   354
  live, attenuated   353
  O139 *Vibrio cholerae*   353
  oral   352
  parenteral   352, 353
  Shigella   353
  toxoids   354
  Ty21a anti-typhoid   353
  WCVs   353
  whole-cell   353
  whooping cough   353
vagina (microflora)   318
vaginosis (bacterial)   383
vancomycin
  action   126
  resistant enterococci   469

VBNC bacteria   315
vector-borne disease   373
vectors   225
  bacterial artificial chromosome   227
  cloning   214, 230
  cosmid   227, 231
  expression   219, 227, 231, 267
  gapped shuttle   228
  phagemid   229, 233
  replacement   227
  retrieval   228
  shuttle   227
  yeast artificial chromosomes   227
verocytotoxic *E. coli*   332
verocytotoxins   332
Vi antigen   354, 523
viable but non-cultivable   315
viable cell count   442
vibrio   6, 7
Vibrio   526
  alginolyticus   102
  anti-cholera vaccines   353
  cholera   331, 383
  cholera toxin   279
  cholerae O139 Bengal   279, 288
  cholerae (genome sequence)   129
  cholerae (smf)   102
  chromosomes   14, 129
  food poisoning   401
  El Tor biotype   279, 288, 353
  fimbriae   37
  fischeri   298
  flagella   31
  multi-drug-resistant O1   469
  parahaemolyticus   129, 401, 499, 510
  TCP adhesin   37
  vulnificus   499
vinegar   395
VirA protein (*Shigella*)   324
VirG protein   337, 353
viridans streptococci   492
virulence
  phage-determined   278
  plasmid-determined   131, 316
virulence factors   358
virulent phages   278
virulon   110
virus   278

VNC bacteria   315
Voges–Proskauer test   488
voltage-regulated channel   20
volutin   16
VP test   488
VSL#3 (probiotic)   380
VT1, VT2   332
VTEC   332

warfare, biological   4
washing powders (biological)   411
waste-water treatment   411
water (for growth)   45
water supplies   415
  aquifers   415
  blooms   417
  breakpoint chlorination   418, 419
  chloramines   418
  chlorination   418
  chlorine demand   418, 419
  clarification   417
  combined residuals   418, 419
  cyanobacterial toxins   417
  dam   416
  disinfection   418, 419
  distribution system   421
  Eijkman test   420
  floc blanket clarifier   417
  free residuals   418
  groundwater   415, 417
  microstraining   417
  nitrates   421
  ozonation   419
  raw water   415
  reservoir   416
  sand filters   417
  small-scale supplies   421
  solar disinfection   422
  superchlorination   418
  tertiary treatment   421
  toxins   417
Waterhouse–Friderichsen syndrome   387
WCVs   353
Weil's disease   386
Western blotting   229
wheat (nitrogen fixation)   308
whey   394
Whipple's disease   392, 526
whole-cell vaccines   353

whooping cough 392
  vaccines 353
wild type 192
Wolbachia 526
Wolinella succinogenes 423
woolsorters' disease 382

X factor (*Haemophilus*) 519
X-rays (in sterilization) 453
xanthan gum 30, 395
Xanthobacter 29
Xanthomonas 30, 310, 395, 526
*Xba*I 223
xenobiotics 304
Xgal 225, 236
  in IVET 360, 361
*Xho*I 223
xis protein 289
XLD medium 428, 429
xylenols 455

xylose–lysine–deoxycholate
  medium 428, 429

YAC 227
yeast artificial chromosome 227
Yersinia 188, 526
  avoiding phagocytosis 359
  coagulase 486
  enterocolitica 399, 495
  food poisoning 401
  invasion of cells 325
  low-temperature growth 399
  pseudotuberculosis 401
  Yop proteins 110
ykuA gene 9
yoghurt 394
Yop proteins 110
  YopE 361
  YopH 361
  YopJ 361

YscN 110

Z ring (FtsZ ring) 54, 55, 74,
  75
Z scheme 97, 98
z value (in food processing)
  397
Ziehl–Neelsen stain 445
zinc chelation 347
zinc-containing enzymes
  de-acetylase 470
  endopeptidase 334
  metalloprotease 328
ZipA 54, 56
zippering 325, 326, 327
Zoogloea ramigera 411, 413,
  527
zoospores 77
Zot toxin 331